Polarimetric Radar Signal Processing

Related titles on radar:

Advances in Bistatic Radar Willis and Griffiths
Airborne Early Warning System Concepts, 3rd Edition Long
Bistatic Radar, 2nd Edition Willis
Design of Multi-Frequency CW Radars Jankiraman
Digital Techniques for Wideband Receivers, 2nd Edition Tsui
Electronic Warfare Pocket Guide Adamy
Foliage Penetration Radar: Detection and characterisation of objects under trees Davis
Fundamentals of Ground Radar for ATC Engineers and Technicians Bouwman
Fundamentals of Systems Engineering and Defense Systems Applications Jeffrey
Introduction to Electronic Warfare Modeling and Simulation Adamy
Introduction to Electronic Defense Systems Neri
Introduction to Sensors for Ranging and Imaging Brooker
Microwave Passive Direction Finding Lipsky
Microwave Receivers with Electronic Warfare Applications Tsui
Phased-Array Radar Design: Application of radar fundamentals Jeffrey
Pocket Radar Guide: Key facts, equations, and data Curry
Principles of Modern Radar, Volume 1: Basic principles Richards, Scheer and Holm
Principles of Modern Radar, Volume 2: Advanced techniques Melvin and Scheer
Principles of Modern Radar, Volume 3: Applications Scheer and Melvin
Principles of Waveform Diversity and Design Wicks *et al.*
Pulse Doppler Radar Alabaster
Radar Cross Section Measurements Knott
Radar Cross Section, 2nd Edition Knott *et al.*
Radar Design Principles: Signal processing and the environment, 2nd Edition Nathanson *et al.*
Radar Detection DiFranco and Rubin
Radar Essentials: A concise handbook for radar design and performance Curry
Radar Foundations for Imaging and Advanced Concepts Sullivan
Radar Principles for the Non-Specialist, 3rd Edition Toomay and Hannan
Test and Evaluation of Aircraft Avionics and Weapons Systems McShea
Understanding Radar Systems Kingsley and Quegan
Understanding Synthetic Aperture Radar Images Oliver and Quegan
Radar and Electronic Warfare Principles for the Non-specialist, 4th Edition Hannen
Inverse Synthetic Aperture Radar Imaging: Principles, algorithms and applications Chen and Martorella
Stimson's Introduction to Airborne Radar, 3rd Edition Baker, Griffiths and Adamy
Test and Evaluation of Avionics and Weapon Sytems, 2nd Edition McShea
Angle-of-Arrival Estimation Using Radar Interferometry: Methods and applications Holder
Biologically-Inspired Radar and Sonar: Lessons from Nature Balleri, Griffiths and Baker
The Impact of Cognition on Radar Technology Farina, De Maio and Haykin
Novel Radar Techniques and Applications, Volume 1: Real Aperture Array Radar, Imaging Radar, and Passive and Multistatic Radar Klemm, Nickel, Gierull, Lombardo, Griffiths and Koch
Novel Radar Techniques and Applications, Volume 2: Waveform Diversity and Cognitive Radar, and Target Tracking and Data Fusion Klemm, Nickel, Gierull, Lombardo, Griffiths and Koch
Radar and Communication Spectrum Sharing Blunt and Perrins
Systems Engineering for Ethical Autonomous Systems Gillespie
Shadowing Function from Randomly Rough Surfaces: Derivation and applications Bourlier and Li
Photo for Radar Networks and Electronic Warfare Systems Bogoni, Laghezza and Ghelfi
Multidimensional Radar Imaging Martorella
Radar Waveform Design Based on Optimization Theory Cui, De Maio, Farina and Li
Micro-Doppler Radar and its Applications Fioranelli, Griffiths, Ritchie and Balleri
Maritime Surveillance with Synthetic Aperture Radar Di Martino and Antonio Iodice
Electronic Scanned Array Design Williams
Advanced Sparsity-Driven Models and Methods for Radar Applications Li
Deep Neural Network Design for Radar Applications Gurbuz
New Methodologies for Understanding Radar Data Mishra and Brüggenwirth
Radar countermeasures for Unmanned Aerial Vehicles Clemente, Fioranelli, Colone and Li
Holographic Staring Radar Oswald and Baker

Polarimetric Radar Signal Processing

Edited by
Augusto Aubry, Antonio De Maio and
Alfonso Farina

The Institution of Engineering and Technology

Published by SciTech Publishing, an imprint of The Institution of Engineering and Technology, London, United Kingdom

The Institution of Engineering and Technology is registered as a Charity in England & Wales (no. 211014) and Scotland (no. SC038698).

First published 2022

The Institution of Engineering and Technology
Futures Place
Kings Way, Stevenage
Hertfordshire, SG1 2UA, United Kingdom

www.theiet.org

British Library Cataloguing in Publication Data
A catalogue record for this product is available from the British Library

ISBN 978-1-83953-402-7 (hardback)
ISBN 978-1-83953-403-4 (PDF)

Typeset in India by Exeter Premedia Services Private Limited

Dedication

To the loves of my life Azzurra and Serena - Augusto
To my mom Imma and my dad Michele - Antonio
To my beloved wife Franca, to my Mom and Dad, always - Alfonso

Contents

About the Authors

Augusto Aubry is an associate professor at the University of Naples Federico II, Italy. His research interests include statistical signal processing and optimization theory, with emphasis on topics related to MIMO communications and radar signal processing. Professor Aubry received the 2022 IEEE Fred Nathanson Memorial Award and is co-recipient of the 2018 IEEE AESS Barry Carlton Best Paper Award. He is an associate editor of the *IEEE Transactions on Signal Processing and* is an IEEE senior member.

Antonio De Maio is a full professor at the University of Naples Federico II, Italy. Professor De Maio received the 2010 IEEE Fred Nathanson Memorial Award. Dr De Maio is a fellow of the IEEE and co-recipient of the 2018 IEEE AESS Barry Carlton Best Paper Award. His research interest is in the field of statistical signal processing, with emphasis on radar detection, optimization theory applied to radar signal processing, and multiple-access communications.

Alfonso Farina is most noted for the development of track while scan techniques for radars and for the development of numerous signal processing techniques used for sensors where tracking plays an essential role. He is author of about 1000 publications. He has won the Fred Nathanson Memorial Radar Award (1987), the IEEE Dennis J. Picard Medal for Radar Technologies and Applications (2010), the IET Achievement Medal (2014) and the IEEE AESS Pioneer Award (2020).

Foreword

"The bells which toll for mankind are – most of them anyway – like the bells on Alpine cattle; they are attached to our necks, and it must be our fault if they do not make a cheerful and harmonious sound" (The Future of Man: The BBC Reith Lectures 1959[1]).
Sir Peter Brian Medawar, 1960 Nobel Prize in Physiology or Medicine.

This foreword starts with a famous quote by Peter Medawar. Quote that is very current in this period of humanity. It is hoped that this book will contribute to the power of culture, of knowledge, and to the wisdom in using them, the ability to cooperate and to seek cross-fertilization.

The Editors invite the Reader to take a breath and immerse themselves in reading the foreword, and therefore encourage themselves to study the chapters with all their necessary technicality. Enjoy the foreword and the whole book, appreciating the sincere effort of the book's Authors.

Definition of radar polarimetry and historical notes

From [1, p. 1] "Radar polarimetry, i.e., the utilization of complete electromagnetic vector wave information, has become an indispensable tool in modern electromagnetic sensor and imaging technology, both in the civil and military sectors, and increasingly more in environmental remote sensing of terrestrial and planetary crusts." From [1, p. 3], "In high resolution polarimetric radar imaging, it is the objective to utilize the complete vector nature of electromagnetic waves, i.e., in addition to amplitude, frequency and phase, also the polarization state information of the transmitted and received waves is incorporated into signal (bin-by-bin) and image (pixel-by-pixel) processing, requiring a 2 by 2 Sinclair matrix and/or a 4 by 4 Mueller matrix description of the scatterer for the coherent and partially polarized cases, respectively." Similar definitions and descriptions are available in [2, p. 245] and [3].

In Table 1 some historical achievements in polarimetry are summarised. It moves from the polarimetric bionics, to the long story of polarization of light, thus the polarization of electromagnetic waves, up to the radar polarimetry of key interest for this book. The table has four columns: events, discoverer of the event, the date of the discovery, main references and comments.

[1]Published February 28, 2021 by Routledge 130 Pages, ISBN 9780367278625

Table 1 Tracing the history of polarimetry (not exhaustive table list)

Events	Discoverer	Date	References & Comments
Polarimetric bionics, i.e., the intrinsic facility of "polarization perception" by insect, birds, fish, and mammals.	Mother Nature		[4], [1].
It has been shown that bees are able to infer the position of the Sun from the polarization pattern of the blue sky and use this to orient themselves. Karl von Frisch first correctly interpreted "The honeybee famous dance communication".	K. R. von Frisch, The Nobel Prize in Physiology or Medicine 1973.	Nobel prize in 1973	https://simple.wikipedia.org/wiki/Karl_von_Frisch
Sky polarimetric navigation.	The Vikings	Late 10th century	[5], [6].
Crystals of the Iceland spar mineral (a transparent, colourless variety of calcite) produce a double image when objects are viewed through the crystals in transmitted light.	Erasmus Bartholin	1669	https://en.wikipedia.org/wiki/Rasmus_Bartholin [7], see also Figure 1 and footnote 2. It is truly remarkable to see how the technology of polarimetry has developed, in such a profound way, from an almost casual observation made in the 17th century.

[2]A. Farina and his colleague Dr. M. Frasca (MBDA, Italy) were invited to deliver talks at "The IQC (Institute of Quantum Computing) Workshop on Quantum Illumination: From Theory to Practice, 3-4 December, 2019, University of Waterloo, Canada, [26]. An interesting presentation by Prof. D. Cory at Waterloo December 2019, "Quantum Showcase" organized by Prof. T. Forrest, Program Director, Transformative Quantum Technologies, University of Waterloo, showed which might have been the earliest quantum sensor. The presentation illustrated the exploitation of birefringence of calcites for late 10th century navigation of the Vikings. They used "the findlings stone (sun-stone), produced from some dichroic mineral (like cordierite and calcite) and were able to navigate in the absence of direct sunlight. Thus, the Vikings during the tenth (10th) century when navigating in the Baltic Sea and Arctic waters prior to traversing the Atlantic, may have been one of the first human beings to utilize polarimetric effects as a tool. This story is narrated in the movie THE VIKINGS (1958), an epic historical fiction movie. Stars K. Douglas, T. Curtis and J. Leigh as well as E. Borgnine.

Table 1 Continued

Events	Discoverer	Date	References & Comments
Each of the two rays arising from refraction by calcite may be extinguished by passing it through a second crystal of the same material if the latter crystal is rotated about the direction of the ray.	Christiaan Huygens	1690	https://en.wikipedia.org/wiki/Christiaan_Huygens [7].
It was discovered that light becomes partially plane polarized when reflected, i.e., its rays vibrate in the same plane. This observation led to a better understanding of the propagation of light.	Étienne-Louis Malus	1808	https://www.britannica.com/biography/Etienne-Louis-Malus https://en.wikipedia.org/wiki/%C3%89tienne-Louis_Malus [7].
Discovered the polarization of light from the clear blue sky.	Dominique Francois Jean Arago	1809	https://en.wikipedia.org/wiki/Fran%C3%A7ois_Arago [7], [4].
Discovery of the law giving the relationship between the refractive index of a medium and the angle of incidence at which reflected light is totally linearly polarized.	David Brewster	1815	https://en.wikipedia.org/wiki/David_Brewster [7].
Investigated the interference of polarized rays of light and discovered that two rays orthogonally polarized never interfere.	Augustin Jean Fresnel	1816	https://en.wikipedia.org/wiki/Augustin-Jean_Fresnel [7].
The first easily usable polarizing filter based on a prism was invented.	William Nicol	1828	https://en.wikipedia.org/wiki/William_Nicol_(geologist) [7].
Discovered the rotation of the plane of linear polarization of light in magnetic fields.	Michael Faraday	1845	https://en.wikipedia.org/wiki/Michael_Faraday [7]. Faraday rotation affects also radar and telecommunication systems. See for instance [8].

Table 1 Continued

Events	Discoverer	Date	References & Comments
Explained theoretically the polarization of scattered skylight.	John William Strutt, alias Lord Rayleigh	1871	https://en.wikipedia.org/wiki/John_William_Strutt,_3rd_Baron_Rayleigh [7].
Formulated the equations of electromagnetism and conjectured that light is made by electromagnetic (e.m.) waves. A rigorous formulation of Faraday's postulates was provided too.	James Clerk Maxwell	1873	https://en.wikipedia.org/wiki/James_Clerk_Maxwell [1], [7].
Confirmed by direct experiments the Maxwell's conjecture, i.e., light is an e.m. wave.	Heinrich Hertz	1888	https://en.wikipedia.org/wiki/Heinrich_Hertz [7]
All possible states of polarization could be represented by points on the Riemann (1872) sphere, where the latitude and longitude of each point are related to the eccentricity and the orientation angle of the polarization ellipse, respectively.	Henri Poincaré	1892	https://en.wikipedia.org/wiki/Henri_Poincar%C3%A9 https://en.wikipedia.org/wiki/Polarization_(waves)#Poincar%C3%A9_sphere [1], [9]. See also Figure 2.
The degree of polarization of rough objects is governed by a simple rule: the darker the surface, the stronger the polarization. Hence, light reflected from concrete is less polarized than that from dark asphalt. Likewise, the diffusely reflected light of wet bricks is more strongly than that from dry bricks[3].	Nikolay Alekseevich Umov	1905	https://en.wikipedia.org/wiki/Nikolay_Umov [7], [4, pp. 35–36]. Umov was the first who introduced in physics such basic concepts as speed and direction of movement of energy, which are now associated with Umov-Poynting vector, and density of energy in a given point of space.

[3]By far the greater part of the light encountered in Nature is scattered (or diffuse) sunlight: consequently the polarization pattern in Nature is very uniform. In this respect, the polarization pattern differs markedly from the colour pattern around us [4].

Table 1 Continued

Events	Discoverer	Date	References & Comments
Properties of partially polarized waves were discovered according to an harmonic analysis in quantum mechanics showing that the coherency matrix is a linear combination of Pauli spin matrices with Stokes parameters.	Norbert Wiener, Wolfgang Pauli	1927–29, 1924–29	https://en.wikipedia.org/wiki/Norbert_Wiener [1, p. 7], [10, p. 200]; see also [11]. Earlier on, during 1924–29, Wolfgang Pauli[4] introduced the concept of the spinor in quantum mechanics which has proven to be also an ideally successful tool in proper description of polarimetric radar problems [1]. The Stokes parameters are a set of values that describe the polarization state of electromagnetic radiation. They were defined by George Gabriel Stokes in 1852, as a mathematically convenient alternative to the more common description of incoherent or partially polarized radiation [9]. See also https://en.wikipedia.org/wiki/Stokes_parameters
Extensive fundamental studies on radar polarization.	George Sinclair at Ohio State University Antenna (later called Electro Science) Laboratory	1946	[1, p. 7]. Sinclair showed that a radar target acts as a polarization transformer and he expressed the properties of a coherent radar target by the 2x2 coherent scattering matrix, the Sinclair matrix.

[4]https://en.wikipedia.org/wiki/Wolfgang_Pauli

Table 1 Continued

Events	Discoverer	Date	References & Comments
The concept of "Optimal Target Polarization" was introduced and leveraged in meteorological radar systems.	M. E. Kennaugh	1952	[12], [1]. Different target polarization properties can be used to boost radar performance, e.g., exploitation of circular polarization for rain clutter rejection or cancellation.
Development of novel methologies to measure radar target scattering matrix.	J. R. Huynen	1965	[13], [1]. This study culminated in remarkable and visionary Huynen's dissertation [14].
Development of advanced weather radar: Joint Doppler Operational Project (JDOP); NEXRAD network; Terminal Doppler Weather Radar (TDWR); Dual polarization weather radar WSR-88D.	U.S. Departments of Commerce, Defense, and Transportation. National Severe Storms Laboratory (NSSL) U.S.	1957–1974, 1976, 1979, 2010	https://en.wikipedia.org/wiki/NEXRAD Polarimetric signal processing improved the estimation accuracy of rain rate and backscattering significantly. Moreover, it allowed the classification of hydrometeors.
Polarimetric multi-notch filter concept and man-machine interface for polarimetric control on radar.	A. J. Poelman	1979	[15], [1].
Development of intra-pulse polarization agile radar (IPPAR) concept.	J. L. Eaves, E. K. Reedy, and co-workers at GIT-EES (Georgia Institute of Technology-Engineering Experiment Station)	1984, 1987	[16], [17], [18], [1].
Target detection algorithms for polarimetric radar	Les Novak, MIT-LL, and co-workers	1989	[19].

Table 1 Continued

Events	Discoverer	Date	References & Comments
Polarimetric signal processing and application to SAR images at CAL-TECH Jet Propulsion Laboratory.	C. Elachi, F. T. Ulaby, J. Van Zyl	1988–2006	[20], [21], [22]. C. Elachi in his book dated 1988 discusses the role of polarization in the spaceborne imaging radars and the multipolarization imagery. More details are in the successive books of 1990 and 2006, co-authored with F. T. Ulaby and J. Van Zyl respectively.
The first spaceborne multifrequency and multipolarization SAR system.	Space Agencies of U.S./ German/ Italy	1994	[23], [24, pp. 18–27 and pp. 18–29]. The SIR-C/X-SAR experiment is a joint U.S./German/ Italian project for space based remote sensing radar with polarimetry. The instrument, flown aboard the space shuttle Endeavour, collected valuable multiparameter SAR data in April and October 1994. Japan's PALSAR is the first operational space-based system to incorporate a quad-pol mode [23].
Full operational use of radar polarimetry for civilian, defence and custom applications enabled by advances in technologies and system engineering.	Across the Planet	Last decades	A wealth of references are available.
6th ESA (European Space Agency) Advanced Course on Radar Polarimetry 2021	ESA	May 10–14 2021	https://eo4society.esa.int/event/6th-advanced-polarimetry-2021/

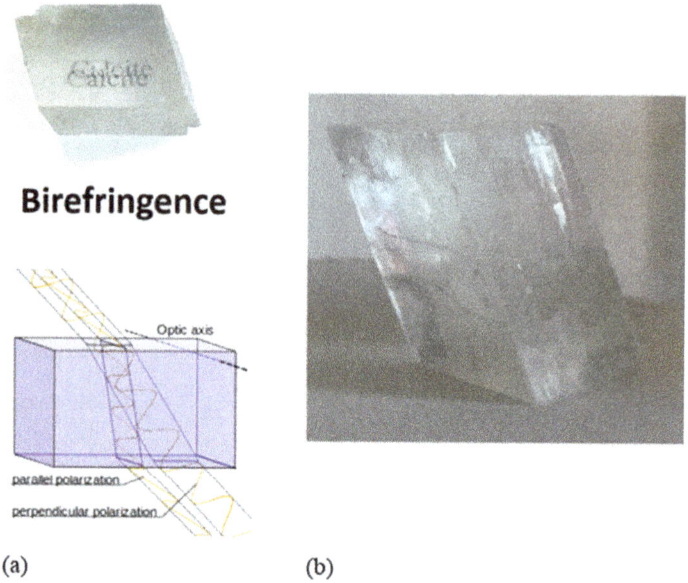

Figure 1 *Birefringence of calcite: a) decomposition of light polarization via calcite properties (by Prof. D. Cory, Waterloo University (CA)); b) a piece of calcite.*

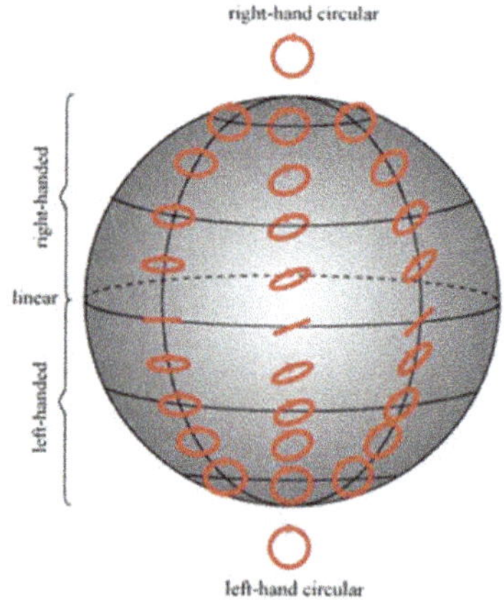

Figure 2 *Poincarè sphere, with all possible states of polarization as points on the Riemann sphere. (from* https://cdn.shopify.com/s/files/1/1026/4509/files/poincaresphere.jpg?2690835184976070306*)*

Description and organization of the book

This book attempts to provide an overview of some advanced techniques and technologies developed for polarimetric radars to meet demanding performance requirements, trying to cover some of the most challenging application fields, including: target detection for either active and passive surveillance systems, mainlobe interference suppression, detection of temporal changes in a given scene, environment classification, automatic target recognition, non-cooperative target imaging, polarimetric coding in radar and SAR systems, pol-SAR ambiguities suppression, space-debris detection, tracking, and classification, estimation of biological and behavioral parameters of insects, precipitations localization as well as type and motion parameters estimation. Specifically, innovative and sophisticated instruments from statistical-signal processing theory such as adaptive detection techniques, covariance matrix estimation, adaptive filtering, invariance theory, independent component analysis, machine learning tools, inverse problem theory, optimization techniques, are framed rigorously and systematically in the context of polarimetric radar signal processing.

It is worth pointing out that there are nowadays in open literature many high-quality books dealing with the use of polarization in radar and SAR, e.g., [10], [9], [27], [28], [29], [30].

Nevertheless, to the best of our knowledge this is the first book containing a very broad range of applications (from active radar detection, passive radar detection, classification, space-debris detection) accounting for the specific topic of polarimetry exploitation. Besides, another important characteristic of the book relies on the use of sophisticated tools form the statistical signal processing theory to address polarization radar problems. The cross-pollination between radar applications and statistical theory is indeed the common thread (leitmotiv) which makes (to the authors' opinion) the present book original.

The book balances a practical point of view with a rigorous mathematical approach corroborated with a wealth of numerical case studies and real experiments. Additionally, the book has a cross-disciplinary approach as it tries to exploit cross-fertilization by the recent and latest research and discoveries in statistical signal processing theory and electromagnetism. The material of the book is organized in fourteen chapters, each one comprising a very comprehensive list of references.

Foreword by the Editors

It contains historical notes, the motivation for the book, discusses its organization, provides the list of used symbols, and introduces the Editors.

Chapter 1: Adaptive Target Detection with Polarimetric Radars
Authors: A. Aubry, A. De Maio, A. Farina
This Chapter is focused on adaptive target detection exploiting polarimetric diversity. The situations of targets embedded in either Gaussian or non-Gaussian interference with unknown covariance matrix are considered, assuming the availability of secondary data (free of signal components and with the same covariance structure of

the cells under test) to infer disturbance statistical properties. To account for target polarimetric features, the perspective target echo from each range bin and polarimetric channel is modelled as a deterministic signal known up to a possibly varying scaling factor. Hence, to solve the resulting composite binary hypothesis testing problem, some adaptive detection schemes are devised according to diverse criteria such as generalized likelihood ratio, Wald, and Rao tests. Along with a discussion on the Constant False Alarm Rate (CFAR) properties shared by the proposed receive architectures, several case studies, both on simulated and measured data, are provided to shed light on the performance gains ensured by polarization diversity as well as to highlight pros and cons of the proposed detection strategies in comparison with some counterparts proposed in the open literature.

Chapter 2: Exploiting Polarimetric Diversity in Passive Radar
Authors: F. Colone and F. Filippini
This Chapter addresses the exploitation of polarimetric diversity in passive radar as a means to improve the surveillance capability of "green sensing systems". Indeed, using third party transmissions to detect and localize targets, passive radar technology may leverage polarimetric diversity via differently polarized antennas on receive. In this Chapter, the target detection task is mainly considered and some practical multi-polarimetric passive radar architectures are illustrated along with the corresponding adaptive signal processing strategies. Detailed comparisons among the proposed approaches are provided in terms of achievable targets detectability and system complexity. In this respect, the performance of the different detection strategies are extensively assessed on measured data, collected by means of experimental passive radar receivers operating at different frequency bands.

Chapter 3: Mainlobe Jamming Suppression for Polarimetric Multi-Channel Radar
Authors: G. Cui, M. Ge, and X. Yu
Electromagnetic interference entering the radar receiver through the antenna mainbeam significantly impairs radar surveillance capabilities, causing false alarm rate increase, miss detection, and last but not least receiver saturation. This Chapter focuses on mainlobe interference suppression for polarimetric multi-channel radar exploiting jointly spatial and polarimetric diversities to discriminate between interference and targets echoes. First, assuming the presence of targets and jammers, the signal model for a polarimetric multi-channel radar is developed and deeply discussed. Then, a suitable approach relying on independent component analysis and using all the available degrees of freedom is proposed to separate target and interference signals, while ensuring mainlobe interference rejection. Finally, some case studies are provided to demonstrate the effectiveness of the new strategy. The results indicate that different kinds of mainlobe interferences, such as noise-like and deceptive jammers via digital radio frequency memory (DRFM), can be significantly suppressed by the proposed method, which outperforms conventional interference cancellation strategies in terms of the achievable Signal to Interference plus Noise Ratio (SINR).

Chapter 4: Coherent Change Detection in Multi-Polarization SAR Images
Authors: V. Carotenuto, A. De Maio, C. Clemente, and A. Farina
Change Detection (CD) refers to the ability of identifying temporal changes within a given scene starting from a pair of co-registered Synthetic Aperture Radar (SAR) images representing an area of interest. Single channel techniques attempt to detect variations between reference and test data, either relying solely on intensity or using jointly amplitude and phase information. This Chapter aims at illustrating the performance improvements that can be attained exploiting polarimetric information. To this end, after presenting the multi-polarization signal model for the SAR systems, the change detection problem is formulated as a binary hypothesis test. Hence, some Coherent CD (CCD) algorithms capitalizing on polarimetric diversity are devised (possibly forcing robustness to intensity mismatches and/or miscalibration) resorting to the elegant theory of invariance in hypothesis testing problems. The effectiveness of the proposed decision rules is illustrated on both simulated and measured data sets.

Chapter 5: Covariance Symmetries Classification in Full-Polarimetric SAR Images
Authors: L. Pallotta, C. Clemente, A. De Maio, and J. Soraghan
Polarimetric covariance (or coherence) matrix captures significant features on the radar-cell scattering behavior, and it is widely exploited in the advanced processing of multi-polarization SAR images. This matrix is typically estimated from the data surrounding the investigated pixel (referred to as looks), and then used in several applications, e.g., scene classification and target detection/identification/recognition. Interestingly, the coherence matrix of a given pixel may exhibit a specific structure as a result of symmetries among co-pol and cross-pol returns, arising from particular geometric configurations. The scope of this Chapter is to describe a systematic framework for detecting covariance symmetries in both homogeneous and heterogeneous polarimetric SAR images. In both cases, the classification problem is formulated as a multiple hypotheses test, solved by means of the well-known Model Order Selection (MOS) rules. The benefits of the proposed signal processing algorithms are demonstrated through case studies involving real SAR polarimetric interferometry (Pol-inSAR) images.

Chapter 6: Polarimetric Information to Enhance SAR ATR Capabilities
Authors: D. Gaglione, C. Clemente, L. Pallotta, C. Ilioudis, A. De Maio, and J. Soraghan
Automatic Target Recognition (ATR) is one of the main applications of modern radar systems. Although it can be used in civilian problems, ATR is an essential requirement in battlefield scenarios, where the correct identification of a target must be guaranteed with a high degree of confidence. In this context, multiple sources of information are often available (i.e. spatial, temporal, frequency, and polarization diversities) and can be potentially combined leading to performance improvements. In particular, polarimetric information can be useful to boost the capabilities of ATR techniques by providing scattering information related to the shape, size, and composing materials of the observed object. In this Chapter the benefits of polarimetric information exploitation to recognize targets are demonstrated by investigating two potential approaches. In the former, the Krogager polarimetric decomposition

is applied to perform moments-based feature extraction and subsequent civilian vehicles classification; while in the latter the polarimetric covariance symmetries extracted from the full polarimetric return are used to feed a Convolutional Neural Network. The results on real-recorded data show the advantages in terms of target recognition accuracy that a polarimetric radar can provide. The final objective of this Chapter is to point the spotlight on the open research questions to be answered in order to unlock the benefits of polarimetry in ATR.

Chapter 7: Polarimetric Inverse Synthetic Aperture Radar Imaging
Authors: A. Lupidi, E. Giusti, and M. Martorella
Fully polarimetric radars have the ability to gather additional information about the target with respect to conventional single polarization radars. Such a richer amount of information can be exploited also in the case of radar imaging, where the received signal is processed to obtain two-dimensional images of targets. In the specific case of non-cooperative target imaging, the combination of Inverse Synthetic Aperture Radar (ISAR) framework and polarimetry brings two main advantages with respect to classic single polarization ISAR. The first is associated with a better ISAR image formation through the capitalization of fully polarimetric ISAR image autofocus algorithms. The second allows for additional target's features to be extracted so as to better support target recognition process. In this Chapter, the fundamentals of Polarimetric ISAR (Pol-ISAR) imaging are first introduced, followed by the definition and assessment of a number of fully polarimetric ISAR algorithms, aimed at demonstrating the superiority of full-polarimetric strategies with respect to single channel techniques. The second part of the chapter looks at ways to extract fully polarimetric features from Pol-ISAR images and their use to classify targets. Results based on measured data accompany the theoretical aspects to validate and substantiate the work done in this area.

Chapter 8: UWB Short-Range Polarimetric Imaging and its Potential for Target Classification.
Authors: J.Wang and A.Yarovoy
Traditionally, the vector nature of an electromagnetic field is ignored in short-range imaging applications, such as ground penetrating radar or body scanner, due to the complexity of polarimetric data acquisition and processing. In this Chapter, different approaches for full-polarimetric data acquisition in short range systems using simple linearly polarized antennas are presented and discussed. Efficient methods for fully-polarimetric image reconstruction are proposed and compared. Finally, the added value of polarimetric approach for image quality improvement and target classification is demonstrated for several applications like land mine detection and underground cable/pipe mapping.

Chapter 9: Robust Transceiver Design for Polarimetric Radars
Authors: X. Cheng, A. Aubry, D. Ciounzo, A. De Maio, and A. Farina
The detection of (possibly low-moving) extended targets is a challenging radar signal processing task and polarization diversity has been proved to be a viable means

to boost radar performance. Indeed, the Target Impulse Response Matrix (TIRM), which contains co-polarized and cross-polarized target information, provides unique object features that can be capitalized to improve clutter mitigation and increase target detectability. The goal of this Chapter is to demonstrate that additional performance benefits can be pursued via polarimetric coding. To this end, a radar receiver equipped with a polarimetric filter bank (with each branch tuned to a specific TIRM, among the possible ones) is supposed and a robust fully-polarimetric transceiver design is proposed, assuming as performance metric the worst-case Signal to Interference plus Noise Ratio (SINR) among the receiver branches. In addition to an energy constraint, a similarity constraint is enforced on the transmit code to bestow desirable features to the probing waveform. Hence, leveraging advanced optimization theory concepts and tools, an iterative optimization procedure, with a polynomial computational complexity and ensured converge features, is devised to optimally synthesize the polarimetric transceiver. Finally, several numerical examples, with measured TIRMs, are provided to assess the effectiveness (also in terms of increased robustness) of the proposed method as compared with existing alternatives, as well as to shed light on the achievable performance improvements with respect to classic single polarization channel radars.

Chapter 10: An Ambiguity Suppression Scheme for Quad-Pol SAR Based on Quasi-Orthogonal Waveforms
Authors: G. Ji, A. Aubry, A. De Maio, R. Wang, and W. Wang
Quad-pol Synthetic Aperture Radar (SAR) constitutes nowadays one of the most promising and appealing remote sensing technologies, with numerous applications in both civilian and defence fields. However, such a sensing system suffers from a severe ambiguity interference, which represent its major shortcoming. This Chapter aims at mitigating ambiguity problems in quad-pol SARs by well-designed quasi-orthogonal waveforms. To this end, the synthesis of Nonlinear Frequency Modulation (NLFM) waveforms with a continuous piecewise linear instantaneous frequency, considering simultaneously the Cross-Correlation Energy (CCE) and the Peak to Side Lobe Ratios (PSLRs) as performance metrics, is formulated. A Coordinate Descent (CD) method is employed to handle the resulting NP-hard problem, where at each step the marginal minimization is tackled via specific optimization tools. Hence, transmission/reception schemes jointly capitalizing on quasi-orthogonal NLFM waveforms and Azimuth Phase Coding (APC) technique are proposed to suppress ambiguity interference. Moreover, a systematic framework for the evaluation of the resulting Azimuth Ambiguity-to-Signal Ratio (AASR) and Range Ambiguity-to-Signal Ratio (RASR) is provided. Finally, detailed simulation experiments are illustrated to assess the ambiguity suppression capabilities of the devised schemes.

Chapter 11: Fully Polarimetric Monopulse SBR for Space Situational Awareness
Authors: M. Maffei, A. Aubry, A. De Maio, and A. Farina
This Chapter delves into a novel Spaceborne Radar (SBR) payload transceiver for augmenting space-based monitoring capabilities for near-Earth Space Situational Awareness (SSA). In order to cope with small-size tumbling debris (which are

orbiting at hyper-velocities in free space interspersed through slabs of plasma media), the aforementioned architectural design comprises polarimetric diversity not only to boost detection and tracking performance per se but also to augment time series analysis on target signatures. In this respect, both the sensor polarimetric monopulse data hyper-cube structure and the timing hierarchies in surveillance mode are addressed along with the definition of a suitable contacts collection strategy. More specifically, a bespoke quad-pol Active Electronically Scanned Array (AESA) based pulse-Doppler radar architecture in the Ka Band is developed to tackle the significant Doppler stress from hyper-velocity debris echoes. Accordingly, the signal processing techniques pertaining to the transceiver functional lineup are thoroughly detailed. Finally, results are illustrated to prove the effectiveness of the proposed debris detection and tracking architecture as well as to show the performance amelioration via polarimetric information diversity as compared with single-pol inference capabilities.

Chapter 12: Polarization Information Processing in Insect Radar
Authors: C. Hu, W. Li, and R. Wang
Insect radar is the most effective tool for monitoring insect migration, which greatly promotes the development of radar entomology. The estimation of biological and behavioral parameters of the insect can be pursued leveraging polarization information. Specifically, based on polarimetric measurements, insect features, such as mass, body length and width, as well as orientation, can be inferred. In this Chapter, after a brief description of the insect radar history, insect parameters estimation problem based on polarimetric radar data is addressed. Hence, a multi-channel fully polarimetric radar system, aimed at directly acquiring the target/insect scatter matrix, is described and advanced fully polarimetric signal processing techniques are provided and tested. Accuracy and robustness improvements with respect to a partial polarimetric radar, in the insect parameters estimation process, are finally proved.

Chapter 13: Polarimetric Weather Radar Signal Processing
Authors: F. Gekat, R. Hannesen, and P. Gölz
Weather/meteorological radar is a key observing system for scientific studies and operational decision-making involving severe weather phenomena and precipitation processes. Recent maturation of dual-polarization methods provides previously unavailable microphysical information that has significantly improved precipitation rate estimates, flood warnings, hail detection, tornado detection, winter weather warnings, and non-meteorological target identification. This Chapter starts with an introduction on the fundamental architectures of polarimetric Doppler weather radars as well as a brief Scattering-matrix-based description of their I-Q signals and polarimetric quantities. Based on this characterization, all steps of the signal processing chain are described, with a special emphasis on polarimetric signal processing aspects. The review starts with the analog reception of the backscattered signal, followed by their digitization, filtering and polarimetric moment estimation. Next signal processing tools for attenuation correction, rain rate estimation, and hydrometeor classification are presented. The Chapter concludes with a review of the calibration schemes required to provide an adequate data quality.

Chapter 14: Meteorological Polarimetric Phased Array Radar
Authors: I. Ivić , D. Conway, S. Torres, J. Herd, D. Zrnić, and M. Weber
Phased array radar (PAR) technology supports methods to adaptively scan only phenomena of interest and rule the surveillance system measurement rate in specific search sectors so as to improve environment awareness with respect to typically employed narrow "pencil beam" rotating antenna systems. With PAR, the measurement of meteorological targets polarimetric properties is more challenging relative to conventional weather radars, that utilize mechanically scanned parabolic reflector antennas, due to several impairments, for instance unequal variation of the horizontal and vertical co-polar patterns along the scanning, the presence of cross-polar patterns, and other system mismatches. As a result, careful calibration procedures leading to bias correction are necessary to achieve polarimetric data quality. In this Chapter, a review of the "state of the art" for meteorological polarimetric PAR (PPAR) for weather observations is provided, with the description of recent methods addressing the mentioned challenges. Plausible meteorological PPAR architectures are illustrated, accounting for array size and geometry, radiated power, as well as level of array digitization, and their comparison in terms of critical PPAR performance requirements is provided. Next, discussions about ongoing research using the "Advanced Technology Demonstrator (ATD)", a 4 m diameter PPAR developed and deployed at NOAA's National Severe Storms Laboratory (NSSL), are provided. In this respect, first novel methods developed for its calibration process are described. Then, meteorological measurements using the ATD are analysed to assess the effectiveness of the described calibration and bias correction methods, as well as pinpoint progress (with respect to a conventional system) and remaining issues. The Chapter concludes with the Authors' point of view on the outlook for PPAR, both for the research community and as a technology basis for the future U.S. national weather radar network.

Each chapter is self-contained and is written by renowned researchers in polarimetric radar signal processing. The emphasis of the book is on both theoretical results and practical applications and clearly show the potential benefits in radar performance using polarimetric diversity in different application domains. Cross referencing and a common notation (reported at the end of this foreword) have been realized so that the related material as well as equations can be easily connected. This significantly enhances the book's value as a reference.

This book is addressed to systems engineers and their managers in civilian as well as defense companies. Technical staff in procurement agencies and their technical advisers. Students at MSc and PhD levels in signal processing, electrical engineering, systems and defense engineering. People interested in applications of polarimetry theory to radar engineering.

Acknowledgements

A. Farina wishes to kindly thank the correspondence with Prof. Dr. L.P. Ligthart FIEEE, FIET, Em. Prof. at the Delft University of Technology, for the exchange of information on the pioneering work of Dr. A.J. Poelman, A. Farina also acknowledges the collaboration with Dr. M. Frasca (MBDA, Italy) for the assistance in recovering some technical references.

A. Farina had the opportunity to be invited, together with Dr. B. Palumbo (Chief of Antenna Laboratory, Selenia SpA), by A. J. Poelman for a technical meeting and to visit the STC (SHAPE Technical Center)'s multipurpose experimental X-band facility having novel polarimetric capabilities. The paper [15], kindly provided by Prof. Dr. L.P. Ligthart, of Dr. Poelman in NERG magazine reflects in a superb way his views on radar polarimetry. Other relevant Dr. Poelman's contributions can be found in IEEE-AESS Vol. 4, 1975, IEEE-AESS Vol. 6, 1976, El. Letters Vol. 5, 1977, and in the NATO Bad Windsheim books edited by W.M. Boerner. Figure 3 from reference [15] shows the polarization control console of the STC's multipurpose experimental X-band facility: an appropriate and clever "man-machine interface" to command to the radar facility on the roof of STC the polarization the radar should radiate. More in general, Dr. Poelman presented on the Poincaré sphere his fundamental polarimetric results connecting measurements, physics and mathematics.

A. Farina had the honour to cooperate with Prof. W. M. Boerner in the field of polarimetric radar. A. Farina contributed with his colleagues with the paper [25] to a major scientific event: NATO - Advanced Research Workshop, on "Direct and Inverse Methods in Radar Polarimetry", Sept. 18–24, 1988, organised by Prof. Boerner.

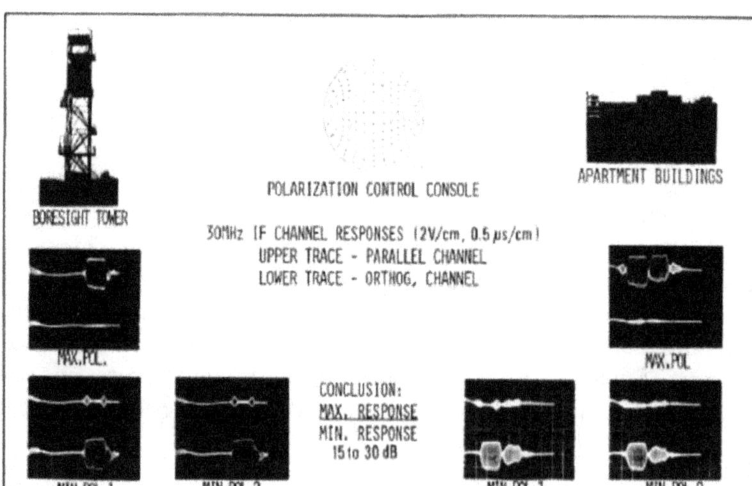

Fig. 19 : Complex radar objects and responses for maximum and minimum polarizations; obtained with STC's experimental X-band facility (1977)

Figure 3 Notional description of the STC's multipurpose experimental X-band facility including the polarization control console (from reference [15])

There were a number of opportunities to learn from Prof. Boerner. A. Farina had the chance to be invited at his seminar on polarimetry at ONERA in May 1987, which was organised by Dr. J. Dorey, Prof. Boerner then was instrumental for A. Farina visit to DFVLR (Deutsche Forschungs- und Versuchsanstalt für Luft- und Raumfahrt) (since 1989 DLR: Deutsches Zentrum für Luft- und Raumfahrt) in Oberpfaffenhofen, Germany in 1987. Dr. A. Schroth showed in operation the Polar 2000, a dual polarization Doppler weather radar.

The co-editors warmly thank all the esteemed Authors for their dedication, their professional skill and hard work in writing the book chapters and taking care of the peer review process. They generously offered their time and competence in carefully reading the submitted chapters, providing their comments and completing the review process in due time.

Co-editor's gratitude goes to IET Team and to Mrs. O. Wilkins, Assistant Editor, so precious in bringing us with her advices and guidance along the avenue that brought us to the completion of the book, starting from a few page proposal.

References

[1] W.M. Boerner, "Historical development of radar polarimetry, incentive for this workshop, and overview of contribution to these proceedings", pp. 1–32 in Part 1 of "Direct and Inverse Methods in Radar Polarimetry", NATO ASI Series, vol. 350, Springer, 1992, edited by W.M. Boerner et al (eds.).

[2] D. Giuli, "Polarization Diversity in Radars," Proc. IEEE, vol. 74, pp. 245–269, February 1986.

[3] https://glossary.ametsoc.org/wiki/Polarimetric_radar

[4] G. P. Konnen, "Polarization in Nature" (invited paper), pp. 33–43 in Part 1 of "Direct and Inverse Methods in Radar Polarimetry", NATO ASI Series, vol. 350, Springer, 1992, edited by W.M. Boerner et al (eds.).

[5] D. Száz and G. Horváth, "Success of Sky-Polarimetric Viking Navigation: Revealing the Chance Viking Sailors Could Reach Greenland from Norway", R. Soc. open sci. 5, 2018.
http://dx.doi.org/10.1098/rsos.172187
https://royalsocietypublishing.org/doi/10.1098/rsos.172187

[6] J. Bohannon, "The Viking Sunstone Revealed? Physicists Show Vikings Could Have Used Calcite Crystals to Navigate to North America", Science. News, Physics, 1 November 2011. https://www.science.org/content/article/viking-sunstone-revealed

[7] G. Horváth, "Brief History of Light Polarization, Brief History of the Discovery of Phenomena Concerning Light Polarization", Chapter 1 of "Polarization Patterns In Nature. Imaging Polarimetry with Atmospheric Optical and Biological Applications", Dr. G. Horváth's D.Sc. Thesis, Hungarian Academy of Sciences, 2003.
https://arago.elte.hu/sites/default/files/DSc-Thesis-2003-GaborHorvath-01.pdf

[8] E. Brookner, W. M. Hall, and R. Hampton Westlake, "Faraday Loss for L-Band Radar and Communications Systems", IEEE Transactions on Aerospace and Electronic Systems, vol. AES-21, no. 4, pp. 459–469, July 1985.
[9] K. Naishadham, "Introduction to Radar Polarimetry", Chap. 13 in "Principles of Modern Radar, Vol II, Advanced Techniques", SciTech Publishing, Edison, NJ. 2013, edited by W. L. Melvin and J. A. Scheer.
[10] W.M. Boerner, L.A. Cram, W.A. Holm, D.E. Stein, and W. Wiesbeck, "Direct and Inverse Methods in Radar Polarimetry", NATO ASI Series, vol. 350, Springer, 1992.
[11] N. Wiener, "Generalized harmonic analysis", Acta Mathematica, vol. 55, pp. 117–258, 1930.
[12] M. E. Kennaugh, "Polarization Properties of Radar Reflectors", M. E. Kennaugh's M. Sc. Thesis, The Ohio State University, 1952.
[13] J.R. Huynen, "Measurement of the Target Scattering Matrix", Proceeding of the IEEE, vol. 53, no. 8, pp. 936–946, August 1965.
[14] J. R. Huynen, "Phenomenological Theory of Radar Targets", J. R. Huynen 's PhD dissertation, Technical University Delft, 1970.
[15] A. J. Poelman, "The Applicability of Controllable Antenna Polarization to Radar Systems", Tijdschrift van het Nederlands Electronica en Radiogenootschap, vol. 44, no. 2, pp. 93–106, 1979.
[16] R. Efurd, M. Cohen, and E. Sjober, "Advanced Intrapulse Polarization Agile Radar Nears Deployment", Microwave Syst. News, vol. 14, pp. 67–76, February 1984.
[17] W. A. Holm, "Polarization Scattering Matrix Approach to Stationary Target/Clutter Discrimination", Proc. Colloque International sur le Radar, pp. 461–465, May 1984.
[18] J. L. Eaves and E. K. Reedy, "Principles of Modern Radar", Van Nostrand Reinhold Co., New York, 1987.
[19] L. M. Novak, M. B. Sectchin, and M. J. Cardullo, "Studies of Target Detection Algorithms that Use Polarimetric Radar Data", IEEE Transactions on Aerospace and Electronic Systems, vol. 25, no. 2, pp. 150–165, March 1989.
[20] C. Elachi, "Spaceborne Radar Remote Sensing: Applications and Techniques", IEEE Press, Piscataway, N. J., 1988.
[21] F. T. Ulaby and C. Elachi, "Radar Polarimetry for Geoscience Applications", Artech House, Inc., 1990.
[22] C. Elachi and J. Van Zyl, "Introduction to the Physics and Techniques of Remote Sensing", 2nd Edition, Wiley Series in Remote Sensing and Image Processing, Wiley-Interscience, 2006.
[23] J. van Zyl and Y. Kim, "Synthetic Aperture Radar", in Encyclopedia of Physical Science and Technology, Third Edition, 2003. https://www.sciencedirect.com/topics/earth-and-planetary-sciences/shuttle-imaging-radar
[24] R.K. Raney, "Space-Based Remote Sensing Radars", Chapter 18 of "Radar Handbook", 3rd Edition, Mc-Graw Hill, Inc., January 2008, edited by M. I. Skolnik.

[25] A. Farina, F. Scannapieco, and F. Vinelli, "Target Detection and Classification with Polarimetric High Range Resolution Radar", pp. 1021–1041 in Part 1 of "Direct and Inverse Methods in Radar Polarimetry", NATO ASI Series, vol. 350, Springer, 1992, edited by W.M. Boerner et al (eds.).
[26] B. Balaji, M. Frasca, and A. Farina, "Quantum Radar Research: A Snapshot in Time", IEEE Aerospace and Electronic Systems Magazine, vol. 35, no. 4, pp. 74–76, April 2020.
[27] J. S. Lee and E. Pottier, "Polarimetric Radar Imaging: From Basics to Applications", CRC Press, 2017.
[28] V. N. Bringi and V. Chandrasekar, "Polarimetric Doppler Weather Radar", Cambridge University Press, 2011.
[29] F. Fabry, "Radar Meteorology: Principles and Practice", Cambridge University Press, 2015.
[30] J. Van Zyl and Y. Kim, "Synthetic Aperture Radar Polarimetry", John Wiley & Sons Inc., 2011.

Notations

$\boldsymbol{a}$	column vector (lower case);
$\boldsymbol{A}$	matrix (upper case);
$a(n)$	n-th element of $\boldsymbol{a}$;
$\boldsymbol{A}(m,n)$	(m,n)-th entry of $\boldsymbol{A}$;
$(\cdot)^T$	transpose operator;
$(\cdot)^*$	conjugate operator;
$(\cdot)^\dagger$	conjugate transpose operator;
$\mathrm{tr}\,(\cdot)$	trace of the square matrix argument;
$\det(\cdot)$	determinant of the square matrix argument;
$\mathrm{Rank}\,(\cdot)$	rank of the square matrix argument;
$\lambda_{max}(\cdot)$	maximum eigenvalue of the square matrix argument;
$\lambda_{min}(\cdot)$	minimum eigenvalue of the square matrix argument;
$\mathbf{diag}(\boldsymbol{a})$	N-dimensional diagonal matrix whose i-th diagonal element is $a(i)$, for $i = 1, \ldots, N$, with $\boldsymbol{a} \in \mathbb{C}^N$;
$\mathbf{Diag}\,(\boldsymbol{A})$	N-dimensional column vector whose i-th entry is $\boldsymbol{A}(i,i)$, $i = 1, \ldots, N$;
$\mathrm{Range}\,(\boldsymbol{X})$	range span of the column vectors of the matrix $\boldsymbol{X}$;
$\boldsymbol{I}$	identity matrix (its size is determined from the context);
$\mathbf{0}$	matrix with zero entries (its size is determined from the context);
$\mathbb{R}^N$	set of N-dimensional vectors of real numbers;
$\mathbb{C}^N$	set of N-dimensional vectors of complex numbers;
$\mathbb{H}^N$	set of $N \times N$ Hermitian matrices;
$\succeq$	for any $\boldsymbol{A} \in \mathbb{H}^N$, $\boldsymbol{A} \succeq \mathbf{0}$ means that $\boldsymbol{A}$ is a positive semidefinite matrix;
$\succ$	for any $\boldsymbol{A} \in \mathbb{H}^N$, $\boldsymbol{A} \succ \mathbf{0}$ means that $\boldsymbol{A}$ is a positive definite matrix;
$\|\boldsymbol{x}\|$	Euclidean norm of the vector $\boldsymbol{x}$;
j	imaginary unit (i.e. $j = \sqrt{-1}$);
i	generic index;
$\mathrm{Re}(x)$	real part of the complex number x;
$\mathrm{Im}(x)$	imaginary part of the complex number x;
$\|x\|$	modulus of the complex number x;
$\arg(x)$	argument of the complex number x;
$\mathbb{E}\,[\cdot]$	statistical expectation;
$\odot$	Hadamard product;
$\otimes$	Kronecker product;

$\dot{y}, \frac{\partial y}{\partial x}, \frac{dy}{dx}$	first derivative of y with respect to variable x;
$\ddot{y}, \frac{\partial^2 y}{\partial x^2}, \frac{d^2 y}{dx^2}$	second derivative of y with respect to variable x;
$\mathbb{P}[\cdot]$	probability measure;
$v(\mathcal{P})$	optimal value of the optimization problem $\mathcal{P}$;
r, θ, ϕ	radius, elevation and azimuth angles in the spherical coordinate system;
f	frequency (Hz);
$\omega = 2\pi f$	radian frequency (rad/s);
κ	wavenumber (1/m);
λ	wavelength (m);
c	speed of light (m/s);
t	time (s);
$\boldsymbol{E}$	electric field intensity (V/m);
(g_0, g_1, g_2, g_3)	Stokes parameters;
d	degree of polarization;
$\boldsymbol{M}_u$	Mueller matrix;
$\boldsymbol{K}_e$	Kennaugh matrix;
$\boldsymbol{S}_t$	target scattering matrix;
$\boldsymbol{S}_d$	diagonalized target scattering matrix;
$\boldsymbol{J}, \boldsymbol{K}, \boldsymbol{L}$	Pauli spin (rotation) matrices;
σ	radar cross section.

Chapter 1
Adaptive target detection with polarimetric radars

A. Aubry[1], A. De Maio[1], and A. Farina[2]

1.1 Introduction

Polarization diversity is a key tool to sensibly boost the detection performance of a radar system, especially when the Doppler discrimination is not entirely possible. This is the case of Doppler ambiguous, slowly moving, or tangentially moving targets, which are masked by the clutter environment in conventional pulse-Doppler radars. In these challenging situations, where angle and Doppler features of the radar returns do not allow to distinguish targets from background clutter, multi polarimetric measurements provide valuable information for target detection.

Not surprisingly, this sensing mode has been widely exploited during the last decades both for defense and civilian applications and has been attracted a lot of interest in the radar research community. In this context, the first studies [1] have been mainly focused on the design of adaptive polarimetric filters and cancellers. In [2], several algorithms capitalizing on polarimetric radar information have been introduced to improve target detection capabilities and to ameliorate the discrimination between different type of targets. Therein, the design of the optimum polarimetric receiver has been addressed, assuming known the target and noise polarimetric scattering matrices. The resulting ideal detector provides an upper bound to the performance achievable by other suboptimal polarimetric schemes. In the same paper, also a new polarimetric target model, used at the analysis stage, has been introduced.

Inspired by the pioneering works [3] and [4] on adaptive target detection for single-polarization channels, in [5], the problem of combined polarization and space-time processing to detect targets (with unknown polarimetric characteristics) embedded in a Gaussian noise with unknown covariance matrix has been addressed. Interestingly, the proposed receivers ensure the Constant False Alarm Rate (CFAR) property with respect to the clutter covariance. Moreover, in [6],

[1]University of Naples "Federico II", Naples, Italy
[2]Selex-ES y (retired; consultant), Rome, Italy

the approach of [5] has been extended to allow the full polarimetric information processing and to capitalize on *a-priori* knowledge of the polarimetric covariance matrix. It is also worth mentioning that in [7], the authors have proposed two adaptive detectors based on the generalized likelihood ratio (GLR) and on the Wald tests, respectively, that exploit the spillover of target energy and also use the polarization diversity to improve the detection and range estimation performance. These decision schemes guarantee the CFAR property with respect to the unknown parameters of the clutter. Some attempts to detect small and slowly moving targets in heavy heterogeneous clutter via polarization diversity have been also pursued in [8–10]. In [9, 10], clutter returns from polarimetric channel have been modeled as spherically invariant random vectors (SIRVs) with possibly correlated textures and speckles. Under these assumptions, in [9], the authors introduced an adaptive receiver designed according to the GLR criterion, which remarkably ensures the CFAR property with respect to the texture statistics. A different GLR procedure with a better computational efficiency than [9] also leading to a polarimetric adaptive receiver ensuring the CFAR property with respect to the texture statistics has been devised and assessed in [10]. In [8], a detector able to establish the presence of a target based only on the data collected from the range cell under test (CUT) (without resorting to secondary data or prior knowledge about the clutter) has been proposed.

The exploitation of the polarization has been investigated also in advanced surveillance systems, i.e., multiple-input multiple-output (MIMO) radars. In this context, the detection of point-like targets against Gaussian interference, with known statistics, has been addressed in [11], where it has been shown that the optimal selection of the antenna polarizations provides improved performance over conventional MIMO systems leveraging only horizontal or vertical polarization. Still assuming Gaussian interference, the detection of moving targets has been considered in [12]. The challenging adaptive target detection in compound-Gaussian clutter has been addressed in [13] and [14], where GLR-, Rao-, and Wald-based tests have been devised. In [15], a partially homogeneous interference environment has been considered and some detection architectures, i.e., the matched subspace detector and the adaptive subspace detector (designed in [16]), have been developed to tackle the detection problem. Therein, it has been also presented a waveform design algorithm that selects the polarization of the transmitted pulses to boost the detection performance of the two detectors.

Finally, in [17], the exploitation of polarimetric diversity in passive surveillance systems as a means to improve the capabilities of so called "green radars" has been investigated. Therein, the disturbance has been modeled as a multichannel autoregressive process allowing a significant reduction on the training data size required for the adaptation process. The interested reader may also refer to Chapter 2 for further details.

In this chapter, the design and analysis of some adaptive architecture exploiting polarimetric diversity for the detection of point-like targets is reviewed. Targets embedded in either Gaussian or non-Gaussian interference with unknown covariance matrix are considered, assuming the availability of secondary data (free of signal components and with the same covariance/covariance structure of the cells under test). To account for target polarimetric features, the perspective target echo from each range bin is modeled as a deterministic signal known up to a possibly varying scaling factor. Hence, to solve the resulting composite binary hypothesis testing problem, some adaptive detection schemes are illustrated according to diverse criteria such as GLR, Wald, and Rao tests. Along with a discussion on the CFAR, several case studies, both on simulated and measured data, are provided to shed light on the performance gains ensured by polarization diversity as well as to highlight pros and cons of the considered detection strategies in comparison with some counterparts from the open technical literature.

The rest of this chapter is organized as follows. In section 1.2, the polarimetric radar system model is introduced along with the formulation of the adaptive target detection problem. In sections 1.3 and 1.4, the design and analysis of some polarimetric adaptive detectors for point-like targets embedded in homogeneous and heterogeneous radar environments are reported. Finally, in section 1.5, concluding remarks are provided.

1.2 Problem formulation & design issues

A radar system transmitting a coherent burst of N pulses with a given Electromagnetic polarization and equipped with two orthogonal polarimetric receive channels is considered. For instance, the radar probes the environment with horizontal polarized waves (or, equivalently, the vertical polarization) and collects information on two linear orthogonal polarizations, denoted as HH and HV (VH and VV). Alternatively, the radar transmits both the horizontal and the vertical polarizations and the HH and the VV polarizations are received.

To proceed further, let us denote by $z_{1,1}$ and $z_{1,2}$ the N-dimensional vectors collecting the baseband signal samples from the range CUT on the two receive polarimetric channels. Hence:

$$z_1 = \left[z_{1,1}^T,\ z_{1,2}^T\right]^T \in \mathbb{C}^{2N} \tag{1.1}$$

represents the primary data vector, which embeds polarimetric echoes backscattered by the prospective target in the CUT. As customary, it is also assumed that a set of $K \geq 2N$ auxiliary data, free of useful signal contributions and sharing the same spectral characteristics as the primary vector (i.e. secondary/training data), is available. The observations in the training set are denoted by z_k, $k = 2, \ldots, K + 1$, where:

$$z_k = \left[z_{k,1}^T, z_{k,2}^T\right]^T, \quad k = 2, \ldots, K + 1 \tag{1.2}$$

with $z_{k,1}$ and $z_{k,2}$ the N-dimensional vectors acquired on the two different polarimetric channels at the kth range cell.

According to the above definitions, the problem of detecting a point-like target can be formulated as the following binary hypotheses test:

$$\begin{cases} H_0: \begin{cases} z_1 = \boldsymbol{n}_1 \\ z_k = \boldsymbol{n}_k \quad k = 2, \ldots, K+1 \end{cases} \\ \\ H_1: \begin{cases} z_1 = \boldsymbol{Pa} + \boldsymbol{n}_1 \\ z_k = \boldsymbol{n}_k \qquad\quad k = 2, \ldots, K+1 \end{cases} \end{cases} \tag{1.3}$$

where

- $\boldsymbol{P} = \begin{pmatrix} \boldsymbol{p} & 0 \\ 0 & \boldsymbol{p} \end{pmatrix} \in C^{2N,2}$, with $\boldsymbol{p}$ the N-dimensional temporal steering vector, which depends on the target Doppler frequency;
- $\boldsymbol{a} = [a(1), a(2)]^T \in C^2$ is a bidimensional complex vector whose elements account for both the target polarimetric characteristics and the channel propagation effects;
- $\boldsymbol{n}_k = [\boldsymbol{n}_{k,1}^T, \boldsymbol{n}_{k,2}^T]^T \in \mathbb{C}^{2N}$, $k = 1, \ldots, K+1$, contains the interference samples on the two polarimetric channels at the kth range cell.

As to the statistical characterization of the disturbance vectors $\boldsymbol{n}_k$, two radar scenarios are considered. The former, known in radar jargon as homogeneous environment, assumes that $\boldsymbol{n}_k$, $k = 1, \ldots, K+1$, are $2N$-dimensional, independent, zero-mean circularly symmetric Gaussian random vectors with an unknown covariance matrix $\boldsymbol{R}$. This represents a valuable model for low-resolution radar systems, where there are a large number of scatterers within each radar cell and the clutter disturbance is indeed well described as Gaussian distributed with the same range-invariant characteristics. The latter, usually known as heterogeneous environment, models $\boldsymbol{n}_{k,1}$ and $\boldsymbol{n}_{k,2}$ as SIRVs [9, 10]. This means that

$$\boldsymbol{n}_{k,1} = \sqrt{\sigma_{k,1}}\boldsymbol{g}_{k,1} \qquad \boldsymbol{n}_{k,2} = \sqrt{\sigma_{k,2}}\boldsymbol{g}_{k,2}$$

where

- the speckle components, namely $\boldsymbol{g}_{k,1}$ and $\boldsymbol{g}_{k,2}$, $k = 1, \ldots, K+1$, are N-dimensional zero-mean circularly symmetric Gaussian random vectors, possibly correlated in the polarimetric domain while statistically independent in the range domain;
- the textures, i.e., $\sqrt{\sigma_{k,1}}$ and $\sqrt{\sigma_{k,2}}$, $k = 1, \ldots, K+1$, are nonnegative random variables, possibly correlated both in the polarimetric and the spatial domains; however, the textures are usually modeled as deterministic but unknown parameters, to circumvent the unavoidable uncertainty about their joint distribution.

By doing so, $\boldsymbol{n}_k$, $k = 1, \ldots, K+1$, are $2N$-dimensional, independent, zero-mean, circularly symmetric Gaussian random vectors whose unknown covariance matrix is

$M_k = \Gamma_k R \Gamma_k$, where $R = \mathbb{E}\left[g_k g_k^\dagger\right]$ (with $g_k = \left[g_{k,1}^T, g_{k,2}^T\right]^T$) is the covariance matrix structure and $\Gamma_k = \textbf{Diag}\left([\sqrt{\sigma_{k,1}}, \sqrt{\sigma_{k,2}}]\right) \otimes I_N$.

Notably, the homogeneous scenario is a special instance of the heterogeneous environment. Specifically, the former is obtained forcing $\sigma_{k,1} = \sigma_{k,2} = 1$, $k = 1, ..., K + 1$, in the heterogeneous setup.

According to the Neyman–Pearson criterion, which aims at maximizing the detection probability under a constraint on the false alarm rate, the optimal decision statistic for the hypotheses testing problem (1.3) is the likelihood ratio test (LRT). Unfortunately, the resulting detector relies on the complete knowledge of the probability density functions (PDFs) under both hypotheses and this requires a perfect knowledge of the target scattering $\boldsymbol{a}$ as well as the primary data covariance matrix. Evidently, these quantities are reasonably unknown in practical situations and, as a result, a uniformly most powerful test (UMP) test for the aforementioned problem is not available. Hence, practically implementable receivers have to be designed resorting to suboptimal criteria, such as GLR, Rao, and Wald tests, which involve maximum likelihood (ML)-based estimates of the unknowns under one or both hypotheses. Thus, the developments reported in the subsequent sections demand specifying the multivariate PDF of the observables. Leveraging previous assumptions, it follows that z_k, $k = 1, ..., K + 1$, are independent random vectors regardless of the hypothesis, with:

$$f(z_k \mid R, \sigma_{k,1}, \sigma_{k,2}; H_0) = \frac{1}{\pi^{2N} det(M_k)} \exp\left[-z_k^\dagger M_k^{-1} z_k\right] \tag{1.4}$$

and

$$f(z_k \mid R, a, \sigma_{k,1}, \sigma_{k,2}; H_1) = \frac{1}{\pi^{2N} det(M_k)} \exp\left[-(z_k - i_k P a)^\dagger M_k^{-1}(z_k - i_k P a)\right] \tag{1.5}$$

where $i_k = 0$ if $k = 2, ..., K + 1$, and $i_k = 1$, for $k = 1$.

1.3 Detection schemes for targets embedded in Gaussian clutter

In this section, the design and analysis of adaptive polarimetric detectors is addressed assuming point-like targets buried in homogeneous clutter. Specifically, the receiver architectures devised according to either the one-step or the two-step GLR design criterion are now discussed.

The one-step GLR test (GLRT) for the hypotheses testing problem (1.3) assuming an homogeneous environment has been proposed in [5]. It has been devised replacing the unknown parameters, appearing in the LRT, with their ML estimates under each hypothesis. The resulting decision rule, denoted as polarimetric GLRT (PGLRT) hereafter, is given by

$$\frac{z^\dagger S^{-1} P (P^\dagger S^{-1} P)^{-1} P^\dagger S^{-1} z}{1 + z^\dagger S^{-1} z} \underset{H_0}{\overset{H_1}{\gtrless}} \gamma \tag{1.6}$$

where γ is the detection threshold set to ensure a desired probability of false alarm* (P_{fa}) and $\boldsymbol{S} = \frac{1}{K}\sum_{n=2}^{K+1} \boldsymbol{z}_n\boldsymbol{z}_n^\dagger$ is the secondary data sample covariance matrix.

The two-step GLRT was synthesized in 18, according to the following two steps:

- it is assumed that the covariance matrix $\boldsymbol{R}$ is known and the GLRT leveraging primary data (i.e. returns from the CUT) and over the remaining unknowns, i.e., a, are computed;
- a fully adaptive detection strategy is conceived plugging the ML estimate of $\boldsymbol{R}$, based on the set of secondary data, into the decision statistic in place of the unknown true covariance matrix.

The resulting adaptive detector, referred to in the following as polarimetric adaptive matched filter (PAMF), is thus,

$$\boldsymbol{z}^\dagger\boldsymbol{S}^{-1}\boldsymbol{P}(\boldsymbol{P}^\dagger\boldsymbol{S}^{-1}\boldsymbol{P})^{-1}\boldsymbol{P}^\dagger\boldsymbol{S}^{-1}\boldsymbol{z} \underset{H_0}{\overset{H_1}{\gtrless}} \gamma \tag{1.7}$$

Notably, both PGLRT and PAMF guarantee CFARness with respect to the covariance matrix of the interference. Besides, both the algorithms involve $O(KN^2)$ floating point operations (flops)†. Nevertheless, the PAMF does not require the evaluation of the term $1 + \boldsymbol{z}^\dagger\boldsymbol{S}^{-1}\boldsymbol{z}$, which is computationally intensive for real-time systems as it must be calculated for each primary data. Thus, the PAMF is faster to implement than the plain GLRT.

1.3.1 Performance assessment

This section is devoted to the performance assessment, in terms of computational complexity and detection capabilities of PGLRT and PAMF receivers. Closed form expressions for P_{fa} and probability of detection P_d (assuming non-fluctuating target) can be found in [5] and [18] for PGLRT and PAMF, respectively. Specifically, for a fixed detection strategy:

- the expression of P_{fa} does not involve the covariance matrix of the disturbance, thus confirming the CFAR property of the proposed algorithm, i.e., the detection threshold can be set regardless of the actual interference scenario;
- for non-fluctuating targets, P_d depends on the detection threshold and the signal to noise ratio (SNR) defined as

$$\text{SNR} = \boldsymbol{a}^\dagger(\boldsymbol{P}^\dagger\boldsymbol{R}^{-1}\boldsymbol{P})\boldsymbol{a} \tag{1.8}$$

*With a slight abuse of notation, for any given detection strategy only one symbol is used to denote the threshold and its possible modification.

†An algorithm is O(n) if its implementation requires a number of flops proportional to n.

- in the case of (fluctuating targets), the P_d can be evaluated averaging the non-fluctuating target P_d expression with respect to the random target strength. Further developments require to specify target and disturbance models. To this end, in the following it is assumed that:

1. $\boldsymbol{a} = \sqrt{\alpha}\boldsymbol{x}$, with $\boldsymbol{x}$ and α statistically independent random quantities. Specifically, $\boldsymbol{x}$ is modeled as a bidimensional, zero-mean, circularly symmetric Gaussian random vector with covariance matrix‡ [2]:

$$\boldsymbol{M}_t = \begin{pmatrix} 1 & \epsilon_t\sqrt{\delta_t} \\ \epsilon_t^*\sqrt{\delta_t} & \delta_t \end{pmatrix} \tag{1.9}$$

Besides, α is modeled as a Gamma random variable, with mean value $\overline{\alpha}$ (related to the average radar cross section) and shape parameter ν;

2. As to the disturbance, a clutter-dominated environment is assumed resulting into a covariance matrix given by

$$\boldsymbol{R} = \boldsymbol{M}_c \otimes \boldsymbol{\Sigma} \tag{1.10}$$

where Σ is an $N \times N$ matrix accounting for the covariance between returns from the same polarimetric channel, and

$$\boldsymbol{M}_c = \begin{pmatrix} 1 & \epsilon_c\sqrt{\delta_c} \\ \epsilon_c^*\sqrt{\delta_c} & \delta_c \end{pmatrix}$$

is the polarimetric scattering matrix of the clutter (see footnote ‡ with reference to typical values of ϵ_c and δ_c).

Under this statistical model, it can be shown that the P_d depends on $\overline{\alpha}\boldsymbol{p}^\dagger\boldsymbol{\Sigma}^{-1}\boldsymbol{p}$ and $\boldsymbol{M}_t^{1/2}\boldsymbol{M}_c^{-1}\boldsymbol{M}_t^{1/2}$ (the interested reader may refer to [5] and [18]); in particular, a key parameter is the average SNR defined as

$$\begin{aligned}
\mathrm{SNR}_{av} &= \mathbb{E}[\boldsymbol{a}^\dagger(\boldsymbol{P}^\dagger\boldsymbol{R}^{-1}\boldsymbol{P})\boldsymbol{a}] \\
&= \mathbb{E}[\alpha]\mathrm{tr}\,(\boldsymbol{M}_t^{1/2}(\boldsymbol{P}^\dagger\boldsymbol{R}^{-1}\boldsymbol{P})\boldsymbol{M}_t^{1/2}) \\
&= \mathbb{E}[\alpha]\mathrm{tr}\,(\boldsymbol{M}_t^{1/2}(\boldsymbol{P}^\dagger(\boldsymbol{M}_c^{-1}\otimes\boldsymbol{\Sigma}^{-1})\boldsymbol{P})\boldsymbol{M}_t^{1/2}) \\
&= \mathbb{E}[\alpha]\boldsymbol{p}^\dagger\boldsymbol{\Sigma}^{-1}\boldsymbol{p}\,\mathrm{tr}\,(\boldsymbol{M}_t^{1/2}\boldsymbol{M}_c^{-1}\boldsymbol{M}_t^{1/2}).
\end{aligned}$$

‡Typical values of et and dt can be found in [2].

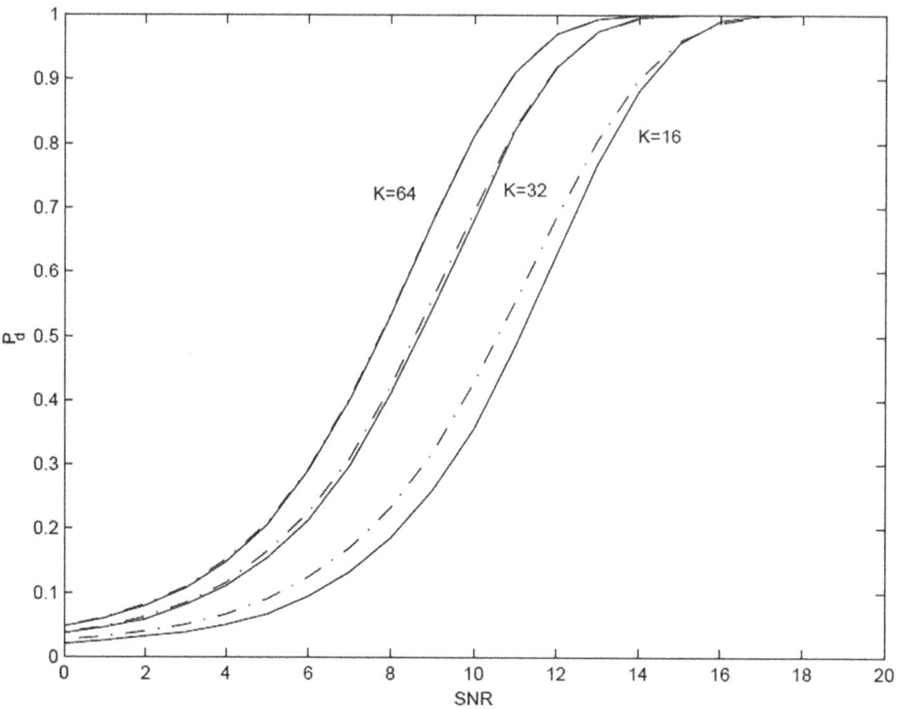

Figure 1.1 *P_d versus SNR of the PAMF (solid curves) and of the PGLRT (dashed-dotted curves) for $P_{fa} = 10^{-4}$, $N = 4$, and different training data sizes*

In Figure 1.1, P_d versus SNR for PGLRT and PAMF is plotted, assuming $P_{fa} = 10^{-4}$, a non-fluctuating target, $N = 4$, and three distinct values of K, i.e., $K \in \{16, 32, 64\}$. Inspection of the figure highlights that the PAMF experiences a performance degradation as compared with the GLRT for $K = 16$. This loss becomes more and more contained as K increases, in particular when $K \geq 32$ no significant P_d reductions can be appreciated. This behavior is not surprising, indeed the GLRT uses all the data to estimate the disturbance covariance matrix, while the PAMF resorts to secondary data only. Thus, the loss of the PAMF for low values of K is mainly due to the reduced sample size, which, however, is less and less relevant as K grows.

On the other hand, it is not difficult to show that the two test statistics are asymptotically equivalent as K increases. Thus, for a high enough value of K they ensure the same performance. Figure 1.1 shows that this is the case when $K \geq 64$.

The situation of a random a, i.e., a fluctuating target, is now analyzed. In Figures 1.2 and 1.3, P_d curves (of the PGLRT and the PAMF) versus the expected value SNR_{av} is reported. Therein, it is assumed $P_{fa} = 10^{-4}$, $N = 4$, $K = 32$, and several values of ν. Besides, Figure 1.2 is a representative of a radar system that processes the HH and the VV channels, while Figure 1.3 refers to a surveillance system

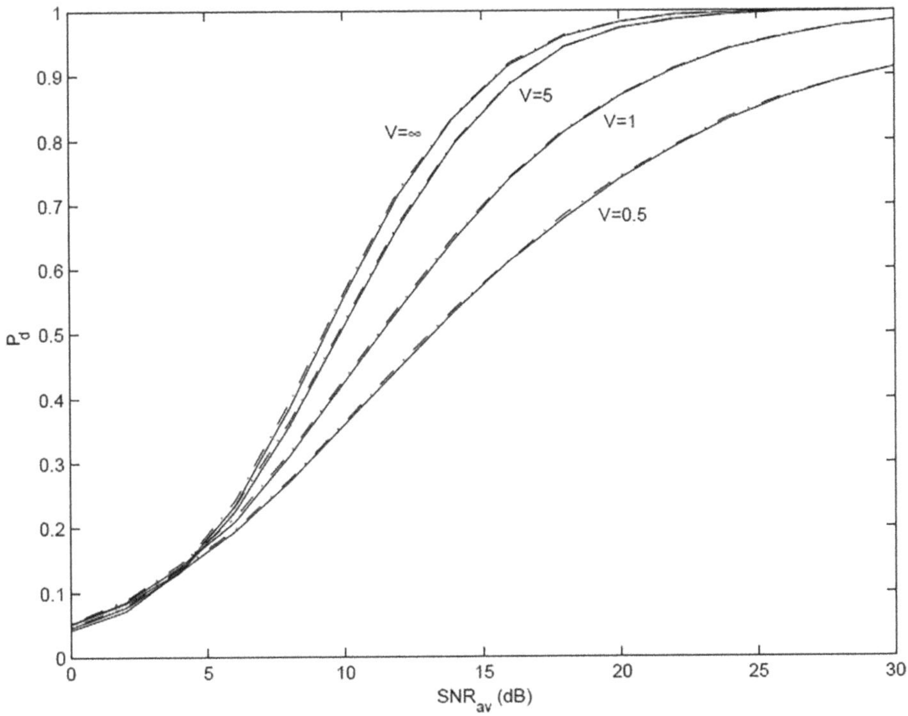

Figure 1.2 P_d *versus* SNR_{av} *of the PAMF (solid curves) and of the PGLRT (dashed-dotted curves) for* $P_{fa} = 10^{-4}$, $N = 4$, $K = 32$, $\epsilon_t = 0.28$, $\delta_t = 1$, $\epsilon_c = 0.5$, *and* $\delta_c = 1.6$. *Each curve refers to a specific value of* ν.

employing HH and HV receive channels. Indeed, in Figure 1.2 $\epsilon_t = 0.28$, $\delta_t = 1$, $\epsilon_c = 0.5$, and $\delta_c = 1.6$, whereas in Figure 1.3 $\epsilon_t = 0$, $\delta_t = 0.19$, $\epsilon_c = 0$, and $\delta_c = 0.18$ are supposed. As a consequence, when HH and HV polarimetric information is acquired, target and clutter returns at the different channels are assumed uncorrelated. Looking over the figures clearly shows that the PAMF and the GLRT achieve almost the same performance also in the presence of fluctuating targets.

1.4 Detection schemes for targets embedded in compound-Gaussian clutter

This section deals with polarimetric detection of point-like targets buried in heterogeneous clutter. As already pointed out, the compound-Gaussian distribution is suitable for polarimetric coherent clutter echoes, received by high-resolution radar operating at low grazing angles [19–21].

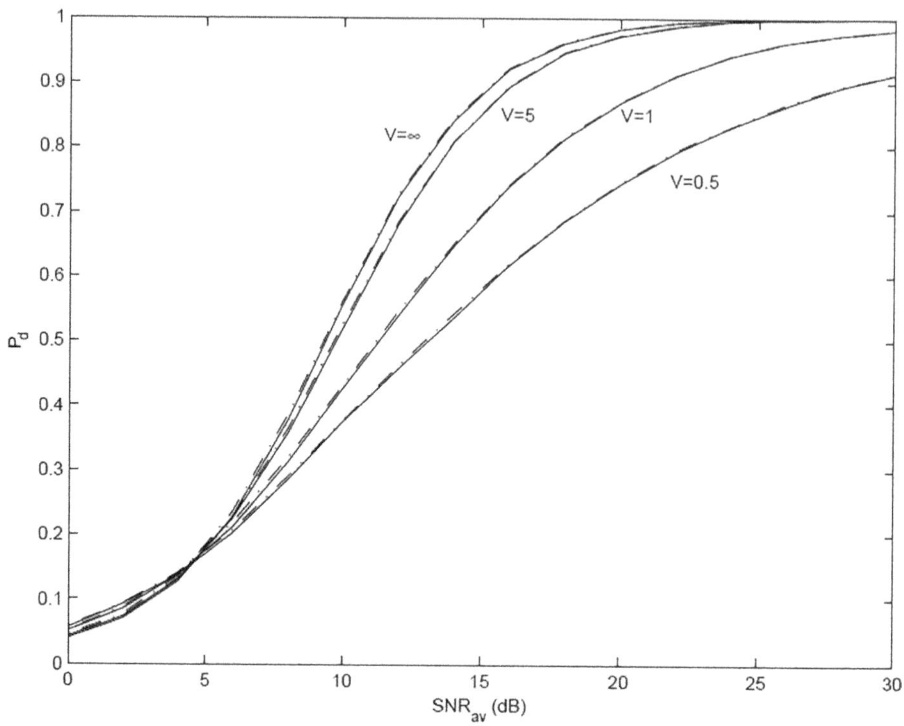

Figure 1.3 *P_d versus* SNR_{av} *of the PAMF (solid curves) and of the PGLRT (dashed-dotted curves) for* $P_{fa} = 10^{-4}$, $N = 4$, $K = 32$, $\epsilon_t = 0$, $\delta_t = 0.19$, $\epsilon_c = 0$, *and* $\delta_c = 0.18$. *Each curve refers to a specific value of* ν.

The reported receivers are synthesized according to a two-step framework: first, the clutter covariance structure is assumed known and detection algorithms based on plain Rao and Wald tests are devised. Then fully adaptivity is provided replacing a suitable estimate of the clutter covariance structure in place of the true one. At the analysis stage, the performance of the considered detection schemes is also assessed in comparison with some alternative processing schemes[§] based on the two-step GLRT design technique [9, 10].

The Rao test for known $\boldsymbol{R}$ has been synthesized in [22] expanding the primary data likelihood ratio in the neighborhood of the ML estimate of the unknown parameters under H_0 [23]. The resulting decision rule is

[§]As discussed in [10], it seems that the decision statistic based on one-step GLR criterion does not admit a closed form expression.

$$z_1^\dagger \boldsymbol{\Gamma}_{1,0}^{-1} \boldsymbol{R}^{-1}\boldsymbol{P}(\boldsymbol{P}^\dagger \boldsymbol{R}^{-1}\boldsymbol{P})^{-1}\boldsymbol{P}^\dagger \boldsymbol{R}^{-1}\boldsymbol{\Gamma}_{1,0}^{-1} z_1 \underset{H_0}{\overset{H_1}{\gtrless}} \gamma \tag{1.11}$$

where

$$\boldsymbol{\Gamma}_{1,0} = \mathbf{Diag}\left(\left[\sqrt{\hat{\sigma}_{1,1,0}}, \sqrt{\hat{\sigma}_{2,1,0}}\right]\right) \otimes \boldsymbol{I}_N$$

with [9]

$$\hat{\sigma}_{1,h,0} = \frac{1}{N}\boldsymbol{\Psi}_0(h,h)\left(1 + \frac{\mathrm{Re}\left(\boldsymbol{\Psi}_0(1,h)\right)}{\sqrt{\boldsymbol{\Psi}_0(1,1)\boldsymbol{\Psi}_0(h,h)}}\right) \quad h = 1,2$$

where $\boldsymbol{\Psi}_0(i,j)$ denotes the (i,j)th entry of the 2×2 matrix $\boldsymbol{\Psi}_0 = \boldsymbol{Z}^\dagger \boldsymbol{R}^{-1}\boldsymbol{Z}$, with:

$$\boldsymbol{Z} = \begin{pmatrix} z_{1,1} & \mathbf{0} \\ \mathbf{0} & z_{2,1} \end{pmatrix}$$

The Wald test for known covariance structure¶ has been devised in [22], exploiting the asymptotic efficiency of the ML estimate [23]. The resulting detector is

$$z_1^\dagger \boldsymbol{\Gamma}_{1,1}^{-1} \boldsymbol{R}^{-1}\boldsymbol{P}(\boldsymbol{P}^\dagger \boldsymbol{R}^{-1}\boldsymbol{P})^{-1}\boldsymbol{P}^\dagger \boldsymbol{R}^{-1}\boldsymbol{\Gamma}_{1,1}^{-1} z_1 \underset{H_0}{\overset{H_1}{\gtrless}} \gamma \tag{1.12}$$

where $\boldsymbol{\Gamma}_{1,1} = \mathbf{Diag}\left(\left[\sqrt{\hat{\sigma}_{1,1,1}}, \sqrt{\hat{\sigma}_{2,1,1}}\right]\right) \otimes \boldsymbol{I}_N$ with

$$\hat{\sigma}_{h,1,1} = \frac{1}{N}\boldsymbol{\Psi}_1(h,h)\left(1 + \frac{\mathrm{Re}\left(\boldsymbol{\Psi}_1(1,h)\right)}{\sqrt{\boldsymbol{\Psi}_1(1,1)\boldsymbol{\Psi}_1(h,h)}}\right) \quad h = 1,2$$

and $\boldsymbol{\Psi}_1(i,j)$ the (i,j)th entry of the 2×2 matrix

$$\Psi_1 = \boldsymbol{Z}^\dagger[\boldsymbol{R}^{-1} - \boldsymbol{R}^{-1}\boldsymbol{P}(\boldsymbol{P}^\dagger \boldsymbol{R}^{-1}\boldsymbol{P})^{-1}\boldsymbol{P}^\dagger \boldsymbol{R}^{-1}]\boldsymbol{Z}$$

In order to make the derived detectors fully adaptive, a suitable estimate of $\boldsymbol{R}$ based upon the secondary data is substituted into the left hand side of (1.11) and (1.12). To this end, it is clear that resorting to the plain sample covariance matrix would lead to adaptive receivers, which do not ensure the CFAR property with respect to the texture statistics.

Thus, a reasonable, even if heuristic, strategy, first proposed in [24], thoroughly analyzed in [25], and extended to the polarimetric case in [9], which allows to come up with an estimator functionally independent of the clutter textures, relies on a normalized sample covariance matrix, i.e.,

¶The covariance structure refers to the covariance matrix of the data up to a multiplicative scalar term. In the SIRVs context, it defines substantially the covariance of the speckles contribution.

$$\hat{\boldsymbol{R}} = \frac{1}{K}\sum_{k=2}^{K+1} \hat{\boldsymbol{\Gamma}}_k^{-1} z_k z_k^\dagger \hat{\boldsymbol{\Gamma}}_k^{-1} \tag{1.13}$$

with

$$\hat{\boldsymbol{\Gamma}}_k = \mathbf{Diag}\left(\left[\sqrt{\frac{1}{N}\|z_{1,k}\|^2}, \sqrt{\frac{1}{N}\|z_{2,k}\|^2}\right]\right) \otimes \boldsymbol{I}_N, \quad k = 2, \ldots, K+1$$

Interestingly, the detectors obtained plugging $\hat{\boldsymbol{R}}$ in place of $\boldsymbol{R}$, ensure the CFAR property with respect to the texture statistics, but, unfortunately, do not guarantee CFARness with respect to the structure of the covariance matrix.

Hence, replacing (1.13) in (1.11) and (1.12), the following adaptive decision rules are obtained:

$$z_1^\dagger \hat{\boldsymbol{\Gamma}}_{1,0}^{-1} \hat{\boldsymbol{R}}^{-1} \boldsymbol{P}(\boldsymbol{P}^\dagger \hat{\boldsymbol{R}}^{-1} \boldsymbol{P})^{-1} \boldsymbol{P}^\dagger \hat{\boldsymbol{R}}^{-1} \hat{\boldsymbol{\Gamma}}_{1,0}^{-1} z_1 \underset{H_0}{\overset{H_1}{\gtrless}} \gamma \tag{1.14}$$

and

$$z_1^\dagger \hat{\boldsymbol{\Gamma}}_{1,1}^{-1} \hat{\boldsymbol{R}}^{-1} \boldsymbol{P}(\boldsymbol{P}^\dagger \hat{\boldsymbol{R}}^{-1} \boldsymbol{P})^{-1} \boldsymbol{P}^\dagger \hat{\boldsymbol{R}}^{-1} \hat{\boldsymbol{\Gamma}}_{1,1}^{-1} z_1 \underset{H_0}{\overset{H_1}{\gtrless}} \gamma \tag{1.15}$$

where $\hat{\boldsymbol{\Gamma}}_{1,0}$ and $\hat{\boldsymbol{\Gamma}}_{1,1}$ are obtained from $\boldsymbol{\Gamma}_{1,0}$ and $\boldsymbol{\Gamma}_{1,1}$, substituting $\hat{\boldsymbol{R}}$ in to $\boldsymbol{R}$. Hereafter, the decision rules (1.14) and (1.15) are denoted as polarimetric Rao (P-Rao) and polarimetric Wald (P-Wald) receivers, respectively. Their block schemes are reported in Figures 1.4 and 1.5, respectively. Both the architectures perform a noncoherent integration of the primary data components through the weight matrices $\boldsymbol{Q}_R$ and $\boldsymbol{Q}_W$, i.e.:

$$\boldsymbol{Q}_R = \hat{\boldsymbol{\Gamma}}_{1,0}^{-1} \hat{\boldsymbol{R}}^{-1} \boldsymbol{P}(\boldsymbol{P}^\dagger \hat{\boldsymbol{R}}^{-1} \boldsymbol{P})^{-1} \boldsymbol{P}^\dagger \hat{\boldsymbol{R}}^{-1} \hat{\boldsymbol{\Gamma}}_{1,0}^{-1}$$

$$\boldsymbol{Q}_W = \hat{\boldsymbol{\Gamma}}_{1,1}^{-1} \hat{\boldsymbol{R}}^{-1} \boldsymbol{P}(\boldsymbol{P}^\dagger \hat{\boldsymbol{R}}^{-1} \boldsymbol{P})^{-1} \boldsymbol{P}^\dagger \hat{\boldsymbol{R}}^{-1} \hat{\boldsymbol{\Gamma}}_{1,1}^{-1}$$

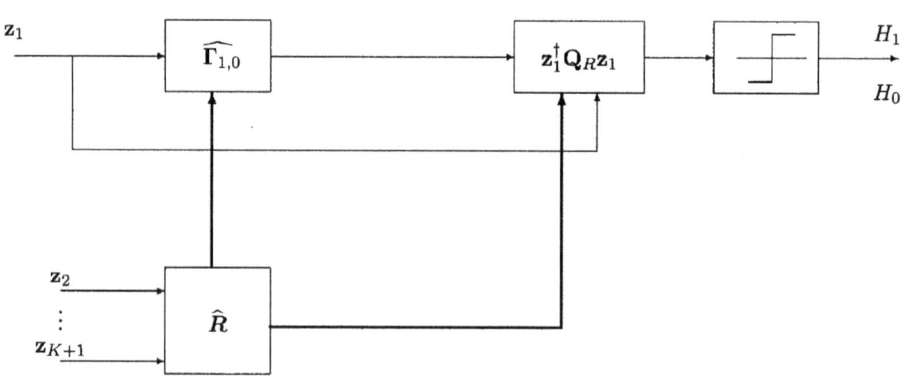

Figure 1.4 Block scheme of the P-Rao detector

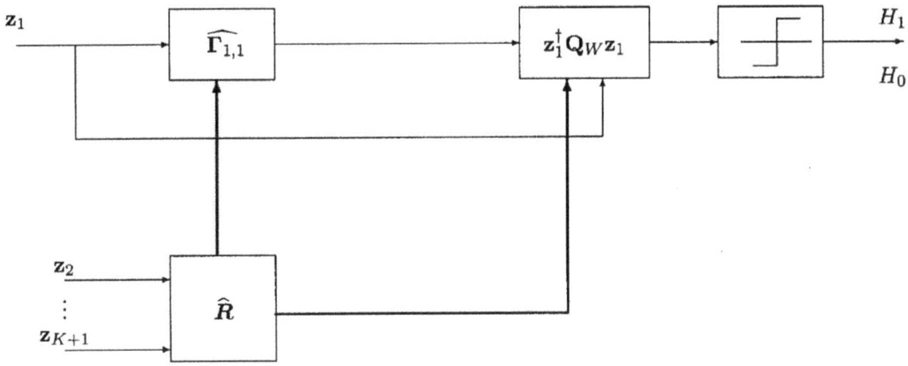

Figure 1.5 *Block scheme of the P-Wald detector*

which require both the primary and the secondary data. Their difference relies on the presence of the matrices $\hat{\Gamma}_{1,0}$ and $\hat{\Gamma}_{1,1}$, due to the fact that the Rao test exploits the texture estimates under H_0, whereas the Wald test employs those under the H_1 hypothesis.

Before proceeding further, it worth pointing out the clutter covariance structure can be also estimated via a recursive procedure able to monotonically improve (along its iterations) the secondary data likelihood function [22]. Remarkably, according to this approach, the estimate (1.13), till now conceived heuristically, can be interpreted as the first instance of the quoted recursion assuming the identity matrix as initial guess.

1.4.1 Performance assessment

This section is devoted to the performance assessment of the P-Rao and the P-Wald receivers in terms of P_{fa} and P_d. Both synthetic and measured data are considered.

1.4.1.1 Performance assessment on simulated data

Since closed form expressions for both the P_{fa} and the P_d are not available, standard Monte Carlo techniques are used, based on $\frac{100}{P_{fa}}$ and 1000 independent trials, respectively. By doing so, reliable estimates are accomplished.

Due to CFAR property with respect to the $\sigma_{i,k}$'s, the threshold can be set independent on the texture statistics. However, as the detectors (1.14) and (1.15) do not guarantee CFARness with respect to the clutter covariance structure, the threshold setting requires knowledge of $\boldsymbol{R}$. Specifically, modeling the covariance structure as (1.10), it is required the knowledge of ϵ_c, δ_c, ρ_c, and $\boldsymbol{\Sigma}$. In this respect, $\boldsymbol{\Sigma}$ is assumed an exponentially shaped matrix with one-lag correlation coefficient ρ_c. Finally, it is assumed that the textures $\sigma_{1,1}$ and $\sigma_{1,2}$ are exponentially correlated gamma random variables with shape parameter ν_c, mean values $\overline{\sigma}_{1,1}$ and $\overline{\sigma}_{1,2}$, and correlation coefficient ρ_σ.

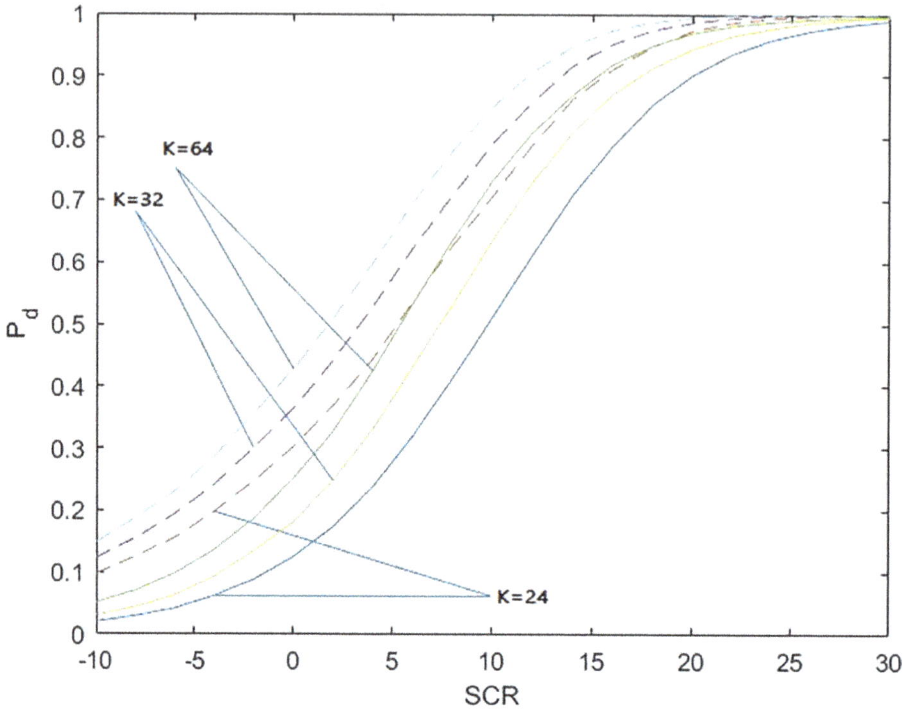

Figure 1.6 P_d versus SCR of the P-Rao (solid curves) and of the P-Wald (dashed curves) receivers, for $P_{fa} = 10^{-4}$, $f_d = 0$, $N = 8$, $\nu_c = 0.5$, $\rho_\sigma = 0$, $\epsilon_t = 0.28$, $\delta_t = 1$, $\epsilon_c = 0.5$, $\delta_c = 1.6$, $\rho_c = 0.9$, and $K \in \{24, 32, 64\}$

As to the target, it is assumed that $\boldsymbol{a}$ is modeled as a bidimensional, zero-mean, circularly symmetric Gaussian random vector with covariance matrix (1.9). Finally, $\boldsymbol{p} = \left[1, e^{j2\pi f_d}, \ldots, e^{j2\pi(N-1)f_d}\right]$, where f_d is the normalized target Doppler frequency.

In Figures 1.6 and 1.7, the P_d of the P-Rao and the P-Wald detectors is plotted versus the signal to clutter ratio (SCR) defined as

$$\mathrm{SCR} = \frac{\overline{\alpha}}{\overline{\sigma}_{1,1} + \overline{\sigma}_{1,2}},$$

for $P_{fa} = 10^{-4}$, $f_d = 0$, $N = 8$, $\nu_c = 0.5$, $\rho_\sigma = 0$, and several values of K. Moreover, Figure 1.6 considers $\epsilon_t = 0.28$, $\delta_t = 1$, $\epsilon_c = 0.5$, and $\delta_c = 1.6$ [2], i.e., it is a representative of a radar that processes the HH and VV channel data. Figure 1.7 instead refers to a system employing the HH and HV channel data and considers $\epsilon_t = 0$, $\delta_t = 0.19$, $\epsilon_c = 0$, and $\delta_c = 0.18$ [2].

The inspection of the plots shows that an increase in the sample support K can lead to a significant performance improvement for both co- and cross-polarized scenarios. Moreover, the P-Rao detector suffers a performance gap with respect to

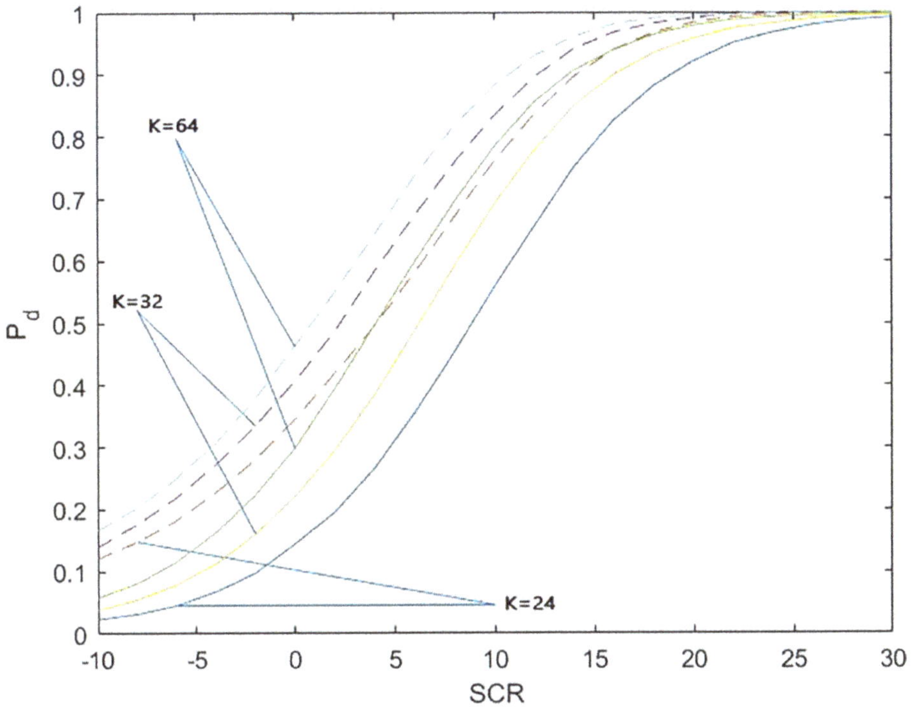

Figure 1.7 P_d *versus SCR of the P-Rao (solid curves) and of the P-Wald (dashed curves) receivers, for* $P_{fa} = 10^{-4}$, $f_d = 0$, $N = 8$, $\nu_c = 0.5$, $\rho_\sigma = 0$, $\epsilon_t = 0$, $\delta_t = 0.19$, $\epsilon_c = 0$, $\delta_c = 0.18$, $\rho_c = 0.9$, *and* $K \in \{24, 32, 64\}$

the P-Wald receiver especially for smaller values of K. Specifically, the SCR-gain of the P-Wald ranges between 2.5 and 4.5 dB at $P_d = 0.9$. This behavior can be justified observing that the P-Rao receiver exploits the estimates of the parameters under the hypothesis H_0, namely, assuming that the target is actually absent into the primary data. As a consequence, a model mismatch may occur, i.e., the receiver suffers a signal contamination problem, which can be reduced asymptotically, namely, for large enough values of N and K. Indeed, asymptotic results ensure that the two detectors achieve the same P_d.

In Figure 1.8, the impact of the clutter shape parameter ν_c on the performance of the P-Rao and the P-Wald architectures is assessed. Therein, it is assumed $P_{fa} = 10^{-4}$, $f_d = 0$, $N = 8$, $K = 32$, $\rho_\sigma = 0$. Besides, the target and clutter covariance parameters as Figure 1.8 are considered. The curves clearly show that ν_c significantly affects the performance of both detectors. Specifically, the P_d curves exhibit a threshold effect as ν_c increases, that is, a sharp transition from almost undetectability to a close-to-one detection rate. Conversely, as ν_c decreases, the threshold effect is progressively smoothed.

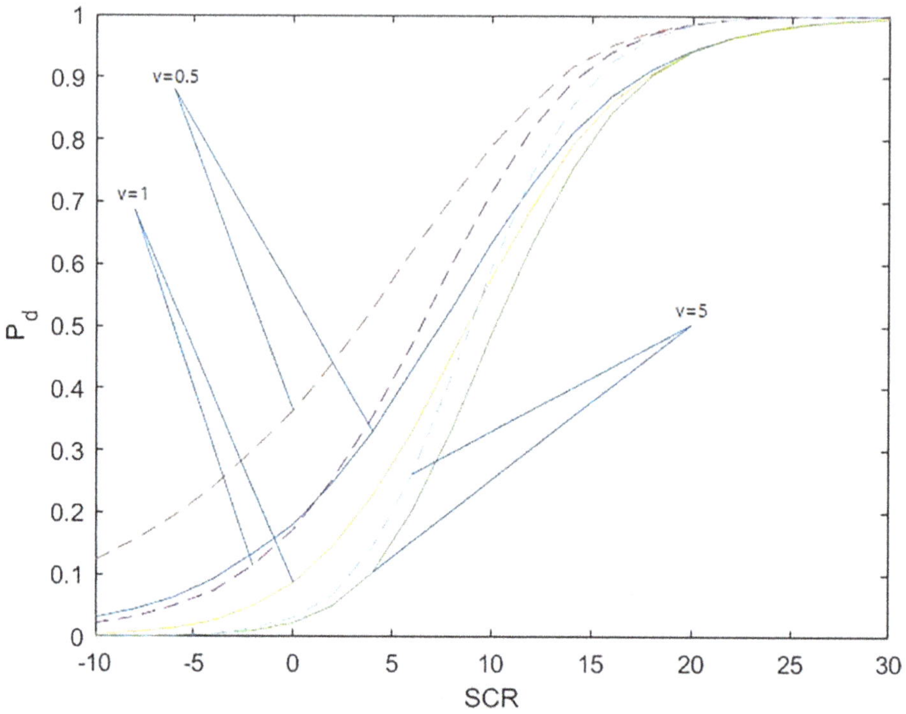

Figure 1.8 P_d *versus SCR of the P-Rao (solid curves) and of the P-Wald (dashed curves) receivers, for* $P_{fa} = 10^{-4}$, $f_d = 0$, $N = 8$, $K = 32$, $\rho_\sigma = 0$, $\epsilon_t = 0.28$, $\delta_t = 1$, $\epsilon_c = 0.5$, $\delta_c = 1.6$, $\rho_c = 0.9$, *and* $\nu_c \in \{0.5, 1, 5\}$

In Figure 1.9, the influence of a correlation between the textures on the two polarimetric channels is analyzed. To this end, the performance of the P-Rao and the P-Wald detectors is plotted for different values of ρ_σ. Therein, $P_{fa} = 10^{-4}$, $f_d = 0$, $N = 8$, $K = 64$, and $\nu_c = 0.5$ are set; besides, the same target and clutter covariance structures as in Figure 1.6 are considered. Looking over the figure highlights that the performance of the P-Wald test significantly improves for vanishingly small ρ_σ. Of course, this fact can be easily justified by noticing that, for small ρ_σ, the receiver benefits more and more of diversity. Conversely, for high values of ρ_σ, the realizations of $\sigma_{1,1}$ and $\sigma_{2,1}$ are very similar and much less advantage can be gained through the adoption of a diversity strategy. As to the curves of the P-Rao receiver, the above considerations apply only for $P_d \geq 0.6$. Moreover, the performance trend shows that this detector is less sensitive to the presence of a possible correlation between the textures of the two processed polarimetric channels. It is conjectured that this unusual behavior is probably due to the signal contamination effect, which is present in $\hat{\boldsymbol{\Gamma}}_{1,0}$ especially for high values of SCR.

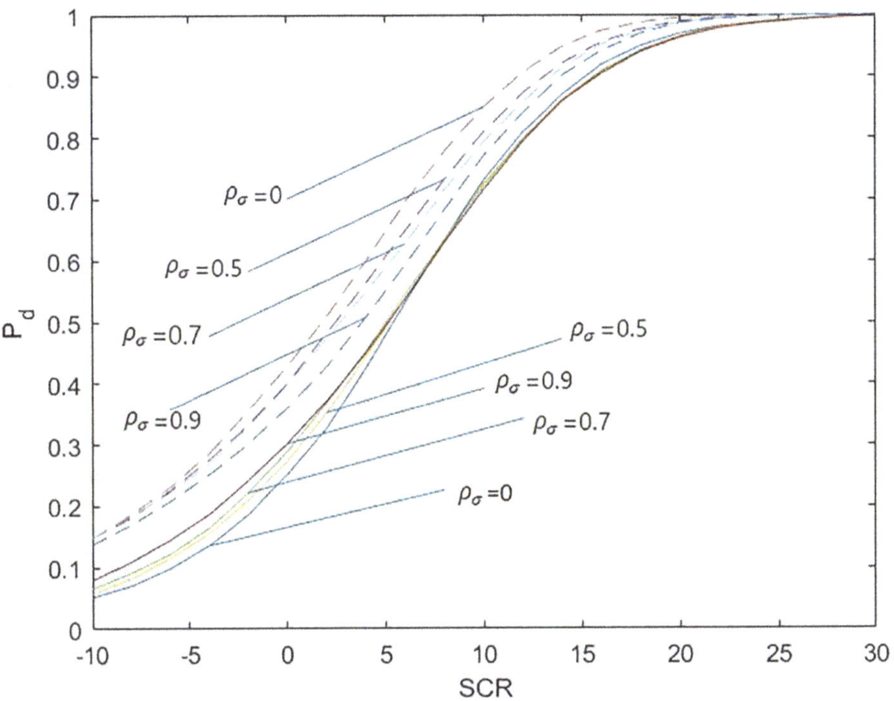

Figure 1.9 P_d *versus SCR of the P-Rao (solid curves) and of the P-Wald (dashed curves) receivers, for* $P_{fa} = 10^{-4}$, $f_d = 0$, $N = 8$, $K = 64$, $\nu_c = 0.5$, $\epsilon_t = 0.28$, $\delta_t = 1$, $\epsilon_c = 0.5$, $\delta_c = 1.6$, $\rho_c = 0.9$, *and* $\rho_\sigma \in \{0, 0.5, 0.7, 0.9\}$

1.4.1.2 Performance assessment on measured clutter data

In this subsection, the performance assessment of the considered adaptive receivers is evaluated exploiting real clutter data. First, the compatibility of the available data set with the compound clutter model is verified; then P_d curves are illustrated. The analysis is conducted also in comparison with the detectors of [9] and [10], referred to in the following as polarimetric-GLRT-1 (P-GLRT-1) and polarimetric-GLRT-2 (P-GLRT-2), respectively. Specifically, P-GLRT-1 corresponds to the decision rule:

$$\frac{\sqrt{\hat{\boldsymbol{\Psi}}_0(1,1)\hat{\boldsymbol{\Psi}}_0(2,2)} + \mathrm{Re}\left(\hat{\boldsymbol{\Psi}}_0(1,2)\right)}{\sqrt{\hat{\boldsymbol{\Psi}}_1(1,1)\hat{\boldsymbol{\Psi}}_1(2,2)} + \mathrm{Re}\left(\hat{\boldsymbol{\Psi}}_1(1,2)\right)} \underset{H_0}{\overset{H_1}{\gtrless}} \gamma \tag{1.16}$$

where $\hat{\boldsymbol{\Psi}}_0$ and $\hat{\boldsymbol{\Psi}}_1$ are obtained from the matrices $\boldsymbol{\Psi}_0$ and $\boldsymbol{\Psi}_1$, introduced with reference to the Rao and Wald tests, replacing R with $\hat{\boldsymbol{R}}$. P-GLRT-2 instead refers to the following detector:

$$\frac{z_1^\dagger \hat{\boldsymbol{\Gamma}}_1^{-1} \hat{\boldsymbol{R}}^{-1} \boldsymbol{P}(\boldsymbol{P}^\dagger \hat{\boldsymbol{R}}^{-1} \boldsymbol{P})^{-1} \boldsymbol{P}^\dagger \hat{\boldsymbol{R}}^{-1} \hat{\boldsymbol{\Gamma}}_1^{-1} z_1}{K + z_1^\dagger (\hat{\boldsymbol{\Gamma}}_1 \hat{\boldsymbol{R}} \hat{\boldsymbol{\Gamma}}_1)^{-1} z_1} \underset{H_0}{\overset{H_1}{\gtrless}} \gamma \tag{1.17}$$

with

$$\hat{\boldsymbol{\Gamma}}_1 = \textbf{Diag}\left(\left[\sqrt{\tfrac{1}{N}\boldsymbol{z}_{1,1}^{\dagger}(\boldsymbol{I}_N - \boldsymbol{p}(\boldsymbol{p}^{\dagger}\boldsymbol{p})^{-1}\boldsymbol{p}^{\dagger})\boldsymbol{z}_{1,1}}, \sqrt{\tfrac{1}{N}\boldsymbol{z}_{2,1}^{\dagger}(\boldsymbol{I}_N - \boldsymbol{p}(\boldsymbol{p}^{\dagger}\boldsymbol{p})^{-1}\boldsymbol{p}^{\dagger})\boldsymbol{z}_{2,1}}\right]\right) \otimes \boldsymbol{I}_N$$

Dataset description and its statistical analysis

A clutter analysis based on measured radar data collected by the IPIX radar in November 1993 and stored in the Dartmouth database [26] is now illustrated. This considered radar, mounted at a site in Dartmouth, Nova Scotia, on the East Coast of Canada, employs a coherent receiver and is capable of operating in a dual-polarized mode**. The results refer to the file 19931117_131609_*stareB0002.cdf*, namely, to the data set 226 of [26]. In [26], additional information concerning the radar parameters, sea state, just to mention a few, can be found.

Let us now focus on the statistical characterization of the clutter amplitude, in terms of cumulative distribution function (CDF). The goal is twice: (1) to provide evidence and results confirming the deviations of the measured radar data from the Rayleigh model and (2) to assess their compatibility with the compound model. Specifically, the empirical CDF (ECDF) is compared with several models available in open literature for describing the amplitude CDF of the clutter returns. Among them, it is worth mentioning the K- and the Weibull distributions whose CDFs are reported in the following, together with the Rayleigh CDF [29].

K-model. The K-CDF with scale parameter $\mu > 0$ and shape parameter $\nu > 0$ is given by

$$F_X(x) = 1 - \frac{1}{2^{\nu-1}\Gamma(\nu)}\left(\sqrt{\frac{2\nu}{\mu}}x\right)^{\nu} K_{\nu}\left(\sqrt{\frac{2\nu}{\mu}}x\right) \qquad x \geq 0$$

This model is amenable to a compound representation and the first two moments are

$$\mathbb{E}[X] = \sqrt{\frac{\pi\mu}{2\nu}}\frac{\Gamma(\nu+0.5)}{\Gamma(\nu)} \qquad \mathbb{E}[X^2] = 2\mu$$

Weibull model. The Weibull-CDF with parameters $b > 0$ and $c > 0$ is given by

This CDF is amenable to a compound representation as long as $c < 2$. Moreover, the first two moments:

$$\mathbb{E}[X] = b\Gamma\left[\frac{(c+1)}{c}\right] \qquad \mathbb{E}[X^2] = b^2\Gamma\left[\frac{(c+2)}{c}\right]$$

Rayleigh model. The Rayleigh-CDF with parameter $b > 0$ is given by

**A complete description of IPIX radar can be found in [27, 28].

$$F_X(x) = 1 - \exp\left(-\frac{x^2}{2b_R^2}\right) \quad x \geq 0$$

This CDF is amenable to a compound representation and its variance is

$$\mathbb{E}[X^2] - \mathbb{E}^2[X] = \left(2 - \frac{\pi}{2}\right) b_R^2$$

The statistical tool usually employed to check if a data set is compatible with a design distribution $F_X(x)$ is the Kolmogorov–Smirnov test [30]. Unfortunately, it cannot be applied for the case at hand as it requires the statistical independence of the samples used for computing the ECDF. In order to circumvent this drawback, a validation procedure relying on the evaluation of a distributional distance between the nominal and the ECDF is employed. According to this framework, a theoretical model fits the empirical data better than another if the distance between the CDF of the former from the ECDF is smaller than the distance between the CDF of the latter from the ECDF. It is worth highlighting that even if the data share a degree of correlation, the mentioned distance still represents a valuable index of similarity between distributions. In the following, the Cramer–von Mises (CV) distance is considered, which was also exploited in [31] to assess the closeness of empirical SAR data to the Weibull model. Hence, given a set of L identically distributed random variables $X_1, \ldots, X_L$, the CV distance (d_{CV}) between the design distribution $F_X(x)$, assumed to be a continuous function, and the empirical distribution $\hat{F}_X(x)$ is defined as [30]:

$$d_{CV}^2 = L \int_{-\infty}^{\infty} \left| F_X(x) - \hat{F}_X(x) \right|^2 dF_X(x). \tag{1.18}$$

Moreover, the integral (1.18) can be evaluated as [30]:

$$d_{CV}^2 = \frac{1}{12L} + \sum_{i=1}^{L} \left| F_X\left(X_{(i)}\right) - \frac{2i-1}{2L} \right|^2$$

where $X_{(i)}$ is the ith-order statistic from the set of random variables.

In Figure 1.10, the ECDF of the measured radar data from the 25th cell of the data set is plotted. Therein, also the K-, Weibull- and Rayleigh-CDFs are reported, whose parameters are obtained via moment matching. The curves, which refer to the HH polarization, highlight that the ECDF is very close to both the K- and the Weibull-CDFs. This behavior is confirmed by Tables 1.1–1.3, where the parameters of the theoretical CDFs that fit the data and the corresponding distance from the ECDF are reported with reference to the four channels HH, VV, HV, and VH.

The values of d_{CV} show that the K-CDF is closer to the ECDF than the others. Moreover, there is a strong deviation from the Rayleigh model for both the co- and cross-polarized channels. Note also that the shape parameter of the K-model, which fits the data, varies from one channel to another and ranges between 0.57 and 1.18. These values confirm the spiky nature of the sea clutter, the agreement of the empirical data with a compound representation, and hence the importance of detectors capable of ensuring the CFAR property with respect to the texture statistics. In Figure 1.11, the CV distance between the Rayleigh-CDF and the ECDF versus the length of the analysis window is reported. The curves highlight that over short periods of the order of 0.5 seconds, the sea clutter amplitude can be fit reasonably

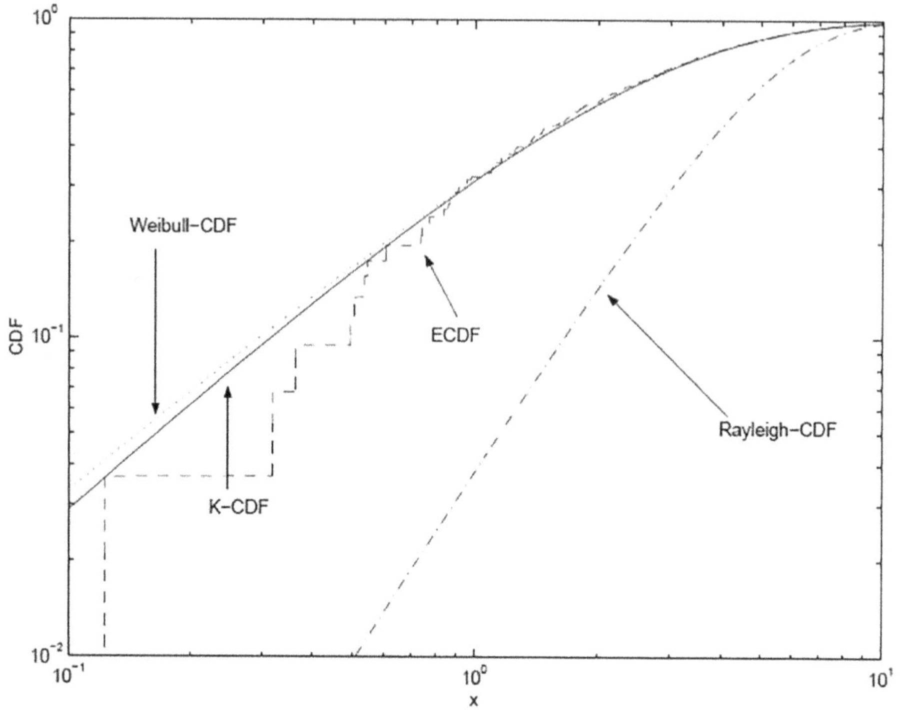

Figure 1.10 *Amplitude ECDF of the returns from the HH channel of the 25th range cell (dashed curve). K-CDF (solid curve) and Weibull-CDF (dotted curve) with the same first- and second-order moments as the ECDF. Rayleigh-CDF (dashed-dotted curve) with the same variance as the ECDF.*

well with a Rayleigh distribution. Moreover, for larger analysis windows, the CV distance increases; thus the Rayleigh assumption appears to be inadequate and the K-CDF better fits the data. This result agrees with the presence of an amplitude

Table 1.1 *Parameters and CV distance (from the amplitude ECDF) of the K -CDF with the same first- and second-order moments as the ECDF*

	K-CDF		
Polarizations	μ	ν	d_{cv}
HH	5.780	0.570	1.462
HV	3.978	0.610	3.046
VV	4.891	1.180	1.035
VH	4.035	0.600	2.947

Table 1.2 Parameters and CV distance (from the amplitude ECDF) of the Weibull-CDF with the same first- and second-order moments as the ECDF

	Weibull-CDF		
Polarizations	***b***	***c***	d_{cv}
HH	2.500	1.050	1.654
HV	2.128	1.080	3.257
VV	2.722	1.340	1.795
VH	2.132	1.070	3.211

modulation of the clutter returns, through a stochastic process whose decorrelation time is higher, as compared to that of the underlying Rayleigh component.

Detection probability analysis

The detection performance of P-Rao, P-Wald, P-GLRT-1, and P-GLRT-2 receivers is now estimated exploiting the IPIX radar data. Due to the limited amount of real clutter data, the detection thresholds (associated with $P_{fa} = 10^{-4}$) are evaluated by means of Monte Carlo simulation as follows. Specifically, the clutter covariance is first estimated from the real clutter data via the sample covariance matrix. Then, exploiting the CFAR property of the receivers with respect to the texture statistics, the detection thresholds have been evaluated through 10^6 independent runs, assuming that the primary and the secondary data are independent zero-mean circularly symmetric Gaussian random vectors whose covariance matrix is the one estimated from the data set. Finally, in order to evaluate the P_d, a synthetic target, simulated according to the statistical model depicted in subsection 1.4.1.1, is injected into the 25th cell of the real data set.

To proceed further, let us define:

$$\mathrm{SCR}_1 = \frac{\overline{\alpha}}{\widetilde{\sigma}_{1,1} + \widetilde{\sigma}_{1,2}},$$

Table 1.3 Parameter and CV distance (from the amplitude ECDF) of the Rayleigh-CDF with the same variance as the ECDF

	Rayleigh-CDF	
Polarizations	b_R	d_{cv}
HH	3.581	23.575
HV	2.922	23.289
VV	2.875	16.511
VH	2.951	23.399

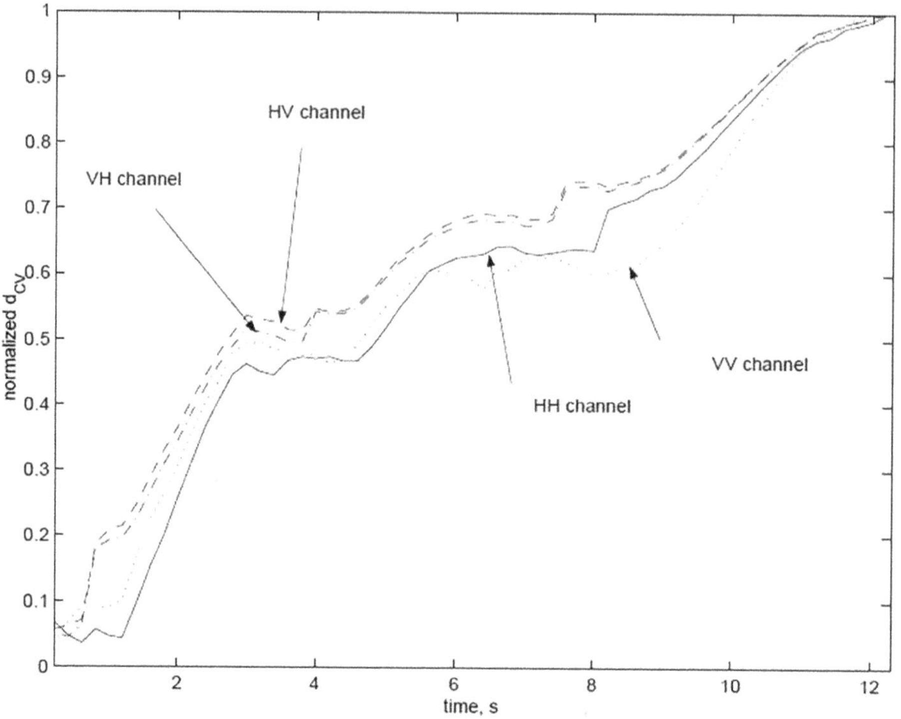

Figure 1.11 Normalized CV distance between the amplitude ECDF of the returns and the Rayleigh-CDF with the same variance

where $\widetilde{\sigma}_{1,1}$ and $\widetilde{\sigma}_{1,2}$ indicate the clutter power values estimated from the real clutter data on the two processed polarimetric channels, respectively. In Figures 1.12 and 1.13, P_d versus SCR_1 of P-Rao, P-Wald, P-GLRT-1, and P-GLRT-2 is displayed, where $P_{fa} = 10^{-4}$, $N = 8$, $K = 24$, $f_d = 0$. Figure 1.12 refers to a radar that processes the co-polarized channels (HH and VV), which thus assumes $\delta_t = 1$ and $\epsilon_t = 0.28$. On the contrary, Figure 1.13 corresponds to a radar which exploits the cross-polarized channels (HH and HV), which accordingly considers $\delta_t = 0.19$ and $\epsilon_t = 0$. The curves highlight that, in the co-polarized case, the P-Wald receiver outperforms the P-Rao, P-GLRT-1, and P-GLRT-2, receivers for all the values of P_d. Moreover, if the HH and HV channels are processed then the P-Wald architecture performs better than the others for $P_d > 0.4$, while it is outperformed by the P-GLRT-2 if $P_d < 0.4$ (which is a detection regime of low practical relevance).

In Figures 1.14 and 1.15, the P_d values estimated from the real data set are compared with those resulting through the use of Monte Carlo techniques. In this last case, the clutter data have been simulated according to the compound-Gaussian model assuming that the covariance structure is that adopted to compute the

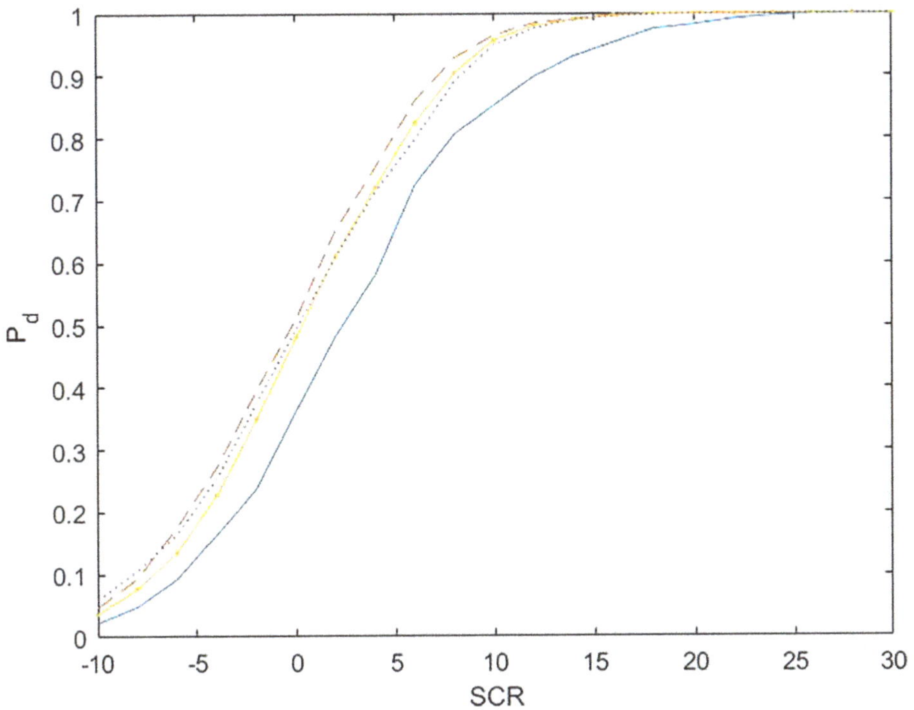

Figure 1.12 P_d *versus* SCR_l *of the P-Rao (solid curve), P-Wald (dashed curve), P-GLRT-1 (dotted-marked curve), and P-GLRT-2 (dotted curve) receivers, for* $P_{fa} = 10^{-4}$, $f_d = 0$, $N = 8$, $K = 24$, *real radar data (HH and VV channels),* $\epsilon_t = 0.28$ *and* $\delta_t = 1$

thresholds, whereas $\sigma_{1,1}$ and $\sigma_{1,2}$ are modeled as independent Gamma random variables with shape parameter $\nu = 1$ and unitary mean square value.

The results show a good agreement between the curves, obtained by processing the real radar data and the P_d values predicted through numerical simulations for both the case of co-polarized and cross-polarized channels. Finally, it is conjectured that the existing gaps between the real and the simulated P_ds are reasonably due to:

- the shape parameters of the simulated textures that do not perfectly match the values estimated from the real data;
- the covariance mismatch existing in the range cells of the processed data set [32–34]. In fact, as in [35], while the covariance matrices estimated from different cells exhibit slight discrepancies, the synthetic data have been simulated by making use of the same covariance matrix;
- the quantization errors of the Analog to Digital converter.

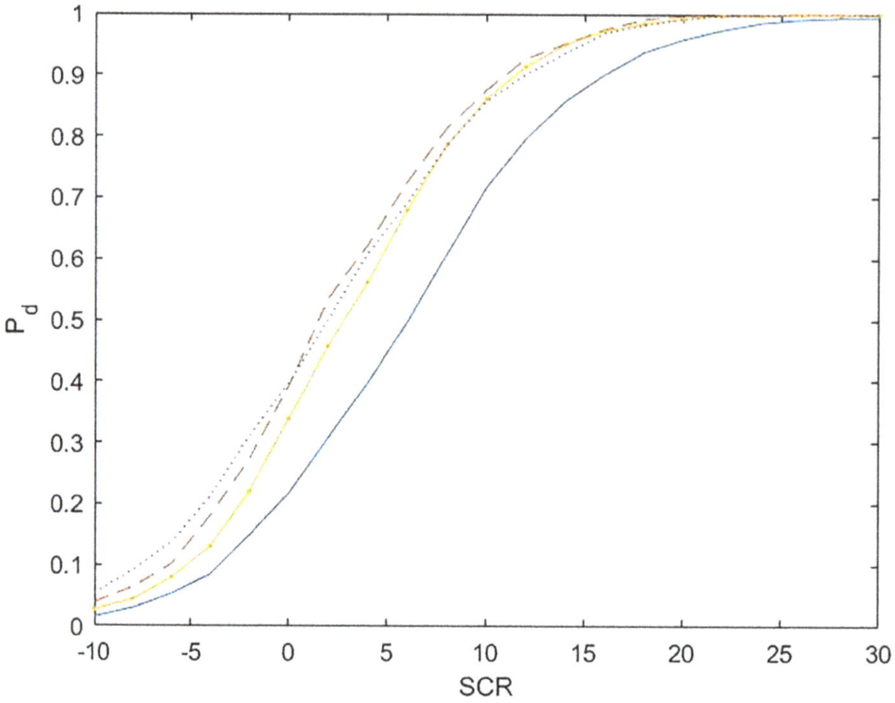

Figure 1.13 P_d versus SCR_1 of the P-Rao (solid curve), P-Wald (dashed curve), P-GLRT-1 (dotted-marked curve), and P-GLRT-2 (dotted curve) receivers, for $P_{fa} = 10^{-4}$, $f_d = 0$, $N = 8$, $K = 24$, real radar data (HH and HV channels), $\epsilon_t = 0$ and $\delta_t = 0.19$

Sensitivity analysis on the P_{fa}

In the following, a sensitivity analysis aimed at assessing the impact on the P_{fa} of a mismatch between the nominal and actual clutter characteristics is conducted. The rationale of this additional study relies on the threshold setting, which requires the knowledge of the clutter covariance matrix as the detectors do not possess CFARness with respect to this unknown quantity matrix parameter. As a consequence, possible discrepancies between the design and the actual covariance might cause an increase in the number of false alarms.

Toward the aforementioned goal, the detection thresholds are set assuming white Gaussian clutter and $P_{fa} = 10^{-4}$. Moreover, the clutter covariance matrix is estimated as the sample covariance of the returns from the available data set. Thus, the actual P_{fa} is evaluated, when compound-Gaussian clutter, with covariance matrix equal to that estimated from the real data, is present. Notice that the CFAR property with respect to the texture statistics makes the P_{fa} fully independent of the actual texture values and their possible correlations.

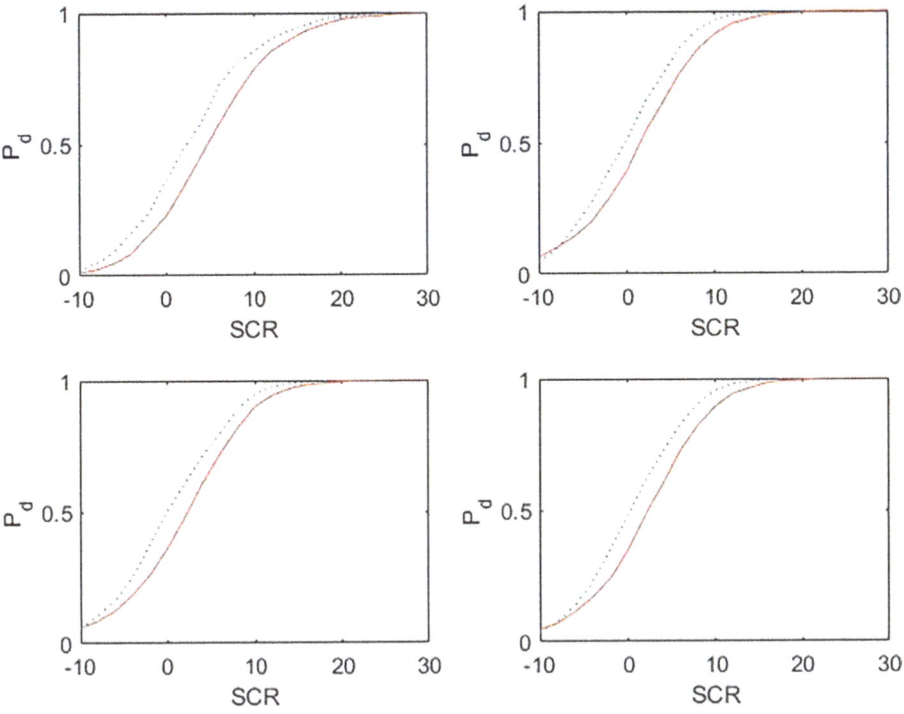

Figure 1.14 *P_d versus SCR_l of the P-Rao, P-Wald, P-GLRT-1, and P-GLRT-2 receivers, for $P_{fa} = 10^{-4}$, $f_d = 0$, $N = 8$ and $K = 24$. Real radar data collected from the HH and VV channels (dotted curves), simulated data with the same covariance matrix of the real data, $\nu = 1$, $\epsilon_t = 0.28$, $\delta_t = 1$, and $\rho_\sigma = 0$ (solid curves).*

The results, reported in Table 1.4, show that a mismatch between the nominal and actual covariance matrix yields a very limited variation of P_{fa}, thus CFAR is substantially achieved both for the cases of co-polarized and cross-polarized channels.

It is worth highlighting, that a further analysis on the P_{fa}, which refers to a nominal $P_{fa} = 10^{-1}$ and which is completely based on real clutter data, has confirmed the robustness of all the detectors. In particular, the P-Rao receiver seems to ensure a more conservative behavior with respect to variations in the clutter covariance.

1.5 Conclusions

This chapter is focused on the design and analysis of receive architectures for coherent polarimetric radars in homogeneous and heterogeneous environments. After an overview on the potential benefits of polarization diversity in target detection, a signal model for the returns of a polarimetric radar assuming a point like targets is

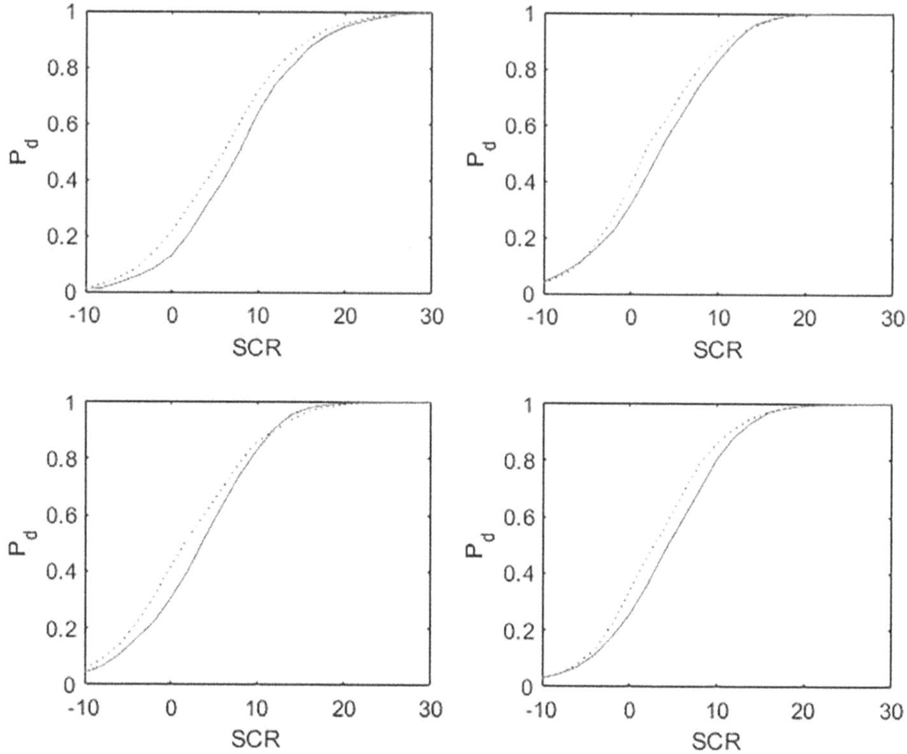

Figure 1.15 P_d *versus* SCR_1 *of the P-Rao, P-Wald, P-GLRT-1, and P-GLRT-2 receivers, for* $P_{fa} = 10^{-4}$, $f_d = 0$, $N = 8$, $K = 24$. *Real radar data collected from the HH and HV channels (dotted curves), simulated data with the same covariance matrix of the real data,* $\nu = 1$, $\delta_t = 0.19$, *and* $\rho_\sigma = 0$ *(solid curves).*

illustrated, along with the resulting target detection problem, formulated in terms of a binary hypothesis test.

Hence, starting from the observation that an UMP test does not exist for the problem at hand, with reference to either homogeneous or heterogeneous environments, valuable adaptive detection schemes are investigated and compared in terms of their performance.

Specifically, when the target is buried in homogeneous environment, both the one-step and the two-step GLRT detectors are presented and their performance assessed. It is shown that both the architectures possess the CFAR property and exhibit almost the same detection capabilities.

As to the compound-Gaussian clutter scenario, adaptive decision rules relying on the Rao and the Wald tests paradigm as well as on a suitable estimate of the clutter covariance structure from the secondary data are developed. It is shown that they ensure the CFAR property with respect to the texture statistics. Besides,

Table 1.4 Sensitivity of the actual P_{fa} for a design $P_{fa} = 10^{-4}$ and $N = 8$

Detector	**Processed channels**	**Actual** P_{fa} $/10^{-4}$	K
P-Rao	HH–VV	1.39	24
P-Rao	HH–VV	1.95	32
P-Rao	HH–HV	1.32	24
P-Rao	HH–HV	1.59	32
P-Wald	HH–VV	1.65	24
P-Wald	HH–VV	1.99	32
P-Wald	HH–HV	1.30	24
P-Wald	HH–HV	1.38	32
P-GLRT-1	HH–VV	1.45	24
P-GLRT-1	HH–VV	1.97	32
P-GLRT-1	HH–HV	1.32	24
P-GLRT-1	HH–HV	1.40	32
P-GLRT-2	HH–VV	4.78	24
P-GLRT-2	HH–VV	14.00	32
P-GLRT-2	HH–HV	3.22	24
P-GLRT-2	HH–HV	7.50	32

their performance, also in comparison with other existing receiver from the open literature, is evaluated both on simulated and measured data. Not surprisingly, an increase in the number of training data leads to a significant performance improvement for both the co- and cross-polarized scenarios. Besides, even if P-Wald and P-Rao achieve asymptotically the same P_d, in general the P-Wald test outperforms the counterparts.

As a last remark, it is pointed out that in the open literature the design of detection schemes for coherent polarimetric radars aimed at detecting range-extended targets in Gaussian and compound-Gaussian disturbance has been also addressed. The interested reader may refer to [22, 36, 37].

References

[1] Giuli D. 'Polarization diversity in radars'. *Proceedings of the IEEE.* 1986, vol. 74(2), pp. 245–69.

[2] Novak L.M., Sechtin M.B., Cardullo M.J. 'Studies of target detection algorithms that use polarimetric radar data'. *IEEE Transactions on Aerospace and Electronic Systems*. 1989, vol. 25(2), pp. 150–65.

[3] Kelly E.J. 'An adaptive detection algorithm'. *IEEE Transactions on Aerospace and Electronic Systems*. 1986, vol. AES-22(2), pp. 115–27.

[4] Robey F.C., Fuhrmann D.R., Kelly E.J., Nitzberg R. 'A CFAR adaptive matched filter detector'. *IEEE Transactions on Aerospace and Electronic Systems*. 1992, vol. 28(1), pp. 208–16.

[5] Park H.-R., Li J., Wang H. 'Polarization-space-time domain generalized likelihood ratio detection of radar targets'. *Signal Processing*. 1995, vol. 41(2), pp. 153–64.

[6] Pastina D., Lombardo P., Bucciarelli T. 'Adaptive polarimetric target detection with coherent radar. I. detection against Gaussian background'. *IEEE Transactions on Aerospace and Electronic Systems*. 2001, vol. 37(4), pp. 1194–206.

[7] Hao C., Gazor S., Ma X., Yan S., Hou C., Orlando D. 'Polarimetric detection and range estimation of a point-like target'. *IEEE Transactions on Aerospace and Electronic Systems*. 2016, vol. 52(2), pp. 603–16.

[8] Hurtado M., Nehorai A. 'Polarimetric detection of targets in heavy inhomogeneous clutter'. *IEEE Transactions on Signal Processing*. 2008, vol. 56(4), pp. 1349–61.

[9] Lombardo P., Pastina D., Bucciarelli T. 'Adaptive polarimetric target detection with coherent radar. II. detection against non-Gaussian background'. *IEEE Transactions on Aerospace and Electronic Systems*. 2001, vol. 37(4), pp. 1207–20.

[10] De Maio A., Alfano G. 'Polarimetric adaptive detection in non-Gaussian noise'. *Signal Processing*. 2003, vol. 83(2), pp. 297–306.

[11] Gogineni S., Nehorai A. 'Polarimetric MIMO radar with distributed antennas for target detection'. *IEEE Transactions on Signal Processing*. 2010, vol. 58(3), pp. 1689–97.

[12] Li N., Cui G., Kong L., Zhang T., Liu Q.H. 'Moving target detection for polarimetric MIMO radar in homogeneous gaussian clutter'. IEEE Radar Conference; Cincinnati, OH, 2014.

[13] Cui G., Kong L., Yang X., Yang J. 'GLRT design for polarimetric MIMO radar in non-Gaussian clutter'. *IEEE Radar Conference*; Kansas City, MO, USA; 2011.

[14] Kong L., Cui G., Yang X., Yang J. 'Rao and Wald tests design of polarimetric multiple-input multiple-output radar in compound-Gaussian clutter'. *IET Signal Processing*. 2011, vol. 5(1), pp. 85–96.

[15] Liu J., Zhang Z.-J., Yang Y. 'Performance enhancement of subspace detection with a diversely polarized antenna'. *IEEE Signal Processing Letters*. 2012, vol. 19(1), pp. 4–7.

[16] Kraut S., Scharf L.L., McWhorter L.T. 'Adaptive subspace detectors'. *IEEE Transactions on Signal Processing*. 2001, vol. 49(1), pp. 1–16.

[17] Colone F., Filippini F. 'Autoregressive model based polarimetric adaptive detection scheme Part I: theoretical derivation and performance analysis'. *IEEE Transactions on Aerospace and Electronic Systems*. 2020, vol. 56(5), pp. 3762–78.

[18] De Maio A., Ricci G. 'A polarimetric adaptive matched filter'. *Signal Processing*. 2001, vol. 81(12), pp. 2583–9.

[19] Ward K.D. 'Compound representation of high resolution sea clutter'. *Electronics Letters*. 1981, vol. 17(16), pp. 561–3.

[20] Sangston K.J., Gerlach K.R. 'Coherent detection of radar targets in a non-Gaussian background'. *IEEE Transactions on Aerospace and Electronic Systems*. 1994, vol. 30(2), pp. 330–40.

[21] Sangston K.J., Farina A. 'Coherent radar detection in compound-Gaussian clutter: Clairvoyant detectors'. *IEEE Aerospace and Electronic Systems Magazine*. 2016, vol. 31(11), pp. 42–63.

[22] Alfano G., De Maio A., Conte E. 'Polarization diversity detection of distributed targets in compound-Gaussian clutter'. *IEEE Transactions on Aerospace and Electronic Systems*. 2004, vol. 40(2), pp. 755–65.

[23] Kay S. *Fundamentals of statistical signal processing, part 2*. Upper Saddle River, NJ: Prentice Hall; 2001.

[24] Conte E., Lops M., Ricci G. 'Adaptive matched filter detection in spherically invariant noise'. *IEEE Signal Processing Letters*. 1996, vol. 3(8), pp. 248–50.

[25] Gini F., Michels J.H. 'Performance analysis of two covariance matrix estimators in compound-Gaussian clutter'. *IEE Proceedings - Radar, Sonar and Navigation*. 1999, vol. 146(3), pp. 133–40.

[26] IPIX Radar *The Mcmaster IPIX radar sea clutter database* [online]. Available from http://soma.crl.mcmaster.ca/ipix/

[27] Krasnor C., Haykin S., Nohara T.J., Currie B.W. 'A new coherent marine radar for ocean-related studies'. Canadian Conference on Electrical and Computer Engineering; Vancouver, Canada, 1988.

[28] Hamburger D., Krasnor C., Haykin S., Nohara T.J., Currie B.W. 'A computer-integrated radar system for research use'. Canadian Conference on Electrical and Computer Engineering; Montreal, Canada, 1989.

[29] Evans M., Hastings N., Peacock B. *Statistical distributions*. New York: Wiley Interscience; 1993.

[30] D'Agostino R.B., Stephens M.A. *Goodness-of-fit Techniques*. New York: Marcel Dekker Inc; 1986.

[31] di Bisceglie M., Galdi C. 'CFAR detection of extended objects in high-resolution SAR images'. *IEEE Transactions on Geoscience and Remote Sensing*. 2005, vol. 43(4), pp. 833–43.

[32] Himed B., Melvin W.L. 'Analyzing space-time adaptive processors using measured data'. Conference Record of the Thirty-First Asilomar Conference on Signals, Systems and Computers; Pacific Grove, CA, 1997.

[33] Blum R.S., McDonald K.F. 'Analysis of STAP algorithms for cases with mismatched steering and clutter statistics'. *IEEE Transactions on Signal Processing*. 2000, vol. 48(2), pp. 301–10.

[34] Richmond C.D. 'Performance of a class of adaptive detection algorithms in nonhomogeneous environments'. *IEEE Transactions on Signal Processing*. 2000, vol. 48(5), pp. 1248–62.

[35] Gini F., Greco M.V., Diani M., Verrazzani L. 'Performance analysis of two adaptive radar detectors against non-Gaussian real sea clutter data'. *IEEE Transactions on Aerospace and Electronic Systems*. 2000, vol. 36(4), pp. 1429–39.

[36] De Maio A. 'Polarimetric adaptive detection of range-distributed targets'. *IEEE Transactions on Signal Processing*. 2002, vol. 50(9), pp. 2152–9.
[37] Cui G., Kong L., Yang X., Yang J. 'Distributed target detection with polarimetric MIMO radar in compound-Gaussian clutter'. *Digital Signal Processing*. 2012, vol. 22(3), pp. 430–8.

Chapter 2

Exploiting polarimetric diversity in passive radar

Fabiola Colone[1] and Francesca Filippini[1]

2.1 Introduction

Radar polarimetry has been widely used in radar systems aiming at the detection, discrimination, and recognition of targets of interest among other interfering sources. It becomes an essential tool in challenging radar scenarios where the achievable performance is not under control of the radar designer. This is certainly the case of passive radar whose performance largely varies with the radiative properties of the transmitter of opportunity as well as with the severity of the electromagnetic (EM) scenario that typically includes many interfering sources.

Passive radar, also known as passive coherent location (PCL), technology has been attracting significant research interest over the past two decades [1–5]. The wide interest received by PCL sensors allowed them to increasingly reach a point of maturity. Nevertheless, by relying on signals emitted by illuminators of opportunity (IOs) to detect and localize targets, the performance of PCL sensors might be largely time-varying and leading to an unreliable surveillance system. The main limitations stem from the lack of control over the exploited waveform structure as well as from the strong direct signal and multipath contributions [6–8]. In addition, significant interference can be experienced due to co-channel or adjacent-channel transmissions, especially when broadcast emitters are exploited as IOs [6–10].

Recently, among the advanced processing strategies devised to overcome these limitations, the exploitation of polarization diversity has been considered as a mean to increase the reliability of passive radar with particular reference to the target detection task [11–23]. Specifically, the availability of multi-polarimetric channels on receive has been demonstrated to provide robustness against the target echo fading that results from the induced variable polarization [11, 12, 15]. Moreover, a proper combination of signals received via differently polarized antennas has been shown to improve the target identification capability against the interfering sources, thanks to the wave polarization diversity [12, 17, 18, 22].

[1]Department of Information Engineering, Electronics and Telecommunications (DIET), Sapienza University of Rome, Rome, Italy

In this chapter, we take this perspective and illustrate practical multi-polarimetric passive radar architectures together with the corresponding signal processing techniques. Different approaches are compared both in terms of achievable performance and based on considerations regarding the resulting system complexity. The aim is to offer the interested reader with an overview of existing solutions and to provide hints for the identification of the most suitable architecture for the application at hand. The discussion is supported by results from real-world tests performed by means of experimental passive radar receivers operating at different frequency bands.

2.1.1 Organization

We discuss the main motivations behind the joint use of different polarimetric channels in passive radar in section 2.2, where we also provide an overview of the different approaches that can be employed, as described in detail in the subsequent sections.

Section 2.3 is devoted to the description of a cascaded polarimetric PCL processing architecture, presenting two different multi-polarimetric detection strategies that operate after the conventional PCL processing scheme.

In section 2.4, we describe an integrated polarimetric PCL processing architecture, where the processing scheme of a passive radar system is modified to exploit the polarimetric diversity at earlier stages in the chain. Also, an extensive comparison of the performance obtained on real data with the two architectures is shown.

Finally, section 2.5 will report some concluding remarks and future outlook.

2.2 Multi-polarimetric PCL architectures

2.2.1 Background and motivations

As well known, a PCL radar system parasitically exploits an existing transmitter as IO to perform the typical radar tasks of target detection, localization, and tracking. The progress on PCL achieved over the last decades has been tremendous. Numerous existing transmitters for telecommunication, radio navigation, and remote sensing applications have been used as IOs that result in the exciting possibility of deploying surveillance sensors with reduced cost of implementation and maintenance, ease of installation, covert operation, and limited impact on the environment.

PCL systems based on transmitters of radio/television broadcast have been the subject of extensive experimentation, and many prototypes have been developed and demonstrated to provide remarkable coverage against both aerial and maritime traffic. Recent studies have looked at the possibility to develop advanced imaging and surveillance capabilities by exploiting passive radar receivers mounted on a moving platform, such as small boats or light aircrafts [24–28]. Moreover, the attractive possibility of local area applications has been investigated, aiming at low-cost pervasive monitoring of public or private areas [29–35]. These studies are fostered also by the progress in commercial, miniaturized, and powerful hardware that makes it feasible to implement compact and cost-affordable multi-channel passive radar systems.

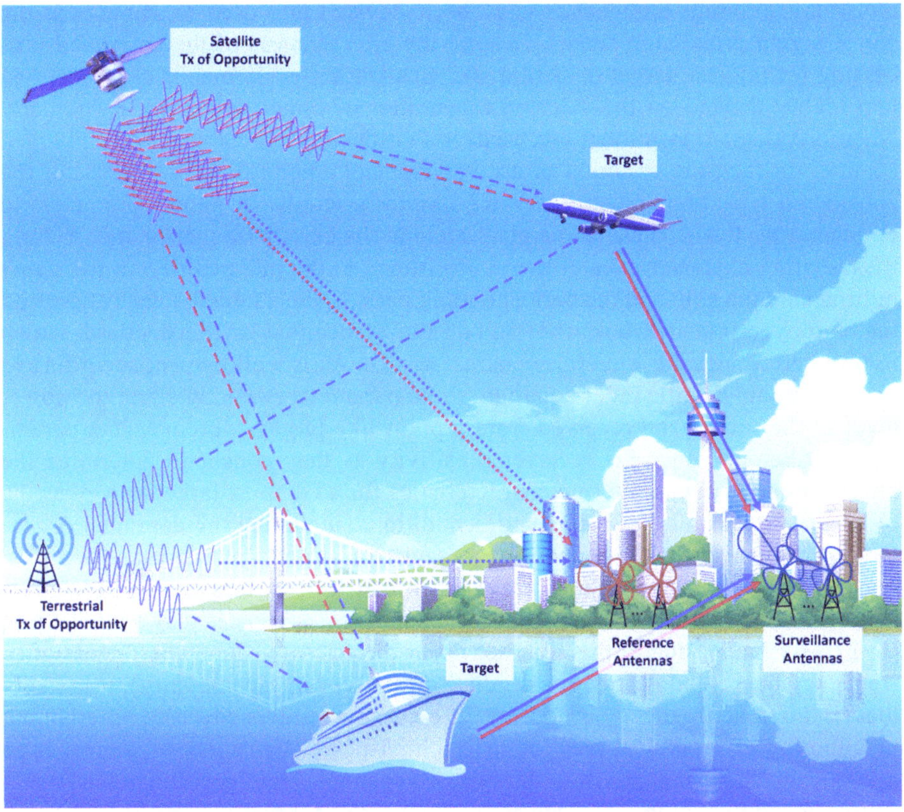

Figure 2.1 Reference scenario for a ground-based PCL receiver

The reader is referred to [36–38] for an overview of recent developments and challenges, new areas of research, and applications.

The acquisition geometry of a ground-based PCL system is sketched in Figure 2.1 for the case of a joint aerial and maritime surveillance application, assuming that the IO can be either a terrestrial or a satellite transmitter. The low power signal reflected from the target is collected by the main receiver – typically known as the surveillance channel – using a directive antenna steered toward the surveillance area. Since the transmitted signal is not known, an auxiliary channel is employed on receive – typically known as the reference channel – and it is usually connected to an additional directive antenna steered toward the IO. This auxiliary signal is exactly used as a reference for both removing disturbance, such as direct signal breakthrough and strong clutter/multipath echoes, at the surveillance channel [6–8, 39–43] and for evaluating the bistatic 2D (range-velocity) cross-ambiguity function (CAF) between the surveillance and the reference signals [5]. Target detection is then performed in the range-velocity domain, according to an appropriate thresholding strategy.

Many different approaches have been devised and used to implement the above signal processing stages. Despite the effectiveness of the employed processing techniques, target detection and localization performance of a PCL system highly depend on the selected transmitter of opportunity and, particularly when broadcast transmitters are used, it significantly varies with the employed channel. In fact, the selection of a given channel among those available for the considered telecommunication service implies a number of radar parameters to be inherently fixed. These parameters include the carrier frequency and, in turn, the radiating characteristics of the transmitter, i.e., the radiated power, the transmitting antenna gain and the corresponding pattern, the adopted polarization, and the employed transmission mode among those specified by the standard, which in turn determines the waveform characteristics. As is well known, according to a conventional bistatic radar equation, these parameters directly affect the power level of the radar echo received from a point-like target. Moreover, the parameters further influence the receiver sensitivity as they concur to determine the aforementioned backscattering characteristics for a complex target as well as the propagation channel electromagnetic conditions, including the dominant propagation mechanism and the possible presence of strong multipath and interfering transmissions. When radio/television broadcast services are considered as source of illumination, co- and adjacent- channel interferences play a crucial role [6–12]. Basically, due to the high density of broadcast channels in the assigned frequency band (and despite the regulations present in each country), it is likely that (i) the same frequency is used by different transmitters covering different but neighboring geographical areas and (ii) a high-power transmission is present on a frequency channel adjacent to the selected channel. This spectrum management causes no significant interference between channels for one-way broadcast reception, for which it is designed. But for two-way reception, which requires working many tens of dB further down in amplitude, the frequency reuse and the broad spectral roll-off of corresponding signals might cause severe interference. Such interference increases the equivalent noise floor of the PCL system in the selected channel bandwidth thus reducing the achievable disturbance cancellation and target detection performance.

Depending on the selected IO and specific broadcast channel, a large diversity could be experienced in term of all the conditions above and these might be time-varying during the radar operation. Therefore, in order to enhance the reliability of the considered PCL system, proper strategies should be adopted aimed at improving the system performance by mitigating its unstable behavior.

Among the recently proposed solutions, the exploitation of information diversity conveyed by multiple receiving channels is often considered as a viable means to improve the system performance against the many effects that are not under the control of the PCL radar designer. The sought diversity can be gained, for example, from the signals collected by displaced antennas [44–47] or at different carrier frequencies [9, 10, 47–51] or exploiting differently polarized antennas [11–23, 52].

A spatial cancellation approach might be exploited by properly combining the signals received by multiple receiving antennas in order to synthesize an equivalent antenna pattern with nulls (or at least strong depressions) in the directions of arrival of strong multipath or interfering transmitters. However, along with the requirement of a higher system complexity, the adaptive application of such approach might be troublesome if *a priori* knowledge on the interfering sources' locations is not available, due to the not extremely high power level of the considered signals. Moreover, synthesizing pattern nulls at given directions would significantly reduce the surveillance area since target returns from those directions would be indiscriminately rejected.

In References 9, 10 and 51, the joint exploitation of the signals of opportunity received at multiple carrier frequencies, i.e., different broadcast channels, was proposed to improve the robustness and the reliability of PCL system against the unpredictable characteristics of both the employed waveforms and the propagation channel. This multi-frequency approach was shown to be able to compensate for the time-varying behavior of the single frequency channel since the instantaneous characteristics of the bad channels are averaged with the good ones thus yielding a limited impact on the final performance. Despite the effectiveness of this approach for both target detection and localization, the resulting performance depends on the number and the quality of the available channels; basically, if a given channel is severely (and continuously) affected by interferences, its contribution to the multi-frequency operation would be negligible or possibly detrimental. Therefore, we expect that the multi-frequency approach will significantly benefit from a preliminary processing stage aimed at the maximization of the performance achievable with the single channel before this information is exploited for the multi-frequency integration.

In this chapter, we focus on the exploitation of different antenna polarizations as a mean to overcome the limitations above. Different approaches have been recently devised to effectively combine the signals received via differently polarized antennas in order to enhance the achievable performance for the resulting polarimetric PCL system. The effectiveness of these approaches has been extensively demonstrated using different IOs and operative scenarios [12, 17, 22, 23]. These approaches leverage polarimetric diversity according to alternative design strategies, which in turn have a different impact on the resulting polarimetric PCL system architecture, as illustrated in the following subsection.

2.2.2 Alternative architectural approaches to polarimetric PCL

Assuming that the PCL system is equipped with at least two receiving antennas with different (e.g., orthogonal) polarizations, many approaches might be envisaged to exploit the conveyed diversity in order to accomplish specific goals at both the surveillance and the reference channels. The following considerations are in order:

1. Due to the reflection of the transmitted signal on the target complex structure, target echoes typically show a random polarization. Hence, the use of a single fixed polarization for the surveillance antenna might result in a significant signal-to-noise (SNR) degradation depending on the target characteristics.

It is, therefore, expected that a proper combination of the signals simultaneously collected with differently polarized antennas might yield a target detection performance improvement [12].

2. Different IOs might exploit different wave polarizations on transmit. This is certainly the case of many terrestrial transmitters of radio and television broadcast that usually employ waves with linear (vertical or horizontal) polarization states; polarization-division multiplexing is used in satellite television downlinks to double the bandwidth by using two orthogonally polarized feed antennas in satellite dishes. Therefore, one of the possible objectives of the use of differently polarized antennas at the reference channel might be to recover a good copy of the transmitted signal; to this purpose, a switching or a selection strategy might be employed to pick out the signal from the preferred reference antenna.
3. For similar reasons, at the surveillance channel, the polarization diversity can be effectively exploited to increase the capability of rejecting the direct signal interference. In fact, assuming a line of sight between the transmitter and the receiver, a better rejection could be obtained by employing a cross-polarized antenna with respect to the polarization employed by the IO.
4. However, the transmitter polarization is not always known or the direct signal polarization might change due to geometric and atmospheric conditions. In this case, an alternative solution is to properly combine a set of reference signals received at differently polarized antennas in order to yield the highest multipath rejection at the surveillance channel [12, 52].
5. Finally, the polarization diversity might be exploited at the surveillance channel to counteract additional interfering contributions such as signals being transmitted in the same or adjacent frequency band or clutter and multipath residuals. Basically, since the target polarimetric characteristics might be very different from the competing disturbance, there is a strong motivation in designing signal processing strategies that are able to exploit the polarimetric differences to further improve the detection capability of the system [12, 22].

The goals above can be accomplished via alternative system and signal processing architectures that can be briefly classified into two main types:

- <u>Cascaded architecture</u> [Figure 2.2(a)]: In this case, the exploitation of the polarization diversity is obtained without modifying the core processing stages of a PCL sensor. This approach might encompass a preliminary selection of the signals to be used in the subsequent processing blocks based on either *a priori* information or appropriate metrics. Then, signals collected at differently polarimetric channels are separately processed according to a conventional single-pol signal processing architecture, using specific techniques. Finally, an appropriate combination of the obtained outputs is performed in order to benefit from the availability of multiple polarimetric channels. As it is apparent, this approach has a backward compatibility with

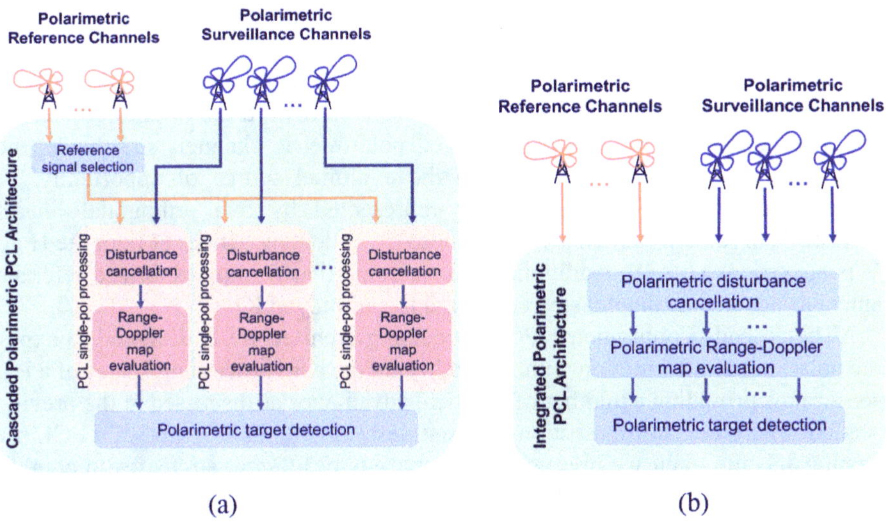

Figure 2.2 Different polarimetric PCL architectures: (a) cascaded architecture and (b) integrated architecture

existing PCL systems since it leverages the tools already developed for a single-pol system. Also, it allows different polarimetric signals to be processed in parallel using different processors as needed. However, due to its limited flexibility, it allows to accomplish only some of the goals above with remarkable performance. Still, it might be a preferred solution in such scenarios where the exploitation of polarization diversity is expected to provide just limited advantages.

- <u>Integrated architecture</u> [Figure 2.2(b)]: A larger flexibility is gained by resorting to an integrated approach where multiple polarimetric signals at both the reference and the surveillance channels are jointly exploited in ad-hoc multi-pol signal processing techniques. This approach requires deep modifications to the signal processing architecture, but it offers the opportunity to globally redesign the whole PCL signal processing chain in order to obtain the best performance in terms of interference rejection and target detection for the resulting polarimetric PCL. This advantage comes with an obvious increase in system complexity since the jointly processed signals should be made available simultaneously to the same processor and this should have more demanding computational requirements in order to allow a real-time operation.

In the following sections, possible signal processing strategies are introduced that could nicely fit into the two types of architectures illustrated above and their performance is comparatively investigated by application against experimental data.

2.3 Cascaded polarimetric PCL processing architecture

Let us consider a multi-polarimetric passive radar system, equipped with L differently polarized surveillance antennas, steered toward the monitored area. As aforementioned, the number of employed polarimetric channels strictly depends on the considered application and on the exploited source of opportunity. For instance, a common configuration is represented by two orthogonal linearly polarized surveillance antennas simultaneously collecting echoes at both the H and V polarizations ($L = 2$). Additionally, we assume that R multi-polarized reference antennas are also available, steered toward the employed IO.

The cascaded polarimetric PCL processing architecture is designed to exploit the polarimetric information without modifying the core processing stages of a PCL sensor thus providing a number of practical advantages as discussed in the previous section. Therefore, in this section, we first describe the main stages of a PCL processing scheme, then we present and compare two different strategies to combine the information provided by different polarimetric channels for the target detection task. Specifically, we focus on the capability to detect moving targets against interference caused by the stationary scene, interference from other transmitters as well as thermal noise.

According to the conventional single-pol PCL processing, see Figure 2.2(a), each surveillance signal first undergoes the temporal disturbance cancellation stage, aimed at removing the direct signal breakthrough and clutter contributions from the stationary scene from each surveillance signal. This stage could be performed with different algorithm, for instance, the well-known extensive cancellation algorithm (ECA) [8], and it is based on the availability of a copy of the transmitted waveform provided by a single reference signal. If the system collects $R > 1$ polarimetric versions of the reference signal, a preliminary selection could be performed of the best available copy of the transmitted signal, for instance, based on the corresponding power levels on receive.

Let $\boldsymbol{s}_l$, $l = 0, \ldots, L-1$ denote the $N \times 1$ vector collecting the samples of the lth surveillance signal, and $\boldsymbol{r}$ denotes the $N \times 1$ vector collecting the samples of the selected reference signal, where N is the number of samples in the considered coherent processing interval (CPI) T_{int}, i.e., $N = \lfloor T_{int} f_s \rfloor$, f_s being the sampling frequency.

Assuming that the multipath and clutter echoes are potentially backscattered from the first N_r range bins, the output of the ECA is evaluated as the difference between the lth surveillance signal and properly scaled and delayed replicas of the reference signal

$$\bar{\boldsymbol{s}}_l = \boldsymbol{s}_l - \boldsymbol{K}\boldsymbol{\varsigma} \tag{2.1}$$

where,

- $\bar{\boldsymbol{s}}_l$ is the lth $N \times 1$ surveillance signal after the ECA has been applied.
- $\boldsymbol{K}$ is a $N \times N_r$ matrix whose columns are the delayed versions of the selected reference signal, i.e., $\boldsymbol{K} = \left[\boldsymbol{r}^{(0)}\ \boldsymbol{r}^{(1)} \ldots\ \boldsymbol{r}^{(N_r-1)}\right]$, where $\boldsymbol{r}^{(l)}$ is the $N \times 1$ vector collecting the samples of the reference signal delayed by l/f_s.

- $\boldsymbol{\varsigma}$ is a $N_r \times 1$ vector containing the adaptive filter weights. These are evaluated according to a least square approach in order to minimize the power level at the output of the filter, thus obtaining $\boldsymbol{\varsigma} = \text{argmin}\{\|\bar{\boldsymbol{s}}_l\|^2\} = (\boldsymbol{K}^\dagger \boldsymbol{K})^{-1}\boldsymbol{K}^\dagger \boldsymbol{s}_l$.

Once this stage has been performed, a CAF is separately evaluated at each polarimetric surveillance channel in order to build up the bistatic range-Doppler map [5]:

$$\Psi_l(i, m) = \sum_{n=0}^{N-1} \bar{s}_l(n) \cdot r^*(n-i) e^{-j2\pi\left(\frac{mn}{N}\right)} \qquad l = 0, \ldots, L-1 \tag{2.2}$$

where $\bar{s}_l(n)$ is the nth sample of the signal at the output of the ECA stage, and (i, m) indicates the generic delay (range) and Doppler (velocity) bin.

At this point, a single-channel conventional PCL processing scheme would foresee a detection stage, typically performed with a cell average constant false alarm (CA-CFAR) detector applied over the bistatic range-Doppler plane. Instead, when multiple polarimetric channels are available, different approaches are possible to implement the detection task by taking advantage of the polarimetric information diversity. The following subsections will be devoted to the description of two polarimetric detection strategies based on different assumptions on the statistical properties of the disturbance affecting the available channels at the output of the processing stages illustrated above.

2.3.1 Polarimetric noncoherent integration detector

Let $\boldsymbol{\psi}$ be the $L\times 1$ vector collecting the complex values extracted from the L CAFs at a given range and Doppler location, considered as the cell under test (CUT). In the presence of a target with complex target amplitude α_l at the lth polarimetric channel, vector $\boldsymbol{\psi}$ is written as $\boldsymbol{\psi} = \boldsymbol{\alpha} + \boldsymbol{\psi}_d$, where $\boldsymbol{\alpha} = \left[\alpha_0, \ldots, \alpha_{L-1}\right]^T$ $(L\times 1)$ and $\boldsymbol{\psi}_d$ is the disturbance component competing against the target echoes in the range-Doppler domain. The latter is typically due to clutter residuals surviving the temporal cancellation stage as well as interfering contributions due to co-/adjacent channel transmissions.

The detector described in this section is based on the following simplifying assumptions:

1. The disturbance contributions affecting the L polarimetric CAFs are statistically independent and identically distributed (IID). Vector $\boldsymbol{\psi}_d$ follows a complex normal distribution with zero mean and covariance matrix $\sigma^2\boldsymbol{I}$, being σ^2 the disturbance power level at the single polarimetric channel.
2. Target amplitudes at different polarimetric channels are deterministic but unknown, so that $\boldsymbol{\psi} \sim \mathcal{CN}\left(\gamma\boldsymbol{\alpha}, \sigma^2\boldsymbol{I}\right)$ under the H_γ hypothesis ($\gamma = 0, 1$).

Under these hypotheses, an effective way to enhance the target SNR is to resort to a simple noncoherent integration (NCI) across the polarimetric maps [11]. More precisely, the outputs obtained at the L polarimetric channels are incoherently summed after square law detector

$$\zeta = \boldsymbol{\psi}^{\dagger}\boldsymbol{\psi} \tag{2.3}$$

Afterward, with the purpose of detecting the target presence, a CA-CFAR can be adopted against the integrated map, assuming that the interference is locally homogeneous on the range-Doppler plane

$$\zeta \underset{H_0}{\overset{H_1}{\gtrless}} \xi_{Pol-NCI} \cdot \sum_{p \in I_\zeta} \zeta_p \tag{2.4}$$

being I_ζ a set of P indices that identify range-Doppler bins surrounding the CUT (ζ_p, $p \in I_\zeta$, $|I_\zeta| = P$). The threshold $\xi_{Pol-NCI}$ is found by numerically inverting the following expression of the theoretical P_{fa} [12]:

$$P_{fa}^{(Pol-NCI)}(\xi) = \sum_{l=0}^{L-1} \binom{PL + l - 1}{l} \left(\frac{\xi}{PL}\right)^{l} \left(1 + \frac{\xi}{PL}\right)^{-PL-l} \tag{2.5}$$

The Pol-NCI detector represents a simple solution for jointly exploiting the available polarimetric channels according to a cascaded architecture, with a number of advantages [11, 12, 15, 23]. The target SNR is enhanced, thanks to the NCI of its contribution across different channels, which in turn is achieved with a limited computational complexity. Moreover, potential interference contributions that severely affect the bad polarimetric channels are moderately reduced due to the capability to average these contributions with those of the good channels, hence reducing their impact on the final performance.

However, the analyses performed against experimental data have revealed the weakness of the simplifying hypotheses made concerning the statistical independence of the disturbance affecting the polarimetric channels, which jeopardize the capability of controlling the false alarm rate [12].

2.3.2 Polarimetric generalized likelihood ratio test detector

The polarimetric detector described in this section is introduced to counteract the limitations identified on the polarimetric NCI strategy. In fact, since the disturbance contributions affecting the received signals are likely to be generated by external sources such as interferences from adjacent frequency channels, residual clutter, and multipath, the assumption of statistically independent disturbance at the different polarimetric channels might not always be realistic.

Therefore, the polarimetric generalized likelihood ratio test (Pol-GLRT) detector was derived based on the following assumptions:

1. The $L \times 1$ vector $\boldsymbol{\psi}_d$ follows a complex normal distribution with zero mean vector and covariance matrix $\boldsymbol{\Upsilon}$ with nonzero off-diagonal elements. Consequently, $\boldsymbol{\psi} \sim \mathcal{CN}(\gamma\boldsymbol{\alpha}, \boldsymbol{\Upsilon})$ under the H_γ hypothesis ($\gamma = 0, 1$), where $\boldsymbol{\alpha}$ is defined as in section 2.3.1.

2. A set of P target-free secondary vectors, $\boldsymbol{\psi}_p$, $p\in I_{\boldsymbol{\psi}}$, $|I_{\boldsymbol{\psi}}| = P$, is available, containing the output of the L maps at range-velocity locations adjacent to the CUT. They are assumed to be IID with the same statistic of $\boldsymbol{\psi}$ under the H_0 hypothesis.

Based on these assumptions and by resorting to the GLRT approach in [53], the detection test is obtained as [12]

$$\boldsymbol{\psi}^{\dagger}\hat{\boldsymbol{\Upsilon}}^{-1}\boldsymbol{\psi} \underset{H_0}{\overset{H_1}{\gtrless}} \xi_{Pol-GLRT} \tag{2.6}$$

where $\hat{\boldsymbol{\Upsilon}} = \frac{1}{P}\sum_{p\in I_\psi} \boldsymbol{\psi}_p\boldsymbol{\psi}_p^{\dagger}$, $p \in I_\psi$ is the estimated covariance matrix and $\xi_{Pol-GLRT}$ is the detection threshold, selected according to the following theoretical P_{fa} expression that accounts for the fluctuations in the estimation of matrix $\boldsymbol{\Upsilon}$ due to finite P [12]

$$P_{fa}^{(Pol-GLRT)}(\xi) = \frac{\left[1-\left(\frac{\xi}{P+\xi}\right)\right]^{P-L+1}}{\Gamma(P-L+1)} \sum_{l=0}^{L-1} \frac{\Gamma(P-l)}{\Gamma(L-l)} \left(\frac{\xi}{P+\xi}\right)^{L-l+1} \tag{2.7}$$

being $\Gamma(\cdot)$ the Gamma function.

The detection test in (2.6) can also be interpreted as the cascade of two subsequent blocks: a whitening transformation followed by an NCI. In other words, this strategy benefits from the target echo integration across the available polarimetric channels as the Pol-NCI but, differently from the previous approach, it has the additional ability to counteract the correlated disturbance contributions in the range-Doppler map, thanks to an adaptive disturbance rejection performed in the polarimetric domain. Therefore, the polarimetric diversity is exploited not only for enhancing the target SNR but also to discriminate target echoes from the competing interference.

Both the Pol-NCI and the Pol-GLRT detection strategies have been tested and compared in different operative scenarios [12, 15, 17, 23]. In the following subsection, the results are reported against experimental passive radar datasets.

2.3.3 Experimental validation

To show the effectiveness of the described cascaded polarimetric PCL processing architecture and to compare the two detectors described above, we consider three experimental datasets, collected with different PCL systems exploiting either terrestrial or satellite transmitters as sources of opportunity and operating in different scenarios.

The three datasets will be referred to as case studies and will be distinguished as follows:

- Case study A features data collected with a FM radio-based PCL system for aerial surveillance.

- Case study B features data collected with a digital video broadcasting-terrestrial (DVB-T) based PCL system for both aerial and maritime surveillance.
- Case study C features data collected with a digital video broadcasting-satellite (DVB-S) based PCL system for short range vehicular traffic monitoring.

In all cases, $L=2$ linearly polarized surveillance antennas (H and V) are employed on receive while the number of available reference channels varies depending on the considered case study. The following subsections will be dedicated to each of the three case studies separately.

2.3.3.1 Case study A: FM radio-based PCL

The dataset employed for case study A has been collected near Fiumicino Airport, Italy, exploiting the FM radio transmission of Monte Cavo (~25 km away from the receiver site) at the frequency channel 94.5 MHz. The receiving system was equipped with two cross-elements Yagi dual-polarized antennas (see Figure 2.3), one employed to collect the H and V versions of the reference signals and the other employed to collect the H and V versions of the surveillance signals. The reference signal collected by the V polarized antenna is generally selected for the subsequent processing by the selection block based on the power level measurement, which is an expected result since the exploited IO uses V polarization at the considered frequency channel.

Case study A will be repeatedly employed in this chapter for an extensive assessment of the described techniques since (i) the dataset is made of more than 2000 datafiles of 1.1 s each, with a temporal spacing of about 2.3 s, for a total acquisition time of 80 minutes; (ii) a reliable air-truth is available for the several targets

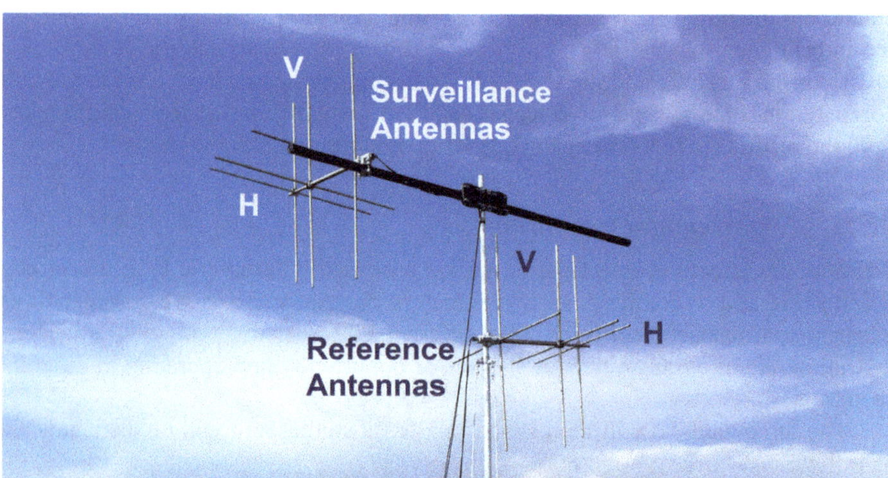

Figure 2.3 Employed multi-polarimetric reference and surveillance antennas

of opportunity (aircraft) that were present in the monitored area at the time of the acquisition, provided by an Automatic Dependent Surveillance–Broadcast (ADS-B) receiver. Further details on the acquisition campaign and the employed receiving system are reported in [12 and 22].

To investigate the performance of the cascaded polarimetric architecture and corresponding detection strategies, we first extract one datafile, obtained for a CPI of 1 s, from the entire dataset, say datafile #1, and we report the obtained results in Figure 2.4. Specifically, Figure 2.4(a,b) shows the results of a single polarimetric conventional PCL processing (referred to as single-pol H or single-pol V), while Figure 2.4(c,d) shows the outcome of the cascaded polarimetric PCL processing where the polarimetric detection stage is performed with the Pol-NCI and Pol-GLRT strategies, respectively.

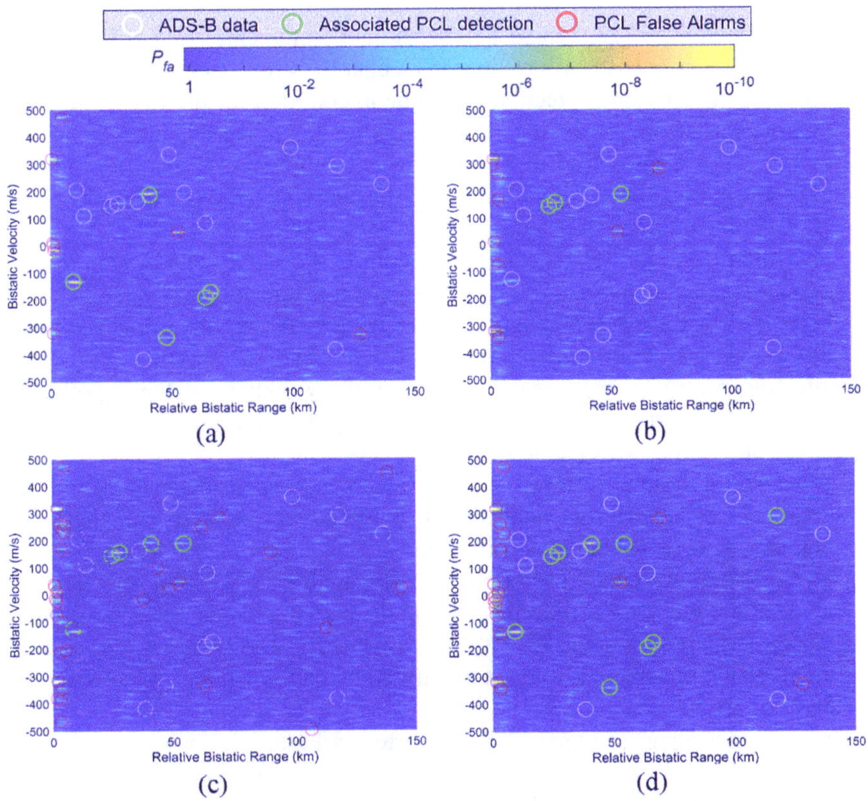

Figure 2.4 *Datafile #1: Range-Doppler map of minimum nominal P_{fa} values to be set in order to detect each bin: (a) single-pol H, (b) single-pol V, (c) cascaded architecture with pol-NCI, and (d) cascaded architecture with pol-GLRT*

In order to directly compare the results obtained with different detectors, each subfigure of Figure 2.4 represents the test statistic over the bistatic range-bistatic velocity plane before the thresholding stage. Specifically, we map each bin into the nominal P_{fa} that would allow the corresponding threshold exceeding. In other words, each pixel has been scaled to represent the minimum value of nominal P_{fa} to be set for that pixel to yield a detection. Consequently, the brighter the pixel in the map is, the more the P_{fa} can be reduced, i.e., the threshold raised, while still ensuring its detection. In all subfigures, the white circles indicate the expected position of the noncooperative targets that were present at the time of this datafile, as provided by the ADS-B receiver; the green circles represent the PCL detection that would be obtained at $P_{fa} = 10^{-5}$ and are correctly associated with the air-truth tracks; finally, red circles represent the PCL detections that have not been associated with any track and thus are likely to correspond to false alarms.

The following considerations are in order:

- As expected from [12, 15 and 17], depending on the employed polarization on receive, target echoes do not appear as equally strong peaks. Specifically, at this datafile, only one detection would be in common for the two orthogonally polarized channels when the corresponding signals are separately used with $P_{fa} = 10^{-5}$ (Figure 2.4(a,b)). This confirms the hypothesis that, depending on the propagation channel conditions and the complex structure of the target, the signal polarization may change along the two-way path.
- When the detection test is performed with the Pol-NCI strategy, the resulting system is not able to recover all the correct detections of the H and V channels. Moreover, it fails in rejecting the disturbance as it shows the highest number of false alarms. As aforementioned, we can attribute the inability to provide an effective control of the false alarm rate to the weakness of the hypothesis of statistically independent disturbance affecting the range-velocity maps at the different polarimetric channels.
- When the polarimetric information is leveraged according to the Pol-GLRT approach, the interfering contribution is effectively rejected and the highest number of correct detections is obtained.
- The reported green and red circles are obtained for a selected P_{fa} value. It is evident that there are additional peaks in the map that do not exceed the current threshold but could do so for a slightly higher P_{fa}. Nevertheless, the considerations remain unchanged and are confirmed by additional analyses not reported here for brevity.
- An enlarged view of Figure 2.4 is reported in Figure 2.5 around the target at approximately 115 km. Observing Figure 2.5, the advantage of the Pol-GLRT strategy applied at the end of the described PCL processing is apparent since the target would be detected up to very low P_{fa} values ($\geq 10^{-8}$).

After analyzing a single datafile, we collect the detection results at $P_{fa} = 10^{-5}$ across 50 consecutive scans. The results are shown on the same bistatic range-velocity map in Figure 2.6, where the same four strategies are compared. In all

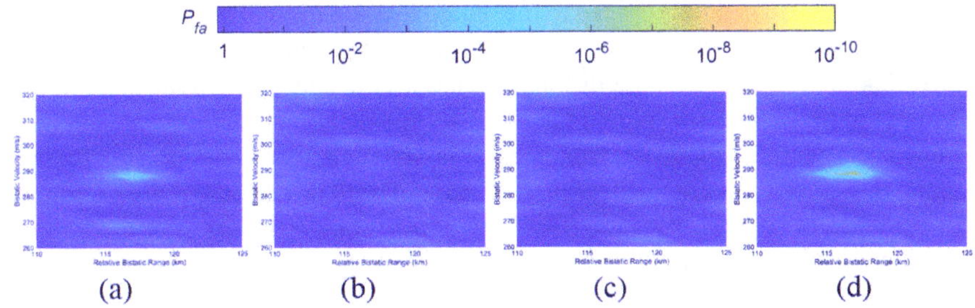

Figure 2.5 *Enlarged view of Figure 2.4 with (a) single-pol H, (b) single-pol V, (c) cascaded architecture with pol-NCI, and (d) cascaded architecture with pol-GLRT*

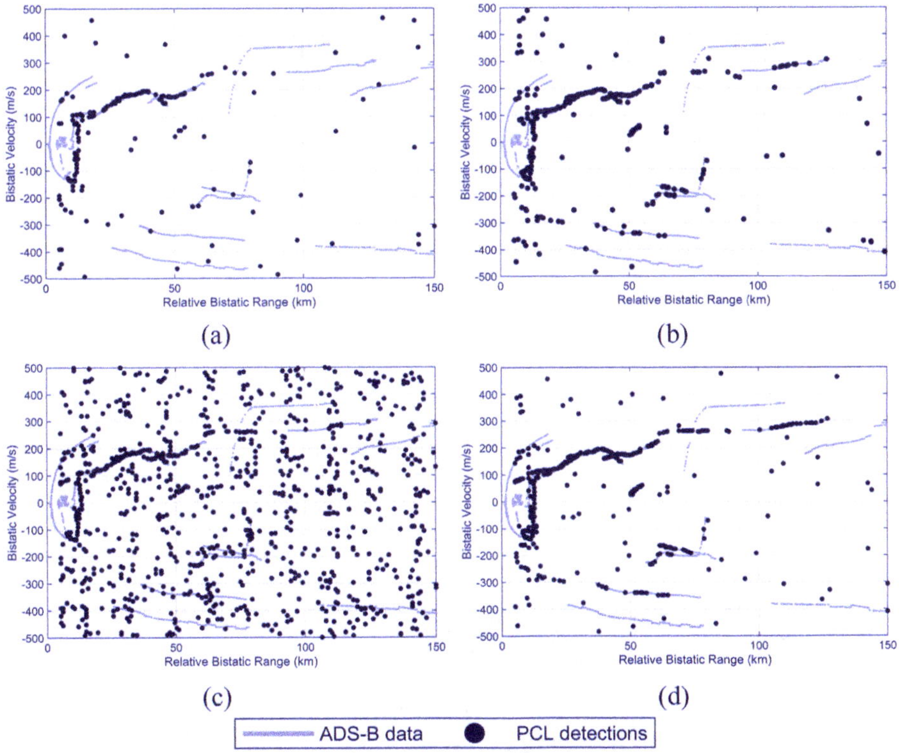

Figure 2.6 *Detection results over 50 consecutive datafiles, with $P_{fa} = 10^{-5}$, using (a) single-pol H, (b) single-pol V, (c) cascaded architecture with Pol-NCI, and (d) cascaded architecture with Pol-GLRT. © [2020] IEEE. Reprinted, with permission from [22].*

subfigures, the gray lines represent the available air-truth while the PCL detections are reported as black dots. Figure 2.6 confirms the considerations made on Figures 2.4 and 2.5. In fact, the different behaviors of the single polarimetric channels are evident on almost every tracks.

On average, the V polarimetric channel is better performing with respect to the H channel, which shows that selecting the channel with orthogonal polarization with respect to the IO is not always the best choice as the performance after the temporal cancellation block might be limited by other interfering sources. However, we have shown in Figure 2.4(a,b) and Figure 2.5(a,b) that this is not always the case and some track portions might be visible only when using the H channel, which in turn highlights the impossibility to select *a priori* the best performing channel and the potential benefits coming from their joint exploitation.

The lack of control of the false alarm rate with the Pol-NCI is evident when looking at Figure 2.6(c) where, although some additional target detections might have been obtained, the disturbance level is unacceptably increased. It is worth mentioning that the selected radio channel was significantly affected by co-/adjacent-channel interference so that the need for a disturbance rejection capability is apparent. In fact, several additional target detections are obtained with the Pol-GLRT in Figure 2.6(d) while maintaining the number of false alarms under control.

2.3.3.2 Case study B: DVB-T based PCL

The dataset employed for case study B has been collected in Eckernförde, Germany, by exploiting the DVB-T IO located in Kiel, approximately 20 km away from the receiver site. The passive radar system PARASOL [54], developed at Fraunhofer FHR, was employed, equipped with two dual polarized (H,V) surveillance antennas.

Three cooperative targets were employed, namely the ultralight Delphin aircraft Figure 2.7(a) and two speedboats as the one in Figure 2.7(b). Further details on the considered data and acquisition campaign are reported in [17]. Their actual trajectories along the test are reported as colored lines in Figure 2.8 mapped on to the bistatic range-velocity plane.

Figure 2.8 also reports the raw PCL detections for 168 consecutive datafiles, with $P_{fa} = 10^{-6}$ when a single-channel PCL processing is performed Figure 2.8(a,b) and when the polarimetric information is exploited at the detection stage using the

(a) (b)

Figure 2.7 Cooperative targets: (a) ultralight Delphin aircraft and (b) speedboat

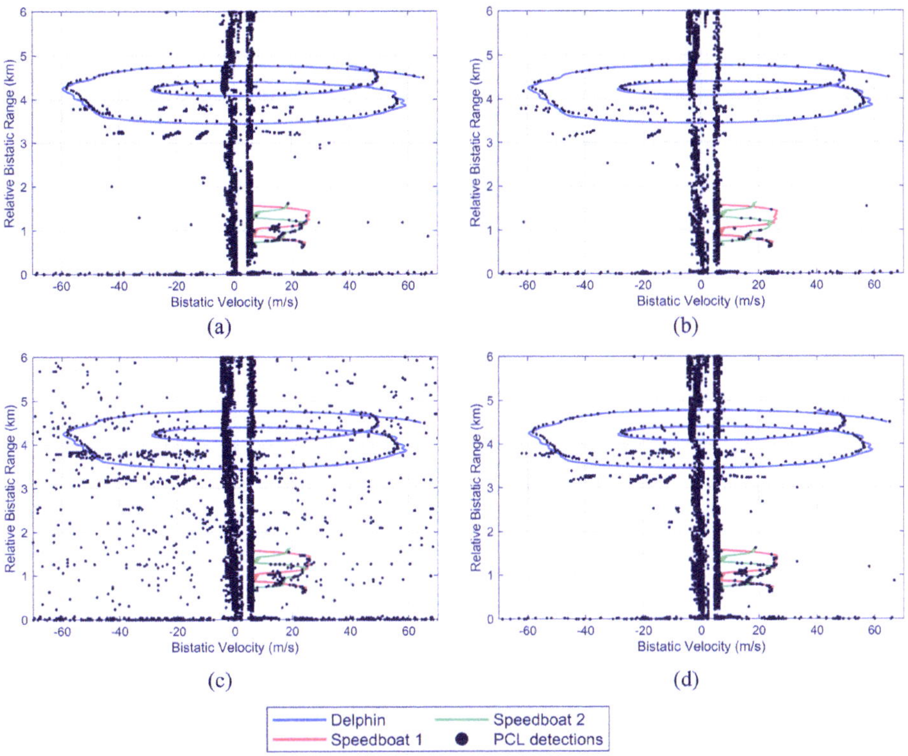

Figure 2.8 Detection results over 168 consecutive datafiles, with $P_{fa} = 10^{-6}$, using (a) single-pol H, (b) single-pol V, (c) cascaded architecture with Pol-NCI, and (d) cascaded architecture with Pol-GLRT.© [2017] IET. Reprinted from [17].

Pol-NCI Figure 2.8(c) and the Pol-GLRT Figure 2.8(d). The considerations made on the results of case study A also apply here. When comparing the results of the H and V polarimetric channel, separately used, it is evident that different portions of the target tracks are better visible in different channels. Moreover, when the two channels are combined according to the Pol-GLRT approach, the improvement is noticeable both with respect to the single polarimetric channels and to the Pol-NCI, thanks to the capability to mitigate the residual background affecting the range-Doppler maps at different polarimetric channels.

2.3.3.3 Case study C: DVB-S based PCL

The dataset employed for case study C has been collected for a test performed in a parking area, when exploiting the DVB-S Astra 1M satellite transmitter operating at frequency channel 11.778 GHz [23]. A parabolic reflector, see Figure 2.9(a), was used to collect a good copy of the reference signal, by receiving the vertical polarized version of the reference signal. A feedhorn with circular flange was employed as

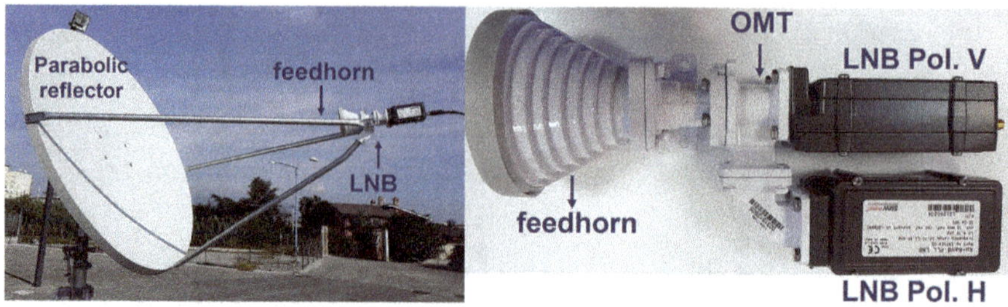

(a) (b)

Figure 2.9 Employed antennas: (a) single pol (V) reference antenna and (b) dual-pol surveillance antenna (H,V)

surveillance antenna in order to guarantee a wider beamwidth for continuously monitoring the area of interest [see Figure 2.9(b)]. To collect both the H and V versions of the surveillance signal, the feedhorn was equipped with an orthomode transducer, with two outputs, each connected to a dedicated low-noise block in order to provide the first low-noise amplification of the received signals and their down-conversion from Ku to L band. A four-channel software-defined radio finally provided the baseband digitized signals for both the reference and the surveillance channels.

A car was used in the trial as cooperative target, moving away from the receiver site and the total acquisition time was of approximately 25 s. The same kind of results as in Figure 2.4 is shown in Figure 2.10 for two datafiles where the car is present. As it is apparent, the target does not appear as an equally strong peak at

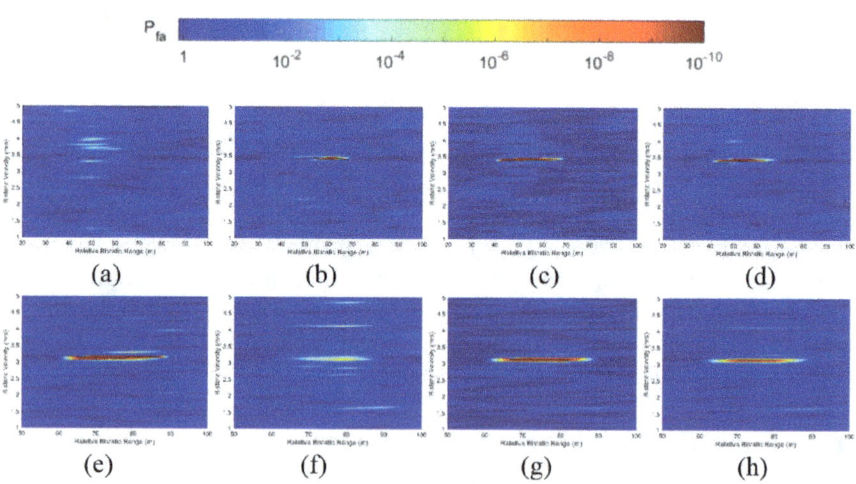

Figure 2.10 Datafiles #1 and #2: Range-Doppler map of minimum nominal P_{fa} values to be set in order to detect each bin: (a,e) single-pol H, (b,f) single-pol V, (c,g) cascaded architecture with Pol-NCI, and (d,h) cascaded architecture with Pol-GLRT

the two single polarimetric channels. Specifically, the V channel outperforms the H channel in datafile #1 Figure 2.10(a–d) while the car appears as a stronger peak in the H map in datafile #2 [Figure 2.10(e–h)]. In both cases, the Pol-NCI and Pol-GLRT strategies would successfully detect the target at both datafiles also with very low P_{fa} values.

We note that, when exploiting DVB-S signals, due to the low power density available on the Earth's surface and the limited cost of the employed receivers, the performance is mostly noise-limited and any other source of interference is well below the receiver noise level.

Therefore, the main advantage of the polarimetric strategies is offered by the integration across different receiving channels, other than a proper cancellation of the disturbance obtained exploiting the polarimetric information. This is also confirmed in Figure 2.11, where we reported the results across 130 signal fragments of 0.5 s each, corresponding to the employed CPI, with an overlap of 0.2 s. Specifically, the raw detections obtained with a $P_{fa} = 10^{-5}$ are reported on the same bistatic range-bistatic velocity plane. In fact, in this case, the result obtained with the Pol-GLRT is only slightly better than the one of the Pol-NCI (+4% of correct detections), while

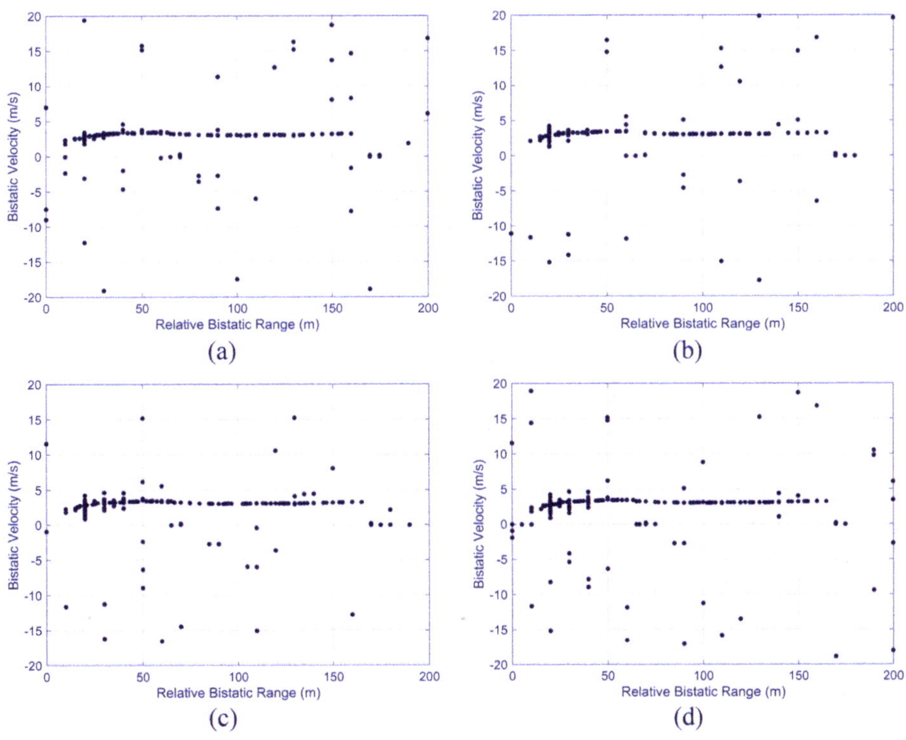

Figure 2.11 Detection results over 130 consecutive datafiles, with $P_{fa} = 10^{-5}$, using (a) single-pol H, (b) single-pol V, (c) cascaded architecture with Pol-NCI, and (d) cascaded architecture with Pol-GLRT

both outperform the single-pol strategies, by providing a higher number of correct detections for the considered target (~+31% and +52% with respect to the H and V channel, respectively).

Based on the analyses reported in this section, the following conclusions apply:

- The use of a single polarization on receive has proved to provide limited performance in all the considered scenarios. This makes the PCL system unreliable since, when operating with a fixed, *a priori* selected, polarization on receive, it is not able to guarantee a continuous coverage on targets of interest.
- The cascaded polarimetric PCL processing has proved to be an effective architectural solution that allows tackling the limitations of a single-pol PCL, while keeping the implementation simple and maintaining a backward compatibility with existing PCL systems by preserving the core processing stages already developed for single-pol sensors. For these reasons, this approach might be an effective solution in PCL systems that are characterized by high computational and implementation complexity, such as, for example, the PCL sensors based on DVB-S and DVB-T signals of opportunity.
- Among the described strategies to leverage the information diversity conveyed by different polarimetric channels when employing a cascaded architecture, the polarimetric GLRT was shown to be robust against severe interference sources and residual disturbance. Depending on the considered IO, the operative scenario, and the severity of the interference, its advantage over a simpler polarimetric NCI might vary, therefore a proper selection of the detector to be adopted will depend on the considered application. For instance, the simpler Pol-NCI should be preferred in such scenarios where the exploitation of polarization diversity is expected to provide just limited advantages, e.g., in DVB-S based PCL sensors.
- Despite the considerable improvement that the described architectural approach can yield, we note that its ability to discriminate target echoes against strong disturbance background may be limited in some specific cases. For instance, this is the case of the FM radio-based PCL considered in Figure 2.6(d) where only isolated detections are available for some of the targets of opportunity, which would probably not be sufficient for a track initiation stage. This might be attributed to the limited flexibility of the cascaded polarimetric PCL processing, and further improvement in the system performance could be expected if the polarimetric information diversity is properly exploited at earlier stages in the processing chain. This possibility will be addressed in the next section.

2.4 Integrated polarimetric PCL processing architecture

According to the integrated polarimetric PCL processing architecture, the multichannel passive radar processing scheme is altered to encompass the exploitation of the polarimetric diversity at the earlier stages of the processing scheme. As mentioned in section 2.2, the information diversity conveyed at both the reference and surveillance channels is leveraged to better reject the interference contribution, counteract the disturbance cancellation residuals, and hence increase the target detection capability.

This approach offers the opportunity to globally redesign the PCL signal processing chain and the signal processing techniques employed therein in order to obtain improved performance for the resulting polarimetric PCL. To this purpose, different approaches might be exploited to implement the tasks listed above. In the following, we refer to the signal processing scheme sketched in Figure 2.12.

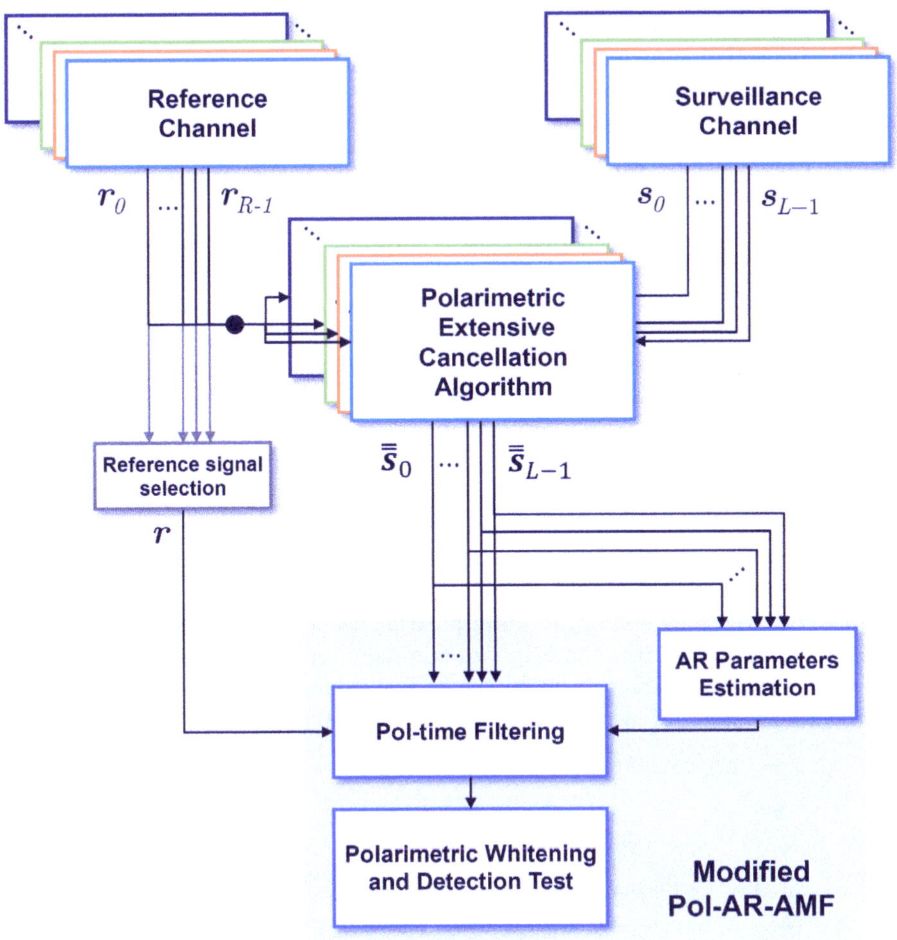

Figure 2.12 Block diagram of the integrated multi-polarimetric PCL processing with the Mod-Pol-AR-AMF detector

According to this scheme, first the R polarimetric versions of the reference signals are jointly exploited to improve the direct signal and multipath removal from the surveillance signals using a polarimetric version of the ECA (Pol-ECA) [12]. Once this stage has been performed at each polarimetric surveillance channel, L sequences of samples are made available that are likely to include interfering contributions, along with target echoes and cancellation residuals. Therefore, an adaptive detection scheme is applied that properly combines the samples of the L signals in order to improve the target discrimination capability against the competing disturbance. Specifically, the proposed detection strategy is derived by modeling the disturbance as a multichannel autoregressive (AR) process and by resorting to a two-stage GLRT approach [19, 20]. The derived detector, referred to as polarimetric AR model based adaptive matched filter (Pol-AR-AMF), has been efficiently adapted to the PCL case in [22] in order to reduce its computational burden while retaining the advantages of exploiting the polarimetric and temporal correlation to counteract the disturbance.

In the next subsections, we illustrate in more detail the techniques mentioned above; then, we test the overall architecture on real data.

2.4.1 Polarimetric extensive cancellation algorithm

With reference to a conventional PCL processing scheme, the availability of a good copy of the transmitted signal can be easily demonstrated to have a larger impact on the direct signal interference and clutter rejection stage rather than on the subsequent matched filtering stage, namely the evaluation of the range-velocity map. Therefore, the polarization diversity provided by the reference channels could be fruitfully exploited with the purpose of enhancing the disturbance cancellation at the surveillance signals. To this purpose, although a selection strategy to pick out the signal from the preferred antenna might represent an effective solution, an improved multipath rejection capability could be obtained if the polarimetric versions of the reference signal are properly combined.

A robust and effective approach was proposed in [11] as a polarimetric extension of the ECA [8] (Pol-ECA). The Pol-ECA is obtained by broadening the disturbance subspace including the multi-polarimetric signals collected with the R reference antennas. In fact, the output of this stage for the lth surveillance channel is evaluated as the difference between the lth surveillance signal and properly scaled and delayed replicas of the reference signals $\boldsymbol{r}_r$, $r = 0, \ldots, R-1$

$$\bar{\bar{\boldsymbol{s}}}_l = \boldsymbol{s}_l - \tilde{\boldsymbol{K}}\tilde{\boldsymbol{\varsigma}} \tag{2.8}$$

where,

- $\bar{\boldsymbol{s}}_l$ is the lth $N \times 1$ surveillance signal after the Pol-ECA has been applied.
- $\tilde{\boldsymbol{K}} = [\boldsymbol{K}_0 \ \ldots \ \boldsymbol{K}_{R-1}]$ and $\boldsymbol{K}_r$, $r = 0, \ldots, R-1$ are $N \times N_r$ matrices whose columns are the delayed versions of the rth reference signal, i.e., $\boldsymbol{K}_r = \left[\boldsymbol{r}_r^{(0)} \ \boldsymbol{r}_r^{(1)} \ldots \ \boldsymbol{r}_r^{(N_r-1)}\right]$, where $\boldsymbol{r}_r^{(l)}$ is the $N \times 1$ vector collecting the samples of the reference signal delayed by l/f_s.

- $\tilde{\varsigma} = \left[\varsigma_0^T \dots \varsigma_{R-1}^T\right]^T$ is a $N_r R \times 1$ vector containing the adaptive filter weights, being ς_r the weights to be applied at the rth reference channel. As for the single-pol ECA, vector $\tilde{\varsigma}$ is found according to a LS approach as $\tilde{\varsigma} = (\tilde{\boldsymbol{K}}^\dagger \tilde{\boldsymbol{K}})^{-1} \tilde{\boldsymbol{K}}^\dagger \boldsymbol{s}_l$.

The results reported in [8] proved that the Pol-ECA yields a disturbance rejection that is always as good as or better than that obtained with a single-pol ECA approach employing a properly selected reference signal. However, the extent of the advantage will strictly depend on the specific application, e.g., the considered scenario or the employed source of opportunity. Moreover, this approach operates with a larger number of degrees of freedom thus requiring a higher computational load, which should be traded for the expected advantage.

2.4.2 Polarimetric autoregressive model based adaptive detector

Although the direct signal, clutter, and multipath contributions are strongly reduced after the disturbance cancellation in (2.8), it is likely for the L output signals to include cancellation residuals and interfering contributions, along with target echoes. At this stage, the L available sequences of samples can be properly modeled to derive a target detection strategy that effectively exploits both the polarimetric and temporal domain to better counteract the residual disturbance, thus further improving the system performance.

2.4.2.1 Polarimetric AR-based adaptive matched filter

At the output of the Pol-ECA stage, we arrange the nth signal snapshot, namely the signal samples $\bar{\bar{s}}_l(n)$, $l = 0, \dots, L-1$, made available at the nth sampling time from the polarimetric surveillance channels, in a L-dimensional vector

$$\boldsymbol{x}(n) = \left[\bar{\bar{s}}_0(n) \dots \bar{\bar{s}}_l(n) \dots \bar{\bar{s}}_{(L-1)}(n)\right]^T \tag{2.9}$$

Furthermore, we arrange N consecutive vectors in a LN-dimensional vector $\boldsymbol{x} = \left[\boldsymbol{x}^T(0) \dots \boldsymbol{x}^T(N-1)\right]^T$, N being the number of samples included in the CPI.

Under the null hypothesis H_0, vector $\boldsymbol{x}$ is solely composed by disturbance contributions $\boldsymbol{d}$, such as thermal noise, interference from transmissions at co-/adjacent channels, and residual clutter and multipath contributions. Conversely, under the alternate hypothesis H_1, vector $\boldsymbol{x}$ is composed by both disturbance and a useful target component $\boldsymbol{z}$, i.e.,

$$\begin{aligned} H_0 &: \boldsymbol{x} = \boldsymbol{d} \\ H_1 &: \boldsymbol{x} = \boldsymbol{z} + \boldsymbol{d} \end{aligned} \tag{2.10}$$

The target component $\boldsymbol{z}$ can be written as

$$\boldsymbol{z} = \boldsymbol{t}(\tau, f_D) \otimes \boldsymbol{\alpha} \tag{2.11}$$

where,

- $\boldsymbol{t}(\tau, f_D)$ is a N–samples fragment of the direct signal waveform $g(t)$, delayed in time by τ and Doppler shifted by f_D, i.e.,

$$\boldsymbol{t}(\tau, f_D) = \begin{bmatrix} g(t_0 - \tau) \\ g\left(t_0 + \frac{1}{f_S} - \tau\right) \\ \vdots \\ g\left(t_0 + \frac{N-1}{f_S} - \tau\right) \end{bmatrix} \odot \begin{bmatrix} 1 \\ e^{j2\pi \frac{f_D}{f_S}} \\ \vdots \\ e^{j2\pi (N-1) \frac{f_D}{f_S}} \end{bmatrix} \tag{2.12}$$

being t_0 the CPI starting time

- $\boldsymbol{\alpha} = [\alpha_0, \ldots, \alpha_{L-1}]^T (L \times 1)$ contains the unknown complex target amplitudes at the different polarimetric channels, assumed constant during the CPI.

We model the disturbance as a L-channel AR process of order $(Q-1)$, denoted as AR $(Q-1)$. Accordingly, the $L \times 1$ vector random process $\boldsymbol{d}(n)$ extracted at the nth time instant satisfies the following relation:

$$\boldsymbol{d}(n) = \sum_{q=1}^{Q-1} \boldsymbol{A}^\dagger(q) \boldsymbol{d}(n-q) + \boldsymbol{\varepsilon}(n) \tag{2.13}$$

where $\{\boldsymbol{A}(q)\}_{q=1}^{Q-1}$ are complex-valued $L \times L$ matrix parameters encoding the regression coefficients at different polarimetric channels, and $\boldsymbol{\varepsilon}(n) \sim \mathcal{CN}(\boldsymbol{0}, \boldsymbol{R})$ is the driving white noise sequence, denoting $\boldsymbol{R}$ as the $L \times L$ polarimetric covariance matrix.

Based on the considered model, the approximate (actually conditional) likelihood function of the data $\boldsymbol{x}$ under the H_γ ($\gamma = 0,1$) hypothesis is written as [19, 55, 56]

$$\begin{aligned} f_\gamma(\boldsymbol{x} \mid \gamma\boldsymbol{\alpha}, \boldsymbol{R}, \boldsymbol{A}) &= (\pi^L |\boldsymbol{R}|)^{-(N-Q+1)} \times \\ &\exp\left\{ -\sum_{n=1}^{N-Q+1} [\tilde{\boldsymbol{x}}(n) - \gamma \tilde{\boldsymbol{z}}(n)]^\dagger \boldsymbol{P} [\tilde{\boldsymbol{x}}(n) - \gamma \tilde{\boldsymbol{z}}(n)] \right\} \end{aligned} \tag{2.14}$$

where $\boldsymbol{P} = \boldsymbol{H}^\dagger \boldsymbol{R}^{-1} \boldsymbol{H}$, with $\boldsymbol{H} = [-\boldsymbol{A}^\dagger \;\; \boldsymbol{I}]$ and we denote as $\tilde{\boldsymbol{x}}(n)$ and $\tilde{\boldsymbol{z}}(n)$ the QL-dimensional subvectors of $\boldsymbol{x}$ and $\boldsymbol{z}$, defined as $\tilde{\boldsymbol{x}}(n) = [\boldsymbol{x}^T(n) \ldots \boldsymbol{x}^T(n+q-1)]^T$ and $\tilde{\boldsymbol{z}}(n) = [\boldsymbol{z}^T(n) \ldots \boldsymbol{z}^T(n+q-1)]^T$, respectively.

The sought polarimetric adaptive detector has been derived in [19] by resorting to a two-stage GLRT approach. First, the disturbance characteristics, namely matrices $\boldsymbol{A}$ and $\boldsymbol{R}$, are assumed known and the clairvoyant detector is derived; then, suitable maximum likelihood (ML) estimates of the AR parameters are plugged into the derived clairvoyant test. This approach has been referred to as Pol-AR-AMF, and the resulting detection test to be applied at a given delay (range)-Doppler CUT (τ, f_D) can be written as

$$\|\chi(\tau, f_D)\|^2 \underset{H_0}{\overset{H_1}{\gtrless}} \xi_{Pol-AR-AMF} \tag{2.15}$$

where $\xi_{Pol-AR-AMF}$ is the detection threshold, $\hat{\boldsymbol{P}}$ is evaluated using the ML estimates of the AR parameters and

$$\chi(\tau, f_D) = \left[\boldsymbol{W}^{-\frac{1}{2}}(\tau, f_D)\right]^{\dagger} \sum_{n=0}^{N-Q} \boldsymbol{\Sigma}^{\dagger}(n, \tau, f_D)\, \hat{\boldsymbol{P}}\, \tilde{\boldsymbol{x}}(n) \tag{2.16}$$

where,

$$\boldsymbol{W}(\tau, f_D) = \frac{1}{2}\sum_{n=0}^{N-Q} \boldsymbol{\Sigma}^{\dagger}(n, \tau, f_D)\, \hat{\boldsymbol{P}}\, \boldsymbol{\Sigma}(n, \tau, f_D) \tag{2.17}$$

and

$$\boldsymbol{\Sigma}(n, \tau, f_D) = \tilde{\boldsymbol{t}}(n, \tau, f_D) \otimes \boldsymbol{I} \tag{2.18}$$

being $\tilde{\boldsymbol{t}}(n, \tau, f_D)$ a Q-dimensional subvector of $\boldsymbol{t}(\tau, f_D)$ in (2.12), starting from the nth sample, i.e., $\tilde{\boldsymbol{t}}(n, \tau, f_D) = [t_n(\tau, f_D) \ldots t_{n+Q-1}(\tau, f_D)]^T$, being v_n the nth element of vector $\boldsymbol{v}$.

In practice, the vector $\boldsymbol{t}(\tau, f_D)$ and its subvectors are built using the best available copy of the transmitted signal provided by a properly selected reference channel $r(t)$, i.e., $g(t) = r(t)$ in (2.12).

Matrix $\hat{\boldsymbol{P}}$ in (2.16) and (2.17) is evaluated by exploiting suitable estimates of matrices $\boldsymbol{A}$ and $\boldsymbol{R}$. To this purpose, the adaptive detection scheme relies on the availability of K secondary data vectors $\boldsymbol{y}_k$, $k = 1, \ldots, K$, assumed independent, IID, target free, and sharing the same distribution of the primary data $\boldsymbol{x}$ under the H_0 hypothesis. Under these hypotheses, proper approximations of the ML estimates of $\boldsymbol{A}$ and $\boldsymbol{R}$ are readily obtained as in [19]

$$\hat{\boldsymbol{A}} = \hat{\boldsymbol{Q}}_{00}^{-1} \hat{\boldsymbol{Q}}_{01} \tag{2.19}$$

and

$$\hat{\boldsymbol{R}} = \frac{1}{K(N-Q+1)}\left(\hat{\boldsymbol{Q}}_{11} - \hat{\boldsymbol{Q}}_{01}^{\dagger}\hat{\boldsymbol{Q}}_{00}^{-1}\hat{\boldsymbol{Q}}_{01}\right) \tag{2.20}$$

where $\hat{\boldsymbol{Q}}_{00}$ $(L(Q-1)\times L(Q-1))$, $\hat{\boldsymbol{Q}}_{01}(L(Q-1)\times L)$, $\hat{\boldsymbol{Q}}_{11}$ $(L\times L)$ are blocks of the following matrix:

$$\hat{\boldsymbol{Q}} = \sum_{k=1}^{K}\sum_{n=0}^{N-Q} \tilde{\boldsymbol{y}}_k(n)\tilde{\boldsymbol{y}}_k^{\dagger}(n) = \begin{bmatrix} \hat{\boldsymbol{Q}}_{00} & \hat{\boldsymbol{Q}}_{01} \\ \hat{\boldsymbol{Q}}_{01}^{\dagger} & \hat{\boldsymbol{Q}}_{11} \end{bmatrix} \tag{2.21}$$

We observe that the $LQ\times LQ$ matrix $\boldsymbol{Q}$ represents an estimate of the pol-time disturbance covariance matrix within a sub-CPI of length Q, being Q the number of temporal degrees of freedom, namely taps, of the employed filter. As is apparent from (2.21), this estimate benefits from a joint average over K secondary data and $(N-Q+1)$ consecutive (overlapped) sub-CPIs within the CPI.

2.4.2.2 Modified Pol-AR-AMF detection scheme for PCL application

In this section, modified versions of the Pol-AR-AMF detector are introduced following the approaches in [20] and [22] with the aim to make the detector robust to the spectral mismatches and to reduce the computational load required by the PCL application.

The analyses in [19] have demonstrated the effectiveness of the Pol-AR-AMF against an input disturbance that exactly matches the employed AR($Q-1$) model in (2.13). In contrast, it was shown in [20] that when the spectral characteristics of the input disturbance differ from the employed AR model, some disturbance residual might arise after the temporal and polarimetric filtering. In other words, differently from the case of a perfect spectral match, the random output vector $\boldsymbol{\chi}(\tau, f_D)$ will be characterized by a covariance matrix $\boldsymbol{D}_{\chi}(\tau, f_D)$ that might have nonzero off-diagonal elements, even in asymptotic conditions (i.e., when the approximating AR parameters are perfectly estimated).

Therefore, in order to make the detector robust with respect to typical spectral mismatches between the input disturbance and the employed model, an additional polarimetric whitening stage is included in the processing scheme based on a proper estimate of matrix $\boldsymbol{D}_{\chi}(\tau, f_D)$. The modified detection test proposed in [20] can be equivalently written as

$$\boldsymbol{w}_0^{\dagger}\left(\tau, f_D\right) \hat{\boldsymbol{D}}_w^{-1}\left(\tau, f_D\right) \boldsymbol{w}_0\left(\tau, f_D\right) \underset{H_0}{\overset{H_1}{\gtrless}} \xi_{Mod-Pol-AR-AMF} \tag{2.22}$$

where

- $\xi_{Mod-Pol-AR-AMF}$ is the detection threshold that can be computed by inverting the following P_{fa} expression

$$P_{fa}^{(Mod-Pol-AR-AMF)}(\xi) = \frac{\left[1-\left(\frac{\xi}{P+\xi}\right)\right]^{P-L+1}}{\Gamma(P-L+1)} \sum_{l=0}^{L-1} \frac{\Gamma(P-l)}{\Gamma(L-l)} \left(\frac{\xi}{P+\xi}\right)^{L-l+1} \tag{2.23}$$

- vector $\boldsymbol{w}_0$ is written as

$$\boldsymbol{w}_0\left(\tau, f_D\right) = \sum_{n=0}^{N-Q} \boldsymbol{\Sigma}^{\dagger}\left(n, \tau, f_D\right) \hat{\boldsymbol{P}}\, \tilde{\boldsymbol{x}}(n) \tag{2.24}$$

- $\hat{\boldsymbol{D}}_w(\tau, f_D)$ is the estimated sample covariance matrix of the filter output, namely $\boldsymbol{D}_w(\tau, f_D) = E[\boldsymbol{w}_0(\tau, f_D)\boldsymbol{w}_0^{\dagger}(\tau, f_D)] = \boldsymbol{W}^{-\frac{\dagger}{2}}(\tau, f_D)\boldsymbol{D}_{\chi}(\tau, f_D)\boldsymbol{W}^{\frac{1}{2}}(\tau, f_D)$. It can be obtained using a set of P target-free training data, $\boldsymbol{w}_p$ ($p = 1, \ldots, P$), that underwent the same filtering stages applied to the primary data, i.e., $\hat{\boldsymbol{D}}_w(\tau, f_D) = 1/P \sum_{p=1}^{P} \boldsymbol{w}_p \boldsymbol{w}_p^{\dagger}$. In this regard, we recall that the secondary data to be exploited for the estimation of matrix $\boldsymbol{D}_w$ should not necessarily coincide with the secondary data exploited to build the first cancellation stage of the detector. For instance, an effective choice is to exploit the filter outputs at a set $I_{(\tau, f_D)}$ of P indices that identify range-Doppler cells surrounding the CUT over the range-Doppler plane.

The detector in (2.22) is referred to as modified Pol-AR-AMF (Mod-Pol-AR-AMF) and a block diagram of the main signal processing stages foreseen in the considered multi-polarimetric processing scheme are shown in Figure 2.12.

Despite this approach operates with a limited number of adaptive degrees of freedom, the long integration time exploited by passive radar typically requires a substantial computational cost and *ad hoc* expedients for its application. Therefore, with the aim to adapt the proposed detector to the passive radar application, the approach presented in [19 and 20] has been revised in [22], and further modifications have been proposed to obtain a computationally effective solution to be exploited in real-world scenarios.

First of all, we revised the strategy adopted to estimate matrix $\hat{\boldsymbol{Q}}$ in (2.21), and consequently matrix $\hat{\boldsymbol{P}}$, and different adaptivity strategies have been considered and compared in [22]. It was shown therein that, since very long CPIs are typically used in passive radar applications, a suitable solution is to estimate the AR parameters based on the primary data itself. More precisely, the filter weights can be updated along the CPI, by repeating the estimate of matrix $\hat{\boldsymbol{P}}$ over subsequent fragments of the employed integration time. The extension of the employed fragments must be carefully selected as (i) short enough to account for possible nonstationary characteristics of the disturbance across the entire CPI and (ii) long enough to guarantee that the target contributions are negligible with respect to the competing disturbance and an asymptotic estimate of the AR parameters is achieved. We refer the interested reader to [22] for a more thorough discussion on the selection of this parameter.

Let us assume that the estimation of the AR parameters is repeated $M = \lceil (N - Q + 1)/J \rceil$ times along the CPI, being J the number of consecutive overlapped sub-CPIs of dimension Q in the considered fragment. At the mth block $(m = 0, \ldots, M{-}1)$ matrix $\hat{\boldsymbol{Q}}_m$ is estimated as follows:

$$\hat{\boldsymbol{Q}}_m = \sum_{n=mJ}^{(m+1)J-1} \tilde{\boldsymbol{x}}(n)\tilde{\boldsymbol{x}}^{\dagger}(n) \tag{2.25}$$

and used to evaluate the AR matrix parameters according to (2.19) and (2.20) and in turn matrix $\hat{\boldsymbol{P}}_m$ to be used at that particular signal fragment. Accordingly, $\hat{\boldsymbol{P}}(n) = \hat{\boldsymbol{P}}_{\lfloor n/J \rfloor}$, $n = 0, \ldots, N - Q$ and (2.24) is modified as

$$\boldsymbol{w}_0(\tau, f_D) = \sum_{n=0}^{N-Q} \boldsymbol{\Sigma}^{\dagger}(n, \tau, f_D)\hat{\boldsymbol{P}}(n)\tilde{\boldsymbol{x}}(n) \tag{2.26}$$

As aforementioned, long CPIs (in the order of seconds) are typically used in passive radar applications to attain desired levels of SNR. In fact, the number N of samples included in the CPI is at least in the order of hundreds of thousands, and this sets quite demanding computational requirements. Moreover, as shown by the explicit dependence on the CUT the operations included therein must be repeated for each bistatic range and Doppler bin in the map where the presence of targets is sought. Consequently, to deal with the unfeasibility of the direct implementation of the Mod-Pol-AR-AMF, a suitable strategy has been investigated in [22] to obtain an efficient implementation of the proposed scheme.

Based on a computational complexity analysis carried out in [22], it was shown that the main contribution to the cost is represented by the pol-time filtering stage in (2.26), i.e., the evaluation of the temporally whitened sequence of L-dimensional vectors $\boldsymbol{y}(n,\tau,f_D) = \boldsymbol{\Sigma}^\dagger(n,\tau,f_D)\hat{\boldsymbol{P}}(n)\tilde{\boldsymbol{x}}(n)$, followed by the summation across consecutive samples to achieve the coherent integration of target echoes in time domain, say $\boldsymbol{w}_0(\tau,f_D) = \sum_{n=0}^{N-Q} \boldsymbol{y}_0(n,\tau,f_D)$.

A cost-effective implementation of the described algorithm, derived in [22], is based on neglecting the Doppler-induced phase variation on the moving target echo within the Q-dimensional sub-CPI, provided that the employed AR model order is assumed sufficiently small since this corresponds to a filter with a limited number of temporal taps.

Under this assumption, vector $\tilde{\boldsymbol{t}}(n,\tau,f_D)$ is approximated as

$$\tilde{\boldsymbol{t}}\,(\mathrm{n},\tau,f_D) \cong \begin{bmatrix} g\left(t_0 + \frac{n}{f_S} - \tau\right) \\ \vdots \\ g\left(t_0 + \frac{n+Q-1}{f_S} - \tau\right) \end{bmatrix} e^{j2\pi n \frac{f_D}{f_S}} = \tilde{\boldsymbol{t}}_0(n,\tau)e^{j2\pi n \frac{f_D}{f_S}} \tag{2.27}$$

where $\tilde{\boldsymbol{t}}_0(n,\tau) = \tilde{\boldsymbol{t}}(n,\tau,0)$. Consequently, by defining $\boldsymbol{\Sigma}_0(n,\tau) = \tilde{\boldsymbol{t}}_0(n,\tau) \otimes \boldsymbol{I}$, matrix $\boldsymbol{\Sigma}(n,\tau,f_D)$ in (2.18) becomes

$$\boldsymbol{\Sigma}(n,\tau,f_D) \cong \boldsymbol{\Sigma}_0(n,\tau)e^{j2\pi n \frac{f_D}{f_S}} \tag{2.28}$$

and the pol-time filter output in (2.26) can be simplified as

$$\boldsymbol{w}_0(\tau,f_D) = \sum_{m=0}^{N-Q} \boldsymbol{\Sigma}_0^\dagger(n,\tau)\hat{\boldsymbol{P}}(n)\tilde{\boldsymbol{x}}(n)e^{j2\pi n \frac{f_D}{f_S}} \tag{2.29}$$

where the computationally intense evaluation of the whitened sequence $\boldsymbol{y}(n,\tau) = \boldsymbol{\Sigma}_0^\dagger(n,\tau)\hat{\boldsymbol{P}}(n)\tilde{\boldsymbol{x}}(n)$ has not to be repeated at each Doppler bin. Instead, the Doppler-dependent phase term can be taken into account at the subsequent coherent integration stage needed to obtain $\boldsymbol{w}_0(\tau,f_D)$. Assuming that the search grid in the Doppler dimension is uniformly spaced and not oversampled, for a given delay bin τ (1.29) corresponds to the evaluation of L discrete Fourier transforms of filtered sequences of $N-Q+1$ samples each, which can be evaluated at FFT speed. By repeating the computations above across an appropriate grid of delays, L range/Doppler maps are obtained, and the final adaptive detector is applied for each CUT.

Further details on the derivation of this simplified algorithm as well as a demonstration of the negligible loss to be accepted with this computationally effective version are reported in [22].

Along with an analysis of the cost reduction obtained with respect to the direct implementation of the algorithm, detailed in [22], it is worth evaluating the surplus cost required by the Mod-Pol-AR-AMF, with the expedients adopted to make it effective in the PCL case, with respect to the Pol-GLRT. To this end, we define the Δ_{FLOPs} as the ratio between the computational cost required by the Mod-Pol-AR-AMF and that needed to implement a Pol-GLRT, deducting from the overall complexity the cost required for the temporal disturbance cancellation stage. The results are reported in Figure 2.13 versus a grid of Q values, using a CPI of 1 s. The Mod-Pol-AR-AMF is assumed to operate with $J = 200$ whereas $P = 32$ is employed in both cases. The dark blue and orange lines represent the case of a PCL system for aerial surveillance exploiting FM radio and DVB-T transmissions, respectively. For both applications, we consider $L = 2$ polarimetric surveillance channels, a maximum relative bistatic range of 200 km and a maximum bistatic velocity of 500 m/s. In the FM radio-based PCL application, we assume that the signal is transmitted at carrier

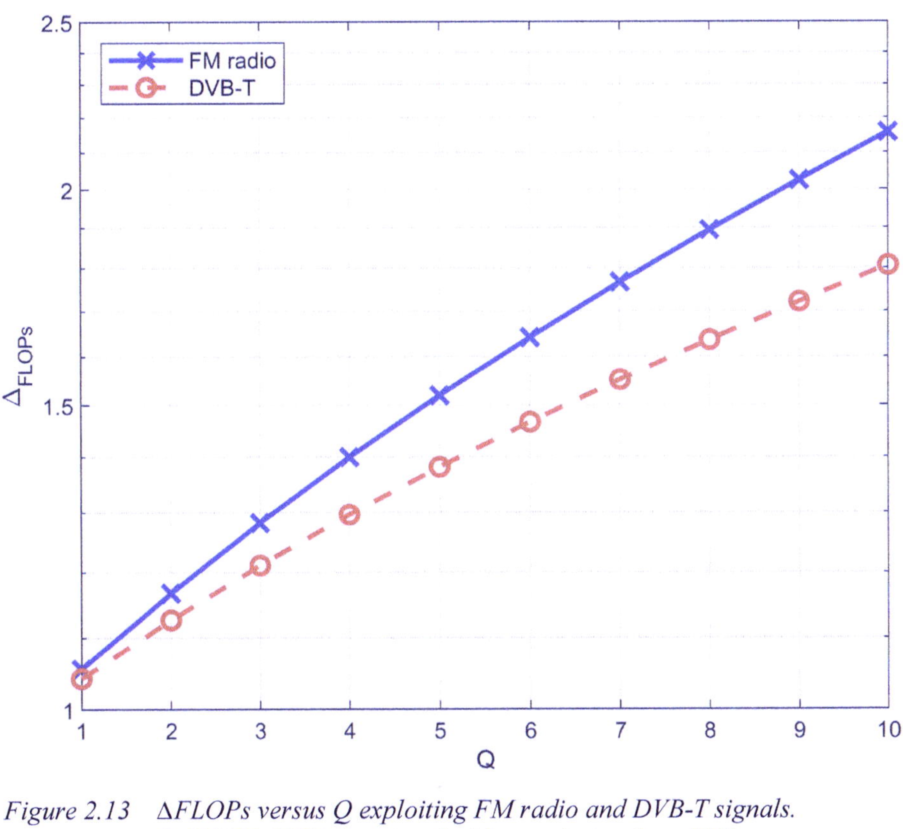

Figure 2.13 ΔFLOPs versus Q exploiting FM radio and DVB-T signals. *© [2020] IEEE. Reprinted with permission from [22].*

frequency $f_c = 94.5$ MHz and received with sampling frequency $f_s = 200$ kHz while in the DVB-T based passive radar case, we assume $f_c = 666$ MHz and $f_s = 64/7$ MHz. The results in Figure 2.13 confirm that a higher computational burden is required by the Mod-Pol-AR-AMF with respect to the Pol-GLRT in the considered scenario. However, the complexity is less than doubled, i.e., $\Delta_{\text{FLOPs}} < 2$, for the typical Q values, i.e., $Q = 3$ or 4, that have been shown to reasonably approximate the disturbance spectral characteristics [22].

2.4.3 Experimental validation

In this section, the effectiveness of the described integrated polarimetric PCL architecture is shown on real data and compared to the alternative architectural scheme described in section 2.3. To this end, we consider the data belonging to case study A, namely that collected with an FM radio-based PCL system for aerial surveillance.

In Figure 2.14, we first consider the same datafile extracted for Figures 2.4 and 2.5 and we map the output values obtained at each bin of the range-Doppler map on to P_{fa} values using the same methodology adopted in section 2.3.3. For immediate comparison, we reported in Figure 2.14(a) the result of the cascaded polarimetric PCL processing with the Pol-GLRT detector applied after a conventional PCL processing [same result as in Figure 2.4(d)]. Figure 2.14(b) shows the outcome of the integrated polarimetric PCL processing, with the Mod-Pol-AR-AMF strategy operating with $Q = 3$; this corresponds to the application of the whole scheme depicted in Figure 2.12. Finally, for a fair comparison, Figure 2.14(c) shows the result of a hybrid solution where the Pol-GLRT detector is used after the PCL processing is applied separately at each surveillance channel, but the temporal disturbance cancellation is performed according to the Pol-ECA approach. This last result allows to understand the impact of the different polarimetric stages on the final results and in particular it allows a direct comparison among the different approaches exploited for polarimetric target detection.

Observing Figure 2.14, the integrated architecture exploiting the Pol-ECA and the Mod-Pol-AR-AMF outperforms both the cascaded Pol-GLRT and the hybrid solution and would enable the discrimination of two additional targets at $P_{fa} = 10^{-5}$ with respect to the alternative approaches. Furthermore, targets that would be detected by all the different strategies at the selected P_{fa} appear as stronger peaks with the integrated scheme, i.e., they would be detected also when further reducing the P_{fa} value. An example is reported in Figure 2.15 where an enlarged view of the target at approximately 115 km and 290 m/s is reported. Finally, at this datafile, the exploitation of the polarimetric information at the reference channels according to the Pol-ECA strategy does not yield evident advantages. In other words, the performance improvement observed for the integrated polarimetric architecture has to be mostly attributed to the Mod-Pol-AR-AMF that apparently yields a better target discrimination capability, thanks to the joint exploitation of polarimetric and temporal correlation characteristics of the competing disturbance.

However, when considering datafile #2, extracted from the same dataset 3 s apart from datafile #1, the difference is evident. Figure 2.16 (and the enlarged view

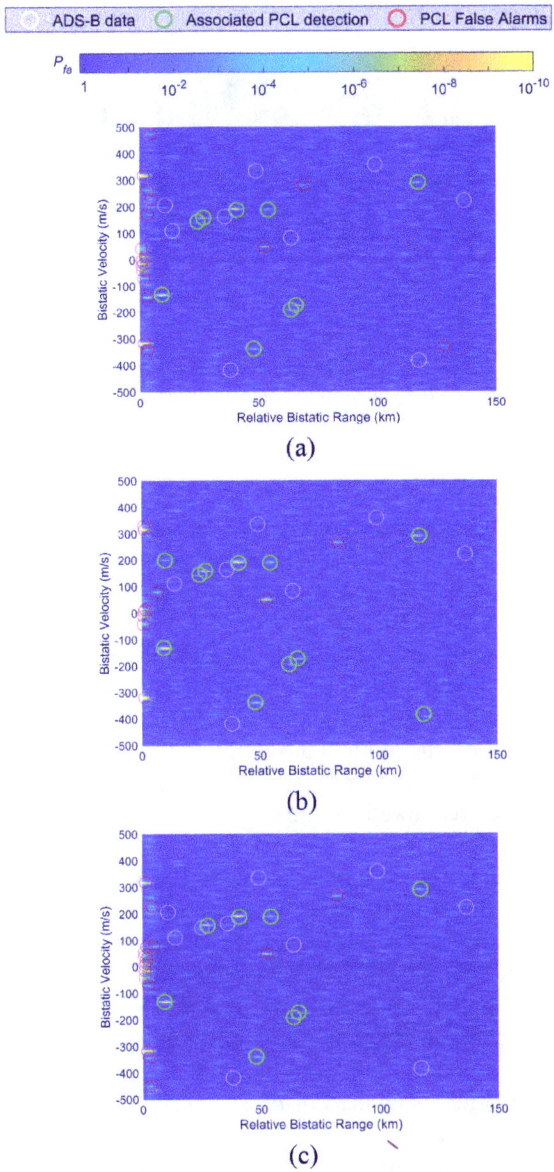

Figure 2.14 *Datafile #1: Range-Doppler map of minimum nominal P_{fa} values to be set in order to detect each bin: (a) cascaded architecture with Pol-GLRT, (b) integrated architecture with Mod-Pol-AR-AMF and Pol-ECA, and (c) Hybrid architecture with Pol-GLRT and Pol-ECA.*

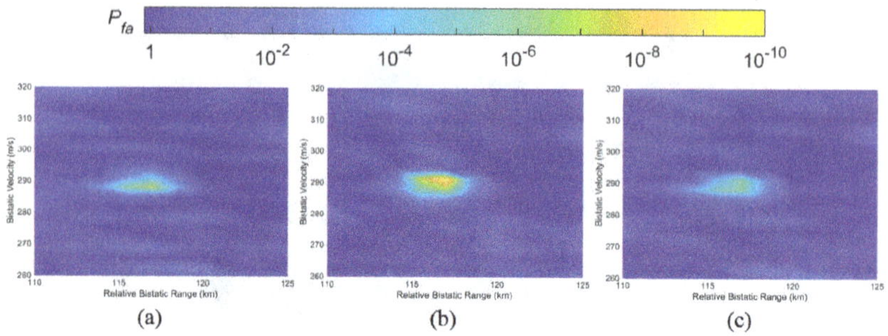

Figure 2.15 *Enlarged view of Figure 2.14 with (a) cascaded architecture with Pol-GLRT, (b) integrated architecture with Mod-Pol-AR-AMF and Pol-ECA, and (c) hybrid architecture with Pol-GLRT and Pol-ECA*

reported in Figure 2.17 around two targets) clearly shows that a simple selection strategy that picks out the preferred reference signal on receive might not always be an effective solution [see Figure 2.16(a) and Figure 2.17(a)]. Instead, an effective way to combine the polarimetric information also at the reference channels could yield a noticeable improvement and allow a better disturbance rejection [see the results for the hybrid solutions in Figure 2.16(c) and Figure 2.17(c)]. Still, when the Mod-Pol-AR-AMF is applied afterward in lieu of the Pol-GLRT, the improvement is noticeable [see Figure 2.16(b) and Figure 2.17(b)].

The performance of the three considered strategies is also investigated in terms of detection capability for 50 consecutive scans in Figure 2.18, which overall confirms what has been remarked for the single datafiles:

1. By properly leveraging the polarization diversity at the reference channels, when available, the detection performance of the resulting PCL system is enhanced [compare Figure 2.18(a) and (c)]. See, for instance, the two tracks between 200 and 300 m/s where the sequence of plots is much denser when the Pol-ECA is applied. Note that this change would not disrupt the cascaded architecture described in section 2.3 as different surveillance signals continue to be separately processed but would require to perform the direct signal and multipath cancellation using the Pol-ECA algorithm when more than one reference signal is available.
2. When the different polarimetric surveillance channels are jointly combined according to an integrated polarimetric PCL architecture [see Figure 2.18(b)], the detection performance is further enhanced. See the farthest track at about −400 m/s that would not be detected with any other combination.

To finally assess the performance of the strategies described in this chapter, we carry out an extensive analysis by considering the entire dataset of case study A. The results are shown in Figures 2.19–2.20. Specifically, Figure 2.19 shows the false alarm rate measured in the range interval [150–200 km] versus the nominal value.

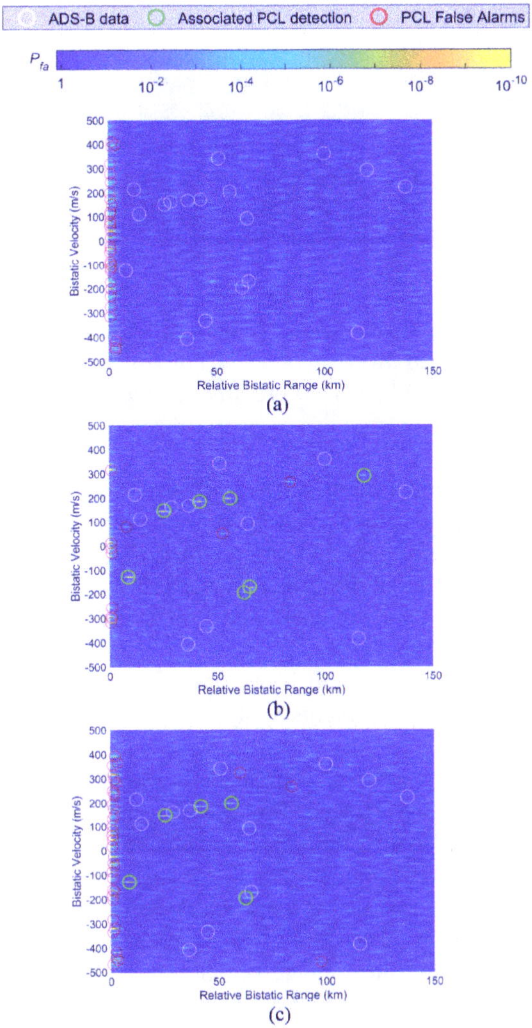

Figure 2.16 Datafile #2: Range-Doppler map of minimum nominal P_{fa} values to be set in order to detect each bin: (a) cascaded architecture with Pol-GLRT, (b) integrated architecture with Mod-Pol-AR-AMF and Pol-ECA, and (c) hybrid architecture with Pol-GLRT and Pol-ECA

Incidentally, note that we measure the false alarm rate in the range interval [150–200 km] where the target detection probability is extremely low to avoid labeling as false alarms detections corresponding to targets not identified by the available air-truth therefore enabling an unsupervised false alarm counting process. On the other hand, Figure 2.20 reports the empirical receiver operating characteristic (ROC) curves. To obtain this figure, at each data file, and for a given considered false alarm rate value, a correct detection is declared when a peak in the range-Doppler plane exceeded the

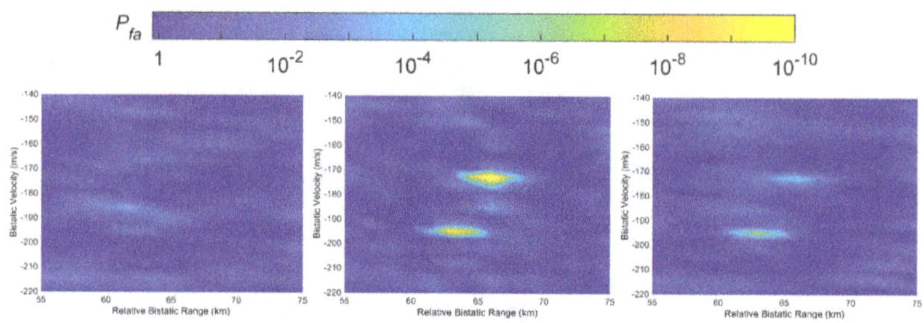

Figure 2.17 Enlarged view of Figure 2.16 with (a) cascaded architecture with Pol-GLRT, (b) integrated architecture with Mod-Pol-AR-AMF and Pol-ECA, and (c) hybrid architecture with Pol-GLRT and Pol-ECA

selected threshold at the radar coordinates predicted by the available air-truth plus a small tolerance. Then, the detection frequency is obtained by dividing the number of correct detections by the maximum number of target occurrences across the entire dataset, namely 9632. Note that the detection analysis is limited to targets laying in the range-band [0–100 km] and included within an angular sector of 90° around the surveillance antenna pointing direction.

From Figures 2.19–2.20, the considerations made in the previous sections are confirmed and the following comments apply:

i. The capability to control the false alarm rate changes with the considered processing scheme. Specifically, the CA-CFAR detection scheme applied to the single-pol channel map yields the best capability to guarantee the nominal P_{fa}. In contrast, the CA-CFAR scheme applied after the Pol-NCI is the worst performing solution, confirming the weakness of the simplifying hypothesis of independent disturbance affecting the L range-velocity maps. The false alarm rate control offered by the remaining multi-pol techniques is comparable and, in all cases, largely acceptable in practical applications.
ii. The polarimetric approaches can effectively improve the performance of the system only if suitable strategies are employed to exploit signals collected via multi-polarized antennas. This is demonstrated by observing the performance of the Pol-NCI strategy. The latter, in fact, does not succeed in improving the target detection performance with respect to the single H and V channels. This is due to the generally higher number of false alarms obtained that confirms the unsuitability of this solution in scenarios that are likely to be affected by severe interference.
iii. The polarimetric strategies that adaptively exploit the information diversity to counteract the disturbance, namely the Pol-GLRT (both with and without polarimetric ECA) and the proposed

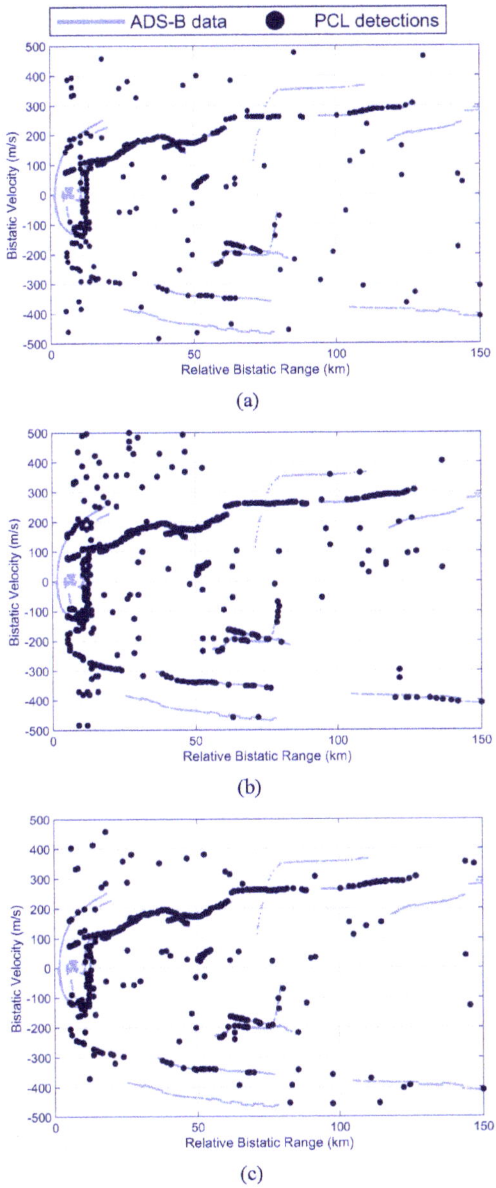

Figure 2.18 *Detection results over 50 consecutive datafiles, with $P_{fa} = 10^{-5}$, using: (a) cascaded architecture with Pol-GLRT, (b) integrated architecture with Mod-Pol-AR-AMF and Pol-ECA, and (c) hybrid architecture with Pol-GLRT and Pol-ECA. © [2020] IEEE. Reprinted with permission from [22]*

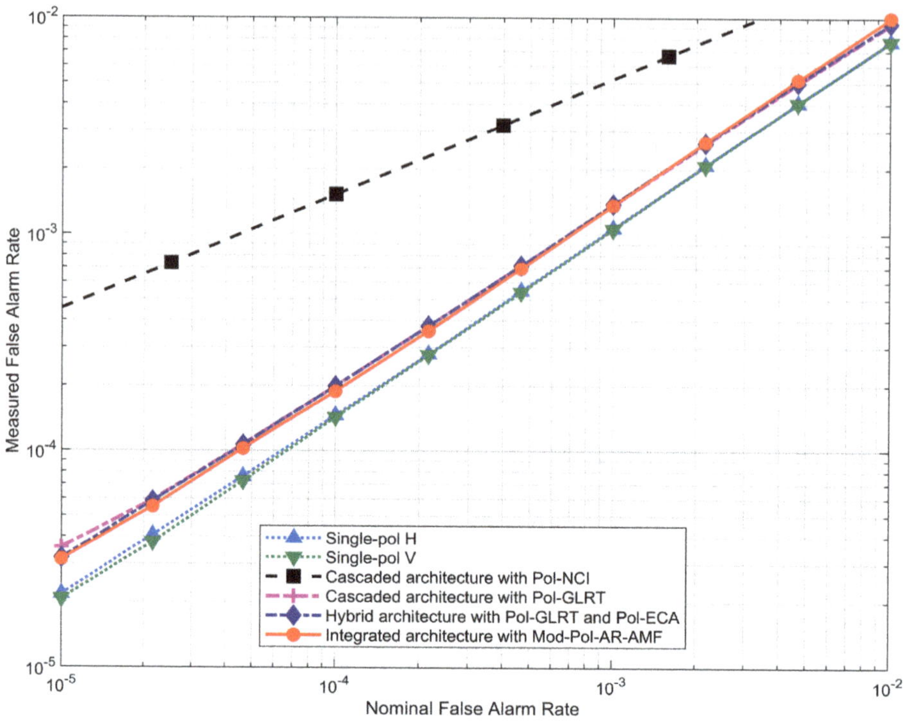

Figure 2.19 Measured versus nominal false alarm rate curves with different polarimetric architectures and signal processing techniques for the bistatic range interval [150–200 km]

Mod-Pol-AR-AMF, substantially increase the target detection performance with respect to the single-pol channels.

iv. The use of the Pol-ECA overall increases the detection performance.

v. The integrated processing architecture featuring the Pol-ECA and the Mod-Pol-AR-AMF is the best performing solution, yielding a tremendous improvement with respect to both the worst and best performing single-pol channel. Moreover, it increases the capability to discriminate targets by approximately 7–8% with respect to the cascaded architecture encompassing the Pol-GLRT and the ECA and by approximately 5–6% with respect to the hybrid architecture encompassing the Pol-GLRT and the Pol-ECA.

2.5 Summary

In this chapter, the exploitation of polarization diversity for PCL systems has been investigated. With the purpose of enhancing the performance of passive radars, two

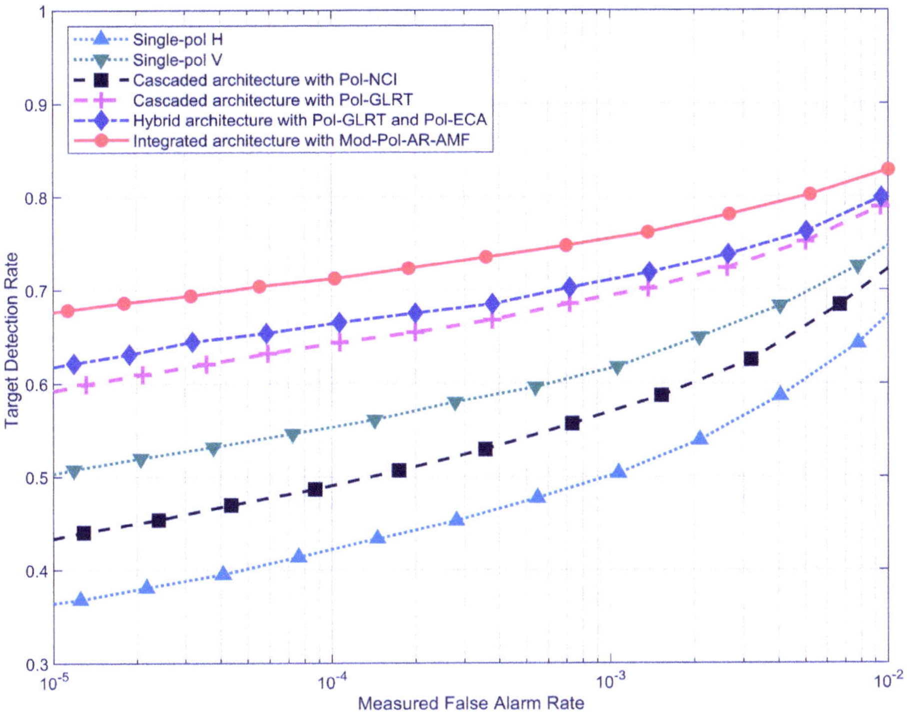

Figure 2.20 *Empirical ROC curves with different polarimetric architectures and signal processing techniques for the bistatic range interval [0–100 km]*

different multi-polarimetric signal processing architectures have been described, that leverage the information diversity conveyed by multi-polarimetric receiving channels according to different approaches. Different signal processing strategies that fit into the described architectural approaches have been presented and compared and the effectiveness of the described strategies has been then extensively demonstrated on real data, collected with PCL systems operating in different scenarios and at different frequency bands.

The reported analyses demonstrated that a suitable combination of the multi-polarimetric signals is useful to recover the target echo that typically shows a variable polarization, to counteract the direct signal interference, as well as to better reject the additional interfering contributions.

We showed that the major benefit is obtained when combining the different polarimetric signals in an integrated architecture where multiple polarimetric signals at both the reference and the surveillance channels are jointly exploited in *ad-hoc* multi-pol signal processing techniques. This architectural approach provides a larger flexibility, which is especially useful in scenarios where the signals are affected by severe interference. This is for instance the case of passive radars exploiting

FM radio signals, where the interference from co-channel or adjacent channel transmission might seriously limit the system performance.

On the other hand, a simpler approach as the cascaded multi-pol architecture has proved to be an effective solution for PCL systems that are less likely to be affected by severe interfering contributions, for example, the PCL sensors based on DVB-S signals, without significantly increasing the computational and implementation complexity. The main advantage of this architectural solution is that it overcomes the limitations of a single-pol PCL, while preserving the core processing stages already developed for single-pol passive radars thus maintaining a backward compatibility with existing PCL systems.

Based on the advantages offered by the joint combination of polarimetric channels, both architectures described in this chapter can be extended to include additional diversity sources. This approach has already been investigated when including frequency diversity, which is typically inherently offered if radio/television broadcast services are used as sources of opportunity [57, 58]. Moreover, while this chapter focuses on the improvement offered by multi-polarimetric PCL systems for target detection purpose, additional radar tasks have been considered in the recent technical literature when exploiting polarization diversity. In particular, target imaging in passive radar might benefit from the availability of polarimetric diversity as described in Chapter 7 [59–61].

The improvement offered by the exploitation of polarimetric diversity in passive radars also suggests further opportunities. Among them, a challenging task is represented by target characterization and classification, which could benefit from the potential diversity of the target signature if observed at different polarizations. Such feature would be extremely useful in low cost and pervasive passive sensors to be employed for the surveillance of critical infrastructure, potentially threatened by small targets such as unmanned aerial vehicles and drones. Further enhanced capabilities can be expected for passive radar sensors exploiting IO of the constantly advancing technologies for mobile communication and networking, where the use of multiple-input and multiple-output (MIMO) dual-pol antennas is foreseen.

Acknowledgements

We gratefully acknowledge the entire Radar Remote Sensing and Navigation research group at Sapienza, in particular Prof. Pierfrancesco Lombardo for the stimulating discussions and insightful comments and for leading the experimental activities on passive radar using different waveforms, Dr. Carlo Bongioanni and Dr. Octavio Cabrera for setting up the experimental prototypes and the acquisition campaigns.

Moreover, we thank Dr. Diego Cristallini and the colleagues at Fraunhofer Institute FHR for the DVB-T based experimental tests.

References

[1] Griffiths H.D., Baker C.J. 'Passive coherent location radar systems. part 1: performance prediction'. *IEE Proceedings - Radar, Sonar and Navigation.* 2005, vol. 152(3), p. 153.
[2] Baker C.J., Griffiths H.D., Papoutsis I. 'Passive coherent location radar systems. part 2: waveform properties'. *IEE Proceedings - Radar, Sonar and Navigation.* 2005, vol. 152(3), p. 160.
[3] Howland P.E., Maksimiuk D., Reitsma G. 'FM radio based bistatic radar'. *IEE Proceedings - Radar, Sonar and Navigation.* 2005, vol. 152(3), p. 107.
[4] Farina A., Kuschel H. 'IEEE aerospace and electronic systems magazine'. *Special Issue on Passive Radar Part I & II.* 2012, vol. 27, pp. 10–11.
[5] Lombardo P., Colone F. 'Advanced processing methods for passive bistatic radar systems' in Melvin W.H., Scheer J.A. (eds.). *Principles of Modern Radar: Advanced Techniques.* Schitech Publisher; 2012. pp. 739–821.
[6] Griffiths H., Baker C. 'The signal and interference environment in passive bistatic radar'. *IEEE Information Decision and Control (IDC '07)*; 2007. pp. 12–14.
[7] Poullin D., Flecheux M. 'Recent progress in passive coherent location (PCL) concepts and technique in France using DAB or FM broadcasters'. *IEEE RADAR Conference 2008 (RADAR '08), May 26–30*; Rome, Italy, 2008.
[8] Colone F., O'Hagan D.W., Lombardo P., Baker C.J. 'A multistage processing algorithm for disturbance removal and target detection in passive bistatic radar'. *IEEE Transactions on Aerospace and Electronic Systems.* 2009, vol. 45(2), pp. 698–722.
[9] Colone F., Bongioanni C., Lombardo P. 'Multifrequency integration in FM radio-based passive bistatic radar. part I: target detection'. *IEEE Aerospace and Electronic Systems Magazine.* 2013, vol. 28(4), pp. 28–39.
[10] Colone F., Bongioanni C., Lombardo P. 'Multifrequency integration in FM radio-based passive bistatic radar. part II: direction of arrival estimation'. *IEEE Aerospace and Electronic Systems Magazine.* 2013, vol. 28(4), pp. 40–47.
[11] Bongioanni C., Colone F., Martelli T., D'Angeli R., Lombardo P. 'Exploiting polarimetric diversity to mitigate the effect of interferences in FM-based passive radar'. *11th International Radar Symposium*; Vilnius, 2010. pp. 1–4.
[12] Colone F., Lombardo P. 'Polarimetric passive coherent location'. *IEEE Transactions on Aerospace and Electronic Systems.* 2015, vol. 51(2), pp. 1079–97.
[13] You J., Wan X., Fu Y., Fang G. 'Experimental study of polarisation technique on multi-FM-based passive radar'. *IET Radar, Sonar & Navigation.* 2015, vol. 9(7), pp. 763–71.
[14] Son I., Yazici B. 'Passive polarimetrie multistatic radar for ground moving target'. *IEEE Radar Conference (RadarConf)*; Philadelphia, PA, 2016. pp. 1–6.

[15] Conti M., Moscardini C., Capria A. 'Dual-polarization DVB-T passive radar: experimental results'. *2016 IEEE Radar Conference (RadarConf16)*; Philadelphia, PA, USA, 2016. pp. 1–5.
[16] Filippini F., Colone F., Cristallini D., Bournaka G. 'Preliminary experimental results of polarimetric detection schemes for DVB-T based passive radar'. *IEEE Radar Conference (RadarConf)*; Seattle, WA, 2017. pp. 0383–88.
[17] Filippini F., Colone F., Cristallini D., Bournaka G. 'Experimental results of polarimetric detection schemes for DVB-T-based passive radar'. *IET Radar, Sonar & Navigation*. 2017, vol. 11(6), pp. 883–91.
[18] Filippini F., Colone F. 'A practical approach to polarimetric adaptive target detection in passive radar'. *International Conference on Radar Systems (Radar 2017)*; Belfast, UK, 2017. pp. 1–6.
[19] Colone F., Filippini F. 'Autoregressive model based polarimetric adaptive detection scheme part I: theoretical derivation and performance analysis'. *IEEE Transactions on Aerospace and Electronic Systems*. 2020, vol. 56(5), pp. 3762–78.
[20] Colone F., Filippini F. 'Autoregressive model based polarimetric adaptive detection scheme part II: performance assessment under spectral model mismatch'. *IEEE Transactions on Aerospace and Electronic Systems*. 2020, vol. 56(5), pp. 3779–95.
[21] Filippini F., Colone F. 'Polarimetric detection scheme for passive radar based on a 2D auto-regressive disturbance model'. *International RADAR Conference (RADAR)*; Toulon, France, 2019. pp. 1–6.
[22] Filippini F., Colone F. 'Polarimetric passive radar: a practical approach to parametric adaptive detection'. *IEEE Transactions on Aerospace and Electronic Systems*. 2020, vol. 56(6), pp. 4930–46.
[23] Cabrera O., Bongioanni C., Filippini F., Sarabakha O., Colone F., Lombardo P. 'Detecting drones and human beings with DVB-S based COTS passive radar for short-range surveillance'. *IEEE International RADAR Conference (RADAR)*; 2020. pp. 37–42.
[24] Dawidowicz B., Kulpa K.S., Malanowski M., Misiurewicz J., Samczynski P., Smolarczyk M. 'DPCA detection of moving targets in airborne passive radar'. *IEEE Transactions on Aerospace and Electronic Systems*. 2012, vol. 48(2), pp. 1347–57.
[25] Brown J., Woodbridge K., Griffiths H., Stove A., Watts S. 'Passive bistatic radar experiments from an airborne platform'. *IEEE Aerospace and Electronic Systems Magazine*. 2012, vol. 27(11), pp. 50–55.
[26] Palmer J., Ummenhofer M., Summers A., Bournaka G., Palumbo S., Cristallini D. 'Receiver platform motion compensation in passive radar'. *IET Radar, Sonar & Navigation*. 2017, vol. 11(6), pp. 922–31.
[27] Wojaczek P., Colone F., Cristallini D., Lombardo P. 'Reciprocal-filter-based STAP for passive radar on moving platforms'. *IEEE Transactions on Aerospace and Electronic Systems*. 2019, vol. 55(2), pp. 967–88.

[28] Blasone G.P., Colone F., Lombardo P., Wojaczek P., Cristallini D. 'Passive radar DPCA schemes with adaptive channel calibration'. *IEEE Transactions on Aerospace and Electronic Systems*. 2020, vol. 56(5), pp. 4014–34.

[29] Chetty K., Smith G.E., Woodbridge K. 'Through-the-wall sensing of personnel using passive bistatic wifi radar at standoff distances'. *IEEE Trans Geosci Remote Sens*. 2012, vol. 50(4), pp. 1218–26.

[30] Pastina D., Colone F., Martelli T., Falcone P. 'Parasitic exploitation of wi-fi signals for indoor radar surveillance'. *IEEE Transactions on Vehicular Technology*. 2015, vol. 64(4), pp. 1401–15.

[31] Colone F., Martelli T., Bongioanni C., Pastina D., Lombardo P. 'WiFi-based PCL for monitoring private airfields'. *IEEE Aerospace and Electronic Systems Magazine*. 2017, vol. 32(2), pp. 22–29.

[32] Samczynski P., Malanowski M., Krawczyk G., Kulpa J., Żywek M. 'Passive radar as a part of critical infrastructure protection system'. *International Conference on Radar*; Brisbane, QLD, 2018. pp. 1–5.

[33] Milani I., Colone F., Bongioanni C., Lombardo P. 'WiFi emission-based vs passive radar localization of human targets'. *IEEE Radar Conference (RadarConf18)*; Oklahoma City, OK, 2018. pp. 1311–16.

[34] Cabrera O., Bongioanni C., Filippini F., Sarabakha O., Colone F., Lombardo P. 'Detecting drones and human beings with DVB-S based COTS passive radar for short-range surveillance'. *IEEE International RADAR Conference (RADAR)*; Washington, DC, 2020. pp. 37–42.

[35] Filippini F., Cabrera O., Bongioanni C., Colone F., Lombardo P. 'DVB-S based passive radar for short range security application'. *IEEE Radar Conference (RadarConf21)*; 2021. pp. 1–6.

[36] Palmer J., Cristallini D., Kuschel H. 'Opportunities and current drivers for passive radar research'. *2015 IEEE Radar Conference*; Johannesburg, South Africa, 2015. pp. 145–50. Available from https://ieeexplore.ieee.org/xpl/mostRecentIssue.jsp?punumber=7402199

[37] Klemm R., Nickel U., Gierull C., Lombardo P., Griffiths H., Koch W. 'Novel radar techniques and applications' in *Part III: passive and multistatic radar*; 2017.

[38] Kuschel H., Cristallini D., Olsen K.E. 'Tutorial: passive radar tutorial'. *IEEE Aerospace and Electronic Systems Magazine*. 2019, vol. 34(2), pp. 2–19.

[39] Kulpa K.S., Czekała Z. 'Masking effect and its removal in PCL radar'. *IEE Proceedings - Radar, Sonar and Navigation*. 2005, vol. 152(3), pp. 174–78.

[40] Cardinali R., Colone F., Ferretti C., Lombardo P. 'Comparison of clutter and multipath cancellation techniques for passive radar'. *2007 IEEE Radar Conference*; Waltham, MA, USA, 2007. pp. 469–74.

[41] Colone F., Palmarini C., Martelli T., Tilli E. 'Sliding extensive cancellation algorithm for disturbance removal in passive radar'. *IEEE Transactions on Aerospace and Electronic Systems*. 2016, vol. 52(3), pp. 1309–26.

[42] Schwark C., Cristallini D. 'Advanced multipath clutter cancellation in OFDM-based passive radar systems'. *IEEE Radar Conference (RadarConf)*; Philadelphia, PA, 2016. pp. 1–4.

[43] Garry J.L., Baker C.J., Smith G.E. 'Evaluation of direct signal suppression for passive radar'. *IEEE Trans Geosci Remote Sens.* 2017, vol. 55(7), pp. 3786–99.

[44] Villano M., Colone F., Lombardo P. 'Antenna array for passive radar: configuration design and adaptive approaches to disturbance cancellation'. *International Journal of Antennas and Propagation.* 2013, vol. 2013(3), 16.

[45] Zemmari R., Broetje M., Battistello G., Nickel U. 'Gsm passive coherent location system: performance prediction and measurement evaluation'. *IET Radar, Sonar & Navigation.* 2014, vol. 8(2), pp. 94–105.

[46] Belfiori F., van Rossum W., Monni S., Hoogeboom P. 'Antenna array characterisation and signal processing for an FM radio-based passive coherent location radar system'. *IET Radar, Sonar & Navigation.* 2012, vol. 6(8), pp. 687–96.

[47] Filippini F., Martelli T., Colone F., Cardinali R. 'Target DOA estimation in passive radar using non-uniform linear arrays and multiple frequency channels'. *IEEE Radar Conference (RadarConf18)*; Oklahoma City, OK, 2018. pp. 1290–95.

[48] Olsen K.E., Woodbridge K. 'FM based passive bistatic radar target range improvement'. *Proceedings of the International Radar Symposium*; Hamburg, Germany, 2009.

[49] Olsen K.E., Woodbridge K., Andersen I.A. 'FM based passive bistatic radar target range improvement: part II'. *Proceedings of the International Radar Symposium*; Vilnius, Lithuania, 2010.

[50] Martelli T., Colone F., Tilli E., Di Lallo A. 'Multi-frequency target detection techniques for DVB-T based passive radar sensors'. *Sensors.* 2016, vol. 16(10), p. 1594.

[51] Martelli T., Colone F., Cardinali R. 'DVB-T based passive radar for simultaneous counter-drone operations and civil air traffic surveillance'. *IET Radar, Sonar & Navigation.* 2020, vol. 14(4), pp. 505–15.

[52] Wojaczek P., Cristallini D., Schell J., O'Hagan D., Summers A. 'Polarimetric antenna diversity for improved reference signal estimation for airborne passive radar'. *IEEE Radar Conference (RadarConf20)*; Florence, Italy, 2020. pp. 1–6.

[53] Kelly E.J. 'An adaptive detection algorithm'. *IEEE Transactions on Aerospace and Electronic Systems.* 1986, vol. AES-22(2), pp. 115–27.

[54] Heckenbach J., Kuschel H., Schell J., Ummenhofer M. 'Passive radar based control of wind turbine collision warning for air traffic PARASOL'. *Proceedings of International Radar Symposium (IRS)*; 2015. pp. 36–41.

[55] Kay S.M., Nagesha V. 'Maximum likelihood estimation of signals in autoregressive noise'. *IEEE Transactions on Signal Processing.* 1994, vol. 42(1), pp. 88–101.

[56] Sheikhi A., Nayebi M.M., Aref M.R. 'Adaptive detection algorithm for radar signals in autoregressive interference'. *IEE Proceedings - Radar, Sonar and Navigation.* 1998, vol. 145(5), pp. 309–14.

[57] Colone F., Lombardo P. 'Non-coherent adaptive detection in passive radar exploiting polarimetric and frequency diversity'. *IET Radar, Sonar & Navigation*. 2016, vol. 10(1), pp. 15–23.
[58] Filippini F., Colone F. 'Multi-carrier and multi-polarimetric model based adaptive target detector for passive radar systems'. *IET Radar, Sonar & Navigation*. 2021, vol. 15(3), pp. 853–66.
[59] Son I., Yazici B. 'Passive imaging with multistatic polarimetric radar'. *IEEE Radar Conference (RadarCon)*; Arlington, VA, 2015. pp. 1584–89.
[60] Pisciottano I., Cristallini D., Pastina D., Santi F. 'Experimental results of polarimetric passive ISAR exploiting DVB-S2 illumination'. *IEEE International RADAR Conference (RADAR)*; Washington, DC, 2020. pp. 518–23.
[61] Pisciottano I., Santi F., Pastina D., Cristallini D. 'DVB-S based passive polarimetric ISAR—methods and experimental validation'. *IEEE Sensors Journal*. 2021, vol. 21(5), pp. 6056–70.

Chapter 3

Mainlobe jamming suppression for polarimetric multi-channel radar

Guolong Cui[1], Mengmeng Ge[2], and Xianxiang Yu[1]

3.1 Introduction

Electronic countermeasure (ECM) techniques are aimed at denying information (detection, position, track initiation, etc.) so as to prevent the victim radar from working effectively [1, 2]. Mainlobe ECM as an effective category of ECM techniques is able to significantly degrade the performance of modern radar systems. It usually utilizes barrage interference or deceptive interference to enter into the main beam of the receiving antenna so as to impair radar ability and reduce radar information reliability.

The mainlobe jamming can be generally classified into two types [3]. The former is the self-defensive jamming (SDJ), where the jammer is installed on the target, and the direction of arrival (DOA) of the jamming signal is equal to that of the target echo. The latter is called the escort-support jamming (ESJ), also known as near-mainlobe jamming, where the jammer is located near the target. In particular, the ESJ signal enters from the main beam of antenna, whereas its DOA is slightly different from the target echo.

The electronic counter-countermeasures (ECCM) techniques, which are devoted to countering the effects of the enemy's ECM and to accomplish the intended mission, have been designed due to the rapid development of ECM. Several mainlobe ECCM strategies have been proposed to thwart mainlobe ECM. Generally, these methods fall into two categories according to the type of jamming. The first category mainly is bent on the deceptive jamming (i.e., false targets) suppression resorting to new radar systems, endowing waveform diversity and signal processing techniques. For example, multiple-input multiple-output radar with frequency diverse array [4–7], distributed radar systems [8–10], pulse agility [11, 12] and adaptive coupled sequential estimation algorithms [13, 14] are investigated to combat mainlobe deceptive jamming.

[1]University of Electronic Science and Technology of China, China
[2]Southwest China Institute of Electronic Technology, China

The other category is focused on addressing suppressive jamming. For example, in [15], a modified block matrix was invoked to cancel suppressive jamming, but requiring an accurate angle of the jamming signal. Aimed at mitigating mainlobe jamming for ground-based radar, an auxiliary array was devised in [16]. However, a large physical space is needed to deploy the large aperture auxiliary array limiting its practical applications. In [17], a logarithm transformation was proposed to suppress constant modulus noise-like jamming. In [18], spatial polarization characteristics of antennas were exploited to suppress mainlobe blanket jamming. However, the performance of the proposed approach will be greatly reduced if the polarization state of jamming cannot be accurately estimated. In [19], an optimized mainlobe jamming cancellation filter was designed to suppress mainlobe jamming for distributed monopulse arrays. Its capability is related to the construction of the covariance matrix of jamming-plus-noise. In practice, the sample covariance matrix of the received signal is usually utilized to replace the covariance matrix of jamming-plus-noise probably usually leading to performance degradation. In this respect, some advanced techniques to estimate the covariance matrix can be used to get better interference suppression performance [20–22].

Recently, the independent component analysis (ICA) belonging to a kind of blind source separation method has been widely adopted in array signal processing, wireless communication, acoustic signal processing and many other fields [23–25]. Not surprisingly, several ICA-based methods have been proposed to solve the problem of jamming cancellation. For instance, in [26], the ICA based on Gaussian moments was leveraged to suppress noise suppressive jamming for the first time. The technique has been extended to deal with the mainlobe jamming and estimate targets' parameters [27–29]. In [30, 31], the multiple mainlobe jammings are also suppressed resorting to the ICA-based space-time multi-channel algorithm. Overall, these works are focused on handling the ESJ suppressive jamming.

In this chapter, the signal model for the polarimetric multi-channel radar is developed accounting for the target and the SDJ (barrage and deceptive interferences) or ESJ jamming in the radar mainlobe. By exploiting polarization characteristics, an approach based on the polarimetric independent component analysis (P-ICA) is presented to separate the target signal and jamming signal while realizing the mainlobe jamming rejection. Finally, the effectiveness and capacities are demonstrated by simulations, and the results indicate that the P-CIA method can suppress well both the SDJ and ESJ, with different signal types of the jamming, e.g., noise-like jamming and digital radio frequency memory (DRFM) jamming.

The rest of the chapter is organized as follows. In section 3.2, the system model of the polarimetric multi-channel radar is introduced briefly. In section 3.3, the P-ICA method is presented to combat SDJ and ESJ suppressive jamming. In section 3.4, the numerical simulations are provided to evaluate the performance. Finally, some conclusions are summarized in section 3.5.

3.2 Signal model

Consider a narrowband polarimetric multi-channel radar system, which consists of a single polarized transmit element and a uniform linear array equipped with L receive elements with the inter-spacing d. Each transmit and receive element has two tunable polarimetric channels, i.e., Horizontal (H) channel and Vertical (V) channel (assuming that the isolation between them is very large), as shown in Figure 3.1.

Assume that the polarization vector of the transmit element is $\boldsymbol{\kappa} = [\kappa_h, \kappa_v]^T$ and the polarization vector of the lth received element is $\boldsymbol{r}_l = [r_{lh}, r_{lv}]^T$, for $l = 1, 2, \cdots, L$, and $\|\boldsymbol{\kappa}\| = 1$, $\|\boldsymbol{r}_l\| = 1$. The transmitter can probe the environment with a train of K pulses with specific polarization, where the polarization is assumed to be fixed between pulses. Denoted by $s(t)$, the complex envelope of each probing pulse signal is given by

$$s(t) = s_T(t)\exp\{j2\pi f_0 t\}, 0 \leq t < T_p \tag{3.1}$$

where f_0 and T_p are, respectively, the carrier frequency and the pulse duration, and $s_T(t)$ denotes the complex baseband narrowband envelope. At each receive branch, signals are captured on the H and V channels and then they are weighted via r_{lh} and r_{lv}, respectively, and further summed before A/D [32]. In particular, the specific values of the polarization vector $\boldsymbol{r}_l, l = 1, \cdots, L$ rely on the specific radar task, and they are usually fixed after the receiving array antenna is fabricated [32].

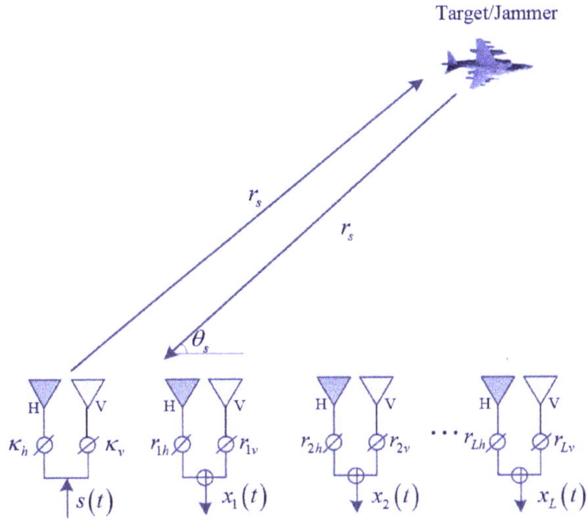

Figure 3.1 The illustration of polarimetric multi-channel radar

3.2.1 Target echo model

Suppose that a far-field target is located at the range r_s and spatial angle θ_s with a radial Doppler frequency f_d. For the lth receive element, the time delay τ_l can be computed as

$$\tau_l = \frac{2r_s}{c} - \frac{d(l-1)\sin\theta_s}{c}, \; l = 1, \ldots, L \tag{3.2}$$

where d is the inter-spacing of receive elements and c is the speed of light. Hence, the received target echo $x_{Tl}(t)$ at the lth element can be expressed as [33]

$$x_{Tl}(t) = \xi \boldsymbol{r}_l^T \boldsymbol{S} \boldsymbol{\kappa} s_T(t-\tau_l) \exp\{j2\pi f_0(t-\tau_l)\} \exp\{j2\pi f_d t\}, \; l = 1, \ldots, L \tag{3.3}$$

where ξ is a complex parameter accounting for both propagation effects as well as the target radar cross section, channel, and other terms involved in the radar range equation. To proceed further, let $\boldsymbol{S}$ be the Sinclair scattering matrix of the target

$$\boldsymbol{S} = \begin{bmatrix} s_{hh} & s_{hv} \\ s_{vh} & s_{vv} \end{bmatrix} \tag{3.4}$$

Note that the Sinclair scattering matrix $\boldsymbol{S}$ describes the relationship between the polarization vectors of the incident wave and the scattered wave, and the values of the elements depend on the carrier frequency, the shape and the material of the targets, and the incident and observation angles of the electromagnetic wave [34].

After down-conversion and some algebraic manipulations, the received signal $x_{Tl}(t)$ can be recast as

$$x_{Tl}(t) = \rho \boldsymbol{r}_l^T \boldsymbol{S} \boldsymbol{\kappa} s_T\left(t - \frac{2r_s}{c}\right) \exp\left\{j2\pi \frac{d(l-1)\sin\theta_s}{\lambda}\right\} \exp\{j2\pi f_d t\} \tag{3.5}$$

where $\rho = \xi \exp\{-j4\pi f_0 r_s/c\}$ and $\lambda = c/f_0$ is the carrier wavelength. Stacking all the target echoes at L received elements, we have

$$\boldsymbol{x}_T(t) = \rho \boldsymbol{R} \odot \boldsymbol{A}(\theta_s) \boldsymbol{S} \boldsymbol{\kappa} s_T\left(t - \tfrac{2r_s}{c}\right) \exp\{j2\pi f_d t\} \tag{3.6}$$

where $\odot$ denotes Hadamard product; $\boldsymbol{R}$ and $\boldsymbol{A}(\theta)$ are, respectively,

$$\boldsymbol{R} = [\boldsymbol{r}_1, \boldsymbol{r}_2, \cdots, \boldsymbol{r}_L]^T \tag{3.7}$$

$$\boldsymbol{A}(\theta) = [\boldsymbol{a}(\theta), \boldsymbol{a}(\theta)] \tag{3.8}$$

where

$$\boldsymbol{a}(\theta) = \left[1, \exp\left\{j2\pi \frac{d\sin\theta}{\lambda}\right\}, \cdots, \exp\left\{j2\pi \frac{d(L-1)\sin\theta}{\lambda}\right\}\right]^T \tag{3.9}$$

3.2.2 Jamming echo model

Suppose that a jammer is located at the radar antenna mainlobe, which intercepts, stores and retransmits the radar probing signals with false information or transmits high power noise-like signals, to reduce the received signal-to-noise ratio (SNR) and degrade detection and tracking performance.

We assume that the jammer continuously transmits high-power jamming signal in the same centre frequency f_0 as the radar working carrier frequency. In other words, the jammer *knows* exactly the radar in frequency domain. Denoted by $j(t)$, the jamming signal is given in

$$j(t) = s_J(t)\exp\{j2\pi f_0 t\} \tag{3.10}$$

where $s_J(t)$ is the complex baseband narrowband envelope.

Note that the jamming waveform $s_J(t)$ may be uncorrelated with the radar probing signal $s_T(t)$, e.g., the noise frequency modulation (NFM) jamming [35] and the noise amplitude modulation (NAM) jamming [36]. Specifically, for the NFM jamming, it is a typical noise-like jamming along with its frequency being modulated by noise, given in

$$s_J(t) = J_p \exp\left(j(2\pi K_{FM}\int_0^t u(t')\,dt' + \phi_f)\right) \tag{3.11}$$

where J_p is the amplitude, $u(t)$ is the Gaussian random process with mean-zero and variance σ_{un}^2, ϕ_f is the initial phase satisfying uniform distribution on $[0, 2\pi)$, and K_{FM} is a constant denoting the frequency modulation rate. The power density spectrum of $s_J(t)$ flats over the radar bandwidth.

For the NAM jamming, its amplitude is modulated by noise, given in

$$s_J(t) = \left(U_0 + U_n(t)\right)\exp\left(j\phi_a\right) \tag{3.12}$$

where U_0 is a constant, $U_n(t)$ is the Gaussian random process with the mean-zero and the variance of σ_{Un}^2, ϕ_a is the initial phase satisfying uniform distribution on $[0, 2\pi)$.

The deceptive jamming waveform $s_J(t)$ is a function of the radar probing signal $s_T(t)$, and it depends on the different jamming strategies; for instance, the SMeared SPectrum (SMSP) jamming [37] contains I_s subwaveforms generated by DRFM jammer [37], specifically,

$$s_J(t) = \sum_{i=0}^{I_s-1} \hat{s}_T\left(t - i\frac{T_p}{I_s}\right) \tag{3.13}$$

where $\hat{s}_T(t)$ is related to the transmit signal $s_T(t)$. For example, if $s_T(t)$ is a linear frequency modulation (LFM) signal with a chirp rate μ, then $\hat{s}_T(t)$ is a LFM waveform with a chirp rate $I_s\mu$ and a during time T_p/I_s.

The jammer also has two polarization antennas, and let $\boldsymbol{\nu} = [\nu_h, \nu_v]^T$ be the jamming transmit polarization vector, where we assume that the jamming polarization state does not change within a short time. The jamming echo received by the lth element, after down-conversion and some algebraic manipulations, can be written in

$$x_{Jl}(t) = \beta\,\boldsymbol{r}_l^T \boldsymbol{\nu} s_J(t)\exp\left\{j2\pi\frac{d(l-1)\sin\theta_J}{\lambda}\right\} \tag{3.14}$$

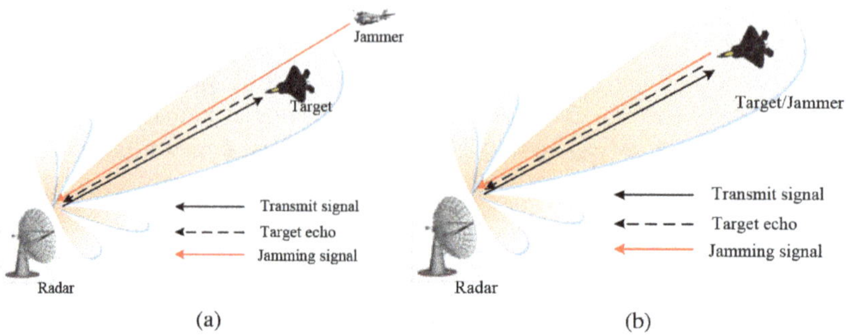

Figure 3.2 Jamming scenario: (a) ESJ; (b) SDJ

where β is the complex parameter of jamming accounting for both the channel propagation effect and other terms involved in the jamming range equation; θ_J is the spatial angle of the jamming. In particular, $|\beta|^2$ denotes the received jamming power. Stacking all the L received jamming echoes, we have

$$\boldsymbol{x}_J(t) = \beta \boldsymbol{R} \odot \boldsymbol{A}(\theta_J)\boldsymbol{v} s_J(t) \tag{3.15}$$

We remark that for the ESJ, the jammer is located near the target, and the ESJ signal enters the radar system also through the antenna mainlobe, but the spatial angle θ_J is usually slightly different from the target spatial angle θ_s, as shown in Figure 3.2(a). For the self-defence jamming (SDJ) where the jammer is often installed on the target, the jamming spatial angle θ_J is exactly equal to the target spatial angle θ_s, as shown in Figure 3.2(b).

3.2.3 Total echo model

Finally, the total received echo is a combination of the target echo, the jamming echo and the noise, given in

$$\begin{aligned} \boldsymbol{x}(t) &= \boldsymbol{x}_T(t) + \boldsymbol{x}_J(t) + \boldsymbol{n}(t) \\ &= \boldsymbol{C}\boldsymbol{g}(t) + \boldsymbol{n}(t) \end{aligned} \tag{3.16}$$

where

- $\boldsymbol{g}(t)$ is the two-component vector for the given t which contains the target echo $s_T\left(t - \frac{2r_s}{c}\right)$ and jamming echo $s_J(t)$, i.e.,

  $$\boldsymbol{g}(t) = \left[s_T\left(t - \frac{2r_s}{c}\right), s_J(t)\right]^T$$

- $\boldsymbol{C}$ is the $L \times 2$ mixing full-column matrix, given in

  $$\boldsymbol{C} = \left[\rho \boldsymbol{R} \odot \boldsymbol{A}(\theta_s)\boldsymbol{S}\boldsymbol{\kappa}, \beta \boldsymbol{R} \odot \boldsymbol{A}(\theta_J)\boldsymbol{v}\right] \in \mathcal{C}^{(L\times 2)}$$

- $\boldsymbol{n}(t) = [n_1(t), n_2(t), \cdots, n_L(t)]^T$ is the complex circular white Gaussian noise with zero-mean and covariance matrix $\sigma^2 \boldsymbol{I}_L$, while $\boldsymbol{I}_L$ denotes the L-dimensional identity matrix.

In the next section, based on the different polarization behaviours of the target and the jamming, we will separate the target signals and jamming signals from the total received echo resorting to P-ICA method.

3.3 Jamming suppression via the P-ICA method

It is usually hard to directly resist the mainlobe jamming using the traditional beamforming techniques. To this end, this section indirectly deals with the jamming suppression via the separation of target and jamming signals. This separation processing can be described as in math,

$$\boldsymbol{y}(t) = \boldsymbol{U}\boldsymbol{x}(t) \approx \boldsymbol{g}(t) \tag{3.17}$$

where $\boldsymbol{U} \in \mathbb{C}^{(2\times L)}$ is defined as the unmixing matrix corresponding to $\boldsymbol{x}(t)$ and $\boldsymbol{y}(t)$ is an estimate of $\boldsymbol{g}(t)$. Thus, the key point of P-ICA is to find the unmixing matrix.

Next, we report the P-ICA procedure for achieving the separation of target and jamming signals. Specifically, the pre-whitening processing is first conducted on the received signal $\boldsymbol{x}(t)$ through a whitening matrix $\boldsymbol{Q} \in \mathbb{C}^{(N\times L)}$ to get $\boldsymbol{z}(t) \in \mathbb{C}^{(N\times N)}$ (i.e., $\boldsymbol{z}(t) = \boldsymbol{Q}\boldsymbol{x}(t)$). Now we are devoted to finding the unmixing matrix $\boldsymbol{W} \in \mathbb{C}^{(2\times N)}$ that satisfies $\boldsymbol{W}\boldsymbol{z}(t) \approx \boldsymbol{g}(t)$. In particular, the fourth-order cumulant matrices with respect to the whitened signal $\boldsymbol{z}(t)$ are estimated. Then, the joint diagonalization algorithm is introduced to accomplish the estimate $\hat{\boldsymbol{W}}$ of the unmixing matrix $\boldsymbol{W}$. Finally, signal separation is performed to obtain the target signal and jamming signal based on the whitened signal $\boldsymbol{z}(t)$ and the estimated unmixing matrix $\hat{\boldsymbol{W}}$. The processing procedure for polarimetric multi-channel radar is shown in Figure 3.3.

3.3.1 Pre-whitening processing

Most ICA methods require that the received signal is whitened in spatial domain [38], which means that the signal components are uncorrelated and their variances are equal to one. To this end, the focus of this subsection is to conduct a pre-whitening process for the observed signal $\boldsymbol{x}(t)$, where the whitening process can be accomplished by resorting to a whitening transformation matrix $\boldsymbol{Q} \in \mathbb{C}^{N\times L}$, i.e.,

$$\boldsymbol{z}(t) = \boldsymbol{Q}\boldsymbol{x}(t) = \boldsymbol{Q}\boldsymbol{C}\boldsymbol{g}(t) + \boldsymbol{Q}\boldsymbol{n}(t), \quad \mathbb{E}\{\boldsymbol{z}(t)\boldsymbol{z}^{\dagger}(t)\} = \boldsymbol{I}_N \tag{3.18}$$

where $\boldsymbol{z}(t)$ is the whitened signal vector, N is equal to 2 being two the components of interest, i.e., target echo and jamming signal (of course, this developed method can be extended to deal with multiple mainlobe jammings), $\boldsymbol{Q}$ is given in

$$\boldsymbol{Q} = \boldsymbol{\Lambda}_s^{-\frac{1}{2}}\boldsymbol{U}_s^{\dagger} \tag{3.19}$$

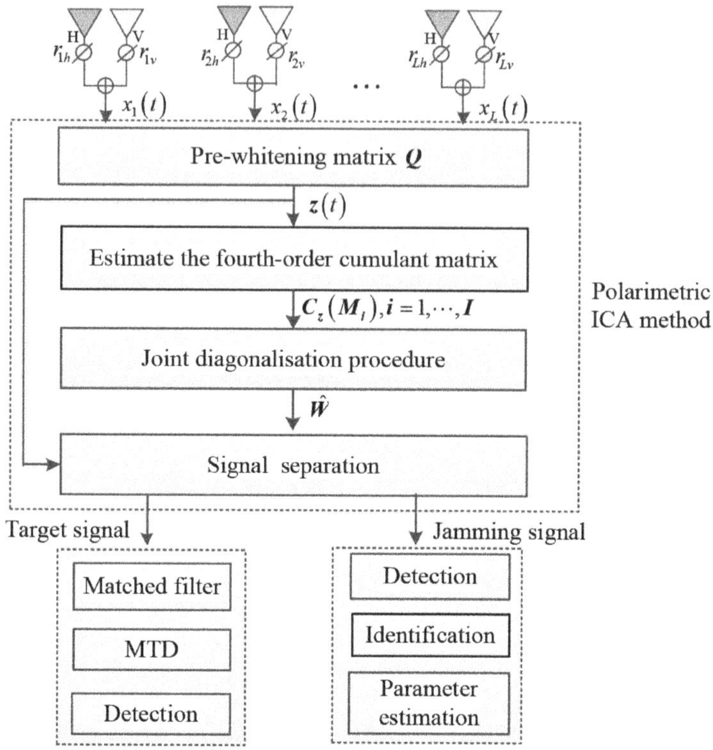

Figure 3.3 The flow chart of jamming suppression

where matrices $\boldsymbol{\Lambda}_S$ and $\boldsymbol{U}_S$, respectively, are composed of the N largest eigenvalues and corresponding eigenvectors of the autocorrelation matrix $\mathbb{E}\{\boldsymbol{x}(t)\boldsymbol{x}^\dagger(t)\}$ of the received signal vector $\boldsymbol{x}(t)$.

3.3.2 Estimation of the fourth-order cumulant matrix

Now our purpose is to find a unmixing matrix $\boldsymbol{W}$ to recover $\boldsymbol{g}(t)$ from $z(t)$. The unmixing matrix $\boldsymbol{W}$ may be obtained via exploiting the fourth-order cumulant matrix of whitened signal $z(t)$. Thus, this subsection focuses on the computation of the fourth-order cumulant matrix. In particular, given an arbitrary non-zero matrix $\boldsymbol{M} \in \mathbb{C}^{N\times N}$, the (p,q)th entry of the fourth-order cumulant matrix $\boldsymbol{C}_z(\boldsymbol{M})$ with respect to the whitened signal $z(t)$ is defined as [39]

$$[\boldsymbol{C}_z(\boldsymbol{M})]_{pq} = \sum_{k=1}^{N}\sum_{l=1}^{N} K_{pqkl}(z)\boldsymbol{M}(k,l) \tag{3.20}$$

where

$$\begin{aligned} K_{pqkl}(z) &= \mathrm{cum}\left(z_p(t), z_k^*(t), z_q(t), z_l^*(t)\right) \\ &= \mathbb{E}\left[z_p(t) z_k(t) z_q^*(t) z_l^*(t)\right] - \mathbb{E}\left[z_p(t) z_k^*(t)\right] \mathbb{E}\left[z_q(t) z_l^*(t)\right] \\ &\quad - \mathbb{E}\left[z_p(t) z_l^*(t)\right] \mathbb{E}\left[z_q(t) z_k^*(t)\right] - \mathbb{E}\left[z_p(t) z_q(t)\right] \mathbb{E}\left[z_k^*(t) z_l^*(t)\right] \end{aligned} \tag{3.21}$$

with cum(·) the cumulant operation and $z_p(t)$ is the pth entry of $z(t)$. Some properties of the matrix $C_z(M)$ are shown in the following propositions.

Proposition 1: Performing some mathematical manipulations to (3.20), $C_z(M)$ can be recast as

$$C_z(M) = \sum_{n=1}^{N} K_n w_n^{\dagger} M w_n w_n w_n^{\dagger} = W \Lambda_M W^{\dagger} \tag{3.22}$$

where $K_n = \mathrm{cum}(g_n(t), g_n^*(t), g_n(t), g_n^*(t))$ is the cumulant of the nth source signal $g(t)$, w_n is the nth column vector of the unmixing matrix $W \in \mathbb{C}^{(2\times N)}$ which is a unitary matrix satisfying $WW^{\dagger} = I$, and

$$\Lambda_M = \mathrm{diag}[K_1 w_1^{\dagger} M w_1, K_2 w_2^{\dagger} M w_2, \cdots, K_N w_N^{\dagger} M w_N]$$

is a diagonal matrix.

Proof. See [39] for details.

Proposition 2: Exploit the cumulants $K_{pqkl}(z), p, q, k, l = 1, \cdots, N$ to construct a $N^2 \times N^2$ dimension matrix H_z, i.e.,

$$H_z(n_1, n_2) = K_{pqkl}(z) \tag{3.23}$$

where $n_1 = p + (q-1)N, n_2 = l + (k-1)N$. Assume that λ_i (sort from largest to smallest) and v_i, $i = 1, \cdots, N^2$ are the eigenvalues and the corresponding orthogonal eigenvectors of H_z [39], respectively, then there are N^2 eigenmatrices $M_i \in \mathbb{C}^{N\times N}$, $i = 1, \cdots, N^2$ that satisfy

$$C_z(M_i) = \lambda_i M_i, \mathrm{Trace}(M_i M_j^{\dagger}) = \delta(i-j) \tag{3.24}$$

where $v_i = \mathrm{vec}(M_i)$ and $\delta(.)$ denotes the Dirac function.

Proof. See [40] for details.

Based on Proposition 1, it can be known that for an arbitrary non-zero matrix $M \in \mathbb{C}^{N\times N}$, the fourth-order cumulant matrix $C_z(M)$ can be diagonalized via the unmixing matrix W. Thus, W is able to make the cumulant matrices set $C_z(M_i), i = 1, 2, \cdots, N^2$ diagonalizable (i.e., $C_z(M_i) = W \Lambda_{M_i} W^{\dagger}$, $i = 1, 2, \cdots, N^2$, where Λ_{M_i} is a diagonal matrix). This paves the way for the development of an estimation procedure of the unmixing matrix W introduced in the following subsection.

3.3.3 Joint diagonalization processing

According to (3.18), to get an estimate of $g(t)$, the unmixing matrix W that satisfies $W(QC)^{\dagger} \approx \Lambda$ is required to obtain, where Λ is a diagonal matrix. At the same time, W is an unitary matrix based on (3.22). Because the joint approximate diagonalization of eigen-matrices (JADE) algorithm is able to deal with complex signals and is asymptotically optimal, while sharing a fast convergence speed [39], this subsection introduces JADE algorithm to accomplish the estimate of W. According to the previous observation in (3.22) of subsection 3.3.2, our goal is to make all $\boldsymbol{C}_z(\boldsymbol{M}_i), i = 1, 2, \cdots, N^2$ diagonalizable by finding a suitable unitary matrix $\boldsymbol{W}$. A possible way is to exploit optimization theory to solve the following minimization problem [40]

$$\min_{\boldsymbol{W}} \sum_{i=1}^{I} \boldsymbol{off}(\boldsymbol{W}^\dagger \boldsymbol{C}_z(\boldsymbol{M}_i)\, \boldsymbol{W}) \tag{3.25}$$

where $\boldsymbol{off}(\boldsymbol{P}) \triangleq \sum_{1 \le m \ne n \le N} |\boldsymbol{P}(m,n)|^2$ for $\boldsymbol{P} \in \mathbb{C}^{N \times N}$, and I is the number of non-zero values among the eigenvalues $\lambda_i, i = 1, \cdots, N^2$, of $\boldsymbol{H}_z$.

Next, we invoke the Jacobi-like technique in [40] to deal with (3.25) by exploiting the special structure on $\boldsymbol{W}$. Specifically, $\boldsymbol{W}$ can be expressed as the product of a set of elementary rotation matrices exploiting the fact that it is a unitary matrix [40], i.e.,

$$W = \prod_{p=1}^{N-1} \prod_{q=p+1}^{N} G_{pq}(c_{pq}, s_{pq}) \tag{3.26}$$

where

$$\boldsymbol{G}_{pq}(c_{pq}, s_{pq}) = \begin{bmatrix} \boldsymbol{I}_{p-1} & \vdots & 0 & \vdots & 0 \\ \cdots & c_{pq} & \cdots & {s_{pq}}^* & \cdots \\ 0 & \vdots & \boldsymbol{I}_{q-p-1} & \vdots & 0 \\ \cdots & -s_{pq} & \cdots & {c_{pq}}^* & \cdots \\ 0 & \vdots & 0 & \vdots & \boldsymbol{I}_{M-q} \end{bmatrix} \tag{3.27}$$

where c_{pq}, s_{pq} are two unknown parameters fulfilling $|c_{pq}|^2 + |s_{pq}|^2 = 1$.

Therefore, the minimization problem with respect to $\boldsymbol{W}$ in Eq. (3.25) can be equivalently transformed into an optimization problem with reference to $C_{pq}, S_{pq}, p = 1, \cdots, N-1, q = p+1, \cdots, N,$ [40], i.e.,

$$\min_{\{c_{pq}\},\{s_{pq}\}} \sum_{i=1}^{I} \boldsymbol{off}\left(\left[\prod_{p=1}^{N-1}\prod_{q=p+1}^{N} \boldsymbol{G}_{pq}(c_{pq}, s_{pq})\right]^\dagger \boldsymbol{C}_z(\boldsymbol{M}_i) \prod_{p=1}^{N-1}\prod_{q=p+1}^{N} \boldsymbol{G}_{pq}(c_{pq}, s_{pq})\right)$$
$$\text{s.t.} \quad |c_{pq}|^2 + |s_{pq}|^2 = 1, p = 1, \cdots, N-1, q = p+1, \cdots, N,$$

Inspection of (3.27) shows that it is hard to solve due to the non-convexity of the objective function and constraints. To this end, resorting to the Jacobi technique, an iteration procedure based on the simultaneous diagonalization algorithm is developed to solve the above problem [40]. In each iteration, it sequentially optimizes $c_{pq}, s_{pq}, p = 1, \cdots, N-1, q = p+1, \cdots, N$ via minimizing

$$\mathscr{P}_{c_{pq}^{(t)}, s_{pq}^{(t)}} \begin{cases} \min\limits_{c_{pq}, s_{pq}} & \sum_{i=1}^{I} \boldsymbol{off}\left(\boldsymbol{G}_{pq}(c_{pq}, s_{pq})^\dagger \boldsymbol{C}^{(t)}{}_z(\boldsymbol{M}_i)\, \boldsymbol{G}_{pq}(c_{pq}, s_{pq})\right) \\ \text{s.t.} & |c_{pq}|^2 + |s_{pq}|^2 = 1 \end{cases} \tag{3.28}$$

for $p = 1, \cdots, N-1, q = p+1, \cdots, N$, where $c_{pq}^{(t)}$ and $s_{pq}^{(t)}$ denote the tth iteration values of c_{pq} and s_{pq}, respectively. $\boldsymbol{C}^{(t)}{}_{z}(\boldsymbol{M}_i)$ denotes the tth iteration value of $\boldsymbol{C}_z(\boldsymbol{M}_i)$ which is related to $\boldsymbol{C}_z^{(t-1)}(\boldsymbol{M}_i)$ (assuming $\boldsymbol{C}_z^{(0)}(\boldsymbol{M}_i) = \boldsymbol{C}_z(\boldsymbol{M}_i)$) and $\boldsymbol{G}_{pq}(c_{pq}^{(t)}, s_{pq}^{(t)})$ corresponding to all the previous index values p, q (not the current values p, q) of the tth iteration.

Next, we focus on the solution of $\mathscr{P}_{c_{pq}^{(t)}, s_{pq}^{(t)}}$ via Jacobi technique. Specifically, letting

$$\tilde{\boldsymbol{C}}_z(\boldsymbol{M}_i) = \boldsymbol{G}_{pq}(c_{pq}, s_{pq})^{\dagger} \boldsymbol{C}^{(t)}{}_{z}(\boldsymbol{M}_i)\, \boldsymbol{G}_{pq}(c_{pq}, s_{pq}) \tag{3.29}$$

the following equation holds true since the unitary transformation using $\boldsymbol{G}_{pq}(c_{pq}, s_{pq})$ preserves the norm [40, p. 2],

$$\boldsymbol{off}(\tilde{\boldsymbol{C}}_z(\boldsymbol{M}_i)) + |\tilde{z}_{i,pp}|^2 + |\tilde{z}_{i,qq}|^2 = \boldsymbol{off}(\boldsymbol{C}_z^{(t)}(\boldsymbol{M}_i)) + |z_{i,pp}|^2 + |z_{i,qq}|^2 \tag{3.30}$$

where $\tilde{z}_{i,pq}$ and $z_{i,pq}$ stand for the (p, q)th entry of the matrices $\tilde{\boldsymbol{C}}_z(\boldsymbol{M}_i)$ and $\boldsymbol{C}_z^{(t)}(\boldsymbol{M}_i)$, respectively.

Hence, the minimization problem in $\mathscr{P}_{c_{pq}^{(t)}, s_{pq}^{(t)}}$ is equivalent to the maximization of $|\tilde{z}_{i,pp}|^2 + |\tilde{z}_{i,qq}|^2$ exploiting the fact that the right-hand side of the equal sign in (3.30) is a constant. In turn, because of $\tilde{z}_{i,pp} + \tilde{z}_{i,qq} = z_{i,pp} + z_{i,qq}$ and invariance of the trace under unitary transforms, it further can be recast as the maximization of $|\tilde{z}_{i,pp} - \tilde{z}_{i,qq}|^2$ [40]. To this end, $\mathscr{P}_{c_{pq}^{(t)}, s_{pq}^{(t)}}$ is tantamount to

$$\begin{aligned} &\max_{c_{pq}, s_{pq}} \quad \sum_{i=1}^{I} |\tilde{z}_{i,pp} - \tilde{z}_{i,qq}|^2 \\ &\text{s.t.} \quad |c_{pq}|^2 + |s_{pq}|^2 = 1 \end{aligned} \tag{3.31}$$

for $p = 1, \cdots, N-1, q = p+1, \cdots, N$.

Substituting (3.27) into (3.29), $\tilde{z}_{i,pp}$ and $\tilde{z}_{i,qq}$ can be, respectively, computed as

$$\tilde{z}_{i,pp} = |c_{pq}|^2 z_{i,pp} - c_{pq} s_{pq}^* z_{i,qp} - c_{pq}^* s_{pq} z_{i,pq} + |s_{pq}|^2 z_{i,qq}$$

and

$$\tilde{z}_{i,qq} = |s_{pq}|^2 z_{i,pp} + c_{pq} s_{pq}^* z_{i,qp} + c_{pq}^* s_{pq} z_{i,pq} + |c_{pq}|^2 z_{i,qq}$$

$\tilde{z}_{i,pp} - \tilde{z}_{i,qq}$ can be further written as

$$\tilde{z}_{i,pp} - \tilde{z}_{i,qq} = |c_{pq}|^2 z_{i,pp} - c_{pq} s_{pq}^* z_{i,qp} - c_{pq}^* s_{pq} z_{i,pq} + |s_{pq}|^2 z_{i,qq} - \tag{3.32}$$

$$\begin{aligned} &(|s_{pq}|^2 z_{i,pp} + c_{pq} s_{pq}^* z_{i,qp} + c_{pq}^* s_{pq} z_{i,pq} + |c_{pq}|^2 z_{i,qq}) \\ &= \boldsymbol{h}_i^T \boldsymbol{v}(c_{pq}, s_{pq}) \end{aligned} \tag{3.33}$$

where

$$\begin{aligned} \boldsymbol{h}_i &= [z_{i,pp} - z_{i,qq}, z_{i,pq} + z_{i,qp}, j(z_{i,qp} - z_{i,pq})]^T \\ \boldsymbol{v}(c_{pq}, s_{pq}) &= \left[|c_{pq}|^2 - |s_{pq}|^2, -c_{pq}^* s_{pq} - c_{pq} s_{pq}^*, j\left(c_{pq} s_{pq}^* - c_{pq}^* s_{pq}\right)\right]^T \end{aligned} \tag{3.34}$$

with $j = \sqrt{-1}$. As a consequence, Equation (3.31) can be rewritten as

$$\begin{aligned} \max_{c_{pq}, s_{pq}} \quad & \boldsymbol{v}^T(c_{pq}, s_{pq})\Big(\sum_{i=1}^{I} \boldsymbol{h}_i^T \boldsymbol{h}_i\Big)\boldsymbol{v}(c_{pq}, s_{pq}) \\ \text{s.t.} \quad & |c_{pq}|^2 + |s_{pq}|^2 = 1 \end{aligned} \tag{3.35}$$

Exploiting the constraint $|c_{pq}|^2 + |s_{pq}|^2 = 1$ and the special structure in $\boldsymbol{v}(c_{pq}, s_{pq})$, the above problem shares an equivalent expression, i.e.,

$$\begin{aligned} \max_{\tilde{\boldsymbol{v}}} \quad & \tilde{\boldsymbol{v}}^T \boldsymbol{H} \tilde{\boldsymbol{v}} \\ \text{s.t.} \quad & \tilde{\boldsymbol{v}}^T \tilde{\boldsymbol{v}} = 1 \end{aligned} \tag{3.36}$$

for $p = 1, \cdots, N-1, q = p+1, \cdots, N$, where $\tilde{\boldsymbol{v}} = \boldsymbol{v}(c_{pq}, s_{pq})$ and $\boldsymbol{H} = \sum_{i=1}^{I} \boldsymbol{h}_i^T \boldsymbol{h}_i$. The closed-form solution $\tilde{\boldsymbol{v}}_{opt}$ to (3.36) is the normalized eigenvector associated with the largest eigenvalue of the matrix $\boldsymbol{H}$.

Then resorting to the constraint $|c_{pq}|^2 + |s_{pq}|^2 = 1$ and $\tilde{\boldsymbol{v}} = \boldsymbol{v}(c_{pq}, s_{pq})$, the solution c_{pq}, s_{pq} to $\mathscr{P}_{c_{pq}^{(t)}, s_{pq}^{(t)}}$ can be calculated as [40]

$$c_{pq}^{(t)} = \sqrt{\frac{\tilde{\boldsymbol{v}}_{opt}(1) + \|\tilde{\boldsymbol{v}}_{opt}\|}{2\|\tilde{\boldsymbol{v}}_{opt}\|}} \tag{3.37}$$

$$s_{pq}^{(t)} = -\frac{\tilde{\boldsymbol{v}}_{opt}(2) - j\tilde{\boldsymbol{v}}_{opt}(3)}{\sqrt{2\|\tilde{\boldsymbol{v}}_{opt}\|(\tilde{\boldsymbol{v}}_{opt}(1) + \|\tilde{\boldsymbol{v}}_{opt}\|)}} \tag{3.38}$$

Based on the discussion, we can update $\boldsymbol{C}_z^{(t)}(\boldsymbol{M}_i) = \boldsymbol{G}_{pq}(c_{pq}^{(t)}, s_{pq}^{(t)})^\dagger \boldsymbol{C}^{(t)}{}_z(\boldsymbol{M}_i)\, \boldsymbol{G}_{pq}(c_{pq}^{(t)}, s_{pq}^{(t)})$ increase the values p, q at the t-th iteration, and continue solving $\mathscr{P}_{c_{pq}^{(t)}, s_{pq}^{(t)}}$. If $c_{pq}^{(t)}, s_{pq}^{(t)}, p = 1, \cdots, N-1, q = p+1, \cdots, N$ are computed (we call this update procedure as a sweep), then the estimate of $\boldsymbol{W}$ at the t-th iteration can be obtained via (3.27), namely,

$$\boldsymbol{W}^{(t)} = \boldsymbol{W}^{(t-1)} \prod_{p=1}^{N-1} \prod_{q=p+1}^{N} G_{pq}(c_{pq}^{(t)}, s_{pq}^{(t)})$$

assuming $\boldsymbol{W}^{(0)} = \boldsymbol{I}_N$ However, from the point of view of practical implementation, one sweep may not ensure to get good diagonalization performance. To this end, several sweeps are generally necessary until convergence [25, 41]. Hence, this will yield an iteration procedure to obtain a good estimate of $\boldsymbol{W}$, where the process is stopped when $s_{(N-1)N} < \zeta$ [40], where ζ is a given threshold. Finally, the whole estimation procedure is reported in Algorithm 1.

The computational complexity of Algorithm 1 is analysed. Specifically, the main computational complexity of the joint diagonalization is connected to the size of $\boldsymbol{C}_z(\boldsymbol{M}_i), i = 1, \cdots, I$. In each iteration, the main computational complexity is associated with the update of $\boldsymbol{C}_z^{(t)}(\boldsymbol{M}_i), i = 1, \cdots, I$, which requires performing $N(N-1)$ times computation of $\boldsymbol{C}_z^{(t)}(\boldsymbol{M}_i)$ along with the computational complexity of $O(N^3)$. Hence, the joint diagonalization shares the computational complexity of $O(\bar{I}(IN^3))$, where $\bar{I}$ is the number of iterations in Algorithm 1.

Algorithm 1 Estimate of $\boldsymbol{W}$ based on joint diagonalization.

Require: $z(t)$, a threshold ζ to control convergence;
Ensure: $\hat{\boldsymbol{W}}$;
Estimate the cumulants $K_{pqkl}(\boldsymbol{z}), p,q,k,l = 1,\cdots,N$ via (1.21), where $\mathbb{E}\left[z_p(t)z_k(t)z_q^*(t)z_l^*(t)\right]$ and $\mathbb{E}\left[z_p(t)z_k^*(t)\right]$ in (1.21) are estimated by their sample averages, respectively;
Exploit the cumulants $K_{pqkl}(\boldsymbol{z}), p,q,k,l = 1,\cdots,N$ to construct the $N^2 \times N^2$ dimension matrix $\boldsymbol{H}_z$ via (1.23);
Exploit **Proposition 2** to construct $\boldsymbol{C}_z(\boldsymbol{M}_i), i = 1,\cdots,I$;
$t = 0; \boldsymbol{W}^{(t)} = \boldsymbol{I}_N, \boldsymbol{C}_z^{(t)}(\boldsymbol{M}_i) = \boldsymbol{C}_z(\boldsymbol{M}_i), i = 1,\cdots,I$;
while $s_{pq} > \zeta$ **do**
  for $1 \leq p \leq N-1$ **do**
    for $p+1 \leq q \leq N$ **do**
      Construct $\boldsymbol{h}_i, i = 1,\cdots,I$ via (1.34) and calculate $\boldsymbol{H} = \sum_{i=1}^{I} \boldsymbol{h}_i^T \boldsymbol{h}_i$;
      Perform the eigenvalue decomposition to $\boldsymbol{H}$ and find its normalized eigenvector $\tilde{\boldsymbol{v}}_{opt}$ corresponding to the maximum eigenvalue;
      Calculate $c_{pq}^{(t)}, s_{pq}^{(t)}$ by Eq. (1.37) and Eq. (1.38) and update the matrix $\boldsymbol{G}_{pq}(c_{pq}^{(t)}, s_{pq}^{(t)})$;
      Update the matrices: $\boldsymbol{C}_z^{(t)}(\boldsymbol{M}_i) = \boldsymbol{G}_{pq}^H(c_{pq}^{(t)}, s_{pq}^{(t)})\boldsymbol{C}_z^{(t)}(\boldsymbol{M}_i)\boldsymbol{G}_{pq}(c_{pq}^{(t)}, s_{pq}^{(t)})$ for $i = 1,2,\cdots,I$;
      Update the unitary matrix: $\boldsymbol{W}^{(t)} = \boldsymbol{W}^{(t)}\boldsymbol{G}_{pq}(c_{pq}^{(t)}, s_{pq}^{(t)})$.
    end for
  end for
  $t = t+1; \boldsymbol{W}^{(t)} = \boldsymbol{W}^{(t-1)}, \boldsymbol{C}_z^{(t)}(\boldsymbol{M}_i) = \boldsymbol{C}_z^{(t-1)}(\boldsymbol{M}_i)$;
end while
Output $\hat{\boldsymbol{W}} = \boldsymbol{W}^{(t)}$.

3.3.4 Signal separation

By exploiting the estimated unmixing matrix $\hat{\boldsymbol{W}}$, the target and jamming signals can be obtained via

$$\hat{\boldsymbol{g}}(t) = \hat{\boldsymbol{W}}\boldsymbol{z}(t) \tag{3.39}$$

where $\hat{\boldsymbol{g}}(t) = [\hat{g}_1(t), \hat{g}_2(t)]^T$ contains the estimate of $\boldsymbol{g}(t)$. Note that it cannot determine which is the target echo in the obtained two components (i.e., $\hat{g}_1(t), \hat{g}_2(t)$). To this end, some radar signal processing operations are exploited to overcome this issue. For example, we can employ the matched filter (MF) [36] to deal with the separated echo

$$u_{mf}(t) = \left| \int \hat{g}_n(t) s_T(-t)\, dt \right|, n = 1, 2 \tag{3.40}$$

In the high SNR case, if the output signal $u_{mf}(t)$ shows the peak in some range bin, the signal $\hat{g}_n(t)$ may be regarded as the target echo. On the contrary, the jamming signal is distinguished. In particular, this discrimination operation can be performed automatically.

3.4 Numerical results

In this section, we evaluate the performance of the P-ICA method through several numerical simulations. In particular, we consider the polarimetric multi-channel radar system with $L = 10$ receiving elements and an inter-element spacing $d = \lambda/2$. Assume that the polarization vector of the transmit element is $\boldsymbol{\kappa} = [\cos 10°, \sin 10° e^{j\frac{\pi}{4}}]^T$, and the polarization vectors of the receiving elements $\boldsymbol{R}$ are defined as

$$\begin{aligned}\boldsymbol{R} &= [\boldsymbol{r}_1, \cdots, \boldsymbol{r}_L] \\ &= \left[\begin{matrix} \cos 30° & \cos 38° & \cos 40° & \cos 59° & \cos 35° \\ \sin 30° e^{j\frac{\pi}{4}} & \sin 38° e^{j\frac{\pi}{2}} & \sin 40° e^{j\frac{\pi}{3}} & \sin 59° e^{j\frac{\pi}{4}} & \sin 35° e^{-j\frac{\pi}{4}} \end{matrix}\right. \\ &\qquad \left.\begin{matrix} \cos 52° & \cos 55° & \cos 65° & \cos 62° & \cos 25° \\ \sin 52° e^{j\frac{\pi}{2}} & \sin 55° e^{-j\frac{\pi}{3}} & \sin 65° e^{-j\frac{\pi}{6}} & \sin 62° e^{j\frac{\pi}{5}} & \sin 25° e^{j\frac{\pi}{3}} \end{matrix}\right]\end{aligned} \tag{3.41}$$

The transmit signal is the LFM waveform, and its complex baseband envelope $s_T(t)$ is given by [36]

$$s_T(t) = a_T \exp(j\mu\pi t^2), 0 \leq t < T_p \tag{3.42}$$

where a_T is the amplitude, T_p is the pulse duration, μ is the chirp rate and $B = \mu T_p$ is the signal bandwidth. In the simulations, unless otherwise stated, we set $T_p = 10\mu s$, $B = 10$MHz and PRT $=50\mu s$. The received baseband echo is discreted with the sample rate $2B$.

Consider that a target is located at angle $\theta_s = 45°$ with respect to the radar boresight, and 300th range bin, and presents a Doppler frequency $f_d = 4$ kHz. Its polarization scattering matrix is set to be

$$\mathbf{S} = \begin{bmatrix} 0.9 & 0.5j \\ -0.8j & 0.9 \end{bmatrix} \tag{3.43}$$

Suppose that the SDJ deploys on the same target, where the jamming signal will enter the radar receiver through the same spatial angle as the target, i.e., $\theta_J = \theta_s$, and is the continuous wave interference which overlaps the target echo signal in time domain. In particular, each PRT echo data are used to perform the separation of target signal and jamming signal.

In Figure 3.4, simulation results in the presence of the NFM jamming are provided for $\sigma_{un}^2 = 1$, $K_{FM} = 10^7$, $f_0 = 1$GHz, $\sigma^2 = 0$ dB, $|\rho|^2 = 10$ dB, and $|\beta|^2 = 50$ dB. The polarization vector of the jammer is set to be $\boldsymbol{\nu} = [\cos 45°, \sin 45° e^{j\frac{\pi}{3}}]^T$. The receive echo $x_1(t)$ for the first received element illustrates that the target echo is covered heavily by the strong jamming both in the raw echo signal in Figure 3.4(a)

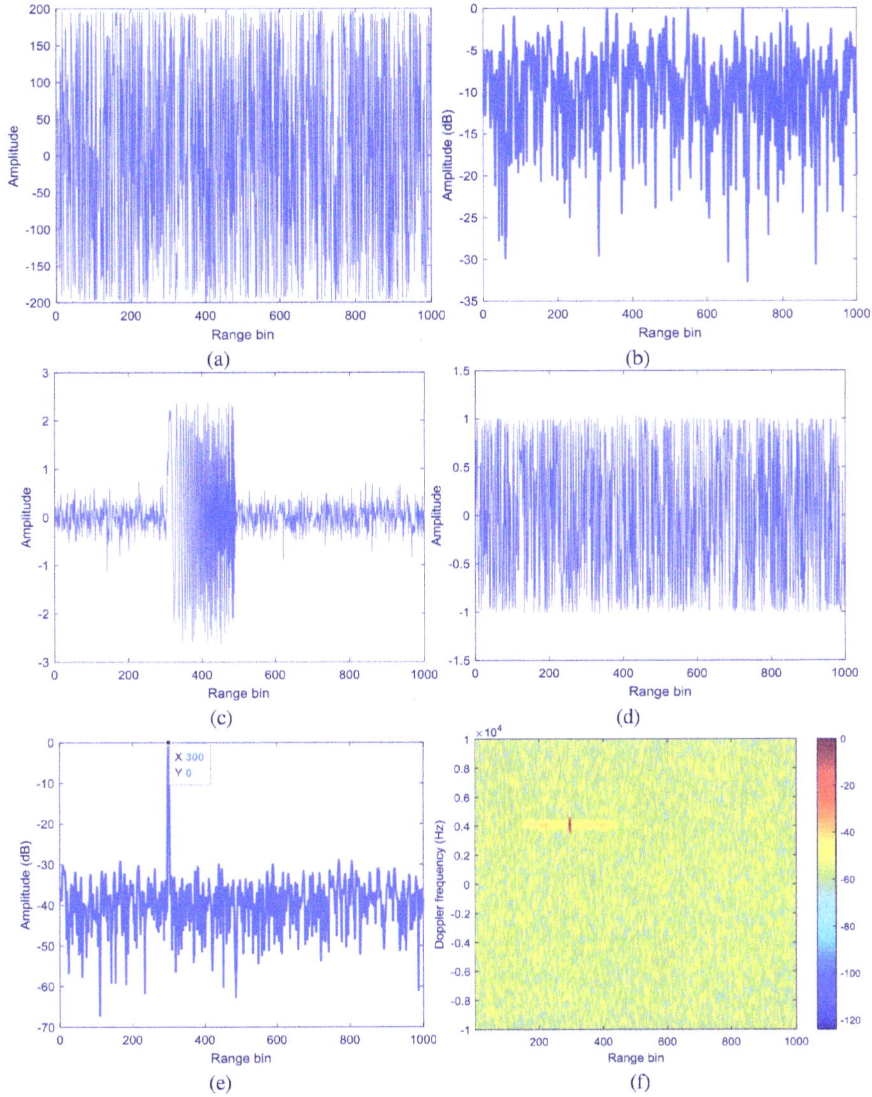

Figure 3.4 Simulation results for the NFM jamming: (a) the real part of $x_1(t)$, where the unit on the vertical axis is "voltage"; (b) MF processing of $x_1(t)$; (c) separated target signal; (d) separated jamming signal; (e) MF processing of separated target signal; (f) range-Doppler plane of separated target signal

and the signal after pulse compression processing in Figure 3.4(b). Moreover, the target signal and the jamming signal can be separated effectively by P-ICA method, as shown in Figure 3.4(c) and (d). When we conduct the pulse compression processing on the separated target signal, as expected, the peak appears in the location of the true target, as shown in Figure 3.4(e), highlighting that the jamming signal is suppressed in comparison with that in Figure 3.4(b). Note that the separated jamming signal can be used to identify and classify the jammer. After performing the separation of target and interference echoes, we conduct the Moving Target Detection (MTD) for all target echoes to estimate the target Doppler. In particular, the performance of the pulse Doppler processing is shown in Figure 3.4(f) with $K = 64$, where the P-ICA separation method is used in each PRI echo. It can be clearly seen an obvious peak corresponding to the 300th range bin and Doppler frequency 4 kHz, which shows the consistency with the true range bin and Doppler frequency of the target. It means that the P-ICA method does not alter the relationship of the relative phases among the coherent pulse train.

Moreover, the NAM jamming and the SMSP jamming are provided to illustrate the performance of different types of the jamming signals. Simulation results in the presence of the NAM jamming in Figure 3.5 and the SMSP jamming in Figure 3.6 are provided for $f_0 = 1$ GHz, $I_s = 4$, $\sigma^2 = 0$ dB, $|\rho|^2 = 10$ dB and $|\beta|^2 = 50$ dB. It is clearly shown that the P-ICA method is capable of separating correctly the target

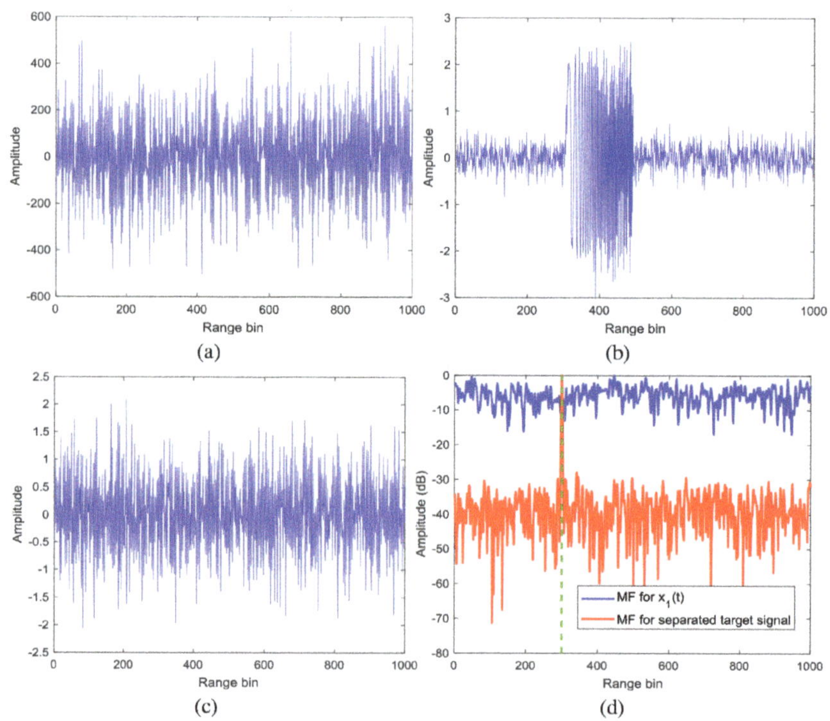

Figure 3.5 Simulation results for the NAM jamming: (a) real part of $x_1(t)$; (b) separated target signal; (c) separated jamming signal; (d) MF processing of separated target signal; and $x_1(t)$, where the units on the vertical axis of (a), (b) and (c) are "voltage"

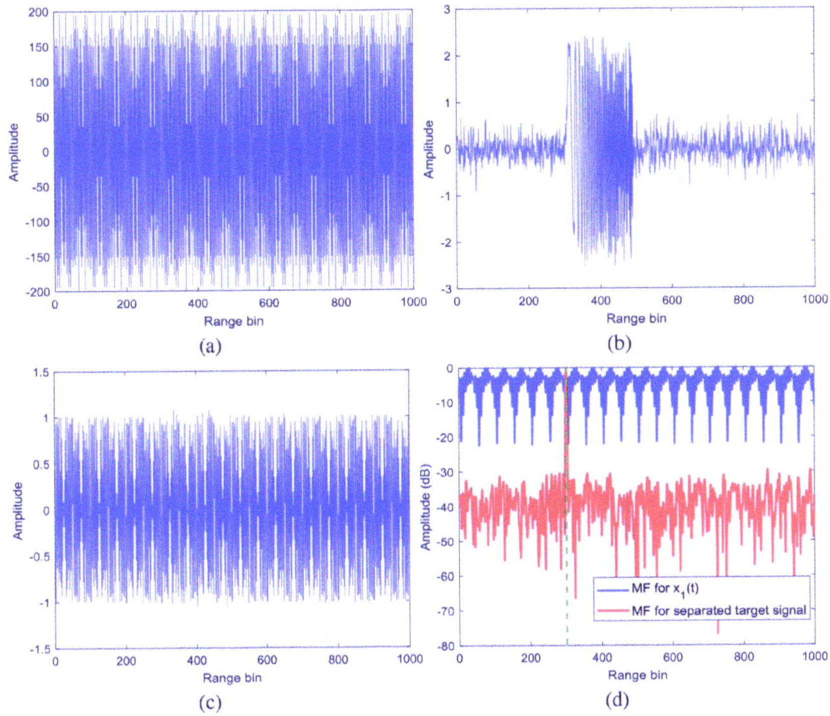

Figure 3.6 Simulation results for the SMSP jamming: (a) real part of $x_1(t)$; (b) separated target signal; (c) separated jamming signal; (d) MF processing of separated target signal; and $x_1(t)$, where the units on the vertical axis of (a), (b) and (c) are "voltage"

signal in the presence of the strong NAM jamming or the SMSP jamming. In other words, the P-ICA method is not sensitive to the signal types of the jamming, which shares very good robustness to tackle different types of jamming signals in practice.

In Figure 3.7, we analyse the performance of the P-ICA method for two different polarization vectors of the jammer, i.e.,

$$\textit{Case } 1: \boldsymbol{v}_1 = [\cos 45°, \sin 45° e^{j\frac{\pi}{3}}]^T \neq \boldsymbol{\kappa} \tag{3.44}$$

$$\textit{Case } 2: \boldsymbol{v}_2 = [\cos 10°, \sin 10° e^{j\frac{\pi}{4}}]^T = \boldsymbol{\kappa} \tag{3.45}$$

For comparison, without considering the polarization diversity, the logarithm cancellation in [17] and traditional signal separation methods, for example, the FastICA method in [26] and JADE method in [31], are also provided. It is shown that FastICA method [26] and JADE method [31] are invalid for the self-defence jammer. It is because the mixing matrix $\boldsymbol{C}$ in the signal separation procedure is not full column rank due to the spatial angle of the jammer being the same as that of the target without the use of polarization information. The log cancellation method [17] can effectively reject both the NFM jamming and the SMSP jamming, while it is failed for the

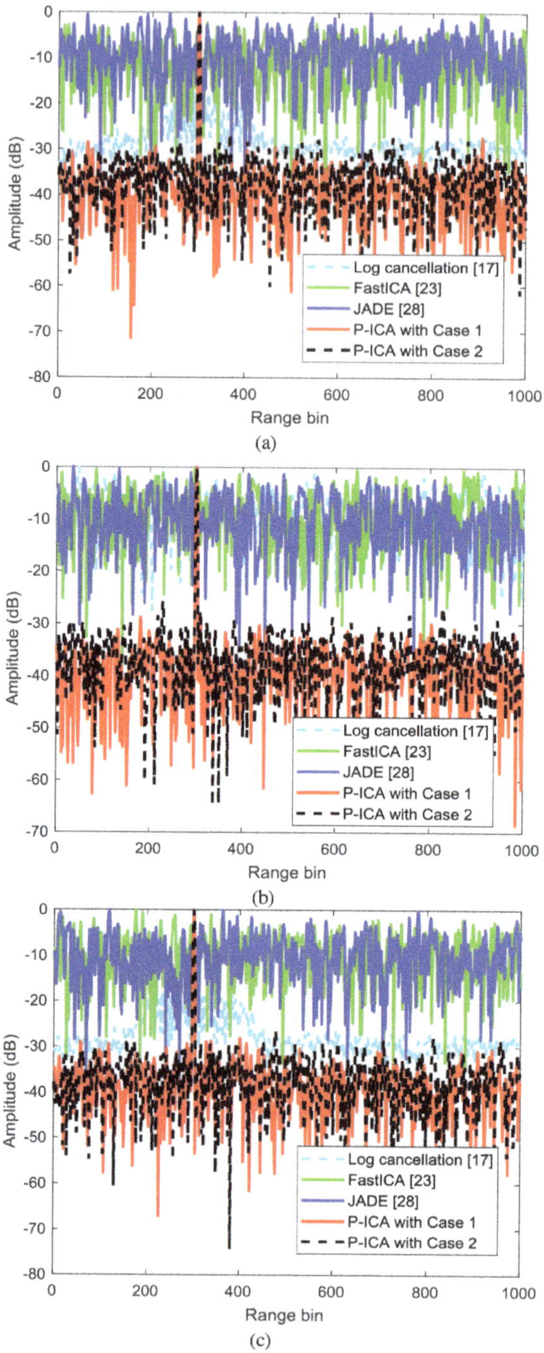

Figure 3.7 Simulation results for jamming suppression: (a) the NFM jamming; (b) the NAM jamming; (c) the SMSP jamming

NAM jamming. On the contrary, the P-ICA method still works for the three types of jamming signals. For *Case 2*, where the transmit polarization vector of the jammer is equal to that of the radar, it is shown that the jamming signals are eliminated well, which means that it is still valid even if the jammer intercepts exactly the transmit polarization vector of the radar.

In Figure 3.8, we plot the averaged output Signal to Jamming plus Noise Ratio (SJNR) over 500 Monte Carlo trials versus $|\rho|^2$ in dB with $|\beta|^2 = 50$ dB, where SJNR is computed as the ratio between target energy and interference and noise energy, i.e.,

$$SJNR = \frac{P_t}{P_{jn}} \tag{3.46}$$

where P_t denotes the energy corresponding to the target range bin obtained via performing the matching filter to the radar echo, and P_{jn} denotes the interference and noise energy which can be estimated via averaging the matching filter results within other range bins. It can be seen that the FastICA and JADE methods attain to almost unchanged output SINR values (less than −2 dB) for the NFM jamming, the NAM jamming and the SMSP jamming. Since the spatial angle of jammer θ_J is the same as that of the target, the FastICA and JADE methods are invalid regardless of target signal power. For both the NFM jamming and the SMSP jamming, the output SJNR values of the P-ICA method are lower than those of log cancellation method for $|\rho|^2$(in dB) < -13 dB for *Case* 1 and $|\rho|^2$(in dB) < -3 dB for *Case* 2, respectively. The P-ICA method exhibits better performance for higher SNR values. In addition, the P-ICA method always outperforms the log cancellation method in [17] for the NAM jamming. Moreover, for $|\rho|^2$(in dB) > -3 dB, the P-ICA method ensures that the output SJNRs increase linearly, which means that the loss in dB of SJNR is a constant in comparison with no jamming case. Note that the output SJNR of the P-ICA method with *Case* 2 is lower than that of *Case* 1, since the transmit polarization vector of the jammer is equal to that of the radar in *Case* 2, and thus the correlation between the columns of matrix $\boldsymbol{C}$ is higher, leading to a worse separation performance.

In Figure 3.9, we plot the output SJNR versus $|\beta|^2$ in dB with $|\rho|^2 = 10$ dB. The FastICA and JADE methods cannot achieve the jamming suppression. Meanwhile, the output SJNRs for the P-ICA method are invariably higher than FastICA and JADE methods. Specifically, for the NFM jamming, the gaps are about 40 dB between the P-ICA method with *Case* 1 and the FastICA or JADE methods, and about 8 dB for the log cancellation method. Compared with the ideal case without jamming, the gap is about 4.3 dB, as shown in Figure 3.9(a). In addition, the SJNR values keep almost the same when increasing $|\beta|^2$. In other words, the P-ICA method is not sensitive to the power of the jamming. Some similar behaviours can be found for the NAM jamming in Figure 3.9(b) and the SMSP jamming in Figure 3.9(c).

In Figure 3.10, we evaluate the performance of the P-ICA method for different spatial angles of the jammer, where the output SJNR values are plotted versus θ_J for the NFM jamming, the NAM jamming and the SMSP jamming. The target is located at $\theta_s = 45°$, and the other simulation parameters are the same as in

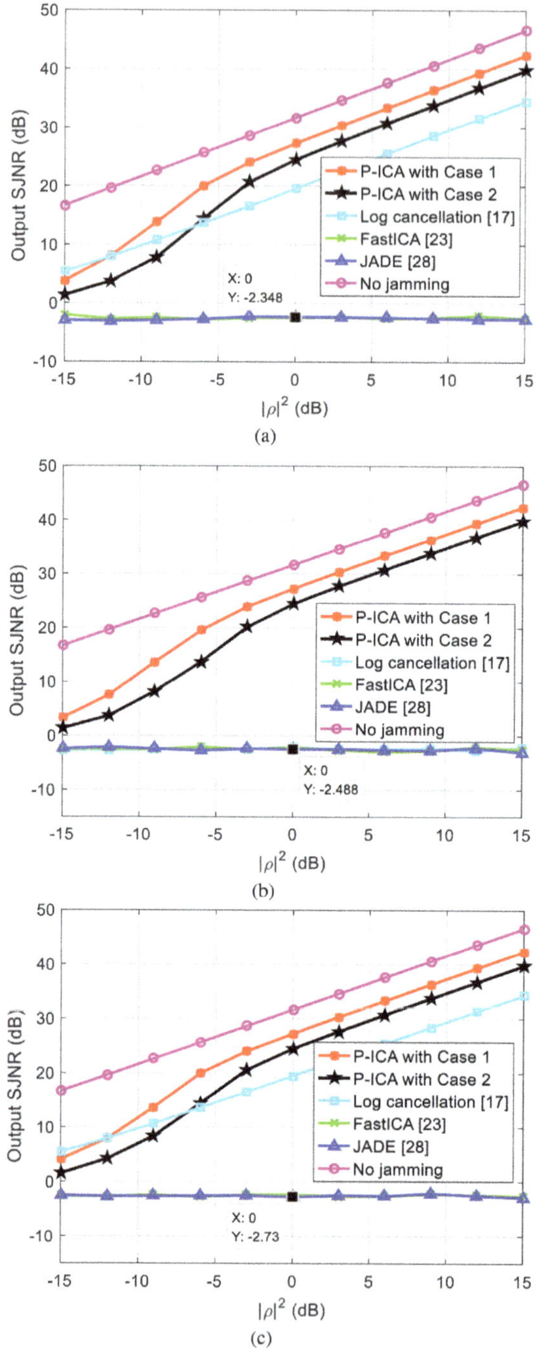

Figure 3.8 Output SJNR versus $|\rho|^2$ in dB: (a) the NFM jamming; (b) the NAM jamming; (c) the SMSP jamming

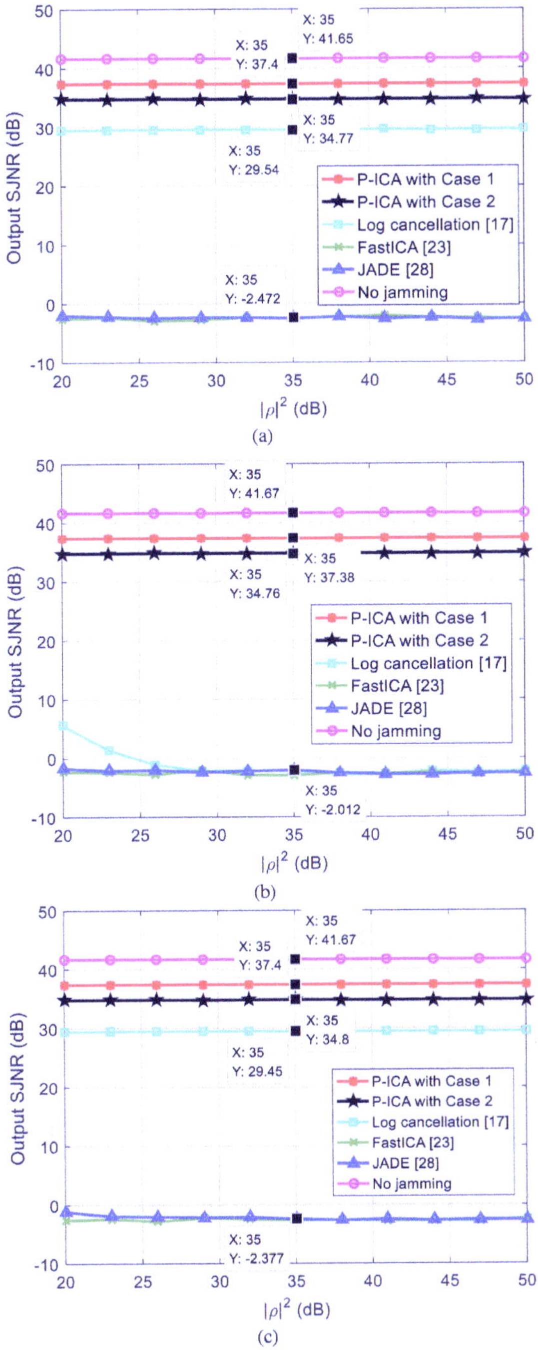

Figure 3.9 Output SJNR versus $|\beta|^2$ in dB: (a) the NFM jamming; (b) the NAM jamming; (c) the SMSP jamming

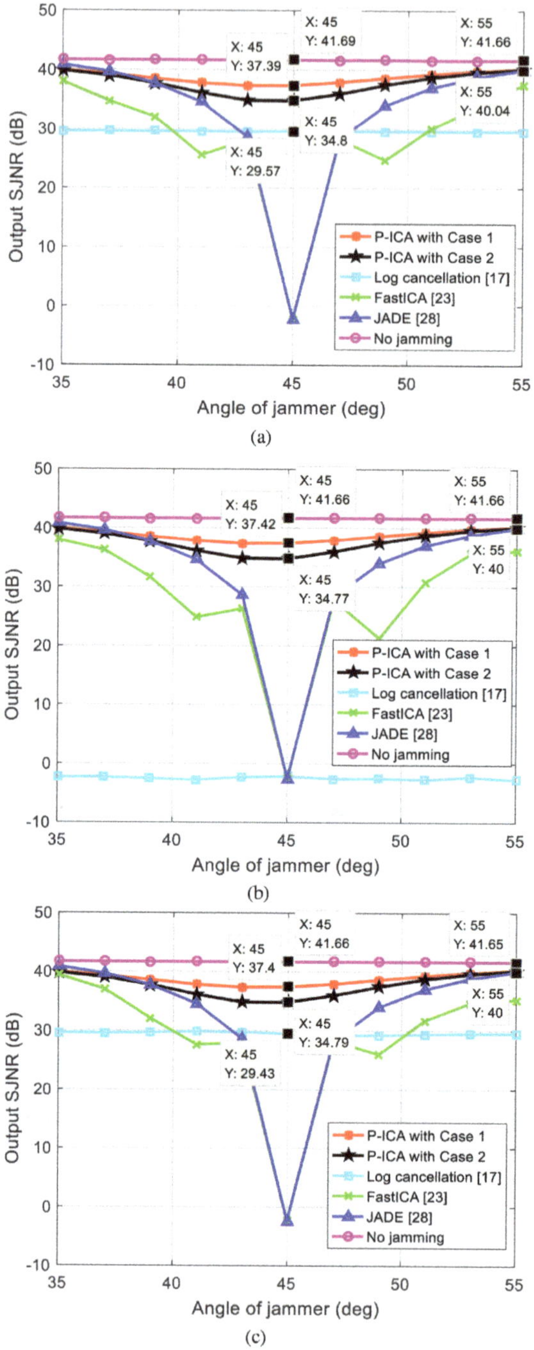

Figure 3.10 Output SJNR versus the spatial angle of jammer: (a) the NFM jamming; (b) the NAM jamming; (c) the SMSP jamming

Figure 3.6. For the P-ICA method, the curves show that the larger the spatial angle difference $|\theta_J - \theta_s|$, the higher the output SJNR values. However, the output SJNR loss is not obvious. Specifically, the gap is about 1.6 dB for $|\theta_J - \theta_s| = 10°$, and they are, respectively, about 4.3 dB for *Case* 1 and about 4.5 dB for *Case* 2 at the worse situation $|\theta_J - \theta_s| = 0°$, compared with the ideal case without jamming. The performance of the FastICA and JADE methods degrade significantly by reducing the spatial angle difference $|\theta_J - \theta_s|$, and they become ineffective leading to the smallest output SJNR values for $|\theta_J - \theta_s| = 0°$, which means that the spatial angle of jammer is identical to that of the target, i.e., the self-defence jammer. The curves show that the log cancellation method works well for the NFM jamming in Figure 3.10(a) and the SMSP jamming in Figure 3.10(c). However, its output SJNR values are always lower than that of the P-ICA method, where the gaps are about 7.8 dB for *Case* 1 and about 5.2dB for *Case* 2. As to the NAM jamming, it does not work, as shown in Figure 3.10(b).

3.5 Conclusions

In this chapter, assuming that the target and self-defence mainlobe jamming share different polarization characteristics; the ICA-based approach for polarimetric multi-channel radar is developed to suppress mainlobe jamming. Specifically, the signal models of polarimetric multi-channel radar accounting for the target and jamming signals were derived. Then, the P-ICA method was utilized to separate the target component and jamming component while achieving the mainlobe jamming suppression. Finally, numerical simulation results have demonstrated that the P-ICA method can suppress many different signal kinds of jammings, such as the NFM jamming, the NAM jamming and the SMSP jamming, and the types of both the SDJ and ESJ. In addition, it can well overcome the disadvantages of the existing methods in [17, 26] which cannot reject NAM jamming and SDJ jamming, respectively.

References

[1] Li N., Zhang Y. 'A survey of radar ECM and ECCM'. *IEEE Transactions on Aerospace & Electronic Systems*. 1995, vol. 31(3), pp. 1110–20.

[2] Morris V., Kastle A. 'Trends in electronic counter-countermeasures'. *Proceedings of National Telesystems Conference (NTC)*; Atlanta, GA, 2002. pp. 265–69.

[3] DeMartino A. *Introduction to modern EW systems*. Norwood, MA: Artech House; 2012.

[4] Xu J., Liao G., Zhu S., So H.C. 'Deceptive jamming suppression with frequency diverse MIMO radar'. *Signal Processing*. 2015, vol. 113, pp. 9–17.

[5] Li S., Zhang L., Liu N., Zhang J., Zhao S. 'Adaptive detection with conic rejection to suppress deceptive jamming for frequency diverse MIMO radar'. *Digital Signal Processing*. 2017, vol. 69(3), pp. 32–40.

[6] Lan L., Liao G., Xu J., Zhang Y., Fioranelli F. 'Suppression approach to main-beam deceptive jamming in FDA-MIMO radar using nonhomogeneous sample detection'. *IEEE Access: Practical Innovations, Open Solutions*. 2018, vol. 6, pp. 34582–97.
[7] Li G. -ming, Zhang Q., Liu Q. -yong, Liang J., Wang D., Zhu F. 'Mainlobe jamming suppression using improved frequency diverse array with MIMO radar'. *Journal of Sensors*. 2019, vol. 2019(3), pp. 1–12.
[8] Zhao S., Zhang L., Zhou Y., Liu N. 'Signal fusion-based algorithms to discriminate between radar targets and deception jamming in distributed multiple-radar architectures'. *IEEE Sensors Journal*. 2015, vol. 15(11), pp. 6697–706.
[9] Zhao S., Liu N., Zhang L., Zhou Y., Li Q. 'Discrimination of deception targets in multistatic radar based on clustering analysis'. *IEEE Sensors Journal*. 2016, vol. 16(8), pp. 2500–08.
[10] Huang D., Cui G., Yu X., Ge M., Kong L. 'Joint range–velocity deception jamming suppression for SIMO radar'. *IET Radar, Sonar & Navigation*. 2019, vol. 13(1), pp. 113–22.
[11] Akhtar J. 'Orthogonal block coded ECCM schemes against repeat radar jammers'. *IEEE Transactions on Aerospace and Electronic Systems*. 2009, vol. 45(3), pp. 1218–26.
[12] Zhang J., Zhu D., Zhang G. 'New antivelocity deception jamming technique using pulses with adaptive initial phases'. *IEEE Transactions on Aerospace and Electronic Systems*. 2013, vol. 49(2), pp. 1290–300.
[13] Cui G., Ji H., Carotenuto V., Iommelli S., Yu X. 'An adaptive sequential estimation algorithm for velocity jamming suppression'. *Signal Processing*. 2017, vol. 134, pp. 70–75.
[14] Cui G., Yu X., Yuan Y., Kong L. 'Range jamming suppression with a coupled sequential estimation algorithm'. *IET Radar, Sonar & Navigation*. 2018, vol. 12(3), pp. 341–47.
[15] Su B., Wang Y., Zhou L. 'A mainlobe interference cancelling method'. *IEEE International Symposium on Microwave*; 2005. pp. 1–4.
[16] Yang X., Yin P., Zeng T., Sarkar T.K. 'Applying auxiliary array to suppress mainlobe interference for ground-based radar'. *IEEE Antennas and Wireless Propagation Letters*. 2013, vol. 12, pp. 433–36.
[17] Du D., Tang B. 'Noise FM jamming suppression via logarithm transform'. *International Conference on Communications, Circuits and Systems*; Xiamen City, China, 2008. pp. 1–4.
[18] Dai H., Wang X., Li Y., Liu Y., Xiao S. 'Main-lobe jamming suppression method of using spatial polarization characteristics of antennas'. *IEEE Transactions on Aerospace and Electronic Systems*. 2012, vol. 48(3), pp. 2167–79.
[19] Sun Q., Zhang Q., Huang X., Gao Q. 'Target detection and localization method for distributed monopulse arrays in the presence of mainlobe jamming'. *EURASIP Journal on Advances in Signal Processing*. 2020, vol. 2020(1), pp. 1–9.

[20] Aubry A., De Maio A., Pallotta L. 'A geometric approach to covariance matrix estimation and its applications to radar problems'. *IEEE Transactions on Signal Processing*. 2017, vol. 66(4), pp. 907–22.

[21] Li J., Aubry A., De Maio A., Zhou J. 'An el approach for similarity parameter selection in KA covariance matrix estimation'. *IEEE Signal Processing Letters*. 2019, vol. 26(8), pp. 1217–21.

[22] Du X., Aubry A., De Maio A., Cui G. 'Toeplitz structured covariance matrix estimation for radar applications'. *IEEE Signal Processing Letters*. 2020, vol. 27, pp. 595–99.

[23] Chabriel G., Kleinsteuber M., Moreau E., Shen H., Tichavsky P., Yeredor A. 'Joint matrices decompositions and blind source separation: a survey of methods, identification, and applications'. *IEEE Signal Processing Magazine*. 2014, vol. 31(3), pp. 34–43.

[24] Raju K., Ristaniemi T., Karhunen J., Oja E. 'Jammer suppression in DS-CDMA arrays using independent component analysis'. *IEEE Transactions on Wireless Communications*. 2006, vol. 5(1), pp. 77–82.

[25] Gong X.-F., Wang X.-L., Lin Q.-H. 'Generalized non-orthogonal joint diagonalization with lu decomposition and successive rotations'. *IEEE Transactions on Signal Processing*. 2015, vol. 63(5), pp. 1322–34.

[26] Huang G., Yang L., Su G. 'Blind source separation used for radar anti-jamming'. *International Conference on Neural Networks*; Signal Processing, Nanjing, China, 2003. pp. 1382–85.

[27] Ge M., Cui G., Yu X. 'Mainlobe jamming suppression via blind source separation'. *IEEE Radar Conference*; Oklahoma City, OK, 2018. pp. 914–18.

[28] Ge M., Cui G., Kong L. 'Mainlobe jamming suppression for distributed radar via joint blind source separation'. *IET Radar, Sonar & Navigation*. 2019, vol. 13(7), pp. 1189–99. Available from https://onlinelibrary.wiley.com/toc/17518792/13/7

[29] Ge M., Cui G., Yu X., Kong L. 'Main lobe jamming suppression via blind source separation sparse signal recovery with subarray configuration'. *IET Radar, Sonar & Navigation*. 2020, vol. 14(3), pp. 431–38. Available from https://onlinelibrary.wiley.com/toc/17518792/14/3

[30] Luo S., Xiong Y., Cheng H. 'An algorithm of radar deception jamming suppression based on blind signal separation'. *International Conference on Computational Problem-Solving (ICCP)*; Chengdu, China, 2011. pp. 2801–06.

[31] Zhou B., Li R., Liu W., Wang Y., Dai L., Shao Y. 'A BSS-based space–time multi-channel algorithm for complex-jamming suppression'. *Digital Signal Processing*. 2019, vol. 87, pp. 86–103.

[32] Xiang Z., Chen B., Yang M. 'Statistical method with dual-polarized MIMO array for target discrimination'. *IEEE Antennas and Wireless Propagation Letters*. 2016, vol. 16, pp. 1313–16.

[33] Ge M., Cui G., Yu X., Kong L. 'Mainlobe jamming suppression with polarimetric multi-channel radar via independent component analysis'. *Digital Signal Processing*. 2020, vol. 106(3), 102806.

[34] Ulaby F.T., Elachi C. 'Radar polarimetry for geoscience applications' in *Geocarto international*. Vol. 5; 1990. pp. 38–38.
[35] Neri F. *Introduction to electronic defense systems*. Raleigh, NC: Artech House; 2006.
[36] Richards M. *Fundamentals of radar signal processing*. Second edition. New York, NY: McGraw-Hill Education; 2014.
[37] Sparrow M.J., Cikalo J. *ECMecm techniques to counter pulse compression radar [online]*. Available from https://patents.google.com/patent/US7081846B1/
[38] Hyvärinen A., Karhunen J., Oja E. *Independent component analysis*. New York, USA: John Wiley Sons Inc; 2001 May 21. Available from http://doi.wiley.com/10.1002/0471221317
[39] Cardoso J.F., Souloumiac A. 'Blind beamforming for non-gaussian signals'. *IEE Proceedings F Radar and Signal Processing*. 1993, vol. 140(6), p. 362.
[40] Cardoso J.-F., Souloumiac A. 'Jacobi angles for simultaneous diagonalization'. *SIAM Journal on Matrix Analysis and Applications*. 1996, vol. 17(1), pp. 161–64.
[41] Li X.-L., Adalı T., Anderson M. 'Joint blind source separation by generalized joint diagonalization of cumulant matrices'. *Signal Processing*. 2011, vol. 91(10), pp. 2314–22.

Chapter 4

Coherent change detection in multi-polarization synthetic aperture radar images

V. Carotenuto[1], *A. De Maio*[1], *C. Clemente*[2], *and A. Farina*[3]

4.1 Introduction

In remote sensing, change detection (CD) is the ability to reveal temporal changes within a given scene. A common approach to tackle this problem relies on the use of a pair of co-registered synthetic aperture radar (SAR) images collected at two different time instants, usually referred to as the reference and test images [1, 2]. The two primary methods adopted in the open literature to process the image pair are incoherent and coherent CD. In the former, the presence of possible changes between the available images is established only by using intensity information (i.e. amplitude) of the reference and test data, whereas the latter jointly uses both amplitude and phase information. Moreover, polarimetric SAR (PolSAR) gives more scattering information than single-polarization channel SAR data, which can be exploited to detect the changes and increase the quality of the CD map.

During the last two decades, the development of CD strategies has attracted the interest of many scientists and engineers. For instance, in [3], a thorough comparison between incoherent and coherent CD strategies, including the maximum likelihood estimate (MLE) of the SAR coherence parameter, is performed based on high-resolution (0.3 m × 0.3 m) SAR images. In [4], several techniques for CD have been presented and compared based on their probability of error as well as on results obtained using repeat-pass ERS-1 SAR data. In [5–7], the multi-polarization signal model for the SAR CD problem is laid down, the detection problem is formulated as a binary hypothesis test, and a decision rule based on the generalized likelihood ratio test (GLRT) is developed. Moreover, the performance analysis [6] of the GLRT is conducted via receiver operating characteristics, namely, detection probability (P_d) versus false alarm probability (P_{fa}), quantifying the benefits of the multi-polarization information in SAR CD. A complementary approach to the GLRT is considered in [8] resorting to the use of perturbation filters. A decision scheme based on canonical

[1]University of Naples "Federico II", Naples, Italy
[2]University of Strathclyde, Glasgow, UK
[3]Selex-ES (retired), Rome, Italy

correlation analysis is applied to scalar Electromagnetics Institute SAR (EMISAR) data in [9, 10], whereas in [11], a mutual information-based framework is developed to address coherent similarities between multi-channel SAR images. In [12], exploiting a proper statistical model for SAR images, a CD method is introduced leveraging a stochastic distance measure between the distributions underlying the multi-look polarimetric data before and after the change. Precisely, the SAR data are modelled according to a product model [13] and closed-form expressions for the Kullback–Leibler (KL), Rényi of order β, Bhattacharyya, and Hellinger distances are provided to compute the sought stochastic measure. In [14], under the assumption of a scaled complex Wishart distribution for the polarimetric data, a determinant ratio test statistic is exploited to measure the similarity of two covariance matrices for unsupervised CD. In [15], the problem of CD for heterogeneous multi-variate SAR images is addressed by modelling the data using a spherically invariant random vector distribution where the heterogeneity of the images is accounted for through a deterministic texture parameter. In [16], a comparison between CD techniques based on the likelihood ratio and statistical methods that depend on information-theoretic measures is carried out. Precisely, modelling the PolSAR data via the scaled complex Wishart distribution, four parametric methodologies for detecting the change are proposed: two considering Shannon and Rényi entropies, one stemming from the KL distance, and the last based on the classic likelihood ratio statistics. In [17], the problem of change discrimination in coherent SAR images is investigated to discriminate between different types of changes. To this end, a feature-based approach is considered by defining a feature vector capable of distinguishing between different types of changes in a scene. Finally, in [18], a CD strategy relying on the generalized eigende composition of the polarimetric covariance matrices associated with the pixels of the SAR images to be compared is designed for agricultural applications.

In this chapter, with reference to multi-polarization SAR coherent CD, some strategies are discussed to identify the possible temporal changes between the reference and test images in the presence of homogeneous unstructured returns [19], homogeneous structured observations [20], and partially homogeneous data [21]. In particular, assuming for each pixel under test the availability of radar returns from two or more polarimetric channels, the detection problem is formulated by testing the equality of the polarimetric covariance matrices associated with the reference and test data. Precisely, for the homogeneous environment, the theory of invariance in hypothesis testing problems is exploited to focus on decision rules which exhibit some natural symmetries leading to important practical properties such as the constant false alarm rate (CFAR) behaviour. Furthermore, the use of invariance determines a data reduction because all invariant tests can be expressed in terms of a statistic, called maximal invariant, which organizes the original data into equivalence classes. Also, the parameter space is usually compressed after reduction by invariance, and the dependence on the original set of parameters becomes embodied into a maximal invariant in the parameter space (induced maximal invariant).

As to the homogeneous structured environment [20], a special structure for the polarimetric covariance matrices is accounted for in the detection problem. This formulation relies on the observation that in remote sensing applications employing polarimetric sensors, different objects in the illuminated area exhibit different reflection symmetry properties [22]. Such symmetries induce special structures on the polarimetric covariance matrix that can be incorporated into the hypothesis testing problem to improve the detection performance.

Lastly, the partially homogeneous environment [21] introduces the capability to account for a possible scale mismatch factor between the reference and test data. This allows the design of scale invariant decision rules, providing advantages in terms of robustness to intensity mismatches and/or miscalibrations and false alarm rejection with respect to the homogeneous environment. This is an important property as images over the same scene can exhibit different intensity scales due to different observation angles and propagation properties. As a matter of fact, there are situations in earth observation where a change detector without scale invariance is likely to fail. A classic example is the sea where different sea states can trigger a detector without scale invariance. Another example is the ground surface that shows very different moisture conditions (due, for instance, to rain). The mentioned effects can often lead to false alarms in a CD structure that does not ensure a robust behaviour with respect to such scale variations.

The rest of this chapter is organized as follows. In section 4.2, the system model for the multi-polSAR sensor is introduced. In sections 4.3, 4.4, and 4.5, the homogeneous, unstructured homogeneous, and partially homogeneous cases are addressed, respectively. In section 4.6, the performance of CD architectures is assessed on recorded live data. Finally, in section 4.7, concluding remarks are given.

4.2 System model

A multi-polarization SAR sensor measure for each pixel of the image under test $N \in \{2, 3\}$ complex returns, collected from different polarimetric channels: for instance, HH (horizontal-horizontal) and VV (vertical-vertical) for $N = 2$, and HH, VV, and HV (horizontal-vertical) with reference to $N = 3$. The N returns from the same pixel are stacked to form the vector $\boldsymbol{X}(l, m)$, where $l = 1, \ldots, L$ and $m = 1, \ldots, M$ (L and M represent the vertical and horizontal size of the image, respectively). Therefore, the sensor provides a 3D data stack $\boldsymbol{X}$ of size $M \times L \times N$ which is referred to in the following as a datacube (Figure 4.1) [19].

For SAR CD applications, let us suppose that two datacubes $\boldsymbol{X}$ (reference data) and $\boldsymbol{Y}$ (test data) of the same geographic area are available; they are collected from two different sensor passes and are accurately pixel aligned (co-registered). Let us focus on the problem of detecting the presence of possible changes in a rectangular neighbourhood $\mathscr{A}$, with size $K = W_1 \times W_2 \geq N$, of a given pixel. To this end, let us denote by $\boldsymbol{R}_X$ ($\boldsymbol{R}_Y$), the matrix whose columns are the vectors of the polarimetric returns from the pixels of $\boldsymbol{X}$ ($\boldsymbol{Y}$) which fall in the region $\mathscr{A}$ and $\boldsymbol{S}_X = \boldsymbol{R}_X \boldsymbol{R}_X^\dagger$ ($\boldsymbol{S}_Y = \boldsymbol{R}_Y \boldsymbol{R}_Y^\dagger$).

The matrices $\boldsymbol{R}_X$ and $\boldsymbol{R}_Y$ are modelled as statistically independent random matrices. Moreover, the columns of $\boldsymbol{R}_X$ ($\boldsymbol{R}_Y$) are assumed statistically independent and

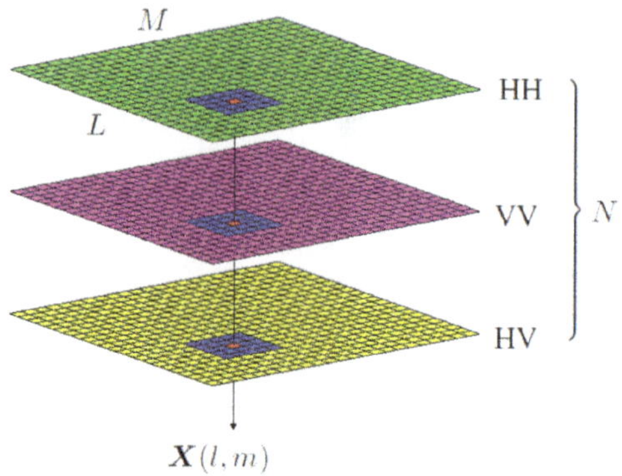

Figure 4.1 Construction of the datacube

identically distributed random vectors drawn from a complex circular zero-mean Gaussian distribution with positive definite covariance matrix $\mathbf{\Sigma}_X$ ($\mathbf{\Sigma}_Y$).

In the following, three different situations are considered for $\mathbf{\Sigma}_X$ and $\mathbf{\Sigma}_Y$:

1. homogeneous unstructured environment [19]
2. homogeneous structured environment [20]
3. partially homogeneous environment [21].

Precisely, in the absence of changes for the homogeneous unstructured scenario, the polarimetric pixels forming the reference and test data share the same covariance matrix. The homogeneous structured case, instead, accounts for a possible special structure of the polarimetric covariance matrices. Lastly, in the partially homogeneous environment, in the absence of changes, the polarimetric pixels share the same covariance structure but can exhibit different power levels.

4.3 Homogeneous unstructured environment

CD problem in the region $\mathscr{A}$ can be formulated in terms of the following binary hypothesis test [19]:

$$\begin{cases} H_0 : \mathbf{\Sigma}_X = \mathbf{\Sigma}_Y \\ H_1 : \mathbf{\Sigma}_X \neq \mathbf{\Sigma}_Y \end{cases} \tag{4.1}$$

where the null hypothesis H_0 of change absence is tested versus the alternative H_1. Under the Gaussian assumption, the joint probability density function (pdf) of $\boldsymbol{R}_X$ and $\boldsymbol{R}_Y$ can be written as:

$$f_{\boldsymbol{R}_X,\boldsymbol{R}_Y}(\boldsymbol{R}_X, \boldsymbol{R}_Y \mid H_1, \mathbf{\Sigma}_X, \mathbf{\Sigma}_Y) = \frac{1}{\pi^{2NK} \det(\mathbf{\Sigma}_X \mathbf{\Sigma}_Y)^K} \exp\left\{-\mathrm{tr}\left(\mathbf{\Sigma}_X^{-1} \boldsymbol{S}_X + \mathbf{\Sigma}_Y^{-1} \boldsymbol{S}_Y\right)\right\} \tag{4.2}$$

A sufficient statistic for (4.1), using the Fisher–Neyman factorization theorem [23], is represented by the two sample Grammian matrices $\boldsymbol{S}_X$ and $\boldsymbol{S}_Y$ which are statistically independent and follow a complex Wishart distribution, i.e. [24].

$$f_{\boldsymbol{S}_X}(\boldsymbol{S}_X \mid H_1, \boldsymbol{\Sigma}_X) = \frac{c_W}{\det(\boldsymbol{\Sigma}_X)^K} \exp\left\{-\mathrm{tr}(\boldsymbol{\Sigma}_X^{-1}\boldsymbol{S}_X)\right\} \det(\boldsymbol{S}_X)^{K-N},\ \boldsymbol{S}_X \succ \mathbf{0} \tag{4.3}$$

$$f_{\boldsymbol{S}_Y}(\boldsymbol{S}_Y \mid H_1, \boldsymbol{\Sigma}_Y) = \frac{c_W}{\det(\boldsymbol{\Sigma}_Y)^K} \exp\left\{-\mathrm{tr}(\boldsymbol{\Sigma}_Y^{-1}\boldsymbol{S}_Y)\right\} \det(\boldsymbol{S}_Y)^{K-N},\ \boldsymbol{S}_Y \succ \mathbf{0} \tag{4.4}$$

with c_W a normalization constant. From the sufficient statistic, the optimum Neyman–Pearson (NP) detector can be evaluated as the likelihood ratio test (LRT):

$$\frac{f_{\boldsymbol{S}_X,\boldsymbol{S}_Y}(\boldsymbol{S}_X, \boldsymbol{S}_Y \mid H_1, \boldsymbol{\Sigma}_X, \boldsymbol{\Sigma}_Y)}{f_{\boldsymbol{S}_X,\boldsymbol{S}_Y}(\boldsymbol{S}_X, \boldsymbol{S}_Y \mid H_0, \boldsymbol{\Sigma}_X)} = \frac{f_{\boldsymbol{S}_X}(\boldsymbol{S}_X \mid H_1, \boldsymbol{\Sigma}_X) f_{\boldsymbol{S}_Y}(\boldsymbol{S}_Y \mid H_1, \boldsymbol{\Sigma}_Y)}{f_{\boldsymbol{S}_X}(\boldsymbol{S}_X \mid H_0, \boldsymbol{\Sigma}_X) f_{\boldsymbol{S}_Y}(\boldsymbol{S}_Y \mid H_0, \boldsymbol{\Sigma}_X)} \underset{H_0}{\overset{H_1}{\gtrless}} T_{0,A}$$

where $T_{0,A}$ is the detection threshold that is set to ensure a given P_{fa} value. Substituting the pdfs of $\boldsymbol{S}_X$ and $\boldsymbol{S}_Y$ in the LRT, and including data-independent terms in the threshold, it can be recast as

$$\mathrm{tr}\left[(\boldsymbol{\Sigma}_X^{-1} - \boldsymbol{\Sigma}_Y^{-1})\boldsymbol{S}_Y\right] \underset{H_0}{\overset{H_1}{\gtrless}} T_0 \tag{4.5}$$

where T_0 is the modification of the original detection threshold. Test (4.5) is not uniformly most powerful (UMP) and is thus not practically implementable because it requires the knowledge (clairvoyant detector) of both $\boldsymbol{\Sigma}_X$ and $\boldsymbol{\Sigma}_Y$ which are usually unknown in realistic applications.

4.3.1 Data reduction and invariance issues

Both hypotheses under test are composite; this means that H_0 and H_1 are equivalent to a partition of the parameter space Θ into the two disjoint sets:

$$\Theta_0 = \{\boldsymbol{\Sigma}_X = \boldsymbol{\Sigma}_Y,\ (\boldsymbol{\Sigma}_X, \boldsymbol{\Sigma}_Y) \in \mathbb{H}^N \times \mathbb{H}^N,\ \boldsymbol{\Sigma}_X \succ \mathbf{0},\ \boldsymbol{\Sigma}_Y \succ \mathbf{0}\}$$
$$\Theta_1 = \{\boldsymbol{\Sigma}_X \neq \boldsymbol{\Sigma}_Y,\ (\boldsymbol{\Sigma}_X, \boldsymbol{\Sigma}_Y) \in \mathbb{H}^N \times \mathbb{H}^N,\ \boldsymbol{\Sigma}_X \succ \mathbf{0},\ \boldsymbol{\Sigma}_Y \succ \mathbf{0}\}$$

where $\mathbb{H}^N$ denotes the set of $N \times N$ Hermitian matrices.

This formulation emphasizes that the specific values of the nuisance parameters are not relevant: what has to be discerned is whether the covariances are equal or not. This means that the data can be clustered considering transformations that leave unaltered:

a. the two composite hypotheses, specifically the partition of the parameter space;
b. the families of distributions under the two hypotheses.

This objective can be accomplished through the *Principle of Invariance* [25]. Accordingly, one look for transformations that keep unaltered the structure of the hypothesis testing problem and, then, for decision rules invariant to them. The principle also acts as a data reduction technique that leads to a reduced observation space of significantly lower dimensionality than the original one.

It is easy to prove that the testing problem at hand is invariant under the group of transformations G acting on the sufficient statistic as [19]:

$$G = \{g : \boldsymbol{S}_X \rightarrow \boldsymbol{B}\boldsymbol{S}_X\boldsymbol{B}^\dagger, \quad \boldsymbol{S}_Y \rightarrow \boldsymbol{B}\boldsymbol{S}_Y\boldsymbol{B}^\dagger, \quad \boldsymbol{B} \in \mathcal{GL}(N)\} \tag{4.6}$$

where $\mathcal{GL}(N)$ denotes the set of $N \times N$ non-singular matrices together with the operation of ordinary matrix multiplication.

4.3.2 Maximal invariant design

The invariance property induces a partition of the data space into orbits (or equivalence classes) where, over each orbit, every point is related to each other through a transformation which is a member of G. Any statistic that identifies different orbits in a one-to-one manner significantly reduces the total amount of data necessary for solving the hypothesis testing problem and constitutes the compressed data set to be used in the design of any invariant detector. This kind of statistics is called maximal invariants since they are constant over each orbit (invariance) while they assume different values on different orbits (maximality).

Formally, a statistic $\boldsymbol{T}(\boldsymbol{S}_X, \boldsymbol{S}_Y)$ is said to be a maximal invariant with respect to the group of transformations G if and only if:

- Invariance: $\boldsymbol{T}(\boldsymbol{S}_X, \boldsymbol{S}_Y) = \boldsymbol{T}[g(\boldsymbol{S}_X, \boldsymbol{S}_Y)]$, $\forall g \in G$.
- Maximality: $\boldsymbol{T}(\boldsymbol{S}_{X_1}, \boldsymbol{S}_{Y_1}) = \boldsymbol{T}(\boldsymbol{S}_{X_2}, \boldsymbol{S}_{Y_2})$ implies that $\exists g \in G$ such that $(\boldsymbol{S}_{X_2}, \boldsymbol{S}_{Y_2}) = g(\boldsymbol{S}_{X_1}, \boldsymbol{S}_{Y_1})$.

Notice that there are many maximal invariant statistics, but they are equivalent in that they yield statistically equivalent detectors. Moreover, all invariant tests can be expressed as a function of the specific maximal invariant statistic [26, 27].

For the problem of interest, the following proposition holds true [19].

Proposition 1: A maximal invariant statistic with respect to the group of transformations (4.6) is the N-dimensional vector of the eigenvalues $\lambda_1, \ldots, \lambda_N$ of $\boldsymbol{S}_X\boldsymbol{S}_Y^{-1}$.

Using the principle of invariance, a significant data reduction is obtained: the maximal invariant statistic is a real N-dimensional vector whereas the original sufficient statistic is composed of two $N \times N$ Grammian matrices $\boldsymbol{S}_X$ and $\boldsymbol{S}_Y$. A useful physical/intuitive meaning of the eigenvalues of $\boldsymbol{S}_X\boldsymbol{S}_Y^{-1}$, which are involved in the computation of the maximal invariant, stems from the observation that they are solutions to suitable optimization problems involving the generalized Rayleigh quotient.

4.3.3 Induced maximal invariant design

By the principle of invariance, the parameter space is also partitioned into orbits and, usually, it involves a reduced set of parameters. The relevant parameters are embodied into any *induced maximal invariant*, namely, any function of the parameters that is constant over each orbit of the parameter space (invariance) but has different values over different orbits (maximality).

For the case at hand, an induced maximal invariant is composed of the eigenvalues $\bar{\delta} = [\delta_1, \ldots, \delta_N]^T$ of the matrix $\mathbf{\Sigma}_X \mathbf{\Sigma}_Y^{-1}$. The induced maximal invariant is an N-dimensional vector while the original parameter space was composed of two covariance matrices $\mathbf{\Sigma}_X$ and $\mathbf{\Sigma}_Y$.

It can be observed that in the reduced parameter space, the partition corresponding to the two composite hypotheses of the test (4.1) is $\Xi_0 = \{\mathbf{1}_N\}$, for $\mathbf{\Sigma}_X = \mathbf{\Sigma}_Y$, and $\Xi_1 = \overline{\{\mathbf{1}_N\}}$, for $\mathbf{\Sigma}_X \neq \mathbf{\Sigma}_Y$, where $\{\mathbf{1}_N\}$ and $\overline{\{\mathbf{1}_N\}}$ are the sets of the N-dimensional column vectors with all entries equal to 1 and with at least one entry different from 1, respectively. The structure of Ξ_0, which now corresponds to a simple H_0 hypothesis, clearly shows that all invariant receivers that process a maximal invariant statistic through a transformation independent of $\delta_1, \ldots, \delta_N$ achieve the CFAR property.

4.3.4 Design of the optimum and sub-optimum invariant detectors

In this section, the design of the most powerful invariant (MPI) receiver is addressed. Moreover, it is shown that no uniformly MPI (UMPI) test exists, namely, that the optimum invariant detector is not implementable. Hence, the focus is moved on sub-optimum invariant receivers based on different functions of the maximal invariant such as arithmetic mean, harmonic mean, and maximum and minimum entries. Remarkably, the GLRT in [6] and [28] belongs to the considered class of receivers.

4.3.4.1 Design of the MPI detector

Based on the NP criterion, the test can be obtained as the LRT based on any maximal invariant, which, as already pointed out, provides sufficient information for constructing any invariant decision rule. In particular, for the considered problem, the LRT can be written as

$$\frac{f_{\mathbf{\Lambda}}(\mathbf{\Lambda}|H_1, \mathbf{\Delta})}{f_{\mathbf{\Lambda}}(\mathbf{\Lambda}|H_0)} \underset{H_0}{\overset{H_1}{\gtrless}} T, \tag{4.7}$$

where $\mathbf{\Lambda} = \mathbf{diag}\left([\lambda_1, \lambda_2, \ldots, \lambda_N]^T\right)$ with $\lambda_1 \geq \lambda_2 \geq \cdots \geq \lambda_N > 0$, $\mathbf{\Delta} = \mathbf{diag}\,(\bar{\delta})$, whereas $f_{\mathbf{\Lambda}}\,(\mathbf{\Lambda} \mid H_0)$ and $f_{\mathbf{\Lambda}}(\mathbf{\Lambda} \mid H_1, \mathbf{\Delta})$ are the pdfs of the devised maximal invariant under the H_0 and the H_1 hypotheses, respectively. Starting from (4.7), it is possible to show that assuming $N = 2$, the MPI test is [19]:

$$\begin{aligned} &\left[\frac{(1+\lambda_1)}{(\delta_1+\lambda_1)}\frac{(1+\lambda_2)}{(\delta_2+\lambda_2)}\right]^{2K}\left[\frac{(\delta_1+\lambda_1)\,(\delta_2+\lambda_2)}{\lambda_1-\lambda_2}\right]+ \\ -&\left[\frac{(1+\lambda_1)}{(\delta_2+\lambda_1)}\frac{(1+\lambda_2)}{(\delta_1+\lambda_2)}\right]^{2K}\left[\frac{(\delta_2+\lambda_1)\,(\delta_1+\lambda_2)}{\lambda_1-\lambda_2}\right]\mathrm{sign}\,(\delta_1-\delta_2) \underset{H_0}{\overset{H_1}{\gtrless}} T_A, \quad \delta_1 \neq \delta_2 \end{aligned} \tag{4.8}$$

$$\frac{(1+\lambda_1)\,(1+\lambda_2)}{(\delta+\lambda_1)\,(\delta+\lambda_2)} \underset{H_0}{\overset{H_1}{\gtrless}} T_B, \quad \delta_1 = \delta_2 = \delta \tag{4.9}$$

with T_A and T_B suitable modifications of the original detection threshold. The case $N = 3$ is available in [19].

The derived expression of the MPI test clearly highlights that it cannot, in general, be implemented unless the induced maximal invariant is known (clairvoyant detector): unfortunately, this hypothesis is not realistic, especially for the SAR CD problem. In other words, the MPI detector, which provides the best performance for a given $\boldsymbol{\Delta}^\star$ is not, in general, optimum for different values of $\boldsymbol{\Delta}$, namely, no UMPI detector exists. But, despite this, it is still important to consider the MPI detector because its performance upper bounds those of any other invariant receiver, operating under the same signal and disturbance models, and hence, it can be used to assess the loss of any implementable invariant test that is sub-optimal.

4.3.4.2 Maximal invariant-based detectors

As the MPI detector is not implementable, it is necessary to investigate invariant decision rules based upon different strategies. However, there is no criterion for an *a priori* choice of a receiver instead of another. An intuitive rule to select invariant tests for the considered problem could be based on the following asymptotic observation: for large values of K, the eigenvalues of $\boldsymbol{S}_X^{-1}\boldsymbol{S}_Y$ tend to δ_i, $i = 1, \ldots, N$. Hence, decision rules

$$h(\lambda_1, \ldots, \lambda_N) \underset{H_0}{\overset{H_1}{\gtrless}} T \tag{4.10}$$

- are very effective to discriminate deviations $\delta_i \gg 1$, when $h(\cdot, \ldots, \cdot)$ is an increasing function of the arguments. However, they perform poor when δ_i are smaller than 1;
- are very effective to discriminate deviations $\delta_i \ll 1$, when $h(\cdot, \ldots, \cdot)$ is a decreasing function of the arguments. Nevertheless they perform poor when δ_i is greater than 1;
- in principle could achieve good detection levels for both $\delta_i \gg 1$ and $\delta_i \ll 1$, when $h(\cdot, \ldots, \cdot)$ complies with $h\left(\frac{1}{\lambda_1}, \ldots, \frac{1}{\lambda_N}\right) = h\left(\lambda_1, \ldots, \lambda_N\right)$.

Among all possible sub-optimum invariant detectors that can be synthesized as a function of the maximal invariant statistic, in the following, the focus is on decision rules based on the GLRT as well as suitable functions of the maximal invariant statistic such as arithmetic/harmonic mean and maximum/minimum entries.

GLRT-based detector: This method is statistically equivalent to replacing the parameters that are not known in the likelihood ratio with their MLEs, under each hypothesis [29]. Under very mild technical assumptions, the GLRT is invariant and it can be shown to be statistically equivalent to [19]

$$\prod_{i=1}^{N} \frac{(1+\lambda_i)^2}{\lambda_i} \underset{H_0}{\overset{H_1}{\gtrless}} T_1 \tag{4.11}$$

where T_1 is the detection threshold. Interestingly, the GLRT complies with condition 3, namely, it is expected to be capable of achieving good detection levels both when $\delta_i \gg 1$ and $\delta_i \ll 1$.

***Arithmetic/Harmonic mean-based detectors*:** These decision rules are, respectively, given by

$$\sum_{i=1}^{N} \lambda_i \underset{H_0}{\overset{H_1}{\gtrless}} T_2 \tag{4.12}$$

$$\sum_{i=1}^{N} \frac{1}{\lambda_i} \underset{H_0}{\overset{H_1}{\gtrless}} T_3 \tag{4.13}$$

where T_2 and T_3 are the detection thresholds. The former complies with condition 1, whereas the latter complies with condition 2 after (4.10). As a consequence, (4.12) is suitable for detecting deviations $\delta_i \gg 1$ while (4.13) for $\delta_i \ll 1$.

From (4.12) and (4.13), it is also possible to construct another decision rule satisfying condition 3 merely summing the decision statistics, i.e.,

$$\sum_{i=1}^{N} \left(\frac{1}{\lambda_i} + \lambda_i \right) \underset{H_0}{\overset{H_1}{\gtrless}} T_4 \tag{4.14}$$

with T_4 the detection threshold.

***Maximum/Minimum eigenvalue-based detectors*:** These tests are, respectively, based on the following comparisons

$$\left(\lambda_1 + \frac{1}{\lambda_N} \right) \underset{H_0}{\overset{H_1}{\gtrless}} T_5 \tag{4.15}$$

$$\max \left(\lambda_1, \frac{1}{\lambda_N} \right) \underset{H_0}{\overset{H_1}{\gtrless}} T_6 \tag{4.16}$$

with T_5 and T_6 the decision thresholds. The former term, λ_1, dominates for large deviations $\delta_i \gg 1$, whereas the latter term $\frac{1}{\lambda_N}$, if $\delta_i \ll 1$. Hence, (4.15) and (4.16) are supposed to perform well for both $\delta_i \gg 1$ and $\delta_i \ll 1$.

4.4 Homogeneous structured environment

Usually, in remote sensing applications employing polarimetric sensors, different objects in the illuminated area exhibit different reflection symmetry properties [22]. Such symmetries induce special structures on the polarimetric covariance matrix that can be incorporated in (4.1) to improve the detection performance. In this respect, the goal of this section is to force a specific structure on the polarimetric covariance matrix by considering:

$$\boldsymbol{\Sigma}_X \in \Xi \quad (\boldsymbol{\Sigma}_Y \in \Xi)$$

where

$$\Xi = \left\{ \boldsymbol{\Sigma} \in \mathbb{H}^N : \boldsymbol{\Sigma} = \begin{pmatrix} \boldsymbol{\Sigma}_1 & \mathbf{0} \\ \mathbf{0} & \sigma^2 \end{pmatrix}, \boldsymbol{\Sigma} \succ \mathbf{0} \right\} \tag{4.17}$$

This block-diagonal structure is used to devise a decision rule based on the GLRT criterion by directly exploiting the received multi-variate complex Gaussian observations.

Exploiting the Gaussian assumption together with the covariance structure, the joint pdf of $\boldsymbol{R}_X$ and $\boldsymbol{R}_Y$ can be written as

$$\begin{aligned} f_{\boldsymbol{R}_X,\boldsymbol{R}_Y}(\boldsymbol{R}_X, \boldsymbol{R}_Y \mid H_1, \boldsymbol{\Sigma}_{X,1}\, \boldsymbol{\Sigma}_{Y,1}, \sigma^2_{X,1}, \sigma^2_{Y,1}) = \\ \frac{1}{\pi^{6K} \det^K(\boldsymbol{\Sigma}_{X,1}) \det^K(\boldsymbol{\Sigma}_{Y,1}) \sigma^{2K}_{X,1} \sigma^{2K}_{Y,1}} \times \\ \exp\left\{ -\mathrm{tr}\left(\boldsymbol{\Sigma}^{-1}_{X,1} \boldsymbol{S}_{X,1} + \boldsymbol{\Sigma}^{-1}_{Y,1} \boldsymbol{S}_{Y,1} \right) - \frac{\widehat{\sigma^2}_{X,1}}{\sigma^2_{X,1}} - \frac{\widehat{\sigma^2}_{Y,1}}{\sigma^2_{Y,1}} \right\} \end{aligned} \tag{4.18}$$

and

$$\begin{aligned} f_{\boldsymbol{R}_X,\boldsymbol{R}_Y}(\boldsymbol{R}_X, \boldsymbol{R}_Y \mid H_0, \boldsymbol{\Sigma}_{X,1}, \sigma^2_{X,1}) = \\ \frac{1}{\pi^{6K} \det^{2K}(\boldsymbol{\Sigma}_{X,1}) \sigma^{4K}_{X,1}} \exp\left\{ -\mathrm{tr}\left[\boldsymbol{\Sigma}^{-1}_{X,1} (\boldsymbol{S}_{X,1} + \boldsymbol{S}_{Y,1}) \right] - \frac{\widehat{\sigma^2}_{X,1} + \widehat{\sigma^2}_{Y,1}}{\sigma^2_{X,1}} \right\} \end{aligned} \tag{4.19}$$

where $\boldsymbol{S}_X$ and $\boldsymbol{S}_Y$ are partitioned as

$$\boldsymbol{S}_X = \begin{bmatrix} \boldsymbol{S}_{X,1} & \boldsymbol{S}_{X,2} \\ \boldsymbol{S}^{\dagger}_{X,2} & \hat{\sigma}^2_{X,1} \end{bmatrix}, \qquad \boldsymbol{S}_Y = \begin{bmatrix} \boldsymbol{S}_{Y,1} & \boldsymbol{S}_{Y,2} \\ \boldsymbol{S}^{\dagger}_{Y,2} & \hat{\sigma}^2_{Y,1} \end{bmatrix} \tag{4.20}$$

and $\widehat{\sigma^2}_{X,1}$ and $\widehat{\sigma^2}_{Y,1}$ are scalars.

Exploiting the considered special covariance structure, namely, by using the pdfs (4.18) and (4.19), it is possible to show that the structured GLRT can be expressed as [20]

$$\frac{\det^2(\boldsymbol{S}_{X,1} + \boldsymbol{S}_{Y,1})}{\det(\boldsymbol{S}_{X,1}) \det(\boldsymbol{S}_{Y,1})} \frac{\left(\widehat{\sigma^2}_{X,1} + \widehat{\sigma^2}_{Y,1} \right)^2}{\widehat{\sigma^2}_{X,1} \widehat{\sigma^2}_{Y,1}} \underset{H_0}{\overset{H_1}{\gtrless}} T_S \tag{4.21}$$

with T_S the detection threshold. It can be proved that (4.21) ensures the CFAR property with respect to both $\boldsymbol{\Sigma}_{X,1}$ and $\sigma^2_{X,1}$. Otherwise stated, the detection threshold ensuring a given false alarm rate can be set independent of the two aforementioned parameters.

4.5 Partially homogeneous environment

Images over the same scene can exhibit different intensity scales due to different observation angles and propagation properties. If not properly accounted for in the system model, such variations could lead to a degradation in terms of CD capabilities between the reference and test images. In the presence of intensity miscalibration, the CD problem in the region $\mathscr{A}$ can be formulated in terms of the following binary hypothesis test [21]:

$$\begin{cases} H_0 : \mathbf{\Sigma}_X = \gamma\, \mathbf{\Sigma}_Y \\ H_1 : \mathbf{\Sigma}_X \neq \gamma\, \mathbf{\Sigma}_Y \end{cases} \tag{4.22}$$

where the null hypothesis H_0 of change absence is tested versus the alternative H_1. The parameter $\gamma > 0$ models possibly received power variations between two different acquisitions from the same scene.

Using the Gaussian assumption previously described, the optimum NP detector is the LRT, which, after standard algebra and statistical equivalences, can be written as

$$\mathrm{tr}\left[\left(\frac{\mathbf{\Sigma}_Y^{-1}}{\gamma} - \mathbf{\Sigma}_X^{-1}\right) \boldsymbol{S}_X\right] \underset{H_0}{\overset{H_1}{\gtrless}} T_0 \tag{4.23}$$

where T_0 is the detection threshold. Evidently, test (4.23) is not UMP, and consequently, it is not practically implementable because it requires the knowledge of $\mathbf{\Sigma}_X$, γ, and $\mathbf{\Sigma}_Y$ which, in realistic applications, are usually unknown.

The approach analyzed in section 4.5.1 is able to produce scale invariant decision rules, providing advantages in terms of robustness to intensity mismatches and/or miscalibrations as well as false alarm rejection with respect to the homogeneous case.

4.5.1 Maximal and induced maximal invariant design

Both hypotheses under test are composite or, otherwise stated, H_0 and H_1 are equivalent to a partition of the parameter space Θ into the two disjoint sets

$$\Theta_0 = \{\mathbf{\Sigma}_X = \gamma \mathbf{\Sigma}_Y,\ (\mathbf{\Sigma}_X, \mathbf{\Sigma}_Y, \gamma) \in \mathbb{H}^N \times \mathbb{H}^N \times \mathbb{R},\ \mathbf{\Sigma}_X \succ \mathbf{0},\ \mathbf{\Sigma}_Y \succ \mathbf{0},\ \gamma > 0\}$$
$$\Theta_1 = \{\mathbf{\Sigma}_X \neq \gamma \mathbf{\Sigma}_Y,\ (\mathbf{\Sigma}_X, \mathbf{\Sigma}_Y, \gamma) \in \mathbb{H}^N \times \mathbb{H}^N \times \mathbb{R},\ \mathbf{\Sigma}_X \succ \mathbf{0},\ \mathbf{\Sigma}_Y \succ \mathbf{0},\ \gamma > 0\}$$

where $\mathbb{R}$ denotes the set of real numbers.

It is not difficult to prove that our testing problem is invariant under the group of transformations G acting on the sufficient statistic as [21]

$$G = \left\{g : \boldsymbol{S}_X \to \boldsymbol{B}\boldsymbol{S}_X\boldsymbol{B}^\dagger,\ \boldsymbol{S}_Y \to a\boldsymbol{B}\boldsymbol{S}_Y\boldsymbol{B}^\dagger,\ \boldsymbol{B} \in \mathscr{G}\mathscr{L}(N),\ a \in \mathbb{R},\ a > 0\right\} \tag{4.24}$$

For the problem (4.22), the following proposition holds true:

Proposition 2: A maximal invariant statistic with respect to the group of transformations (4.24) is the $(N-1)$-dimensional vector [21]

$$\left(\frac{\lambda_2}{\lambda_1}, \frac{\lambda_3}{\lambda_1}, \ldots, \frac{\lambda_N}{\lambda_1}\right) \tag{4.25}$$

where $\lambda_1 \geq \lambda_2 \geq \ldots \geq \lambda_N$ are the eigenvalues of $\boldsymbol{S}_X \boldsymbol{S}_Y^{-1}$ [21].

As desired, the principle of invariance produces a significant data reduction: the maximal invariant statistic is a real $(N-1)$-dimensional vector whereas the original sufficient statistic is composed of the two $N \times N$ Grammian matrices $\boldsymbol{S}_X$ and $\boldsymbol{S}_Y$.

On the other hand, for the considered case, an induced maximal invariant is given by $\left(\frac{\delta_2}{\delta_1}, \ldots, \frac{\delta_N}{\delta_1}\right)$, where $\bar{\boldsymbol{\delta}} = [\delta_1, \ldots, \delta_N]^T$, $\delta_1 \geq \delta_2 \geq \ldots \geq \delta_N$, are the eigenvalues of the matrix $\boldsymbol{\Sigma}_X \boldsymbol{\Sigma}_Y^{-1}$ [21]. The physical interpretation of the induced maximal invariant components follows from the observation that they are related to the spread among the backscattering ratios associated with two different scattering mechanisms.

The previous observation highlights that the principle of invariance also yields a significant reduction of the number of the parameters: in fact, the induced maximal invariant is an $(N-1)$-dimensional vector while the original parameter space was composed of the two covariance matrices $\boldsymbol{\Sigma}_X$, $\boldsymbol{\Sigma}_Y$, and γ.

Let us explicitly observe that in the reduced parameter space, the partition corresponding to the two composite hypotheses of the test (4.22) is $\Xi_0 = \{\mathbf{1}_{N-1}\}$, relative to $\boldsymbol{\Sigma}_X = \gamma\, \boldsymbol{\Sigma}_Y$, and $\Xi_1 = \overline{\{\mathbf{1}_{N-1}\}}$, relative to $\boldsymbol{\Sigma}_X \neq \gamma\, \boldsymbol{\Sigma}_Y$, where $\overline{\{\mathbf{1}_{N-1}\}}$ is the set of the $(N-1)$-dimensional column vectors with positive elements and at least one entry different from 1. The structure of Ξ_0, which now corresponds to a simple H_0 hypothesis, clearly shows that all invariant receivers that process a maximal invariant statistic through a transformation independent of $(\omega_1, \ldots, \omega_{N-1}) = \left(\frac{\delta_2}{\delta_1}, \ldots, \frac{\delta_N}{\delta_1}\right)$ achieve the CFAR property with respect to both γ and $\boldsymbol{\Sigma}_X$.

4.5.2 GLRT derivation

This section is dedicated to the derivations of the GLRT detector for the considered problem. The following decision rule is considered

$$\frac{\max\limits_{\boldsymbol{\Sigma}_X}\max\limits_{\boldsymbol{\Sigma}_Y} \frac{1}{(\pi)^{2NK}\det^K(\boldsymbol{\Sigma}_X)\det^K(\boldsymbol{\Sigma}_Y)} \exp\left[-\mathrm{tr}\left(\boldsymbol{\Sigma}_X^{-1}\boldsymbol{S}_X + \boldsymbol{\Sigma}_Y^{-1}\boldsymbol{S}_Y\right)\right]}{\max\limits_{\gamma>0}\max\limits_{\boldsymbol{\Sigma}_Y} \frac{1}{(\pi)^{2NK}\gamma^{NK}\det^{2K}(\boldsymbol{\Sigma}_Y)} \exp\left[-\mathrm{tr}\left(\boldsymbol{\Sigma}_Y^{-1}\left(\frac{\boldsymbol{S}_X}{\gamma} + \boldsymbol{S}_Y\right)\right)\right]} \underset{H_0}{\overset{H_1}{\gtrless}} T \tag{4.26}$$

which, after the optimizations over $\boldsymbol{\Sigma}_X$ and $\boldsymbol{\Sigma}_Y$ at the numerator and over $\boldsymbol{\Sigma}_Y$ at the denominator, can be recast (after some algebra and statistical equivalences) as,

$$\frac{\min\limits_{\gamma>0}\left[\gamma^N \det^2\left(\frac{\boldsymbol{S}_X}{\gamma} + \boldsymbol{S}_Y\right)\right]}{\det(\boldsymbol{S}_X)\det(\boldsymbol{S}_Y)} \underset{H_0}{\overset{H_1}{\gtrless}} T_1 \tag{4.27}$$

or equivalently as

$$\frac{\min\limits_{\gamma>0}\left[\gamma^N \det^2\left(\frac{\boldsymbol{S}_Y^{-\frac{1}{2}}\boldsymbol{S}_X\boldsymbol{S}_Y^{-\frac{1}{2}}}{\gamma} + \boldsymbol{I}\right)\right]}{\det(\boldsymbol{S}_Y^{-\frac{1}{2}}\boldsymbol{S}_X\boldsymbol{S}_Y^{-\frac{1}{2}})} \underset{H_0}{\overset{H_1}{\gtrless}} T_1 \tag{4.28}$$

where T_1 is a suitable modification of the original threshold, T, in (4.26). In order to proceed further, it is necessary to distinguish between the cases of two- ($N = 2$) and three- ($N = 3$) polarimetric channels.

For $N = 2$, the decision rule turns out equivalent to [21]:

$$\frac{\lambda_1}{\lambda_2} \underset{H_0}{\overset{H_1}{\gtrless}} T_2 \tag{4.29}$$

with T_2 the detection threshold. Two important comments have to be made. First, test (4.29) is equivalent to comparing the condition number of the matrix $\boldsymbol{S}_Y^{-\frac{1}{2}} \boldsymbol{S}_X \boldsymbol{S}_Y^{-\frac{1}{2}}$ with a detection threshold to establish the presence of changes in the considered scene. Second, the GLRT statistic is a maximal invariant.

For $N = 3$, the GLRT can be computed as [21]:

$$\frac{\left[\gamma_{opt,3}^3 \left(\frac{\lambda_1}{\gamma_{opt,3}} + 1\right)^2 \left(\frac{\lambda_2}{\gamma_{opt,3}} + 1\right)^2 \left(\frac{\lambda_3}{\gamma_{opt,3}} + 1\right)^2\right]}{\lambda_1 \lambda_2 \lambda_3} \underset{H_0}{\overset{H_1}{\gtrless}} T \tag{4.30}$$

where $\gamma_{opt,3}$ is the real and positive solution of the following third-order equation

$$\gamma^3 + \frac{1}{3}(\lambda_1 + \lambda_2 + \lambda_3)\gamma^2 - \frac{1}{3}(\lambda_1\lambda_2 + \lambda_1\lambda_3 + \lambda_2\lambda_3)\gamma - \lambda_1\lambda_2\lambda_3 = 0 \tag{4.31}$$

In fact, Descartes' rule of signs implies that (4.31) shares one positive real root which necessarily coincides with the desired optimal point for γ. Additionally, Tartaglia's formula can be exploited to obtain the analytic expression of the optimizer $\gamma_{opt,3}$. Precisely, denoting by

- $a = \frac{1}{3}(\lambda_1 + \lambda_2 + \lambda_3)$,
- $b = -\frac{1}{3}(\lambda_1\lambda_2 + \lambda_1\lambda_3 + \lambda_2\lambda_3)$,
- $c = -\lambda_1\lambda_2\lambda_3$,
- $p = -a^2/3 + b$,
- $q = 2a^3/27 - ab/3 + c$,

the optimal point is the real positive number chosen among the three roots of (4.31) given by

$$-\frac{a}{3} + \sqrt[3]{-\frac{q}{2} + \sqrt{\frac{q^2}{4} + \frac{p^3}{27}}} + \sqrt[3]{-\frac{q}{2} - \sqrt{\frac{q^2}{4} + \frac{p^3}{27}}}$$

where $\sqrt[3]{(\cdot)}$ and $\sqrt{(\cdot)}$ are complex roots*.

*Given a complex number $y = \rho \exp(j\theta)$ the n-th roots are $u_k = \rho^{\frac{1}{n}} \left[\cos\left(\frac{\theta}{n} + \frac{2\pi k}{n}\right) + j \sin\left(\frac{\theta}{n} + \frac{2\pi k}{n}\right)\right]$, $k = 0, 1, \ldots, n-1$.

An equivalent expression in terms of the maximal invariant can be also obtained. From (4.31), it can be shown that

$$\gamma_{opt,3} = \lambda_1 \bar{h}\left(\frac{\lambda_2}{\lambda_1}, \frac{\lambda_3}{\lambda_1}\right) \tag{4.32}$$

with $\bar{h}(\cdot,\cdot)$ a suitable bi-dimensional real function of two real variables. Hence, the GLRT statistic as a function of the maximal invariant can be obtained by substituting (4.32) in (4.30).

4.5.3 Sub-optimum invariant detectors

In this section, additional sub-optimum invariant detectors with reference to the fully polarimetric processing (i.e. $N = 3$) are designed. They exploit suitable combinations of the maximal invariant components and, based on extensive numerical analysis, are seen to achieve satisfactory detection performances. Two chosen combination rules are the standard arithmetic and geometric means which, respectively, lead to the tests

$$\frac{\lambda_1}{\lambda_2} + \frac{\lambda_1}{\lambda_3} \underset{H_0}{\overset{H_1}{\gtrless}} T_a \tag{4.33}$$

$$\frac{\lambda_1}{\lambda_2}\frac{\lambda_1}{\lambda_3} \underset{H_0}{\overset{H_1}{\gtrless}} T_b \tag{4.34}$$

where T_a and T_b are the detection thresholds. Moreover, based on the observations that unbalances among the eigenvalues of $\boldsymbol{\Sigma}_Y^{-\frac{1}{2}} \boldsymbol{\Sigma}_X \boldsymbol{\Sigma}_Y^{-\frac{1}{2}}$ are representative of changes between the reference and test images, and an index to quantify eigenvalues' unbalances is the deviation of the arithmetic mean from the geometric mean, the following decision rule is also considered

$$\frac{\operatorname{tr}\left(\boldsymbol{S}_Y^{-\frac{1}{2}}\boldsymbol{S}_X\boldsymbol{S}_Y^{-\frac{1}{2}}\right)}{\det^{\frac{1}{3}}\left(\boldsymbol{S}_Y^{-\frac{1}{2}}\boldsymbol{S}_X\boldsymbol{S}_Y^{-\frac{1}{2}}\right)} = \left(\frac{\lambda_2}{\lambda_1}\frac{\lambda_3}{\lambda_1}\right)^{-\frac{1}{3}}\left(1+\frac{\lambda_2}{\lambda_1}+\frac{\lambda_3}{\lambda_1}\right) \underset{H_0}{\overset{H_1}{\gtrless}} T_c \tag{4.35}$$

where T_c is the detection threshold.

4.6 Performance assessment

In this section, the performance analysis of the decision rules discussed for the unstructured homogeneous, structured homogeneous, and partially homogeneous cases is presented. The study is conducted using the X-band coherent CD challenge data set acquired by the Air Force Research Laboratory [30]. The airborne SAR sensor used to acquire the data employed a coherent receiver with 640 MHz bandwidth and dual-polarized mode. The data are in the form of focused complex images with a range and cross-range resolution of 0.3 m. The overall data set is composed of ten complex images, acquired the same day, for each of the three available polarizations (HH, VV, and HV). The size of the images is 4501 × 4501 pixels and they have been coherently aligned to a single reference (per polarization) with the help of digital elevation map information [30].

The effectiveness of the considered approaches is assessed by selecting two acquisitions from the entire data set being the reference and test images. Unfortunately, the actual changes between the different acquisitions, namely, the ground truths of the data, are not available. As a consequence, the two passes for performance validation have been selected so as to create a ground truth that is sufficiently accurate. To this end, the acquisition named "FP0124" is used as a reference pass, while the acquisition "FP0121" is used as a test pass. From the two passes, the area with the higher activity (changes) between the passes has been selected. It is represented by a sub-image of 1000 × 1000 pixels (i.e. $L = M = 1000$) and is composed of numerous parked vehicles in various parking lots. Figure 4.2 shows the reference and test sub-images for the HH mode. For this particular scenario, the changes between the reference and test images (denoted by $\boldsymbol{X}$ and $\boldsymbol{Y}$, respectively) occurred during the time interval between the two acquisitions can be classified in two cases

1. A vehicle is present in $\boldsymbol{X}$ but is not present in $\boldsymbol{Y}$, i.e. the vehicle has left from its parking space (pixels relative to this kind of event will be referred to in the following as departures).
2. A vehicle is not present in $\boldsymbol{X}$ but is present in $\boldsymbol{Y}$, i.e. the vehicle has arrived in an empty parking space (pixels relative to this kind of event will be referred to in the following as arrivals).

A total of 34 changes between $\boldsymbol{X}$ and $\boldsymbol{Y}$ can be visually identified (by flickering the two images). The obtained ground truth is shown in Figure 4.3, where the black

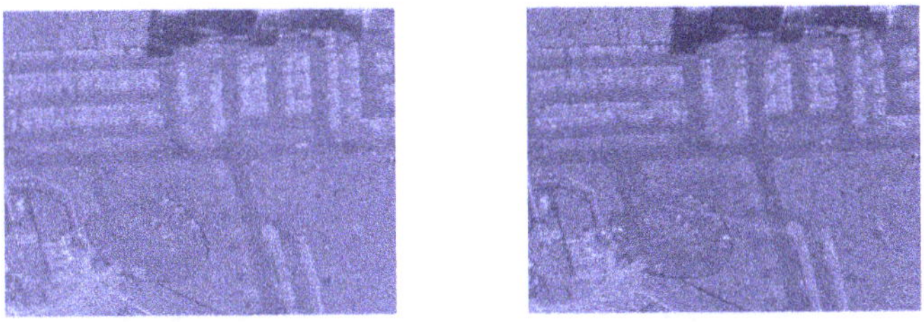

(a) Reference image. (b) Test image.

Figure 4.2 Reference and test images gathered in HH mode

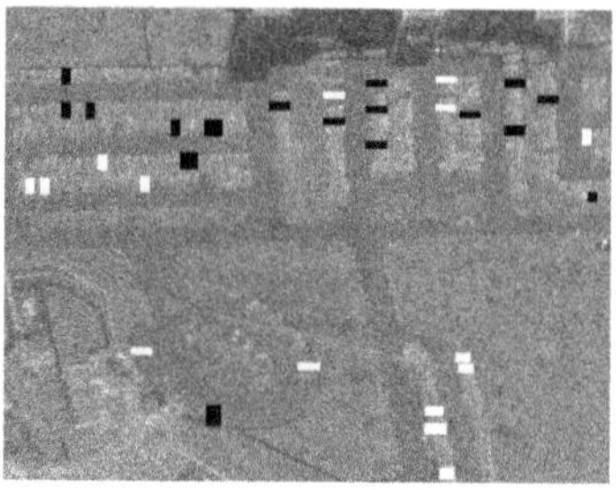

Figure 4.3 Ground truth superimposed to the reference image

and the white regions represent the departures and the arrivals, respectively. In particular, denoting by $\mathscr{K}$ the set of pixels that correspond to changes, the ground truth can be represented as a matrix $\boldsymbol{G}_T$ whose entries are given by

$$\boldsymbol{G}_T(l,m) = \begin{cases} 1 & \text{if } (l,m) \in \mathscr{K} \\ 0 & \text{otherwise} \end{cases}, \quad l = 1,\ldots,L, \quad m = 1,\ldots,M \tag{4.36}$$

Although the acquisitions were performed during the same day, the returns from a scatterer contribute differently to neighbour pixels. For example, slightly different aspect angles can produce a different amount of energy spill-over. These relative differences in the imaged data can lead to false alarms in the CD results. Hence, to prevent such issue, a guard area around each arrival-departure is considered. This permits the definition of an extended ground truth which is used in the following to assess the performance of the considered detection algorithms. The ground truth and the extended ground truth masks are shown in Figure 4.4.

In the subsequent analyses, the performance is assessed using both the number of detected changes and the CD maps. Precisely, for a specific receiver, the corresponding map of changes, $\boldsymbol{C}$, is a $L \times M$ matrix whose (l, m)-th entry is the decision statistic considering the $N \times N$ sample Grammian matrix $\boldsymbol{S}_{X_{l,m}}$ $(\boldsymbol{S}_{Y_{l,m}})$ evaluated

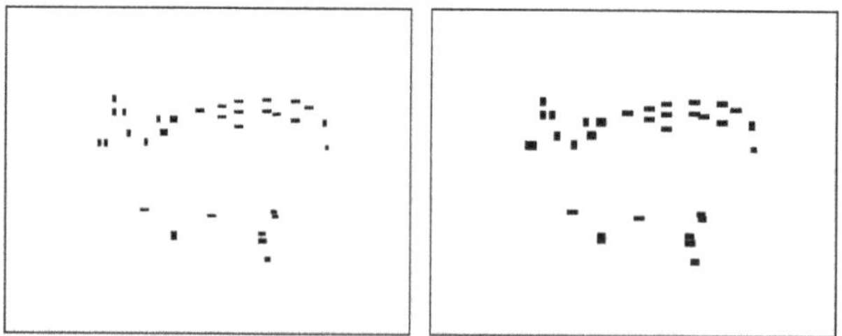

(a) Ground truth. (b) Ground truth with guard cells.

Figure 4.4 Ground truth before and with the addition of guard cells using a square neighbourhood with size $W \times W$ of the pixel (l, m) of $X(Y)$. The detection map corresponding to $\boldsymbol{C}$ is then defined as

$$\boldsymbol{D}(l, m) = \begin{cases} 1 & \text{if } \boldsymbol{C}(l, m) > T \\ & \\ 0 & \text{otherwise} \end{cases}, \quad l = 1, \ldots, L, \quad m = 1, \ldots, M \tag{4.37}$$

where T is the detection threshold associated with the considered detector. In particular, for each of the considered decision rule, the corresponding detection threshold is set to ensure a given P_{fa} in the complement of the extended ground truth area, namely, in the region where no changes occur (there are no true positives). This means that for each detector, after computing the decision statistics for each pixel belonging to the complement of the extended ground truth, the threshold has been selected such that

$$P_{fa} \times \text{total number of available statistics (trials)}$$

are greater than the threshold. This ensures that all the comparisons refer to the same P_{fa} level, namely, the number of threshold crossings in the complement of the extended ground truth is exactly the same for all the analyzed decision rules. Finally, for each detector, starting from $\boldsymbol{D}$, and denoting by $\mathscr{K}_{\bar{G}}$ the set of the pixels that correspond to changes in the extended ground truth area, performance is also given in terms of the number of detections belonging to $\mathscr{K}_{\bar{G}}$ which corresponds to the cardinality of the set

$$\mathscr{D} = \{\boldsymbol{D}(l, m) : \boldsymbol{D}(l, m) \in \mathscr{K}_{\bar{G}}\}$$

where $\bar{G}$ represents the number of guard cells used to generate the extended ground truth map. In the subsequent analyses, $\bar{G} = 5$ is selected meaning that around each set of pixels representing a change, a 5-pixel frame is included. Thus, for a given P_{fa} level, the best detector is the one ensuring the highest number of detections within the extended ground truth region (number of pixels in the extended ground truth region where the decision statistic is over the threshold).

4.6.1 Homogeneous unstructured environment

Table 4.1 summarizes the number of detections for each receiver corresponding to an actual change present in the extended ground truth considering $P_{fa} = 10^{-3}$ for different values of W and N. For comparison purposes, the performance of the detector proposed in 6 (which is an adaptive implementation of the LRT) is also reported, whose expression in terms of the eigenvalues $\lambda_1, \ldots, \lambda_N$ is [6, eq. (B-5)]

$$\sum_{i=1}^{N} \left(\frac{1}{\lambda_i} - \ln \frac{1}{\lambda_i} \right) \underset{H_0}{\overset{H_1}{\gtrless}} T_H \tag{4.38}$$

As expected, the common trend is that the performance improves by increasing W. Furthermore, as to the effects of polarization, the following remarks are in order.

Table 4.1 Homogeneous unstructured environment: number of correct detections in the extended ground truth for some values of W and N

Detector			**(4.11)**	**(4.12)**	**(4.13)**	**(4.14)**	**(4.15)**	**(4.16)**	**(4.38)**
$W=5$	$N=1$	HH	4512	3977	2301	4512	4512	4512	2301
		VV	5190	3941	2485	5190	5190	5190	2759
		HV	5557	4043	2796	5557	5557	5557	2818
	$N=2$	HH-VV	5372	5129	2337	5103	4877	4980	2300
		HH-HV	6088	4932	3019	5969	5597	5778	2946
		VV-HV	6425	4962	3396	6540	6165	6553	3274
	$N=3$	HH-VV-HV	6492	5513	2884	5901	5463	5644	2819
$W=9$	$N=1$	HH	5671	5426	2765	5671	5671	5671	2797
		VV	7712	5627	3684	7712	7712	7712	5564
		HV	7854	5653	3798	7854	7854	7854	5055
	$N=2$	HH-VV	6544	6591	2664	5852	5334	5433	2601
		HH-HV	8034	6381	3692	7525	6889	7140	3593
		VV-HV	9009	6773	4580	8994	8355	9005	5086
	$N=3$	HH-VV-HV	8300	7049	3389	7120	6233	6386	3255

- Processing either VV or HV channel leads in general to better performances than processing the HH polarization; the cross-polarized returns provide the highest number of correct detections for the single-channel processing (the sole exception is receiver (4.38) for $W = 5$; in this case, the VV channel leads to the best choice). The high discrimination capability of the cross-polarized returns can be explained observing that in the presence of a car target, there are dihedrals which significantly depolarize the incident radiation. On the contrary, the depolarization effect is much limited in the absence of car since the parking area is much more similar to a flat plane. Moreover, the intuitive reason for the better performances achievable with the VV polarization compared with those obtained with the HH one resides in the dependence of the polarization return on the specific change considered (absence or presence of the car object in the image) and on its geometry (height extension and acquisition geometry).
- With reference to two polarimetric channels, jointly processing the pair VV-HV seems to be the best choice for almost all the considered decision rules (with exceptions represented by detectors (4.12), for $W = 5$, for which the HH-VV pair seems to be the most suitable choice, and detector (4.38), for $W = 9$, for which polarization diversity does not seem to provide detection improvements).
- A suitable two-channel polarization diversity can outperform the single-channel processing (the sole exception is detector (4.38) for $W = 9$).
- Processing three-polarimetric channels does not always improve the results obtained using two polarizations; the possible improvement depends both on the parameter W as well as on the specific decision rule. In particular for $W = 9$, it appears more convenient the use of a polarization diversity of order two with the pair VV-HV (the sole exception is detector (4.12) whose correct detections increase exploiting the full polarimetric information).

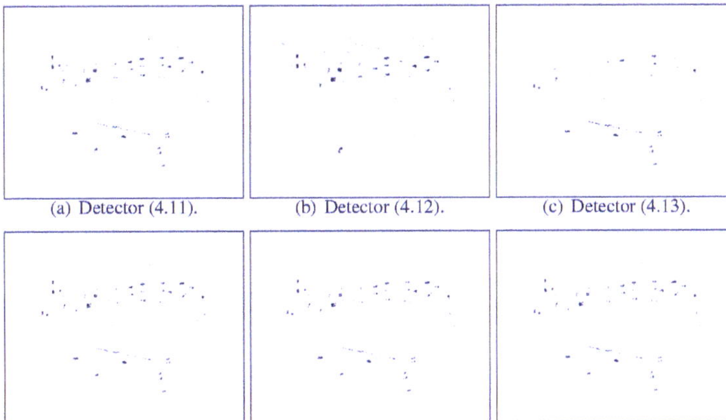

(a) Detector (4.11). (b) Detector (4.12). (c) Detector (4.13).

(d) Detector (4.14). (e) Detector (4.15). (f) Detector (4.16).

Figure 4.5 Homogeneous unstructured environment: detection maps for $W = 5$ and $N = 3$

Figure 4.5 shows the detection maps obtained for the case with $W = 5$ and $N = 3$ for the detectors (4.11)–(4.16). Based on them as well as on the results in Table 4.1, it can be observed that detector (4.16) ensures the best performance for $W = 5$, whereas detector (4.11) outperforms the counterparts for $W = 9$. Moreover, for $W = 3$, additional results (not reported in the table for conciseness) highlight that the best performance is obtained exploiting detector (4.16) processing the pair VV-HV.

Interestingly, inspection of the detection maps for tests (4.12) and (4.13) reveals that the former identifies the departures, whereas the latter identifies the arrivals. This behaviour is also clearly visible in the maps of changes shown in Figure 4.6. In fact, for the detector (4.12), the departures present the highest

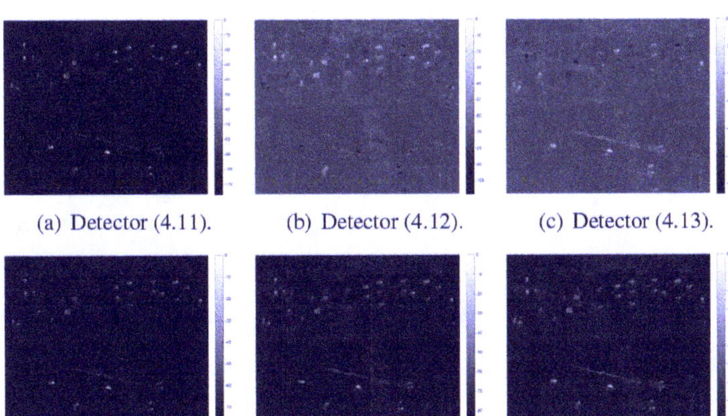

(a) Detector (4.11). (b) Detector (4.12). (c) Detector (4.13).

(d) Detector (4.14). (e) Detector (4.15). (f) Detector (4.16).

Figure 4.6 Homogeneous unstructured environment: maps of changes (in dB) for $W = 5$ and $N = 3$

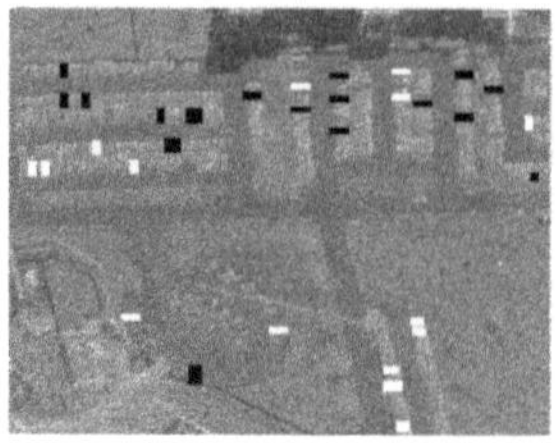

(a) Ground Truth.

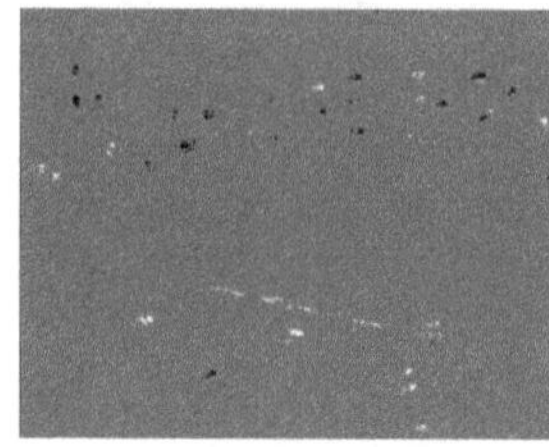

(b) Detector (4.16).

Figure 4.7 Homogeneous unstructured environment: ground truth and enhanced detection map considering $W = 5$ and $N = 3$ for detector (4.16)

values of the decision statistic and the arrivals the lowest, while the dual behaviour is obtained for the detector (4.13). Moreover, although the detector (4.16) is outperformed by (4.11) and (4.14), it enables the discrimination between the departures and the arrivals based on the argument of the max in (4.16). In particular, if the first argument in (4.16) is selected, the change corresponds to a departure, while it corresponds to an arrival if the second argument is selected. By considering the additional information coming from the maximum argument in (4.16), it is thus possible to discriminate the departures from the arrivals detected by the receiver (4.16). In Figure 4.7(b), this enhanced detection map is shown together with the ground truth for the case $W = 5$ and $N = 3$. Herein, departure and arrivals are shown in black and white, respectively.

4.6.2 Homogeneous structured environment

Table 4.2 shows the number of correct changes detected using receivers (4.11) and (4.21) assuming $P_{fa} = 10^{-3}$ for the cases with $W = 3$, $W = 5$, and $W = 7$. From Table 4.2, it is possible to see that the structured GLRT outperforms the unstructured GLRT for the smaller window sizes ($W = 3$ and 5) whereas the unstructured GLRT outperforms the structured GLRT for the larger window size of $W = 7$. This last result can be justified in terms of a covariance model mismatch in the sense that the off-diagonal entries (1, 3) and (3, 1) of the polarimetric covariance matrix which in the theoretical model have been set to zero might not be exactly zero in reality (even if very close to that value). Additionally, there might be some

Table 4.2 Homogeneous structured environment: number of correct detections for $W = 3$, 5, and7

		W	
Detector	3	5	7
Unstructured GLRT (4.11)	3802	6492	7533
Structured GLRT (4.21)	4949	6655	7387

other deviations from the theoretical model for instance due to environmental non-homogeneities.

Figure 4.8, instead, shows detection maps for detectors (4.11) and (4.21) for the case of $W = 3$. From the detection maps, the higher number of detections achievable with the detector (4.21) is appreciable.

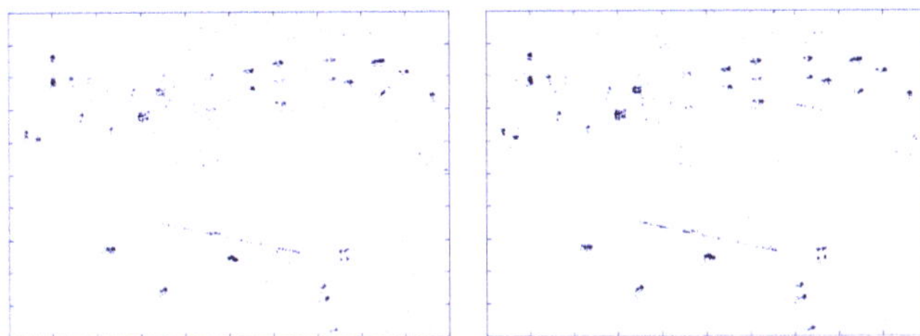

(a) Detector (4.11). (b) Detector (4.21).

Figure 4.8 Homogeneous structured environment: detection maps for $W = 3$

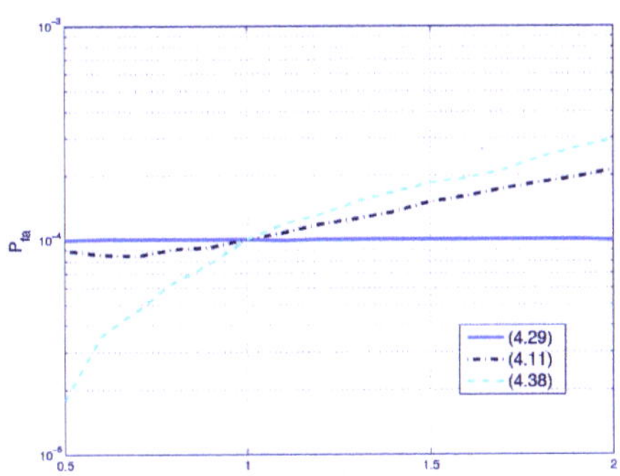

(a) $N = 2, W = 3$

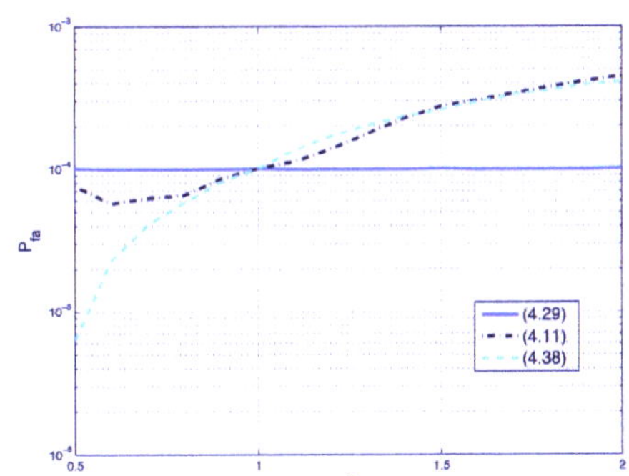

(b) $N = 2, W = 5$

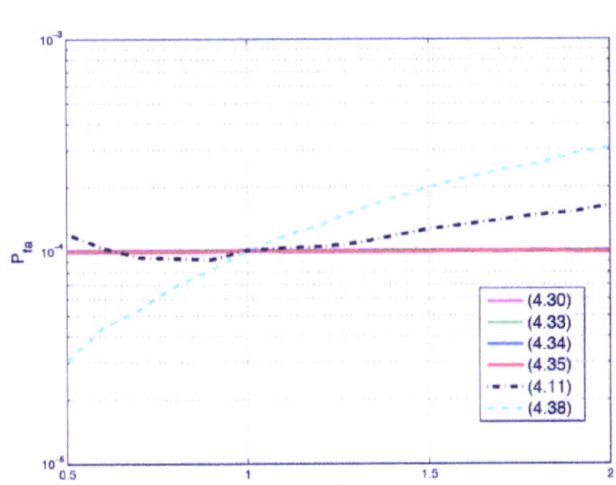

(c) $N = 3, W = 3$

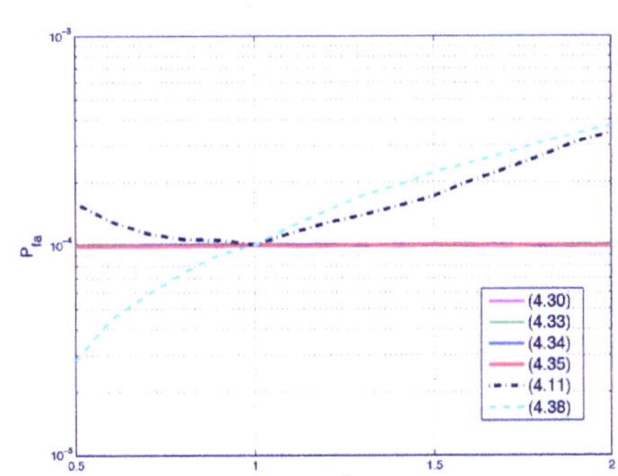

(d) $N = 3, W = 5$

Figure 4.9 Partially homogeneous environment: P_{fa} versus α

4.6.3 Partially homogeneous environment

The capability of the considered detectors to provide scale invariance is assessed via a CFAR analysis. For each decision rule, the threshold is set to ensure a nominal $P_{fa} = 10^{-4}$. Then, exploiting this threshold, the actual P_{fa} is estimated applying a scaling factor equal to $\sqrt{\alpha}$ to the test image with $\alpha \in [0.5, 2]$.

Figure 4.9 shows the actual P_{fa} versus α for $N = 2$ and $N = 3$ with inspection window sizes of $W = 3$ and $W = 5$. Herein, for comparison purposes, the behaviour of the detectors (4.38) and (4.11) is also reported. As expected, in all the situations, the invariant detectors provide a stable probability of false alarm. The detectors (4.38) and (4.11), instead, highlight a poor ability to keep a CFAR in the presence of scale variations between the reference and test images. Specifically, they exhibit an actual P_{fa} different from the nominal value when a scale variation is present ($\alpha \neq 1$).

For the same scenarios, Figure 4.10 shows the number of correct detections in the extended ground truth. As expected, the scale invariant detectors provide

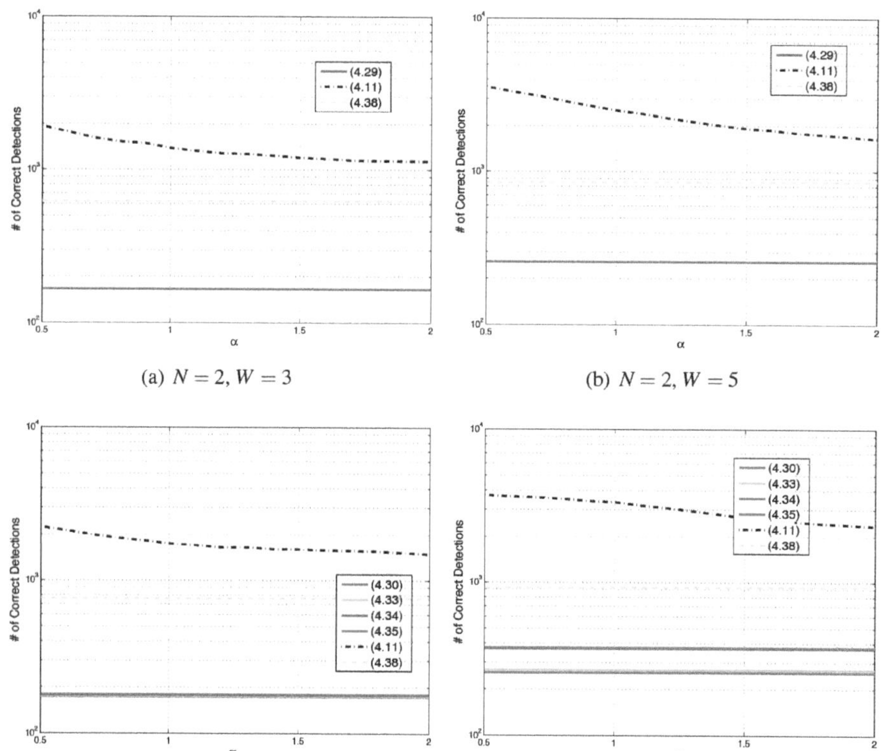

(a) $N = 2$, $W = 3$ (b) $N = 2$, $W = 5$

(c) $N = 3$, $W = 3$ (d) $N = 3$, $W = 5$

Figure 4.10 Partially homogeneous environment: number of correct detections versus α constant performance for different values of α, but in this case, they are outperformed by the detectors (4.38) and (4.11).

4.7 Conclusions

This chapter has addressed the problem of the multi-polarization SAR coherent CD in the presence of homogeneous unstructured returns, homogeneous structured observations, and partially homogeneous data. For the mentioned environments, to identify the possible temporal changes between the reference and test images, the detection problem has been formulated as a specific binary hypothesis test involving the comparison of the polarimetric covariance matrices associated with the reference and test data. For the homogeneous environment, some CFAR decision rules designed by the means of the theory of invariance in hypothesis testing have been discussed. Subsequently, to improve the detection performance in the presence of homogeneous structured data, a special structure for the polarimetric covariance matrices in the hypothesis testing problem has been accounted for. Finally, the capability to account for a possible scale mismatch factor between the reference and test data has been addressed and some scale invariant decision rules have been discussed.

The performance of the considered strategies has been assessed using measured PolSAR data which revealed the capability of the considered decision rules to correctly identify the temporal changes as well as to provide a good CFAR behaviour even in the presence of power mismatches among the different acquisitions.

Possible future research tracks might consider the extension of the framework relaxing the Gaussian requirement for the data as well as the analysis on other datasets acquired under different environmental and operating conditions.

References

[1] Preiss M., Stacy N.J.S. *Coherent Change Detection: Theoretical Description and Experimental Results*. Intelligence, Surveillance and Reconnaissance Division, Defence Science and Technology Organisation, DSTO-TR-1851; 2006.

[2] Touzi R., Lopes A., Bruniquel J., Vachon P.W. 'Coherence estimation for SAR imagery'. *IEEE Trans Geosci Remote Sens*. 1999, vol. 37(1), pp. 135–49.

[3] Stojanovic I., Novak L. 'Change detection experiments using gotcha public release SAR data'. Proceedings of Algorithms for Synthetic Aperture Radar Imagery XX, SPIE Defense, Security, and Sensing; Baltimore, MD, 2013.

[4] Rignot E.J.M., van Zyl J.J. 'Change detection techniques for ERS-1 SAR data'. *IEEE Trans Geosci Remote Sens*. 1993, vol. 31(4), pp. 896–906.

[5] Conradsen K., Nielsen A.A., Schou J., Skriver H. 'A test statistic in the complex Wishart distribution and its application to change detection in polarimetric SAR data'. *IEEE Trans Geosci Remote Sens*. 2003, vol. 41(1), pp. 4–19.

[6] Novak L.M., Zelnio E.G., Garber F.D. 'Change detection for multi-polarization multi-pass SAR (invited paper)'. Defense and Security; Orlando, FL, 2005.

[7] Reigber A., Scheiber R., Jager M, *et al.* 'Very-high-resolution airborne synthetic aperture radar imaging: signal processing and applications'. *Proc IEEE*. 2013, vol. 101(3), pp. 759–83.

[8] Marino A., Cloude S.R., Lopez-Sanchez J.M. 'A new polarimetric change detector in radar imagery'. *IEEE Trans Geosci Remote Sens*. 2013, vol. 51(5), pp. 2986–3000.

[9] Nielsen A.A., Larsen R., Skriver H. 'Change detection in bi-temporal EMISAR data from Kalø, Denmark, by means of canonical correlations analysis'. *3rd Int Air Rem Sens Conf Exhibit*; 1997.

[10] Nielsen A.A. 'Change detection in multispectral bi-temporal spatial data using orthogonal transformations'. 8th Austral-Asian Sensing Conference; Canberra, ACT, 1996.

[11] Erten E., Reigber A., Ferro-Famil L., Hellwich O. 'A new coherent similarity measure for temporal multichannel scene characterization'. *IEEE Trans Geosci Remote Sens*. 2012, vol. 50(7), pp. 2839–51.

[12] Bouhlel N., Meric S. 'Multilook polarimetric SAR change detection using stochastic distances between matrix-variate G_d^0 distributions'. *IEEE Trans Geosci Remote Sens*. 2020, vol. 58(10), pp. 6823–43.

[13] Yue D.-X., Xu F., Frery A.C., Jin Y.-Q. 'Synthetic aperture radar image statistical modeling: part one-single-pixel statistical models'. *IEEE Trans Geosci Remote Sens*. 2021, vol. 9(1), pp. 82–114.

[14] Bouhlel N., Akbari V., Meric S. 'Change detection in multilook polarimetric SAR imagery with determinant ratio test statistic'. *IEEE Trans Geosci Remote Sens*. 2020, vol. 60(5200515), pp. 1–15.

[15] Mian A., Ginolhac G., Ovarlez J.-P., Atto A.M. 'New robust statistics for change detection in time series of multivariate SAR images'. *IEEE Trans Signal Process*. 2019, vol. 67(2), pp. 520–34.

[16] Nascimento A.D.C., Frery A.C., Cintra R.J. 'Detecting changes in fully polarimetric SAR imagery with statistical information theory'. *IEEE Trans Geosci Remote Sens*. 2019, vol. 57(3), pp. 1380–92.

[17] West R.D., Riley R.M. 'Polarimetric interferometric SAR change detection discrimination'. *IEEE Trans Geosci Remote Sens*. 2019, vol. 57(6), pp. 3091–104.

[18] Alonso-Gonzalez A., Lopez-Martinez C., Papathanassiou K.P., Hajnsek I. 'Polarimetric SAR time series change analysis over agricultural areas'. *IEEE Trans Geosci Remote Sens*. 2020, vol. 58(10), pp. 7317–30.

[19] Carotenuto V., De Maio A., Clemente C., Soraghan J. 'Invariant rules for multipolarization SAR change detection'. *IEEE Trans Geosci Remote Sens*. 2015, vol. 53(6), pp. 3294–311.

[20] Carotenuto V., De Maio A., Clemente C., Soraghan J. 'Unstructured versus structured GLRT for multipolarization SAR change detection'. *IEEE Geosci Remote Sens Lett*. 2015, vol. 12(8), pp. 1665–69.

[21] Carotenuto V., De Maio A., Clemente C., Soraghan J.J., Alfano G. 'Forcing scale invariance in multipolarization SAR change detection'. *IEEE Trans Geosci Remote Sens*. 2016, vol. 54(1), pp. 36–50.

[22] Pallotta L., Clemente C., De Maio A., Soraghan J.J. 'Detecting covariance symmetries in polarimetric SAR images'. *IEEE Trans Geosci Remote Sens.* 2017, vol. 55(1), pp. 80–95.

[23] Kay S.M. *Fundamentals of statistical signal processing: estimation theory.* Vol. 1. Prentice-Hall PTR; 1998.

[24] James A.T. 'Distributions of matrix variates and latent roots derived from normal samples'. *Ann. Math. Stat.* 1964, vol. 35(2), pp. 475–501.

[25] Lehmann E.L. *Testing statistical hypotheses.* 2nd ed. New York, NY: Springer-Verlag; 1986.

[26] Muirhead R.J. *Aspects of multivariate statistical theory.* Hoboken, NJ: Wiley; 1982.

[27] Scharf L.L. *Statistical signal processing: detection, estimation, and time series analysis.* Addison-Wesley; 1991.

[28] Conradsen K., Nielsen A.A., Schou J., Skriver H. 'A test statistic in the complex wishart distribution and its application to change detection in polarimetric SAR data'. *IEEE Trans Geosci Remote Sens.* 2003, vol. 41(1), pp. 4–19.

[29] Van Trees H.L. *Detection, estimation, and modulation theory.* John Wiley & Sons; 1968.

[30] Scarborough S., Gorham L., Minardi M.J. 'A challenge problem for SAR change detection and data compression'. Proceedings Algorithms for Synthetic Aperture Radar Imagery XVII, SPIE Defense, Security, and Sensing; Orlando, FL, 2010.

Chapter 5

Classification of covariance symmetries in full-polarimetric SAR images

Luca Pallotta[1], Carmine Clemente[2], Antonio De Maio[3], and John J. Soraghan[2]

5.1 Introduction

The exploitation of the information extracted by all channels of modern polarimetric synthetic aperture radar (PolSAR) systems is becoming one of the main approaches in the development of advanced SAR image processing for both detection and classification purposes. In particular, the polarimetric covariance, coherence, and Muller matrix are typically used [1–4] to fully exploit the polarimetric information. In fact, PolSAR sensors provide four quantities composing the scattering or Sinclair matrix [1, 3], which (when expressed in specific vectorial forms) are used to derive their respective second-order moment-based metrics (viz., covariance, coherence, and Muller matrices). These matrices, in turn, are exploited to make inference about the scattering mechanisms characterizing the scene of interest. In addition, the aforementioned matrices are typically estimated from the available data surrounding the investigated pixel (indicated as looks), and then used in several processes spanning from scene classification to target detection and recognition. Consequently, by using the available polarimetric data [5–9], it is possible to automatically classify the observed scene by assigning a label to each pixel of the illuminated area.

Currently, there is a plethora of work published in the literature related to the exploration of PolSAR-based techniques for scene classification purposes. In particular, such techniques belong to two competing groups referred to as unsupervised and supervised techniques. The first group entails a clustering approach in which the pixels are automatically grouped on the basis of some common features. On the other hand, supervised approaches comprise a training phase that exploits the pre-labeled sampled data followed by a classification stage. The interested readers can consider the following references for further insights on the topic [10–28].

[1]University of Roma Tre, Italy
[2]University of Strathclyde, UK
[3]University "Federico II", Italy

Studying the behavior of the polarimetric covariance matrix (or equivalently its corresponding coherence matrix), it has been observed that it assumes some special structures under symmetrical properties of the returns associated to the pixels under observation [2]. These symmetries, if correctly detected, can be properly exploited to improve the performance of applications such as knowledge-based ground moving target indicator, oil spill detection, as well as land cover classification.

In this chapter, some polarimetry-based methodologies are described in detail. Namely, the detection of covariance structure associated with the particular symmetry experienced by the pixel returns in both homogeneous and heterogeneous environments as well as for the challenging polarimetric interferometry SAR (PolInSAR) applications. Specifically, the problem is formulated as a multiple hypothesis test comprising both nested and non-nested hypotheses. By doing so, it is immediately observed that the classic generalized maximum-likelihood (GML) approach [29] cannot properly work since it always selects the instance with the highest degree of uncertainty [29, 30]. This drawback can be solved using modified versions of the GML, the so-called model order selection (MOS) rules [31–34], to properly manage the involved nested signal models [35, 36]. Specifically, MOS techniques consist in adding a penalty term, namely, a function of the unknown parameters, to the GML under each hypothesis. As to PolInSAR [37], it is based on the use of two spatially separated polarimetric antennas that allow for the joint acquisition of the coherence associated with the four polarimetric channels together with the elevation of the polarimetric phase centers. This finds application in many situations where it would be of interest to not only detect the covariance symmetry but also associate it to the corresponding height of the scattering target (e.g., a tree in a forest or a building in a city). Consequently, a framework aimed at detecting covariance symmetries together with the estimation of the elevation in each resolution cell in three-dimensional (3-D) PolInSAR data is also described [38, 39]. The validity of the methodologies explained in this chapter is demonstrated through numerical simulation. Moreover, their effectiveness is validated on L-band real-recorded data.

The chapter is organized as follows. Section 5.2 provides some definitions and the procedure to build up the PolSAR datacube. Section 5.3 describes the possible symmetries that could occur in nature together with their corresponding covariance/coherence structure. Then, section 5.4 formalizes the symmetry detection problem as a multiple hypothesis testing problem and provides the GML under each hypothesis for both the homogeneous and heterogeneous scenarios. Section 5.5 derives the MOS rules to solve the classification problem, which also includes a description of the application to PolInSAR. In section 5.6, a description of the use of detected covariance structures for the classification of PolInSAR scenes is given. Results of numerical simulations and tests on L-band measured PolSAR data are described and commented in section 5.7. Finally, the chapter ends with some conclusions drawn in section 5.8.

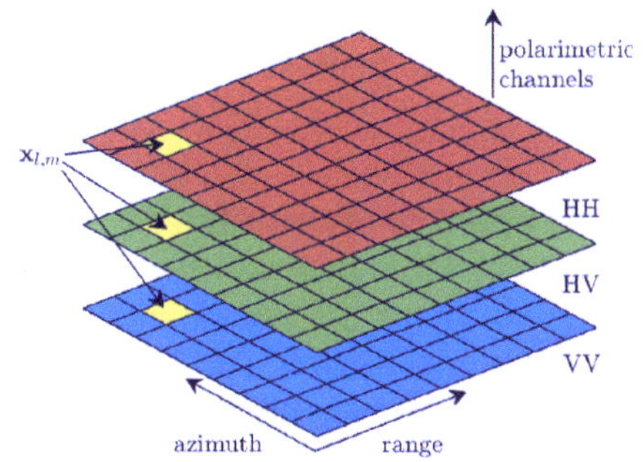

Figure 5.1 Polarimetric datacube acquired through a PolSAR system, where each level is a single-polarization SAR image (viz., HH, HV, and VV for a reciprocal medium)

5.2 Preliminary definitions

In a multi-polarization SAR sensor, each pixel of the image under test measures $N = 3$ complex returns (in case of a reciprocal medium), collected from three different polarimetric channels (viz., HH, HV, and VV). The N returns corresponding to the same pixel in the image are organized in the specific order HH, HV, and VV so as to constitute the vector $\boldsymbol{x}_{l,m}$, $l = 1, \ldots, L$ and $m = 1, \ldots, M$, with L and M representing the vertical and horizontal (or azimuthal and range) size of the SAR image, respectively. Therefore, the sensor provides a 3D data stack $\boldsymbol{X}$ of size $L \times M \times N$, which is indicated by a datacube throughout this chapter, and whose pictorial representation is illustrated in Figure 5.1.

In this context, without loss of generality, the vectors $\boldsymbol{x}_{l,m}$, $l = 1, \ldots, L$ and $m = 1, \ldots, M$, are modeled as statistically independent and identically distributed (i.i.d.) zero-mean complex circularly symmetric Gaussian vectors with positive definite covariance matrix $\xi_{l,m}\boldsymbol{C} \in \mathbb{C}^{3\times 3}$, i.e., $\boldsymbol{x}_{l,m} \sim \mathcal{CN}\left(\boldsymbol{0}, \xi_{l,m}\boldsymbol{C}\right)$, where $\xi_{l,m} > 0$ is a scaling factor accounting for the reflectivity (also indicated as power level) of the (l, m)-th pixel and $\boldsymbol{C} > \boldsymbol{0}$ is the positive definite covariance structure. Under this assumption, the probability density function (pdf) of $\boldsymbol{x}_{l,m}$ can be written as follows:

$$f\left(\boldsymbol{x}_{l,m}; \xi_{l,m}, \boldsymbol{C}\right) = \frac{1}{(\pi \xi_{l,m})^3 \det(\boldsymbol{C})} \exp\left\{-\xi_{l,m}^{-1} \mathrm{tr}\left[\boldsymbol{C}^{-1}\boldsymbol{x}_{l,m}\boldsymbol{x}_{l,m}^{\dagger}\right]\right\} \tag{5.1}$$

Note that the different pixels share the same covariance structure, but, in general, they can possibly have different unknown power levels represented by $\xi_{l,m}$. An important comment follows next: the general model for the pdf given in (5.1) comprises both the homogeneous (viz., $\xi_{l,m} = 1$, $\forall l, m$) and the heterogeneous [40] scenarios, namely, the quantities $\xi_{l,m}$ may vary from pixel to pixel. Therefore, in the next section, the focus will be on these two different scenarios that can, however, be recast into a single general framework. More precisely, starting from the overall datacube $\boldsymbol{X}$ of the scene illuminated by the PolSAR, for each pixel under test, a rectangular neighborhood $\mathcal{A}$ of size $K = W_1 \times W_2 \geq N$ is extracted by means of a moving window. Then, the matrix whose columns are the vectors of the polarimetric returns from the pixels of $\boldsymbol{X}$ that fall in the region $\mathscr{A}$ is denoted by $\boldsymbol{R}$.

5.3 Polarimetric covariance symmetries

This section describes the covariance/coherence matrix structures arising when important scattering symmetric properties [1, 2] become predominant in the return associated with the pixel under investigation. In particular, in the following subsections, the polarimetric covariance matrix for a reciprocal medium is considered in four different situations, viz., absence of symmetry, reflection symmetry, rotation symmetry, and azimuthal symmetry.

5.3.1 Absence of symmetry

The first case studied in this chapter considers the polarimetric covariance matrix in the presence of a reciprocal medium and specifically in the absence of symmetry [1,2], which is a 3×3 Hermitian matrix:

$$\boldsymbol{C} = \begin{bmatrix} c_{hhhh} & c_{hhhv} & c_{hhvv} \\ c_{hhhv}^{*} & c_{hvhv} & c_{hvvv} \\ c_{hhvv}^{*} & c_{hvvv}^{*} & c_{vvvv} \end{bmatrix} \tag{5.2}$$

that is completely described by $n = 9$ real scalar values. The diagonal entries of this matrix are the conventional backscattering coefficients, which are real quantities, whereas the off-diagonal elements are complex values. Recalling that a PolSAR image comprises all the polarimetric channels (viz., S_{HH}, S_{VV}, S_{HV}, and S_{VH}), in many data processing schemes, a calibration is applied to reduce the mismatch between HV and VH channels [41]. For instance, in [42], the fully polarimetric image is reduced to three channels fusing together the HV and VH through their coherent average. During the acquisition process, some targets/media might be mismatched due to statistical fluctuations, Faraday rotations [43], or measurement errors related to sensor non-idealities. Nevertheless, to evaluate the impact of these mismatches (assuming reciprocity), a statistical test on the reciprocity assumption for each pixel in the polarimetric image can be applied [44, 45].

5.3.2 Reflection symmetry

The second case analyzes the presence of a reflection symmetry with respect to a vertical plane. This effect is usually observed over horizontal natural environments and produces complete decorrelations between the co-polarized and the cross-polarized elements (the interested reader can find details regarding the proof of this covariance structure in [2]). As a consequence, the covariance matrix under a dominant reflection symmetry assumes the following special form

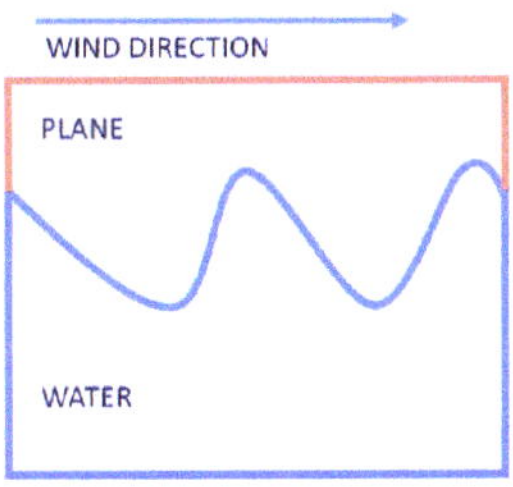

Figure 5.2 Example of reflection symmetry in nature

$$C = \begin{bmatrix} c_{hhhh} & 0 & c_{hhvv} \\ 0 & c_{hvhv} & 0 \\ c^*_{hhvv} & 0 & c_{vvvv} \end{bmatrix} \quad (5.3)$$

which is described by $n = 5$ real scalar values. From a physical point of view, this result is valid for volume scattering, surface scattering, or volume–surface interactions to all scattering orders or to the total scattering effects independently of how dense is the medium, or how rough is the surface [2]. In scenarios of practical interest, as pictorially represented in Figure 5.2, the reflection symmetry can be observed on water surfaces in the upwind or downwind direction. Further examples include ploughed fields in the directions perpendicular to the row structure, as well as forest, snow, or sea ice [2]. In addition, it can also be observed on horizontal and vertical wires or horizontal dihedrals.

5.3.3 Rotation symmetry

The third case studied in this chapter is the dominant presence of a rotation symmetry in the pixel returns, characterized by a covariance matrix that is invariant under rotation around an axis by any considered angle [2]. In such a situation, the corresponding polarimetric covariance matrix assumes the following special structure described by $n = 3$ real scalar values:

$$C = \begin{bmatrix} c_{hhhh} & c_{hhhv} & c_{hhvv} \\ -c_{hhhv} & c_{hvhv} & c_{hhhv} \\ c_{hhvv} & -c_{hhhv} & c_{hhhh} \end{bmatrix} \quad (5.4)$$

where c_{hhhv} is a purely imaginary quantity, c_{hhvv} is a purely real value, and $c_{hvhv} = (c_{hhhh} - c_{hhvv})/2$. The rotation symmetry assuming reciprocity property can be found, for example, in a chiral medium made by embedding helixes in an isotropic background [2]. Further examples of rotation symmetries, schematically represented in Figure 5.3, can be found in a gyrotropic random medium such as the Earth's ionosphere, which is a plasma magnetized by the geomagnetic field, containing rod-like density irregularities along the field direction [2].

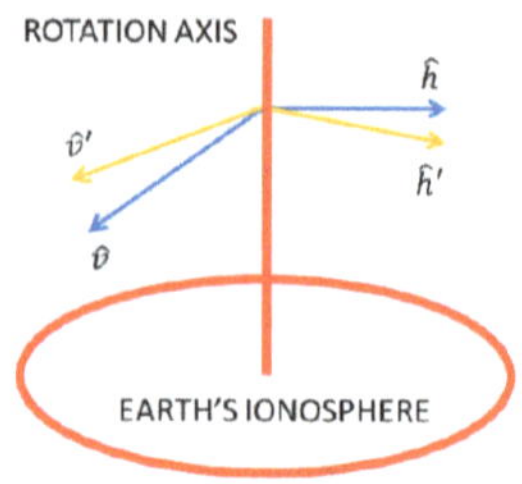

Figure 5.3 Example of rotation symmetry in nature

5.3.4 Azimuthal symmetry

The final case studied in this chapter arises as the combination of a rotation symmetry with a reflection symmetry in any plane which contains the rotation symmetry axis, as highlighted in Figure 5.4. This is typically referred to as azimuthal symmetry and can be observed in nature for dense volumetric environments [3]. Its corresponding polarimetric covariance matrix is described by $n = 2$ real scalar values and shares the form:

$$\boldsymbol{C} = \begin{bmatrix} c_{hhhh} & 0 & c_{hhvv} \\ 0 & c_{hvhv} & 0 \\ c_{hhvv} & 0 & c_{hhhh} \end{bmatrix} \tag{5.5}$$

where $c_{hvhv} = (c_{hhhh} - c_{hhvv})/2$. An example of azimuthal symmetry can be observed in vegetated areas at low frequencies and normal incidence. In fact, for low-frequency SAR, the foliage canopy is penetrated by the electromagnetic wave and it is scattered back by the horizontal branches or the vertical tree trunks, producing an azimuthal symmetry [2].

5.3.5 Transformed matrix domain

Let us now consider a transformed matrix domain (as per Lemma 1 in [35]) where some redundant information (contained in the functionally dependent coefficients of the covariance) translates into zeros of the corresponding transformed matrix.

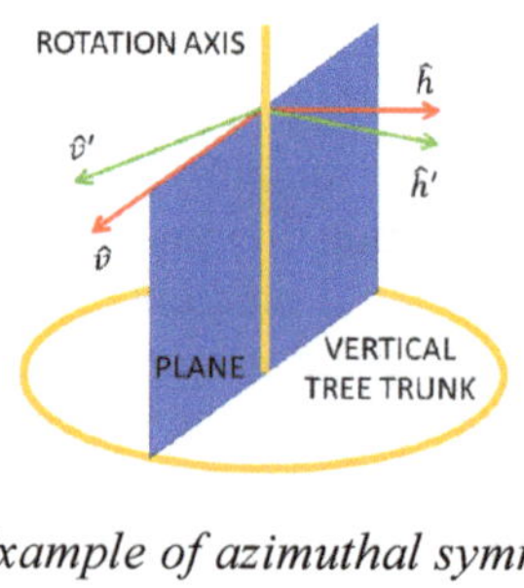

Figure 5.4 Example of azimuthal symmetry in nature

Lemma 1 of [35]

Let us denote by

$$U = \begin{bmatrix} 1 & 0 & 0 \\ 0 & 0 & 1 \\ 0 & 1 & 0 \end{bmatrix}, \quad E = \begin{bmatrix} 1 & 0 & 0 \\ 0 & 1/\sqrt{2} & 0 \\ 0 & 0 & 1 \end{bmatrix}, \quad T = \frac{1}{\sqrt{2}} \begin{bmatrix} 1 & 0 & 1 \\ 1 & 0 & -1 \\ 0 & \sqrt{2} & 0 \end{bmatrix}$$

$$V = \begin{bmatrix} 1 & 0 & 0 \\ 0 & 0 & j \\ 0 & 1 & 0 \end{bmatrix}$$

four transformation matrices, with $\boldsymbol{U}$ orthogonal and $\boldsymbol{V}$ and $\boldsymbol{T}$ unitary. Then, under the reflection symmetry hypothesis

$$\boldsymbol{UCU}^{\dagger} = \begin{bmatrix} c_{hhhh} & c_{hhvv} & 0 \\ c_{hhvv}^{*} & c_{vvvv} & 0 \\ 0 & 0 & c_{hvhv} \end{bmatrix} = \begin{bmatrix} \boldsymbol{C}_1 & \boldsymbol{0} \\ \boldsymbol{0} & c \end{bmatrix} \tag{5.6}$$

where $\boldsymbol{C}_1 \in \mathbb{H}^2, \boldsymbol{C}_1 \geq \boldsymbol{0}$ [i.e., a 2×2 positive semi-definite (psd) Hermitian matrix] and c is a positive real number. Under the rotation symmetry hypothesis

$$\boldsymbol{VETCT}^{\dagger}\boldsymbol{EV}^{\dagger} = \begin{bmatrix} c_{hhhh} + c_{hhvv} & 0 & 0 \\ 0 & c_{hvhv} & \mathrm{Im}\{c_{hhhv}\} \\ 0 & \mathrm{Im}\{c_{hhhv}\} & c_{hvhv} \end{bmatrix} = \begin{bmatrix} a & \boldsymbol{0} \\ \boldsymbol{0} & \boldsymbol{C}_2 \end{bmatrix}$$

where $\boldsymbol{C}_2 \in \mathbb{P}^2, \boldsymbol{C}_2 \geq \boldsymbol{0}$ (i.e., a 2×2 real psd persymmetric matrix) and a is a positive real number.

Finally, under the azimuthal symmetry hypothesis

$$\boldsymbol{ETCT}^{\dagger}\boldsymbol{E} = \begin{bmatrix} c_{hhhh} + c_{hhvv} & 0 & 0 \\ 0 & c_{hvhv} & 0 \\ 0 & 0 & c_{hvhv} \end{bmatrix} = \begin{bmatrix} a & 0 & 0 \\ 0 & b & 0 \\ 0 & 0 & b \end{bmatrix}$$

where b is a positive real number.

5.4 Problem formulation and GML computation

The framework described previously recognizes the special structures assumed by the covariance and, consequently, by the corresponding coherence matrix of objects that scatter with a specific symmetry property. Now, on the basis of the considerations and models for the polarimetric covariance matrix structure in section 5.3.4, the focus is on the problem of data classification exploiting the scattering properties of the pixel under test and of a specific set of neighboring pixels. Specifically, an *average* or *dominant* symmetry property is associated to each pixel on the basis of the specific structure assumed by its covariance matrix. As a consequence, the considered classification problem is proposed as the following multiple hypothesis test comprising both nested and non-nested hypotheses:

$$\begin{cases} H_1: & \text{absence of symmetry;} \\ H_2: & \text{reflection symmetry;} \\ H_3: & \text{rotation symmetry;} \\ H_4: & \text{azimuth symmetry.} \end{cases} \tag{5.7}$$

Before providing a solution to the above-defined multiple hypothesis test, the GML as well as the covariance maximum likelihood estimates (MLEs) is derived under each hypothesis. Hence, sections 5.4.1 and 5.4.2 focus on the homogeneous and on the more general heterogeneous case, respectively.

5.4.1 Homogeneous case

Starting from (5.1), and setting $\xi_{l,m} = 1$ (a homogeneity condition), we can consider the complex multivariate pdf of the observable matrix $\boldsymbol{R} = [\boldsymbol{r}_1, \ldots, \boldsymbol{r}_K]$ in the region $\mathscr{A}$:

$$f_R(\boldsymbol{R}|\boldsymbol{C}) = \frac{1}{\pi^{3K}[\det(\boldsymbol{C})]^K} \exp\left\{-\text{tr}\left(\boldsymbol{C}^{-1}\boldsymbol{R}\boldsymbol{R}^\dagger\right)\right\} \tag{5.8}$$

Now, it is worth recalling that the Fisher–Neyman factorization theorem [29] implies that $\boldsymbol{S}_0 = \boldsymbol{R}\boldsymbol{R}^\dagger$ is a sufficient statistic for $\boldsymbol{C}$. The corresponding log-likelihood function of $\boldsymbol{R}$ can be also derived to be

$$l(\boldsymbol{C};\boldsymbol{R}) = -K\left[\log\det(\boldsymbol{C}) + \text{tr}\left(\boldsymbol{C}^{-1}\boldsymbol{S}\right)\right] - 3K\log\pi \tag{5.9}$$

where $\boldsymbol{S} = (1/K)\boldsymbol{S}_0$ is the classic sample covariance matrix. The MLE of $\boldsymbol{C}$ can be then obtained as the optimal solution to the following optimization problem:

$$\max_{C} \log\left(f_R(\boldsymbol{R}|\boldsymbol{C})\right) = -K\min_{C}\left[\log\det(\boldsymbol{C}) + \text{tr}\left(\boldsymbol{C}^{-1}\boldsymbol{S}\right)\right] - 3K\log\pi \tag{5.10}$$

The next proposition, derived and demonstrated in [35], shows the form assumed by the optimal value of problem (5.10), as well as the corresponding ML optimizer, in the presence of the four different scattering symmetries previously described.

Proposition 1. The optimal value of

$$\min_{C\in\mathbb{H}^3, C\succeq 0}\left[\log\det(\boldsymbol{C}) + \text{tr}\left(\boldsymbol{C}^{-1}\boldsymbol{S}\right)\right] \tag{5.11}$$

and the corresponding optimal solution $\hat{\boldsymbol{C}}$ are, respectively, given in

1. Absence of symmetry:

$$\log\det(\boldsymbol{S}) + 3$$

$$\hat{\boldsymbol{C}} = \boldsymbol{S}$$

2. Reflection symmetry:

$$\log\det\left(\bar{\boldsymbol{S}}_{1,1}\right)+\log\left(\bar{S}_{3,3}\right)+3$$

$$\hat{\boldsymbol{C}}=\boldsymbol{U}^{\dagger}\begin{bmatrix}\bar{\boldsymbol{S}}_{1,1} & \boldsymbol{0}\\ \boldsymbol{0} & \bar{S}_{3,3}\end{bmatrix}\boldsymbol{U}$$

where

$$\bar{\boldsymbol{S}}=\boldsymbol{U}\boldsymbol{S}\boldsymbol{U}^{\dagger}=\begin{bmatrix}\bar{\boldsymbol{S}}_{1,1} & \bar{\boldsymbol{S}}_{1,3}\\ \bar{\boldsymbol{S}}_{3,1} & \bar{S}_{3,3}\end{bmatrix}$$

3. Rotation symmetry:

$$\log\det\left(\tfrac{1}{2}\left(\tilde{\boldsymbol{S}}_{2,2}+\boldsymbol{J}\tilde{\boldsymbol{S}}_{2,2}\boldsymbol{J}\right)\right)+\log\left(\tilde{S}_{1,1}\right)+3+\log 2$$

$$\hat{\boldsymbol{C}}=\boldsymbol{T}^{\dagger}\boldsymbol{E}^{-1}\boldsymbol{V}^{\dagger}\begin{bmatrix}\tilde{S}_{1,1} & \boldsymbol{0}\\ \boldsymbol{0} & \tfrac{1}{2}\left(\tilde{\boldsymbol{S}}_{2,2}+\boldsymbol{J}\tilde{\boldsymbol{S}}_{2,2}\boldsymbol{J}\right)\end{bmatrix}\boldsymbol{V}\boldsymbol{E}^{-1}\boldsymbol{T}$$

where

$$\tilde{\boldsymbol{S}}=\boldsymbol{V}\boldsymbol{E}\boldsymbol{T}\boldsymbol{S}\boldsymbol{T}^{\dagger}\boldsymbol{E}\boldsymbol{V}^{\dagger}=\begin{bmatrix}\tilde{S}_{1,1} & \tilde{\boldsymbol{S}}_{1,2}\\ \tilde{\boldsymbol{S}}_{2,1} & \tilde{\boldsymbol{S}}_{2,2}\end{bmatrix}$$

4. Azimuthal symmetry:

$$\log\left(\hat{S}_{1,1}\right)+2\log\left(\frac{\hat{S}_{2,2}+\hat{S}_{3,3}}{2}\right)+3+\log 2$$

$$\hat{\boldsymbol{C}}=\boldsymbol{T}^{\dagger}\boldsymbol{E}^{-1}\mathbf{diag}\left(\hat{S}_{1,1},\frac{\hat{S}_{2,2}+\hat{S}_{3,3}}{2},\frac{\hat{S}_{2,2}+\hat{S}_{3,3}}{2}\right)\boldsymbol{E}^{-1}\boldsymbol{T}$$

where $\hat{\boldsymbol{S}}=\boldsymbol{E}\boldsymbol{T}\boldsymbol{S}\boldsymbol{T}^{\dagger}\boldsymbol{E}$ and $\hat{S}_{1,1}, \hat{S}_{2,2}, \hat{S}_{3,3}$ are its diagonal entries.

5.4.2 Heterogeneous case

Differently from the homogeneous environment treated in section 5.4.1, in the heterogeneous case, the pixels share different power factors even if they have the same covariance structure. In such a situation, there is a prevalence of nuisance parameters (represented by the scaling factors $\xi_{l,m}$) whose number grows as a function of the amount of data contained in the sliding window $\mathscr{A}$. In this case, the ratio between the number of unknown parameters and the number of observations K does not reduce as K increases, thus leading to the loss of some regularity conditions used to derive the MOS rules [32]. A possible solution to remove this limitation resides in the principle of invariance [46]. This has been explored in [36], where the principle of invariance is used to replace the starting polarimetric data vectors with a suitable function of them so as to remove their statistical dependence on the parameters $\xi_{l,m}$. More precisely, as derived in [36], for the generic (l,m)-th pixel, the considered transformation is given in

$$z_k = \frac{\boldsymbol{r}_k^{l,m}}{\|\boldsymbol{r}_k^{l,m}\|}, \quad k = 1, \ldots, K$$

Due to the above transformation, the data z_k, $k = 1, \ldots, K$, are still i.i.d. and their joint pdf is referred to as angular Gaussian as described in [47]

$$f(z_1, z_2, \ldots, z_K; \boldsymbol{C}) = \Phi(N)\left[\det(\boldsymbol{C})\right]^{-K} \prod_{k=1}^{K} \left\{\text{tr}\left(\boldsymbol{C}^{-1}\boldsymbol{S}_k\right)\right\}^{-3} \tag{5.12}$$

where $\boldsymbol{S}_k = z_k z_k^{\dagger}$ and $\Phi(N)$ is a suitable normalization constant. In section 5.5, we derive decision schemes for covariance structure classification exploiting the transformed data, which are invariant to the scaling operation.

As in the homogeneous case, the MLEs of $\boldsymbol{C}$, under each hypothesis, can be found exploiting the corresponding log-likelihood function of the matrix $\boldsymbol{Z} = [z_1, \ldots, z_K]$, whose expression is

$$l(\boldsymbol{C}; \boldsymbol{Z}) = K\left[\frac{1}{K}\log \Phi(N) - l_1(\boldsymbol{C}; \boldsymbol{Z})\right] \tag{5.13}$$

where

$$l_1(\boldsymbol{C}; \boldsymbol{Z}) = \log\det(\boldsymbol{C}) + \frac{3}{K}\sum_{k=1}^{K} \log \text{tr}\left(\boldsymbol{C}^{-1}\boldsymbol{S}_k\right) \tag{5.14}$$

and maximizing it with respect to $\boldsymbol{C}$. Since maximizing (5.13) is equivalent to minimizing the term $l_1(\boldsymbol{C}; \boldsymbol{Z})$, the following proposition, derived and proved in [36], provides the expressions of the optimal values of $l_1(\boldsymbol{C}; \boldsymbol{Z})$ and the associated MLEs. This accounts for the presence of the four different scattering symmetries described by the structures of $\boldsymbol{C}$. Note that in the first general case, MLE is the classic Tyler's estimator [40, 47–50], whereas in the other three cases, MLE can be found exploiting the symmetric Tyler's estimator introduced in [51]. More details about these derivations can be found in [36].

Similarly, to Proposition 1, Proposition 2 shows the form assumed by the optimal value of problem (5.14), as well as the corresponding ML optimizer, in the presence of the four considered scattering symmetries.

Proposition 2. The optimal value of $l_1(\boldsymbol{C}; \boldsymbol{Z})$ and the corresponding optimal solution $\hat{\boldsymbol{C}}$ are, respectively, given in

1. Absence of symmetry:

$$\log\det\left(\hat{\boldsymbol{C}}\right) + \frac{3}{K}\sum_{k=1}^{K} \log \text{tr}\left(\hat{\boldsymbol{C}}^{-1}\boldsymbol{S}_k\right) \tag{5.15}$$

with $\hat{\boldsymbol{C}} \in \mathbb{H}^3, \hat{\boldsymbol{C}} \succeq \boldsymbol{0}$ a fixed point of

$$\boldsymbol{C} = \frac{3}{K}\sum_{k=1}^{K} \frac{\boldsymbol{S}_k}{\text{tr}\left(\boldsymbol{C}^{-1}\boldsymbol{S}_k\right)} \tag{5.16}$$

2. Reflection symmetry:

$$\log\det\left(\hat{\boldsymbol{C}}_1\right) + \log\hat{c} + \frac{3}{K}\sum_{k=1}^{K}\log\left[\operatorname{tr}\left(\hat{\boldsymbol{C}}_1^{-1}\bar{\boldsymbol{S}}_{1,1,k}\right) + \frac{\bar{S}_{2,2,k}}{\hat{c}}\right] \tag{5.17}$$

where $\bar{\boldsymbol{S}}_k = \boldsymbol{U}\boldsymbol{S}_k\boldsymbol{U}^\dagger = \begin{bmatrix} \bar{\boldsymbol{S}}_{1,1,k} & \bar{\boldsymbol{S}}_{1,2,k} \\ \bar{\boldsymbol{S}}_{2,1,k} & \bar{S}_{2,2,k} \end{bmatrix}$, with $\bar{\boldsymbol{S}}_{1,1,k} \in \mathbb{H}^2(\bar{\boldsymbol{S}}_k(1,1) \geq \boldsymbol{0})$, $\hat{\boldsymbol{C}}_1 \in \mathbb{H}^2(\hat{\boldsymbol{C}}_1 \geq \boldsymbol{0})$ and $\hat{c} \in \mathbb{R}(\hat{c} > 0)$ (strictly positive real number) are fixed points of

$$\boldsymbol{C}_1 = \frac{3}{K}\sum_{k=1}^{K}\frac{\bar{\boldsymbol{S}}_{1,1,k}}{\operatorname{tr}\left(\boldsymbol{C}_1^{-1}\bar{\boldsymbol{S}}_{1,1,k}\right) + \frac{\bar{S}_{2,2,k}}{c}} \tag{5.18}$$

and

$$c = \frac{3}{K}\sum_{k=1}^{K}\frac{\bar{S}_{2,2,k}}{\operatorname{tr}\left(\boldsymbol{C}_1^{-1}\bar{\boldsymbol{S}}_{1,1,k}\right) + \frac{\bar{S}_{2,2,k}}{c}} \tag{5.19}$$

3. Rotation symmetry:

$$\log\det\left(\hat{\boldsymbol{C}}_2\right) + \log\hat{a} + \log 2 + \frac{3}{K}\sum_{k=1}^{K}\log\left\{\operatorname{tr}\left[\frac{\hat{\boldsymbol{C}}_2^{-1}}{2}\left(\tilde{\boldsymbol{S}}_{2,2,k} + \boldsymbol{J}\tilde{\boldsymbol{S}}_{2,2,k}\boldsymbol{J}\right)\right] + \frac{\tilde{S}_{1,1,k}}{\hat{a}}\right\}$$

where $\tilde{\boldsymbol{S}}_k = \boldsymbol{V}\boldsymbol{E}\boldsymbol{T}\boldsymbol{S}_k\boldsymbol{T}^\dagger\boldsymbol{E}\boldsymbol{V}^\dagger = \begin{bmatrix} \tilde{S}_{1,1,k} & \tilde{\boldsymbol{S}}_{1,2,k} \\ \tilde{\boldsymbol{S}}_{2,1,k} & \tilde{\boldsymbol{S}}_{2,2,k} \end{bmatrix}$, with $\tilde{\boldsymbol{S}}_{2,2,k} \in \mathbb{H}^2(\tilde{\boldsymbol{S}}_k(2,2) \geq \boldsymbol{0})$, $\hat{a} \in \mathbb{R}(\hat{a} > 0)$ and $\hat{\boldsymbol{C}}_2 \in \mathbb{S}^2 \cap \mathbb{P}^2, \hat{\boldsymbol{C}}_2 \geq \boldsymbol{0}$ (with $\mathscr{S}_2$ indicating a 2×2 symmetric matrix) are fixed points of

$$a = \frac{3}{K}\sum_{k=1}^{K}\frac{\tilde{S}_{1,1,k}}{\operatorname{tr}\left[\frac{\boldsymbol{C}_2^{-1}}{2}\left(\tilde{\boldsymbol{S}}_{2,2,k} + \boldsymbol{J}\tilde{\boldsymbol{S}}_{2,2,k}\boldsymbol{J}\right)\right] + \frac{\tilde{S}_{1,1,k}}{a}} \tag{5.20}$$

and

$$\boldsymbol{C}_2 = \frac{3}{K}\sum_{k=1}^{K}\frac{\frac{1}{2}\left(\tilde{\boldsymbol{S}}_{2,2,k} + \boldsymbol{J}\tilde{\boldsymbol{S}}_{2,2,k}\boldsymbol{J}\right)}{\operatorname{tr}\left[\frac{\boldsymbol{C}_2^{-1}}{2}\left(\tilde{\boldsymbol{S}}_{2,2,k} + \boldsymbol{J}\tilde{\boldsymbol{S}}_{2,2,k}\boldsymbol{J}\right)\right] + \frac{\tilde{S}_{1,1,k}}{a}} \tag{5.21}$$

4. Azimuthal symmetry:

$$\log(\hat{a}) + 2\log(\hat{b}) + \log(2) + \frac{3}{K}\sum_{k=1}^{K}\log\left(\frac{\hat{S}_{1,1,k}}{\hat{a}} + \frac{\hat{S}_{2,2,k} + \hat{S}_{3,3,k}}{\hat{b}}\right) \tag{5.22}$$

where $\hat{\boldsymbol{S}}_k = \boldsymbol{E}\boldsymbol{T}\boldsymbol{S}_k\boldsymbol{T}^\dagger\boldsymbol{E} \in \mathscr{H}_3(\hat{\boldsymbol{S}}_k \geq \boldsymbol{0})$ and the i-th diagonal entry $\hat{S}_{i,i,k} \in \mathbb{R}, \hat{S}_k(i,i) > 0$, $i = 1, \ldots, 3$, with $\hat{a} \in \mathbb{R}, \hat{a} > 0$ and $\hat{b} \in \mathbb{R}, \hat{b} > 0$ fixed points of

$$a = \frac{3}{K}\sum_{k=1}^{K}\frac{\hat{S}_{1,1,k}}{\frac{\hat{S}_{1,1,k}}{a} + \frac{\hat{S}_{2,2,k}+\hat{S}_{3,3,k}}{b}} \tag{5.23}$$

and

$$b = \frac{3}{K}\sum_{k=1}^{K}\frac{\frac{1}{2}\left(\hat{S}_{2,2,k}+\hat{S}_{3,3,k}\right)}{\frac{\hat{S}_{1,1,k}}{a}+\frac{\hat{S}_{2,2,k}+\hat{S}_{3,3,k}}{b}} \tag{5.24}$$

It is worth underlying that the expressions given in (5.16), (5.18), (5.19), (5.20), (5.21), (5.23), and (5.24) must be applied iteratively using some initial guesses related to the sample covariance matrix.

5.5 Model order rules for classification of symmetries

The previous sections focused on a multiple hypothesis testing problem that includes both nested and non-nested instances. In such a situation, the GML approach does not work well and may fail because the likelihood tends to always assume the highest value that is under the H_1 hypothesis. Nevertheless, this drawback can be solved considering specific modified versions of the GML that are capable of accommodating nested instances. In particular, under each hypothesis, the GML is modified by employing the addition of a penalty term depending on the number of unknown parameters. This modification takes the name of MOS rules [31–34], which allows estimating the correct structure (namely, the model order) starting from the available observables (represented by the data matrix $\boldsymbol{R}$ in this context). Following this line of reasoning, the order is selected as the one corresponding to the minimum among four decision statistics, each of them evaluated under one of the above hypotheses.

In particular, a generic MOS rule derived in [31] and [32] can be expressed through the corresponding decision statistics in the following compact form

$$H_{\hat{i}} = \underset{\substack{H_i \\ i=1,\ldots,4}}{\arg\min}\left\{-2l(\hat{\boldsymbol{C}}^{(i)};\boldsymbol{R}) + n\,\eta(K)\right\} \tag{5.25}$$

where in the heterogeneous case $\boldsymbol{R}$ should be replaced by $\boldsymbol{Z}$ together with its corresponding log-likelihood $l(\hat{\boldsymbol{C}}^{(i)};\boldsymbol{Z})$. Note that, in (5.25), the quantity $\hat{\boldsymbol{C}}^{(i)}$ indicates the MLE of $\boldsymbol{C}$ under the i-th hypothesis comprising n parameters. Moreover, the term $n\,\eta(K)$ is generally referred to as penalty coefficient and its role can be intuitively explained by observing that the first term in (5.25) decreases with increasing n (for nested models), whereas the second term increases. As a consequence, $n\,\eta(K)$ in (5.25) penalizes overfitting [32]. Therefore, several MOS rules can be defined by simply changing the penalty term, for instance,

- AIC (Akaike information criterion): $\eta(K) = 2$
- GIC (generalized information criterion): $\eta(K) = \rho + 1$, with ρ an integer number greater than or equal to 2,
- BIC (Bayesian information criterion): $\eta(K) = \log(K)$.

Finally, considering the number of unknown parameters n in (5.25), it is not difficult to show that it is equal to:

- H_1(absence of symmetry): $n = 9$ (or $n = 8$),
- H_2(reflection symmetry): $n = 5$ (or $n = 4$),
- H_3(rotation symmetry): $n = 3$ (or $n = 2$),
- H_4(azimuthal of symmetry): $n = 2$ (or $n = 1$),

for the homogeneous and heterogeneous cases, respectively.

Before concluding this list of MOS rules, it is worth noting that an additional decision criterion can be considered [referred to as exponentially embedded families (EEF) approach], whose general theoretical formulation is provided in [34]. However, in this chapter, it has been omitted because this decision criterion requires the computation of the generalized likelihood ratio test implying that a suitable H_0 hypothesis must be defined. Nevertheless, the interested readers can find all the required details in [35].

In summary, for each pixel under test, the selected structure is the one associated to $H = H_{\hat{i}}$. The block schemes describing the covariance symmetries classification procedures for both the homogeneous and heterogeneous cases are shown in Figure 5.5. Both schemes, starting from the polarimetric images, are capable of providing a classified image using as few steps as described before. In particular, for each pixel under test, both schemes extract a pixel neighborhood by applying a sliding window. This neighborhood is then processed to obtain the covariance estimates

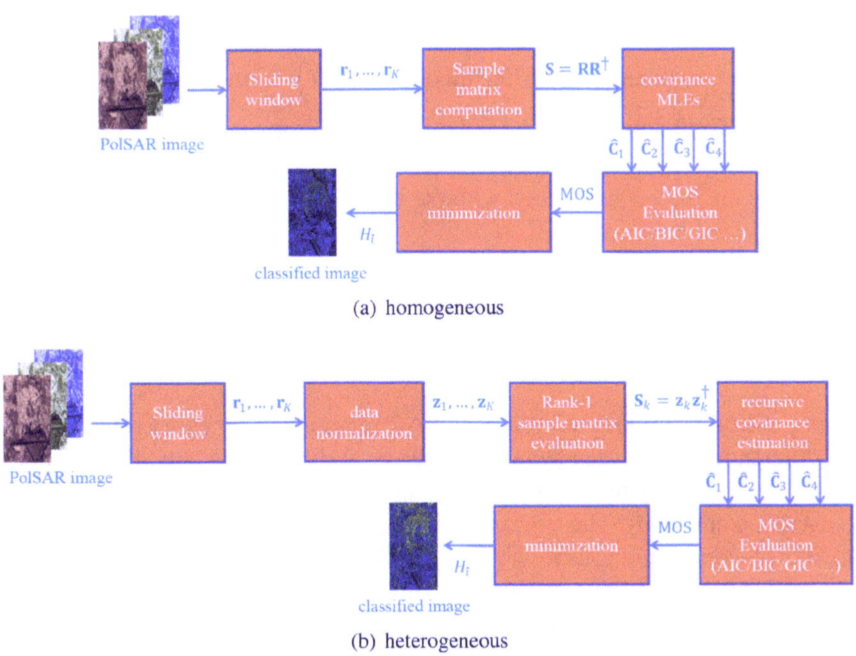

Figure 5.5 Block scheme of the covariance symmetries classification procedure for the (a) homogeneous and (b) heterogeneous cases

(for either a homogeneous or a heterogeneous scenario) under each hypothesis to feed the MOS-based classifier. Once the MOS values under each hypothesis have been computed, the hypothesis minimizing it is selected as an estimation of the covariance structure for the selected pixel. As to choice of the scheme to be applied, it would be useful to observe that the homogeneous case is a particularization of the heterogeneous one, with a reduced computational complexity. As a consequence, if the execution time is a precious resource, it is suggested to use the homogeneous scheme (for, otherwise, the heterogeneous one is preferred for its higher accuracy). Moreover, for the sake of completeness and to help the reader to acquire all the necessary information to implement the described procedures, in Algorithm 1, all the steps required to classify each pixel of the image are explicitly reported for both the homogeneous and heterogeneous cases. It indistinctly refers to one of the model order selectors AIC, BIC, and GIC. However, to use the EEF, it is sufficient to substitute the maximum in place of the minimum, as reported in [35].

Algorithm 1 Covariance Symmetries Detection

Require: $K = W \times W$, $\boldsymbol{R}$ for the homogeneous scenario or $\boldsymbol{Z}$ for the heterogeneous case.

Ensure: $H_{\hat{i}}$.

1: Compute the sample matrix $\boldsymbol{S}$ (homogeneous) or the rank-1 sample matrices $\boldsymbol{S}_k$ (heterogeneous), $k = 1,\ldots,K$, and the matrices $\bar{\boldsymbol{S}}$, $\tilde{\boldsymbol{S}}$, and $\hat{\boldsymbol{S}}$ as defined in Proposition 1 and Proposition 2 if the homogeneous or heterogeneous case is considered, respectively.

2: Compute the maximum log-likelihoods $l\left(\widehat{\boldsymbol{C}}, \boldsymbol{R}\right)$ (homogeneous), $i = 1,\ldots,4$, or $l\left(\widehat{\boldsymbol{C}}, \boldsymbol{Z}\right)$ (heterogeneous), $i = 1,\ldots,4$, from Proposition 1 and Proposition 2, respectively.

3: Choose the index $\hat{i}$ associated with the minimum AIC, BIC, or GIC as given in (1.25).

4: Associate the label $H_{\hat{i}}$ to the pixel under test.

5.6 Detection of symmetries in PolInSAR applications

The use of PolInSAR instead of just PolSAR represents a potential solution to provide differentiation between symmetry sources at different heights. For example, it allows deducing if a detected azimuthal symmetry originates at the ground level due to vegetation or canopy. Thus, it can be derived a technique to discriminate between the different contributions enabling the estimation also of the elevation in each resolution cell [38, 39]. The grand advantage of this framework is that it allows extracting more information from PolInSAR data compared to polarimetric data only.

In monostatic PolInSAR systems, the imaging area is scanned at least twice from slightly different angles, thus yielding two scattering vectors [37], namely, $\boldsymbol{k}_m$, and $\boldsymbol{k}_s$, where the subscript m stands for master and s for slave:

$$\boldsymbol{k}_m = \frac{1}{\sqrt{2}}\begin{pmatrix} S_{hh_m} \\ \sqrt{2}S_{hv_m} \\ S_{vv_m} \end{pmatrix} \quad \boldsymbol{k}_s = \frac{1}{\sqrt{2}}\begin{pmatrix} S_{hh_s} \\ \sqrt{2}S_{hv_s} \\ S_{vv_s} \end{pmatrix} \tag{5.26}$$

The subscripts *hh*, *vv*, and *hv* identify the transmitter/receiver polarimetric SAR channels. The complete scattering phenomenon representing one pixel can be described using a (6 × 6) covariance matrix $\boldsymbol{C}^6$, formed using the superposition of the scattering vectors (5.26):

$$\boldsymbol{C}^6 = \left\langle \begin{bmatrix} \boldsymbol{k}_m \\ \boldsymbol{k}_s \end{bmatrix} \begin{bmatrix} \boldsymbol{k}_m^H & \boldsymbol{k}_s^H \end{bmatrix} \right\rangle = \begin{bmatrix} \boldsymbol{C}(1,1) & \boldsymbol{C}(1,2) \\ \boldsymbol{C}(2,1) & \boldsymbol{C}(2,2) \end{bmatrix}$$

where $\boldsymbol{C}(1,1)$ and $\boldsymbol{C}(2,2)$ are the 3 × 3 conventional Hermitian polarimetric covariance matrices of each image separately; $\boldsymbol{C}(1,2)$ or $\boldsymbol{C}(2,1)$ is the 3 × 3 cross-covariance matrix that contains polarimetric information and related interferometric phases. Following the same line of reasoning of the previous sections, the covariance structures can also be found in this case. Once, the covariance structure is properly detected as well as the elevation of each pixel is derived, they can be used for the classification of PolInSAR scenes.

5.7 Performance assessment

In this section, the effectiveness of the classification rules described in this chapter is assessed. More precisely, a quantitative analysis over simulated data is first performed, and then the algorithms are tested on measured fully-polarimetric SAR data.

5.7.1 Analysis of simulated data

In this section, an in-depth performance analysis is conducted resorting to simulated data adopting as performance metric the correct classification of each hypothesis. The metric is estimated over a total of $M_C = 10^4$ Monte-Carlo independent trials. The described classifiers derived in both homogeneous and heterogeneous environments are compared assuming data modeled as spherically invariant random vectors (SIRVs) [52–54], namely

$$\boldsymbol{r}_k = \sqrt{\tau_k}\,\boldsymbol{g}_k, \quad k = 1, \ldots, K \tag{5.27}$$

where τ_k is a positive real random variable (usually known as texture) statistically independent of $\boldsymbol{g}_k$, which are statistically independent N-dimensional zero-mean complex circular Gaussian vectors with covariance matrix $\bar{\boldsymbol{C}}^{(i)}$, $i = 1, \ldots, 4$. The nominal covariance matrices associated to the considered four scenarios (viz., absence of symmetry, reflection, rotation, and azimuthal symmetry) are, respectively, given in

$$\bar{C}^{(1)} = \begin{bmatrix} 1 & 0.2+0.3j & 0.5-0.3j \\ 0.2-0.3j & 0.25 & -0.2-0.2j \\ 0.5+0.3j & -0.2+0.2j & 0.8 \end{bmatrix} \tag{5.28}$$

$$\bar{C}^{(2)} = \begin{bmatrix} 1 & 0 & 0.5-0.3j \\ 0 & 0.25 & 0 \\ 0.5+0.3j & 0 & 0.4 \end{bmatrix} \tag{5.29}$$

$$\bar{C}^{(3)} = \begin{bmatrix} 1 & 0.3j & 0.2 \\ -0.3j & 0.4 & 0.3j \\ 0.2 & -0.3j & 1 \end{bmatrix} \tag{5.30}$$

and

$$\bar{C}^{(4)} = \begin{bmatrix} 1 & 0 & 0.5 \\ 0 & 0.25 & 0 \\ 0.5 & 0 & 1 \end{bmatrix} \tag{5.31}$$

Moreover, the scaling factors $\tau_1, \ldots, \tau_K$ are also modeled to be statistically independent and follow the Gamma distribution whose pdf is

$$f(x) = \frac{1}{\Gamma(\nu)} \frac{1}{\mu^\nu} x^{\nu-1} e^{-x/\mu} u(x)$$

where $\Gamma(\cdot)$ is the Eulerian Gamma function and μ and $\nu > 0$ are the scale and shape parameters, respectively (the considered setting assumes $\mu = 1/\nu$ to have a Gamma distribution with unit mean). By doing so, the component amplitudes of $\boldsymbol{r}_1, \ldots, \boldsymbol{r}_K$ are K-distributed. It is now also worth underlining that for large values of ν, data distribution tends toward the Gaussian distribution. Finally, the number of iterations to estimate the covariance matrix in the heterogeneous case is set equal to 5.

Figure 5.6 displays the confusion matrix of the AIC, BIC, and GIC (with $\rho = 2$ in all the considered tests) selectors (subplots from top to bottom) for a number of looks inside the sliding windows equal to $K = 9$. The analyses refer to the aforementioned SIRV environment, having set the scale parameter $\nu = 2$. Figures on the left refer to the classifier in the homogeneous environment, whereas figures on the right refer to those in the heterogeneous case. As expected, the MOS rules in the heterogeneous environment are not sensitive to the shape parameter due to the invariance under scaling. For this reason, they are capable of ensuring better performance in terms of structure classification than the homogeneous counterparts. Interestingly, under the H_1 hypothesis, all the considered MOS rules share almost the same performance. In fact, it represents the case in which all the detectors are capable of correctly detecting the current symmetry. Conversely, the azimuthal symmetry is the most challenging structure to be revealed (as clearly emphasized by the confusion matrices) due to the fact that it can be seen as a special case of all the other structures.

(a) homogeneous AIC (b) heterogeneous AIC

(c) homogeneous BIC (d) heterogeneous BIC

(e) homogeneous GIC (f) heterogeneous GIC

Figure 5.6 Confusion matrices of the AIC, BIC, and GIC rules for a number of looks K = 9 in a SIRV environment with a scale parameter ν = 2. Figures on the left refer to the classifier in homogeneous environment, whereas figures on the right refer to those in the heterogeneous case.

Similar to the results in Figures 5.6 and 5.7 shows the confusion matrices for all the selectors considering a larger number of looks, viz., $K = 25$. This second analysis also refers to a SIRV environment with the parameter $\nu = 2$. The evidence from

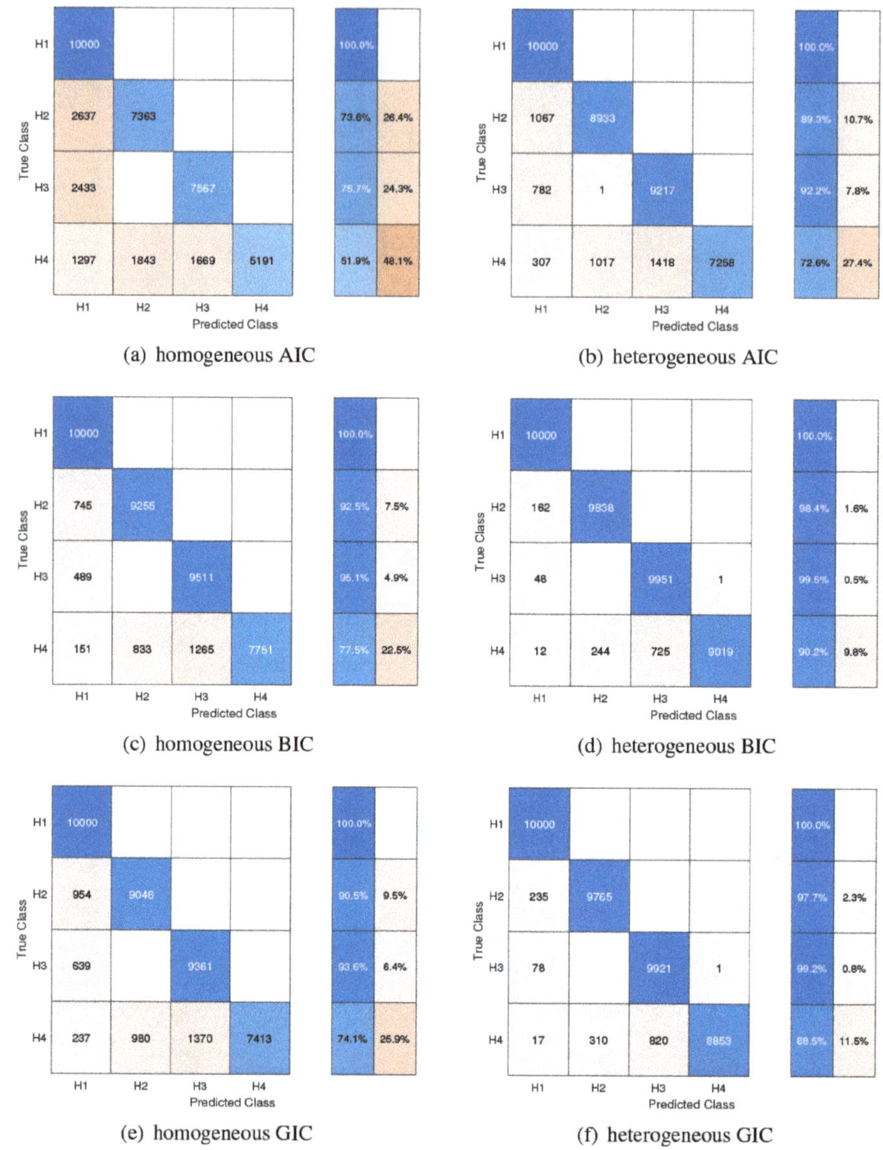

(a) homogeneous AIC (b) heterogeneous AIC

(c) homogeneous BIC (d) heterogeneous BIC

(e) homogeneous GIC (f) heterogeneous GIC

Figure 5.7 Confusion matrices of the AIC, BIC, and GIC rules for a number of looks K = 25 in a SIRV environment with a scale parameter ν = 2. Figures on the left refer to the classifier in homogeneous environment, whereas figures on the right refer to those in the heterogeneous case.

these confusion matrices is again that the algorithm developed in the heterogeneous case works better than its homogeneous counterpart. Again, the H_4 hypothesis represents the more challenging case. Nevertheless, this second study case demonstrates that increasing the number of looks in the sliding window allows improving the classification performance of the covariance structure. Finally, the GIC rule is the one that reaches the higher classification scores slightly overcoming the BIC.

5.7.2 *Test on real-recorded SAR images*

In this section, the classification capabilities of the proposed methods are tested over a measured SAR dataset. More precisely, an L-band (1.3 GHz) coherent fully polarimetric sample dataset*, acquired by the airborne Experimental SAR (E-SAR) system of the German Aerospace Center (DLR) on July 20, 1999, is herein considered. The nominal one-look spatial resolution is ~3×0.4 m. The image (2816×1540 pixels), representing a scene of the Oberpfaffenhofen Test Site Area (DE), contains mixed urban areas, forests, and bare areas that contain an aircraft runway. Its span and the Pauli decomposition [1, 16] in the Red-Green-Blue color space are reported in Figure 5.8a and b, respectively. The corresponding optical image drawn from Google Earth©, and compressed in one dimension to better match with the SAR image, is shown in Figure 5.9.

In section 5.7.1, the quantitative analysis conducted on simulated data has pointed out that for high values of K, the considered classifiers share about the same performance. Thus, in order to avoid redundant examples, in the following study,

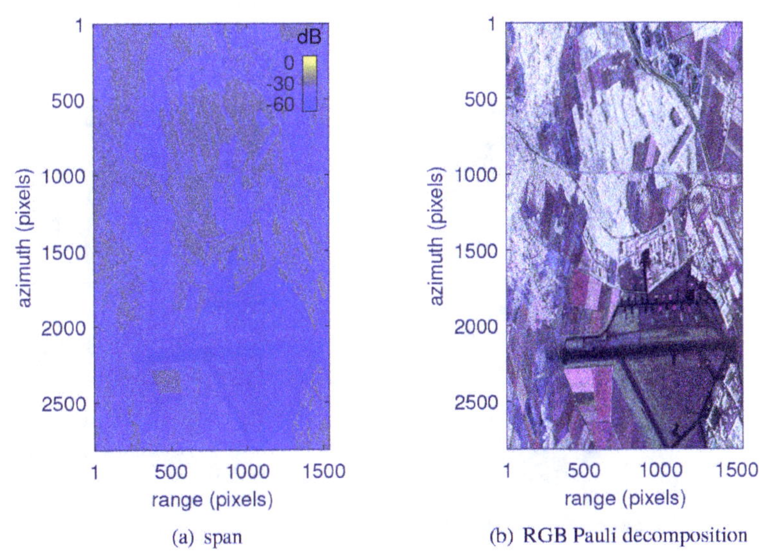

Figure 5.8 Oberpfaffenhofen Test Site Area (DE) (E-SAR L-band 1998) data. (a) Span (expressed in dB). (b) RGB of the Pauli decomposition for the reference image of size 2816 × 1540 pixels.

*Data can be downloaded at https://earth.esa.int/web/polsarpro/data-sources/sample-datasets thanks to Dr Kostas Papathanassiou.

only two classification schemes along with high values of K will be considered. Specifically, the behavior of BIC and GIC for $K = 49$ is investigated.

In Figure 5.10, the classification results are shown by comparing the BIC rule in the homogeneous (left subplot) and heterogeneous (right subplot) cases. The image

Figure 5.9 *Optical image of the observed scene of Oberpfaffenhofen Test Site Area (DE). The image is drawn from Google Earth©.*

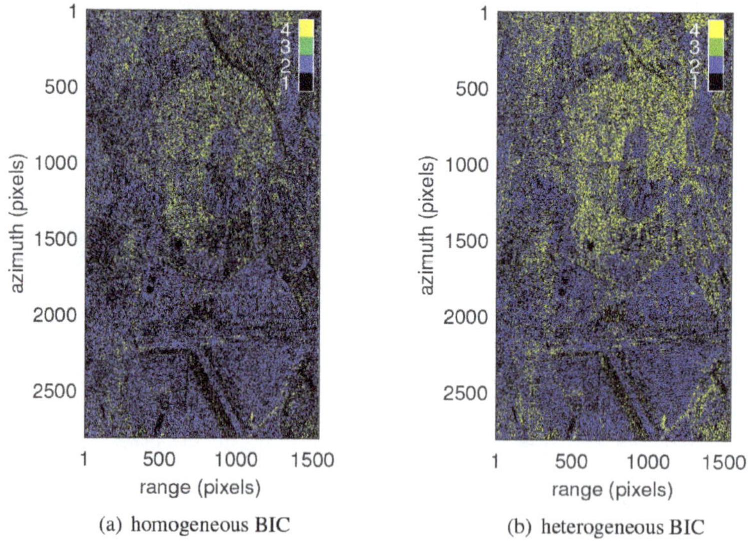

Figure 5.10 *Detected symmetries within the reference image (L-band E-SAR data of Oberpfaffenhofen Test Site Area, DE) with the BIC-based selectors in homogeneous and heterogeneous scenarios, for $K = 49$ looks*

Table 5.1 Number (expressed in percentage) of detected symmetries for the pixels of the L-band E-SAR image of Oberpfaffenhofen Test Site Area (DE) with the BIC- and GIC-based detectors for both homogeneous and heterogeneous cases

	Homogeneous case		Heterogeneous case	
Detected symmetry	**BIC (%)**	**GIC (%)**	**BIC (%)**	**GIC (%)**
H_1 (no-symmetry)	52.73	51.30	37.30	35.80
H_2 (reflection)	37.27	38.18	45.51	46.24
H_3 (rotation)	3.59	3.71	5.17	5.28
H_4 (azimuthal)	6.41	6.81	12.02	12.68

is represented in pseudo-color, namely, a specific color is associated to each of the classification outcomes. Therefore, each pixel can assume one of the following four colors corresponding to the same hypotheses:

- black for H_1 (absence of symmetry),
- blue for H_2 (reflection symmetry),
- green for H_3 (rotation symmetry),
- yellow for H_4 (azimuthal symmetry).

By doing so, areas sharing the same features are classified with the same symmetry and, hence, displayed with the same color. Specifically, it is revealed that reflection symmetry (blue pixels) is predominant over the crops and bare fields, whereas forests are classified as azimuthal symmetry (yellow pixels). Moreover, there exist a few green pixels associated to the scattering of some buildings present within the scene. Finally, the absence of symmetry (black pixels) essentially arises over the streets and the aircraft runway. Comparing left and right subplots highlights some advantages of the heterogeneous BIC over the homogeneous one. For instance, with a focus on the forest area that is reasonably characterized by azimuthal symmetry, the former labels a more extended area of such pixels with the correct symmetry than the latter. The same remark holds also for cultivated and bare areas, whose pixels are almost everywhere better classified by the heterogeneous BIC. These results are also confirmed by the values reported in Table 5.1, where the percentage of pixels classified as belonging to each hypothesis is indicated. In fact, the percentage of pixels assigned to the class of absence of symmetry is higher in the homogeneous case despite the large extent of vegetation and fields in the observed scene. Similarly, in Figure 5.11, the classification performance of the homogeneous and heterogeneous GIC is also reported. The figure confirms the benefits provided by the classifier developed in the heterogeneous case with respect to its homogeneous counterpart, despite an additional computational complexity due to the recursive nature of the involved covariance estimators of the heterogeneous classifier.

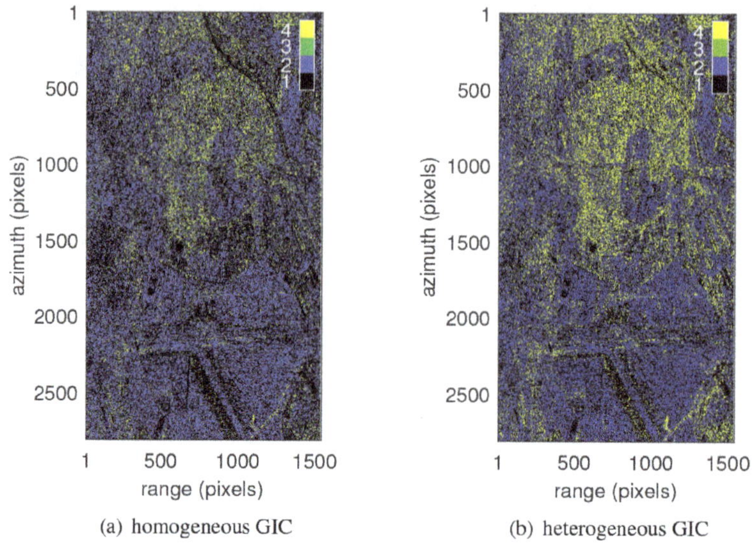

Figure 5.11 Detected symmetries within the reference image (L-band E-SAR data of Oberpfaffenhofen Test Site Area, DE) with the GIC-based selectors in homogeneous and heterogeneous scenarios, for K = 49 looks

5.8 Conclusion

This chapter has dealt with the problem of covariance matrix classification in PolSAR images on the base of the special structures assumed under symmetrical properties of the returns associated to the pixels under test. In particular, the chapter has focused on both homogeneous and heterogeneous SAR images' classification, including a description of the symmetry classification within the PolInSAR imagery. For all the described frameworks, the problem has been formulated as a multiple hypothesis test comprising both nested and non-nested hypotheses. For this reason, it has been solved by resorting to the well-known MOS rules to overcome the limitations of the classic GML approach.

Results conducted on both simulated and L-band real-recorded PolSAR data have proven the effectiveness of the described methodologies, thus paving the way for further applications, e.g., as a preliminary step of a more sophisticated two-stage algorithm aimed at, for instance, classifying the scene.

References

[1] Lee J.S., Pottier E. *Polarimetric radar imaging: from basics to applications*. Boca Raton, FL: CRC Press; 2009.

[2] Nghiem S.V., Yueh S.H., Kwok R., Li F.K. 'Symmetry properties in polarimetric remote sensing'. *Radio Science*. 1992, vol. 27(5), pp. 693–711.

[3] Ferro-Famil L., Pottier E. 'Dual frequency polarimetric SAR data classification and analysis'. *Progress In Electromagnetics Research.* 2001, vol. 31, pp. 247–72.
[4] Curlander J.C., McDonough R.N. *Synthetic aperture radar*. New York: John Wiley & Sons; 1991.
[5] Cloude S.R., Pottier E. 'A review of target decomposition theorems in radar polarimetry'. *IEEE Transactions on Geoscience and Remote Sensing*. 1996, vol. 34(2), pp. 498–518.
[6] Krogager E. 'New decomposition of the radar target scattering matrix'. *Electronics Letters*. 1990, vol. 26(18), pp. 1525–7.
[7] Cameron W.L., Leung L.K. 'Feature motivated polarization scattering matrix decomposition'. *IEEE International Radar Conference*; Arlington, VA, 1990. pp. 549–57.
[8] Simone G., Farina A., Morabito F.C., Serpico S.B., Bruzzone L. 'Image fusion techniques for remote sensing applications'. *Information Fusion*. 2002, vol. 3(1), pp. 3–15.
[9] Eltoft T., Doulgeris A.P. 'Model-based polarimetric decomposition with higher order statistics'. *IEEE Geosci Remote Sens Lett*. 2019, vol. 16(6), pp. 992–96.
[10] van Zyl J.J. 'Unsupervised classification of scattering behavior using radar polarimetry data'. *IEEE Trans Geosci Remote Sens*. 1989, vol. 27(1), pp. 36–45.
[11] Novak L.M., Burl M.C., Chaney R.D. 'Optimal processing of polarimetric synthetic-aperturesynthetic-aperture radar imagery'. *The Lincoln Laboratory Journal*. 1990, vol. 3, pp. 273–90.
[12] Toma M.R., Vinelli F., Farina A. 'Processing of polarimetric and multifrequency SAR data recorded by maestro-1 campaign'. *Proceedings of the Remote Sensing: Global Monitoring for Earth Management.* vol. 2; 1991. pp. 349–51.
[13] Verbout S.M., Netishen C.M., Novak L.M. 'Polarimetric techniques for enhancing SAR imagery'. *Synthetic Aperture Radar*. 1992, vol. 1630, pp. 141–73.
[14] Larson V.J., Novak L.M. 'Polarimetric subspace target detector for SAR data based on the Huynen dihedral model'. *Algorithms for Synthetic Aperture Imagery II*. 1995, vol. 2487, pp. 235–50.
[15] Lemoine G.G., de Grandi G.F., Sieber A.J. 'Polarimetric contrast classification of agricultural fields using MAESTRO 1 AIRSAR data'. *International Journal of Remote Sensing*. 1994, vol. 15(14), pp. 2851–69.
[16] Cloude S.R., Pottier E. 'An entropy based classification scheme for land applications of polarimetric SAR'. *IEEE Trans Geosci Remote Sens*. 1997, vol. 35(1), pp. 68–78.
[17] Lee J.S., Papathanassiou K.P., Hajnsek I, *et al*. 'Applying polarimetric SAR interferometric data for forest classification'. *IEEE International Geoscience and Remote Sensing Symposium (IGARSS)*. vol. 7; 2005.
[18] Zhang L., Zou B., Zhang J. 'Classification of polarimetric SAR image based on support vector machine using multiple-component scattering model and

texture features'. *EURASIP Journal on Advances in Signal Processing*. 2010, vol. 1, pp. 1–9.

[19] Dabboor M., Shokr M. 'A new likelihood ratio for supervised classification of fully polarimetric SAR data: an application for sea ice type mapping'. *ISPRS Journal of Photogrammetry and Remote Sensing*. 2013, vol. 84, pp. 1–11.

[20] Uhlmann S., Kiranyaz S. 'Evaluation of classifiers for polarimetric SAR classification'. Presented at IGARSS - IEEE international geoscience and remote sensing symposium; 2013. Melbourne, Australia,

[21] Lee J.S., Grunes M.R., Ainsworth T.L. 'Unsupervised classification using polarimetric decomposition and the complex Wishart classifier'. *IEEE Trans on Geoscience and Remote Sensing*. 1999, vol. 37(5), pp. 2249–58.

[22] Lee J.S., Grunes M.R., Pottier E. 'Unsupervised terrain classification preserving polarimetric scattering characteristics'. *IEEE Trans Geosci Remote Sens*. 2004, vol. 42(4), pp. 722–31.

[23] De Maio A., Orlando D., Pallotta L., Clemente C. 'A multifamily GLRT for oil spill detection'. *IEEE Trans Geosci Remote Sens*. 2016, vol. 55(1), pp. 63–79.

[24] Zhong N., Yang W., Cherian A., Yang X., Xia G.-S., Liao M. 'Unsupervised classification of polarimetric SAR images via riemannian sparse coding'. *IEEE Trans Geosci Remote Sens*. 2017, vol. 55(9), pp. 5381–90.

[25] Pallotta L., Orlando D. 'Polarimetric covariance eigenvalues classification in SAR images'. *IEEE Geosci Remote Sens Lett*. 2018, vol. 16(5), pp. 746–50.

[26] Biondi F., Clemente C., Orlando D. 'An eigenvalue-based approach for structure classification in polarimetric SAR images'. *IEEE Geosci Remote Sens Lett*. 2019, vol. 17(6), pp. 1003–07.

[27] Addabbo P., Biondi F., Clemente C., Orlando D., Pallotta L. 'Classification of covariance matrix eigenvalues in polarimetric SAR for environmental monitoring applications'. *IEEE Aerospace and Electronic Systems Magazine*. 2019, vol. 34(6), pp. 28–43.

[28] Addabbo P., Han S., Biondi F., Giunta G., Orlando D. 'Adaptive radar detection in the presence of multiple alternative hypotheses using Kullback-Leibler information criterion-part II: applications'. *IEEE Transactions on Signal Processing*. 2021, vol. 69, pp. 3742–54.

[29] Kay S.M. *Fundamentals of statistical signal processing: estimation theorystatistical signal processing: estimation theory*. Vol. 1. Hoboken: Prentice Hall; 1993.

[30] Kay S.M. 'The multifamily likelihood ratio test for multiple signal model detection'. *IEEE Signal Processing Letters*. 2005, vol. 12(5), pp. 369–71.

[31] Selén Y. *Model selection*. [PhD Dissertation in Electrical Engineering]. Sweden: Uppsala University; 2004.

[32] Stoica P., Selén Y. 'Model-order selection: a review of information criterion rules'. *IEEE Signal Processing Magazine*. 2004, vol. 21(4), pp. 36–47.

[33] Stoica P., Selen Y., Li J. 'On information criteria and the generalized likelihood ratio test of model order selection'. *IEEE Signal Processing Letters*. 2004, vol. 11(10), pp. 794–97.

[34] Kay S. 'Exponentially embedded families-new approaches to model order estimation'. *IEEE Transactions on Aerospace and Electronic Systems*. 2005, vol. 41(1), pp. 333–45.

[35] Pallotta L., Clemente C., De Maio A., Soraghan J.J. 'Detecting covariance symmetries in polarimetric SAR images'. *IEEE Trans Geosci Remote Sens*. 2017, vol. 55(1), pp. 80–95.

[36] Pallotta L., De Maio A., Orlando D. 'A robust framework for covariance classification in heterogeneous polarimetric SAR images and its application to L-band data'. *IEEE Trans Geosci Remote Sens*. 2019, vol. 57(1), pp. 104–19.

[37] Cloude S.R., Papathanassiou K.P. 'Polarimetric SAR interferometry'. *IEEE Trans Geosci Remote Sens*. 1998, vol. 36(5), pp. 1551–65.

[38] Tahraoui S., Clemente C., Pallotta L., Soraghan J.J., Ouarzeddine M. 'Covariance symmetries detection in polinsar data'. *IEEE Trans Geosci Remote Sens*. 2018, vol. 56(12), pp. 6927–39.

[39] Tahraoui S., Clemente C., Pallotta L., Soraghan J.J., Ouarzeddine M. 'Applying symmetries detection to P-band POLinSAR data'. *International Radar Conference*; Toulon, France, 2019.

[40] Vasile G., Ovarlez J.-P., Pascal F., Tison C. 'Coherency matrix estimation of heterogeneous clutter in high-resolution polarimetric SAR images'. *IEEE Trans Geosci Remote Sens*. 2010, vol. 48(4), pp. 1809–26.

[41] Ulaby F.T., Long D.G. *Microwave radar and radiometric remote sensing*. Ann Arbor, MI: University of Michigan Press; 2014. Available from http://hdl.handle.net/2027/mdp.39015093623307

[42] ESA *PolSARPro (the polarimetric SAR data processing and educational tool) [online]; 2000-2019*. Available from https://earth.esa.int/web/polsarpro/

[43] Rignot E.J.M. 'Effect of faraday rotation on L-band interferometric and polarimetric synthetic-aperture radar data'. *IEEE Trans Geosci Remote Sens*. 2000, vol. 38(1), pp. 383–90.

[44] Aubry A., Carotenuto V., De Maio A., Pallotta L. 'Assessing reciprocity in polarimetric SAR data'. *IEEE Geosci Remote Sens Lett*. 2019, vol. 17(1), pp. 87–91.

[45] Pallotta L. 'Reciprocity evaluation in heterogeneous polarimetric SAR images'. *IEEE Geosci Remote Sens Lett*. 2022, vol. 19, pp. 1–5.

[46] Lehmann E.L. 'Testing statistical hypotheses' in *Testing statistical hypotheses*. 2nd ed. New York, NY: Springer-Verlag; 1986.

[47] Ollila E., Tyler D.E., Koivunen V., Poor H.V. 'Complex elliptically symmetric distributions: survey, new results and applications'. *IEEE Transactions on Signal Processing*. 2012, vol. 60(11), pp. 5597–625.

[48] Soloveychik I., Wiesel A. 'Tyler's covariance matrix estimator in elliptical models with convex structure'. *IEEE Transactions on Signal Processing*. 2014, vol. 62(20), pp. 5251–59.

[49] Chitour Y., Pascal F. 'Exact maximum likelihood estimates for SIRV covariance matrix: existence and algorithm analysis'. *IEEE Transactions on Signal Processing*. 2008, vol. 56(10), pp. 4563–73.

[50] Conte E., De Maio A., Ricci G. 'Recursive estimation of the covariance matrix of a compound-gaussian process and its application to adaptive CFAR detection'. *IEEE Transactions on Signal Processing*. 2002, vol. 50(8), pp. 1908–15.

[51] Soloveychik I., Trushin D., Wiesel A. 'Group symmetric robust covariance estimation'. *IEEE Transactions on Signal Processing*. 2016, vol. 64(1), pp. 244–57.

[52] Conte E., Longo M. 'Characterisation of radar clutter as a spherically invariant random process'. *IEE Proceedings F Communications, Radar and Signal Processing*. 1987, vol. 134(2), pp. 191–7.

[53] Greco M.S., Gini F. 'Statistical analysis of high-resolution SAR ground Clutter data'. *IEEE Transactions on Geoscience and Remote Sensing*. 2007, vol. 45(3), pp. 566–75.

[54] Oliver C., Quegan S. *Understanding synthetic aperture radar images*. Raleigh, NC: SciTech Publishing; 2004.

Chapter 6

Polarimetric information to enhance synthetic aperture radar automatic target recognition capabilities

Domenico Gaglione[1], *Carmine Clemente*[1], *Luca Pallotta*[2], *Christos Ilioudis*[1], *Antonio De Maio*[3], *and John Soraghan*[1]

6.1 Introduction

Automatic target recognition (ATR) represents the capability of a system to reliably distinguish different targets. Radar systems are particularly suitable for this function being used in any weather and light conditions, effective for long-range operations, and capable of almost instantaneous response to threats. ATR technologies find use in several civilian scenarios, although the military sector represents the main driver for recent and future developments. Several systems are currently used for the recognition of cooperative targets, such as the Identification Friend or Foe system, also called secondary surveillance radar, used by military and civilian authorities for the identification of aircraft. However, the ability to automatically distinguish between friend, hostile, and neutral targets in a non-cooperative environment is a much more valuable feature for any radar. This also presents significant design challenges. Specifically, in a battlefield scenario, if a target can be accurately classified, an appropriate countermeasure can be deployed in order to successfully neutralise the threat. Reliability of the decision is also key to avoid potential casualties caused by friendly fire. The ATR capability of radars derives from the peculiar and characterising interactions between the target and the electromagnetic wave transmitted by the systems. The reflected signal from the target contains unique information that can be extracted through signal processing techniques and exploited to perform the classification task. The nature of this information varies, and in addition to the nature of the target, it also depends on the actual characteristics of the radar system that is employed, such as acquisition mode or bandwidth. For example, radar imaging systems, such as synthetic aperture radars (SARs), provide images of potential

[1]University of Strathclyde, Centre for Signal and Image Processing (CESIP), UK
[2]University of Roma Tre, Italy
[3]University of Naples "Federico II", Italy

targets from which discriminative features, such as shape, size, and reflectivity, can be extracted. Advanced methods also combine multiple images acquired with sensors on board multiple platforms or acquired in different instants of time and/or with polarimetric or multi-frequency sensors. The terminology used in the field of ATR may be sometimes ambiguous. Several terms, such as classification, recognition, identification and characterisation, are often used interchangeably, but sometimes they are also meant to indicate different levels of discrimination. Throughout this chapter, the used terminology is in accordance with the NATO AAP-6 Glossary Terms and Definitions [1]:

- *recognition* describes the ability to give a type/category to a target, e.g. car or truck, tank or personnel carrier, fighter aircraft or helicopter;
- *identification* regards the capability to assign the target to a subclass, e.g. compact car or SUV, Eurofighter Typhoon or F-16 Fighting Falcon;
- *characterisation* takes into account the class variants or target configurations, e.g. Porsche 911 Carrera coupé or Porsche 911 Carrera cabriolet.

The term classification, instead, is used in this chapter to describe the general process of assigning objects to categories, whatever the detail level. Common to all the discrimination levels is the dependence of the performance from key parameters, such as the signal-to-noise ratio (SNR), resolutions, accuracy, cardinality of the classification problem and correlation between the targets in the observed domain. A way to enhance performance is to identify domains, physical or target representations, that allow a higher degree of target separability; one of this domain is the polarimertic one.

Among the various imaging modes, polarimetric SAR imaging can offer useful insight on the scattering mechanism of a target, thus allowing for the extraction of more characterising information that can be useful for ATR purposes. Polarimetric decomposition can support the extraction of useful information by expressing the polarimetric returns as a combination of returns from canonical objects, which allow for an easier physical interpretation of a fully polarimetric SAR image.

The capability of polarimetric decompositions to represent the scattering mechanism of the illuminated objects makes their application highly effective in target recognition from radar imagery. Comparisons between different methods, both coherent and incoherent, were carried out in [2, 3]. In [2], Pauli, Cameron and Krogager decompositions were tested on the classification of several urban and non-urban areas (water, houses, roads, trees, grass and crops) from real fully polarimetric SAR images acquired with the German E-SAR system, and compared with incoherent target decomposition (ICTD) techniques. Results showed that the Krogager and Cameron decompositions outperformed all the other assessed techniques, with an overall accuracy of 87% and 76%, respectively. On the other hand, the Pauli decomposition did not result in better performance than using the original elements of the scattering matrix. In the later work, classification performance of the Krogager decomposition was evaluated on Radarsat-2 fully polarimetric images and compared with ICTD techniques. The data set consisted of images of crops acquired in different days of the year; once again, the Krogager decomposition gave the best results, with an overall accuracy of 78%. The authors of both the comparison studies ascribed better performance of the Krogager decomposition over the ICTD techniques to the fact that, being a coherent decomposition, it maintains full resolution during the feature extraction process. Recent advances in wide-band fully polarimetric SAR system have further revived the employment of coherent decomposition techniques for target recognition, with particular focus on the classification of civilian vehicles [4].

In this chapter, an example of integration of polarimetric information for ATR purposes is shown. In particular, an approach aimed at integrating Krogager decomposition with image-moments-based feature extraction is presented with the aim to highlight the benefits of polarimetic-based physical interpretation in ATR. The results of fully polarimetric SAR data show the advantages that a polarimetric radar could provide in terms of recognition accuracy and training data requirements. However, due to limitations in data availability to the research community, there are still a number of open research questions to be answered to fully unlock the benefits of the use of polarimetry in ATR, and the final goal of this chapter is to introduce some of these outstanding questions.

6.2 Target decompositions

The additional information provided by the polarisation of the scattered electromagnetic wave can be exploited for classification purposes, as objects interact differently based on the polarisation on the incident wave. A fully polarimetric SAR image is completely characterised by the scattering matrix $\boldsymbol{S}$, defined as follows [5]:

$$\boldsymbol{S}_{(H,V)} = \begin{bmatrix} S_{HH} & S_{HV} \\ S_{VH} & S_{VV} \end{bmatrix} \in \mathbb{C}^2 \tag{6.1}$$

where $\mathbb{C}$ is the set of complex numbers, S_{HH} and S_{VV} are the linear co-polarised terms, and S_{HV} and S_{VH} are the linear cross-polarised terms*. Accordingly, in the case of circular polarised waves, the scattering matrix can be expressed as:

$$\boldsymbol{S}_{(R,L)} = \begin{bmatrix} S_{RR} & S_{RL} \\ S_{LR} & S_{LL} \end{bmatrix} = \frac{1}{2}\begin{bmatrix} 1 & j \\ 1 & -j \end{bmatrix}\begin{bmatrix} S_{HH} & S_{HV} \\ S_{VH} & S_{VV} \end{bmatrix}\begin{bmatrix} 1 & 1 \\ j & -j \end{bmatrix} \in \mathbb{C}^2 \tag{6.2}$$

where S_{RR} and S_{LL} are the circular co-polarised terms, and S_{RL} and S_{LR} are the circular cross-polarised terms, respectively.

The interpretation of a fully polarimetric SAR image can be a challenging task. For this reason, decomposition techniques are employed in order to obtain a different representation of the data that has a direct physical interpretation. These polarimetric decomposition techniques can be divided into two main classes. ICTD methods

*Note that in the monostatic case, the cross-polarised terms are equal assuming that electromagnetic reciprocity holds.

are used to represent *distributed targets* through a second-order characterisation of the scattering matrix (i.e. coherency matrix or covariance matrix). Distributed targets, in fact, are highly affected by speckle noise†, and therefore, their characterisation cannot merely rely on intensity values picked from the fully polarimetric SAR image. However, ICTD techniques rely on the estimation of either the coherency matrix or the covariance matrix through $\boldsymbol{S}$, and for the same reason, they cannot maintain the full resolution of the SAR image [2, 3]‡. On the other hand, coherent target decomposition (CTD) methods are suitable for the representation of point-like targets. Here, the objective is to express the scattering matrix as a combination of canonical objects which can then be interpreted more easily from a physical point of view. Furthermore, as CTDs are directly computed from $\boldsymbol{S}$, they maintain the full resolution of the SAR image. ICTD techniques are generally used to characterise natural targets, such as forests, agriculture areas, water, etc., while CTD methods are suitable for the extraction of features from man-made targets [7, 8]. The remainder of this chapter will focus on CTD class of polarimetric decompositions that are more suited for ATR.

6.2.1 Coherent target decompositions

A CTD can be expressed as [6]:

$$\boldsymbol{S} = \sum_{n=0}^{N_S-1} c_n \boldsymbol{S}_n \tag{6.3}$$

where $\boldsymbol{S}_n$ is the response of the n-th canonical object, c_n indicates the weight of $\boldsymbol{S}_n$ in the combination leading to $\boldsymbol{S}$ and N_S is the number of components. The most common polarimetric decomposition, which also represents the basis for the coherency matrix formulation, is the *Pauli decomposition*, defined as [6]:

$$\boldsymbol{S}_{(H,V)} = \frac{c_0^{(P)}}{\sqrt{2}}\boldsymbol{I} + \frac{c_1^{(P)}}{\sqrt{2}}\boldsymbol{J} + \frac{c_2^{(P)}}{\sqrt{2}}\boldsymbol{K} + \frac{c_3^{(P)}}{\sqrt{2}}\boldsymbol{L} \tag{6.4}$$

$$= \frac{c_0^{(P)}}{\sqrt{2}}\begin{bmatrix} 1 & 0 \\ 0 & 1 \end{bmatrix} + \frac{c_1^{(P)}}{\sqrt{2}}\begin{bmatrix} 1 & 0 \\ 0 & -1 \end{bmatrix} + \frac{c_2^{(P)}}{\sqrt{2}}\begin{bmatrix} 0 & 1 \\ 1 & 0 \end{bmatrix} + \frac{c_3^{(P)}}{\sqrt{2}}\begin{bmatrix} 0 & -j \\ j & 0 \end{bmatrix} \tag{6.5}$$

where

†Speckle noise is the pixel-to-pixel variation in intensity caused by the phase fluctuations of the waves reflected by the large number of scatterers that are contained in a resolution cell [6].
‡Generally, the degradation of the resolution is traded with the accuracy of estimation of the second-order characterisation.

$$c_0^{(P)} = \frac{S_{HH} + S_{VV}}{\sqrt{2}}, \quad c_1^{(P)} = \frac{S_{HH} - S_{VV}}{\sqrt{2}} \tag{6.6}$$

$$c_2^{(P)} = \frac{S_{HV} + S_{VH}}{\sqrt{2}}, \quad c_3^{(P)} = j\frac{S_{HV} - S_{VH}}{\sqrt{2}} \tag{6.7}$$

The four components represent the scattering mechanism from four canonical objects. The first one is interpreted as sphere, plate or trihedral, as it takes into account single- or odd-bounce scattering. The second and the third components are seen as non-rotated and 45°-rotated dihedrals, respectively, representing double or even-bounce scattering. The last one includes all antisymmetric components of the scattering matrix $\boldsymbol{S}$, representing a non-reciprocal target. A basis proportional to the Pauli matrices is also the starting point for the definition of the *Cameron decomposition*, defined as [9]

$$\boldsymbol{S}_{(H,V)} = c_0^{(C)}\boldsymbol{S}_{\text{sym}}^{\max} + c_1^{(C)}\boldsymbol{S}_{\text{sym}}^{\min} + c_2^{(C)}\boldsymbol{S}_{\text{non-rec}} \tag{6.8}$$

where $\boldsymbol{S}_{\text{sym}}^{\max}$ and $\boldsymbol{S}_{\text{sym}}^{\min}$ are the maximum and minimum symmetry components, respectively; $\boldsymbol{S}_{\text{non-rec}}$ is the non-reciprocal component:

$$c_0^{(C)} = \left\|\boldsymbol{S}_{(H,V)}\right\| \cos\theta_{\text{rec}} \cos\tau_{\text{sym}} \tag{6.9}$$

$$c_1^{(C)} = \left\|\boldsymbol{S}_{(H,V)}\right\| \cos\theta_{\text{rec}} \sin\tau_{\text{sym}} \tag{6.10}$$

$$c_2^{(C)} = \left\|\boldsymbol{S}_{(H,V)}\right\| \sin\theta_{\text{rec}} \tag{6.11}$$

The Cameron decomposition is based on the concepts of reciprocity and symmetry [6]: a scatterer is reciprocal if it obeys the reciprocity principle such that its scattering matrix is symmetric, while it is symmetric if it has an axis of symmetry in the plane orthogonal to the radar line of sight. The angle θ_{rec} separates the reciprocal ($\theta_{\text{rec}} \leq \pi/4$) and non-reciprocal ($\theta_{\text{rec}} > \pi/4$) components of the scattering matrix, while τ_{sym} further divides the former one into a symmetric ($\tau_{\text{sym}} \leq \pi/8$) and an asymmetric ($\tau_{\text{sym}} > \pi/8$) component.

Of particular interest in this chapter is the Krogager decomposition, which is applicable in case the reciprocity principle is satisfied, and thus, the scattering matrix is symmetric. It is another example of CTD and it is defined by means of the circular polarisation scattering matrix $\boldsymbol{S}_{(R,L)}$, introduced in (6.2), as [10]:

$$\boldsymbol{S}_{(R,L)} = e^{j\theta_0}\left\{k_s e^{j\theta_s}\begin{bmatrix} 0 & j \\ j & 0 \end{bmatrix} + k_d\begin{bmatrix} e^{j2\eta_0} & 0 \\ 0 & -e^{-j2\eta_0} \end{bmatrix} + k_h\begin{bmatrix} e^{j2\eta_0} & 0 \\ 0 & 0 \end{bmatrix}\right\} \tag{6.12}$$

where k_s, k_d and k_h are real-valued quantities which can be interpreted as scattering coefficients from a sphere, a diplane and a helix, respectively[§]. Moreover, θ_0 is the absolute phase term that depends on the distance between the scatterer and the radar, θ_s is the displacement of the sphere with respect to the diplane and the helix components, while their orientation angle is expressed by η_0. The scattering coefficients k_s, k_d and k_h are computed as follows [2]:

$$k_s = |S_{RL}| \tag{6.13}$$

$$k_d = \min\left(|S_{RR}|, |S_{LL}|\right) \tag{6.14}$$

$$k_h = \left||S_{RR}| - |S_{LL}|\right| \tag{6.15}$$

where $|\cdot|$ represents the magnitude of a complex number. An important property of this decomposition is the roll invariance of the components k_s, k_d and k_h, demonstrated in [11]. This important property of the Krogager decomposition means that the Krogager scattering coefficients do not depend on the orientation of the target in roll (meaning invariance with respect to observations from platforms with different incidence angles). This is a particularly relevant property for ATR purposes as this native invariance of the decomposition reduces the requirements of data acquisitions for training with different incidence angles.

For the capabilities of CTDs to be physically interpreted, to satisfactorily represent man-made target and for the above-mentioned properties of the Krogager decomposition, the latter has been selected to be used in the ATR framework presented in this chapter.

6.3 Feature-based ATR techniques

Feature-based ATR algorithms rely on the capability to extract effective target features from the SAR image that are able to represent well the target itself. The features are not directly related to physical or statistical characteristics of the object but are rather obtained through the application of a transform or a projection onto a selected basis.

The algorithm developed by Srinivas *et al.* [12] belongs to this category since the features of the targets were extracted by means of the Wavelet transform[¶]. First, it decomposes the SAR image into four sub-images, containing low- and high-frequency components in both horizontal and vertical directions; then, a probabilistic graphical model, able to represent conditional dependencies between the features, is used for redundancy reduction and for the actual classification.

Exploiting recent advances in *machine learning*, Chen *et al.* [13] and Wagner [14] proposed two approaches for automatically learning hierarchies of features from a large amount of data. This means that the type of feature was not selected

[§]This decomposition is also known as sphere, diplane and helix decomposition.

[¶]However, authors stated that their proposed framework readily generalised to any other suitable choice of feature sets that offered complementary benefits.

a priori, and this task was assigned to the algorithm itself. Chen *et al.* used a convolutional neural network (CNN), with sparsely connected layers to avoid overfitting due to limited training images, that could be directly applied to SAR images. Wagner, instead, proposed the combination of a CNN for the automatic generation of the features and a support vector machine (SVM) for decision-making.

A class of feature-based ATR algorithms for SAR images employs image moments as a tool for the extraction of targets' characteristics. To this end, image moments are scalar quantities defined as weighted averages of the image pixels' intensities.

The first example of the use of image moments for target recognition from SAR is given in [15]. Yan *et al.* used seven Hu moments [16] in combination with an SVM for the classification of planes and tanks. The authors exploited the rotational, translational and scaling invariance of Hu moments [17], achieving an overall recognition accuracy of 96% on noise-free simulated images. Hu moments were also used in [18] in combination with intensity, texture and scattering features. The feature extraction phase was followed by a feature-selection stage, aimed at removing those with low stability, low discriminability and low correlation, in cascade with a k-NN classifier. The approach was tested on real SAR images of three types of ships from TerraSAR-X, obtaining an overall accuracy of 91.54%.

Amoon *et al.* proposed in [19] the use of Zernike moments, obtained as a projection of an SAR image on Zernike polynomials with different orders. Zernike polynomials are orthogonal on the unit circle, and this orthogonality makes the moments independent, thus decreasing the intrinsic redundancy of the features. Moreover, Zernike moments are rotational-invariant but not translational- and scale-invariant. This means that a variation in the pose or dimension of the target modifies the feature value, thus negatively affecting the classification results or introducing additional training data requirements; therefore, position and size normalisation were needed in [19]. After these pre-processing steps, moments were extracted from both the shape (binary image) and the intensity image of the target and used with an SVM. The algorithm was tested on the MSTAR data set, reaching an average accuracy of 96.48%.

Clemente *et al.* presented in [20, 21], a novel algorithm for the classification of civilian vehicles from SAR based on the use of pseudo-Zernike (pZ) moments [22]. The pZ moments inherit the orthogonality of the Zernike moments [23], and similarly to the Hu moments, they own valuable geometric properties, such as invariance to scale [24], rotation and translation [25, 26]. Moreover, pZ moments are less sensitive to noise [23, 27] compared to the Zernike moments**, and this makes them reliable features in recognition systems for different kind of applications, both in image processing [28–31] and radar processing [32].

**With sensitivity to noise, it is meant the variance of the moments when noise is added to an image. Therefore, low sensitivity means that moments computed on a noise-free and a noisy version of the same image are similar in a defined metric.

6.3.1 PZ moments with applications to Polarimetric SAR classification

In this section, the focus will be on the pZ moments as they represent one of the fundamental blocks of the processing framework presented in this chapter. The complex-valued pZ moment $w_{n,l}$ of an image $f(x,y) \in \mathbb{R}$, with $f(x,y) \geq 0$ and (x,y) indices of the pixels, is defined as its projection on the pZ polynomial $W_{n,l}(r,\vartheta)$ of degree $n \geq |l|$, $n \in \mathbb{N}_0$, $l \in \mathbb{N}_0$ with $\mathbb{N}_0$ the set of natural numbers, 0 included:

$$w_{n,l} = \frac{n+1}{\pi} \int_0^{2\pi} \int_0^1 W_{n,l}^* (r,\vartheta) f(r\cos\vartheta, r\sin\vartheta)\, r dr d\vartheta \tag{6.16}$$

where $r = \sqrt{x^2+y^2}$, $\vartheta = \tan^{-1}(y/x)$ and

$$W_{n,l}(r,\vartheta) = \sum_{m=0}^{n-|l|} \frac{r^{n-m}(-1)^m (2n+1-m)!}{m!\,(n+|l|+1-m)!\,(n-|l|-m)!} e^{jl\vartheta} \tag{6.17}$$

The orthogonality condition of the pZ polynomials applies on the unit disc, that is, for $r \leq 1$, and it can be written as:

$$\int_0^{2\pi} \int_0^1 W_{n,l}^* (r,\vartheta)\, W_{m,k}(r,\vartheta)\, r dr d\vartheta = \frac{\pi}{n+1} \delta_{n,m} \delta_{l,k} \tag{6.18}$$

where

$$\delta_{n,m} = \begin{cases} 1 & n = m \\ 0 & n \neq m \end{cases} \tag{6.19}$$

is the Kronecker delta. Therefore, as the number of orthogonal polynomials is $(n+1)^2$ (once n is given), this is also the number of independent pZ moments[††].

The block diagram of the algorithm presented in [20, 21] is shown in Figure 6.1. The SAR image to process, of dimensions $R_{\text{im}} \times C_{\text{im}}$ containing a detected target[‡‡],

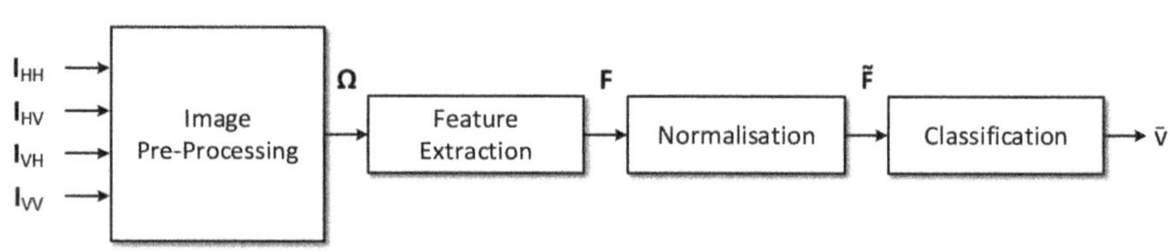

Figure 6.1 Block diagram of the pseudo-Zernike (pZ)-based SAR ATR algorithm proposed in References [20, 21]

[††]In this context, with independence is meant a low level of information redundancy among the moments [33].

[‡‡]The dimensions $\boldsymbol{R}_{\text{im}}$ and $\boldsymbol{C}_{\text{im}}$ are chosen according to the resolution of the SAR image and to the physical dimensions of the expected target.

is assumed to be fully polarimetric with complex-valued components $\boldsymbol{I}_{HH}(x,y)$, $\boldsymbol{I}_{HV}(x,y)$, $\boldsymbol{I}_{VH}(x,y)$ and $\boldsymbol{I}_{VV}(x,y)$. The pre-processing block consists of two steps. First, $\tilde{\boldsymbol{\Omega}} \in \mathbb{R}^{R_{\text{im}} \times C_{\text{im}}}$ was obtained as a scaled sum of the magnitude of the four polarimetric components:

$$\tilde{\boldsymbol{\Omega}}(x,y) = \log_{10}\left(\left|\boldsymbol{I}_{HH}(x,y)\right| + \left|\boldsymbol{I}_{HV}(x,y)\right| + \left|\boldsymbol{I}_{VH}(x,y)\right| + \left|\boldsymbol{I}_{VV}(x,y)\right|\right) \tag{6.20}$$

where the logarithm is performed to highlight weaker contributions in the final image. Then, $\tilde{\boldsymbol{\Omega}}$ is normalised such that its intensity is restricted in the range $[0, 1]$, leading to features that are independent of different intensity levels:

$$\boldsymbol{\Omega}(x,y) = \frac{\tilde{\boldsymbol{\Omega}}(x,y) - \min_{(x,y)} \tilde{\boldsymbol{\Omega}}(x,y)}{\max_{(x,y)}\left[\tilde{\boldsymbol{\Omega}}(x,y) - \min_{(x,y)} \tilde{\boldsymbol{\Omega}}(x,y)\right]} \tag{6.21}$$

The feature extraction consists in the projection of the image $\boldsymbol{\Omega}(x,y) \in \mathbb{R}$, with $\boldsymbol{\Omega}(x,y) \geq 0$, onto the pZ polynomials as per (6.16), which could be precomputed, stored and used when necessary, as the polynomials depend on the size of the image to process only $R_{\text{im}} \times C_{\text{im}}$[§§]. The feature vector $\boldsymbol{F} \in \mathbb{R}^{(N_{\text{ord}}+1)^2}$ finally contains the magnitude of the pZ moments up to the order N_{ord}, that is,

$$\boldsymbol{F} = \left[\left|w_{0,0}\right|, \ldots, \left|w_{N_{\text{ord}},-N_{\text{ord}}}\right|\right]^T \tag{6.22}$$

Note that in order to avoid information loss due to the fact that the pZ polynomials in (6.17) are defined in the unit disk, the support of $\boldsymbol{\Omega}(x,y)$ is scaled before computing the pZ moments. This scaling consists in inscribing the rectangular image in the unit circle, while setting to zero the pixels between the image edges and the unit circumference. The feature vector is then normalised to zero-mean and unit variance; this step ensures that the classification is not affected by polarised features [34]:

$$\tilde{F} = \frac{\boldsymbol{F} - \mu_F}{\sigma_F} \tag{6.23}$$

where μ_F and σ_F are the mean and standard deviation of $\boldsymbol{F}$, respectively. The final step is the classification, performed with a k-NN classifier, that returns the score vector $\boldsymbol{s} = [s_1, \ldots, s_L] \in \mathbb{R}^L$ whose elements are the occurrences (normalised to k) of each class among the k-nearest neighbours to $\tilde{F}$, and L is the number of possible classes. Clemente *et al.* [20, 21] also proposed a fusion rule when more than one SAR images of the target under test are available, as in a multi-sensor system. Let $\boldsymbol{s}_i$ be the score vector obtained by testing the features extracted from the i-th image, with $i = 0, \ldots, T-1$. The proposed fusion rule simply consists of adding up the score vectors and deciding for the class that presented the largest overall score, that is,

[§§]Without loss of generality, this framework can also be applied to single polarisation SAR images. In fact, the feature extraction is applied on just one real-valued image, which may be the magnitude of a single-polarisation image.

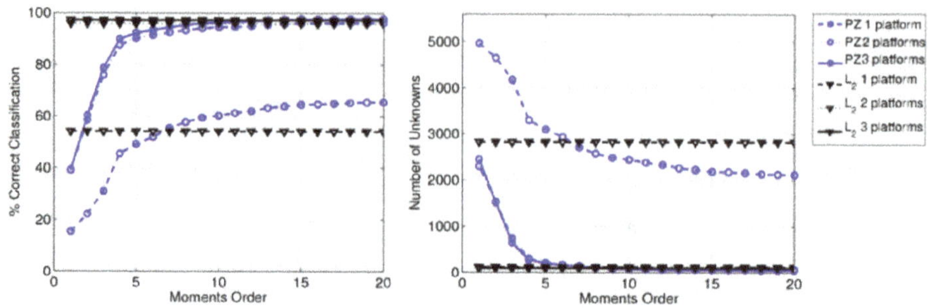

Figure 6.2 *From [21], performance comparison between (blue) the pseudo-Zernike (pZ)-based SAR ATR algorithm on varying the moments order and (black) the ℓ_2-norm-based algorithm. Images from 30 aspect angles were selected for the training of the classifiers. Results for the cases in which classification results derived from the images acquired from 1, 2 and 3 platforms are fused.*

$$\bar{\nu} = \begin{cases} \arg\max\limits_{\nu \in \{1,\ldots,L\}} \sum\limits_{i=0}^{T-1} \boldsymbol{s}_i & \text{if } \exists! \max\limits_{\nu \in \{1,\ldots,L\}} \left(\sum\limits_{i=0}^{T-1} \boldsymbol{s}_i \right) > \delta_{\text{fus}} \\ \text{unknown} & \text{otherwise} \end{cases} \tag{6.24}$$

where δ_{fus} is a fixed threshold. When more than one classes present the same largest score, the decision is not taken and an *unknown target* is declared. The algorithm in [20, 21] was compared with the ℓ_2-norm-based algorithm reported in [35] and was tested on the GOTCHA data set (see section 6.4.1.1 for details), a set composed by eight fully polarimetric circular SAR images acquired with an airborne X-band sensor. The results are shown in Figures 6.2 and 6.3, where the blue lines refer to the pZ moments-based approach and the black lines refer to the ℓ_2-norm-based algorithm. In the first case, the classifiers are trained with observations of the targets chosen from 30 different equally spaced aspect angles, while in the second case, 10 aspect angles are selected. Moreover, dashed lines refer to the single sensor case (i.e. only one SAR image of the target was available), while dotted lines and solid lines refer to 2 and 3 sensors cases, respectively. Given the number of sensors, the pZ moments-based algorithm outperforms the ℓ_2-norm-based algorithm when fewer training images are used, both in terms of accuracy and unknowns. When the training is performed with images selected from 30 aspect angles, comparable results are obtained with moments order equal to or greater than 10; however, as highlighted by the authors, the dimension of the feature vector is 20 times smaller.

6.4 Krogager-enhanced pseudo-Zernike-based algorithm

An enhanced version of the algorithm proposed in [20, 21] was introduced in [36], with the aim to exploit better the polarimetric characteristics of the target. In particular, in the approach in [36], both the fully polarimetric image of the target and its

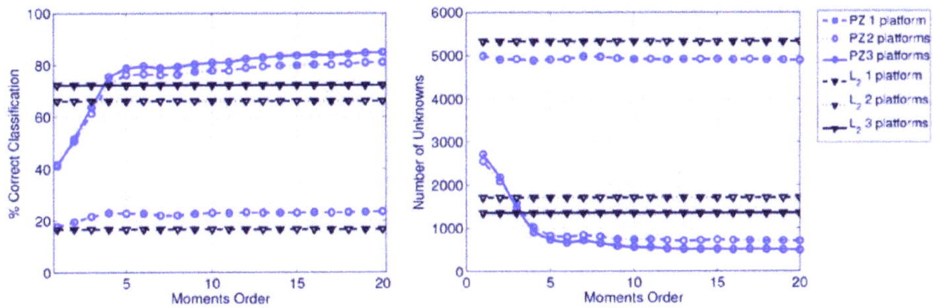

Figure 6.3 *From [21], performance comparison between (blue) the pseudo-Zernike (pZ)-based SAR ATR algorithm on varying the moments order and (black) the ℓ_2-norm-based algorithm. Images from 10 aspect angles were selected for the training of the classifiers. Results for the cases in which classification results derived from the images acquired from 1, 2 and 3 platforms are fused.*

Krogager components are used in order to improve the capability of the algorithm to correctly discriminate targets. A block diagram of the proposed approach is illustrated in Figure 6.4.

It is possible to recognise two branches of the pZ algorithm presented in Figure 6.1: one applied on the polarimetric components of the SAR image and another one applied on its Krogager components, $\boldsymbol{k}_s(x,y)$, $\boldsymbol{k}_d(x,y)$ and $\boldsymbol{k}_h(x,y)$, which, according to equation (6.13), are computed as follows:

$$\boldsymbol{k}_s(x,y) = |\boldsymbol{I}_{RL}(x,y)| \tag{6.25}$$

$$\boldsymbol{k}_d(x,y) = \min\left(|\boldsymbol{I}_{RR}(x,y)|, |\boldsymbol{I}_{LL}(x,y)|\right) \tag{6.26}$$

$$\boldsymbol{k}_h(x,y) = \left||\boldsymbol{I}_{RR}(x,y)| - |\boldsymbol{I}_{LL}(x,y)|\right| \tag{6.27}$$

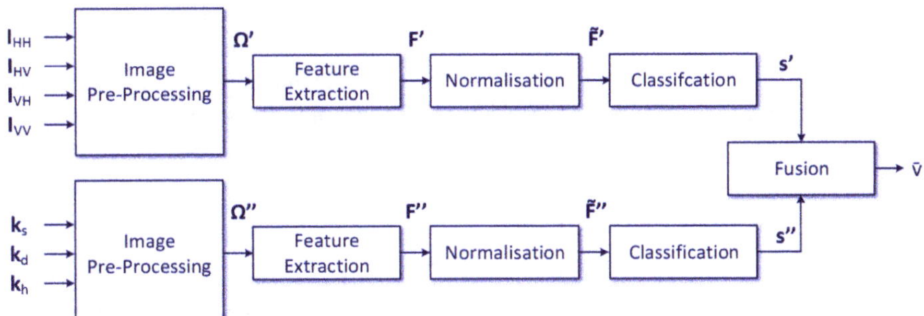

Figure 6.4 *Block diagram of the Krogager pseudo-Zernike (Kr/pZ)-based SAR ATR algorithm*

with $\boldsymbol{I}_{RR}(x,y)$, $\boldsymbol{I}_{LL}(x,y)$ and $\boldsymbol{I}_{RL}(x,y)$ being the circular polarimetric components of the SAR image of the target, obtained from the linear polarimetric components $\boldsymbol{I}_{HH}(x,y)$, $\boldsymbol{I}_{VV}(x,y)$ and $\boldsymbol{I}_{HV}(x,y)$ in accordance to (6.2):

$$\boldsymbol{I}_{RR}(x,y) = \frac{\boldsymbol{I}_{HH}(x,y) - \boldsymbol{I}_{VV}(x,y)}{2} + j\boldsymbol{I}_{HV}(x,y) \tag{6.28}$$

$$\boldsymbol{I}_{LL}(x,y) = \frac{\boldsymbol{I}_{HH}(x,y) - \boldsymbol{I}_{VV}(x,y)}{2} - j\boldsymbol{I}_{HV}(x,y) \tag{6.29}$$

$$\boldsymbol{I}_{RL}(x,y) = \frac{\boldsymbol{I}_{HH}(x,y) + \boldsymbol{I}_{VV}(x,y)}{2} \tag{6.30}$$

The feature extraction blocks always compute the pZ moments of the input images $\boldsymbol{\Omega}' \in \mathbb{R}^{R_{\text{im}} \times C_{\text{im}}}$ and $\boldsymbol{\Omega}'' \in \mathbb{R}^{R_{\text{im}} \times C_{\text{im}}}$. The first one is obtained as in (6.20) and (6.21), while similarly $\boldsymbol{\Omega}''$ is:

$$\boldsymbol{\Omega}''(x,y) = \frac{\tilde{\boldsymbol{\Omega}}''(x,y) - \min_{(x,y)} \tilde{\boldsymbol{\Omega}}''(x,y)}{\max_{(x,y)}\left[\tilde{\boldsymbol{\Omega}}''(x,y) - \min_{(x,y)} \tilde{\boldsymbol{\Omega}}''(x,y)\right]} \tag{6.31}$$

where

$$\tilde{\boldsymbol{\Omega}}''(x,y) = \log_{10}\left(\boldsymbol{k}_s(x,y) + \boldsymbol{k}_d(x,y) + \boldsymbol{k}_h(x,y)\right) \tag{6.32}$$

The feature vectors, $\boldsymbol{F}' \in \mathbb{R}^{(N_{\text{ord}}+1)^2}$ and $\boldsymbol{F}'' \in \mathbb{R}^{(N_{\text{ord}}+1)^2}$, are then normalised as in (6.23) and used to classify the target with two different k-NN classifiers. The fusion rule used for the overall classification is the one already presented in (6.24) and adapted to the proposed approach, that is,

$$\bar{v} = \begin{cases} \arg\max\limits_{v \in \{1,\dots,L\}} (\boldsymbol{s}' + \boldsymbol{s}'') & \text{if } \exists! \max\limits_{v \in \{1,\dots,L\}} (\boldsymbol{s}' + \boldsymbol{s}'') > \delta_{\text{fus}} \\ \text{unknown} & \text{otherwise} \end{cases} \tag{6.33}$$

where $\boldsymbol{s}' = [s'_1, \dots, s'_L] \in \mathbb{R}^L$ and $\boldsymbol{s}'' = [s''_1, \dots, s''_L] \in \mathbb{R}^L$ are the score vectors obtained from the two branches of the framework. Note that, as for the pZ moments-based algorithm, this framework can be easily extended to multiple sensors (provided the availability of several images from multiple sensors), by simply modifying the fusion rule to accommodate for score vectors coming from multiple branches.

6.4.1 Performance evaluation

The proposed integrated Krogager pseudo-Zernike (Kr/pZ) algorithm is tested and compared with the original one [20, 21], solely based on the extraction of the features from only the polarimetric components of the SAR images. The analysis is performed using the images of the GOTCHA data set. Furthermore, several noise conditions are considered in order to demonstrate the robustness of the approach.

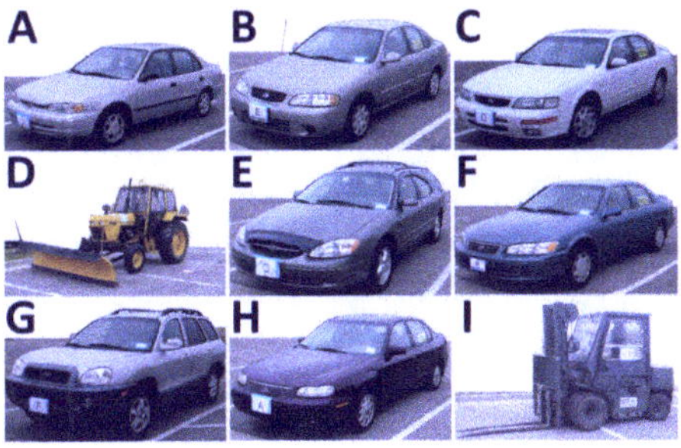

(a) Vehicles.

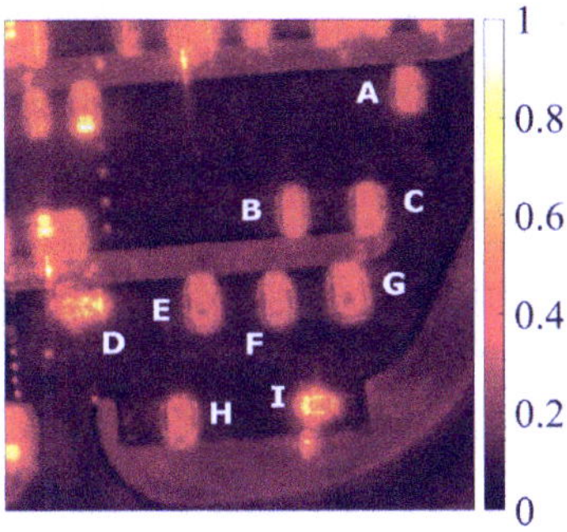

(b) SAR image.

Figure 6.5 *GOTCHA data set: (a) images of the targets present in the scene; (b) full-azimuth and fully polarimetric magnitude SAR image of the imaged car park with the nine vehicles labelled*

6.4.1.1 GOTCHA data set

The algorithm is tested using the GOTCHA data set [37], which is a set of eight fully polarimetric circular SAR images acquired with an airborne X-band SAR, with carrier frequency 9.6 GHz and 640 MHz bandwidth at eight different elevation angles. Figure 6.5 shows nine different civil vehicles used in this evaluation, whereas Table 6.1 reports their classification at different detail levels, that is, for recognition, identification and characterisation purposes.

The fully synthetic aperture (360°) is split into 90 sub-apertures of 4° in azimuth each, leading to approximately equal range-azimuth resolution cells of 23 cm; thus, the set used is made of 720 fully polarimetric images for each of the nine commercial vehicles considered as target. Moreover, equal-sized sub-images, with $R_{im} = 51$ and $C_{im} = 46$, containing each vehicle are selected.

Four different analyses are presented, as a combination of two configurations for the training set and two configurations for the test set. The training set is formed by images acquired using the lowest altitude pass; either 10 or 30 images for each

Table 6.1 *GOTCHA data set*

Target	Description	Recognition	Identification	Characterisation
Nissan Sentra	Compact car	R1	I1	C1
Chevy Prizm	Compact car			C2
Chevy Malibu	Mid-size car		I2	C3
Toyota Camry	Mid-size car			C4
Ford Taurus	Full-size car		I3	C5
Nissan Maxima	Full-size car			C6
Hyundai Santa Fe	SUV		I4	C7
Case Tractor	Industrial truck	R2	I5	C8
Fork Lift	Industrial truck			C9

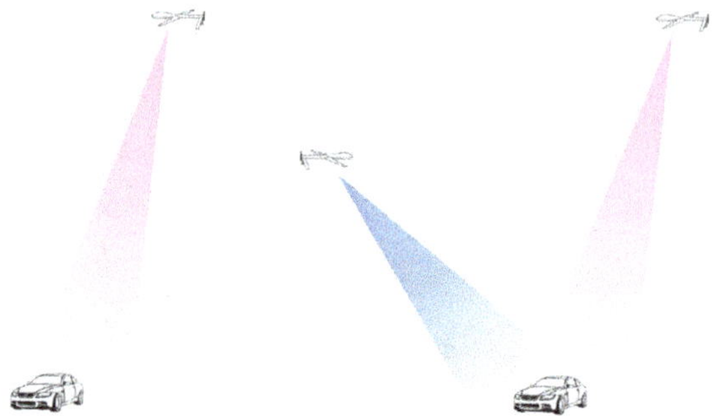

(a) One sensor case. (b) Two sensors case.

Figure 6.6 *Test configurations: (a) one- and (b) two-sensor cases*

vehicles are used to train the classifiers, selected with a separation of 36° or 12° in azimuth, respectively. The test set is composed by all but the images used for the training. Two scenarios are investigated: in the first scenario, shown in Figure 6.6(a), one image is used to classify the target; in the other scenario, depicted in Figure 6.6(b), two images are used to show the benefits of a multi-sensor approach.

6.4.1.2 Noise conditions

The images of the GOTCHA data set are real-recorded SAR images affected by noise. However, in order to demonstrate the robustness of the approach, additional noise is also considered, both additive and multiplicative. In the first case, the noise is modelled as a compound-Gaussian random variable with a gamma texture [38, 39], which can be written in the form

$$\xi_c = \sqrt{\gamma}\,\xi \tag{6.34}$$

where ξ is a complex symmetric zero-mean Gaussian variable and γ is a gamma variable, whose distribution is:

$$f_\gamma\left(x\right) = \frac{1}{\Gamma\left(\nu_s\right)}\frac{1}{\mu_s^{\nu_s}}x^{\nu_s-1}e^{-x/\mu_s}, \quad x > 0 \tag{6.35}$$

where $\Gamma\left(\cdot\right)$ is the gamma function, and μ_s and ν_s are the scale and shape parameters, respectively. To this end, $\nu_s = 1/\mu_s$ to have a gamma distribution with unit mean; specifically, ν_s is chosen equal to 0.5. Equations (6.34) and (6.35) ensure that the amplitude probability density function of ξ_c is K-distributed. In the multiplicative noise case, instead, each pixel is multiplied with a square root of a gamma random variable.

6.4.1.3 Results

The assessment of the performance is carried out by means of a Monte-Carlo simulation with 100 tests. The results are presented in terms of characterisation

Table 6.2 Fusion rule threshold δ_{fus}

	One sensor	**Two sensors**
pZ	1/3	2/3
Kr/pZ	2/3	4/3

and identification accuracies¶, defined as the number of vehicles correctly classified divided by the total number of targets under test and expressed as percentage. Moreover, the percentage of unknowns is also shown, defined as the ratio of the number of targets declared unknown as per (6.33) to the total number of tests (the chosen thresholds δ_{fus} are summarised in Table 6.2).

For the one-sensor case, all the available images are used for the testing, meaning that a total of 710 images for each target are used if the training is performed with 10 images, while 690 images are used for 30 training images case. For the two-sensor case, the total number of randomly selected couples is 71000 or 69000, depending on the training configurations. Moreover, the analysis is performed for different values of the degree of the pZ polynomial, different SNRs and using a 3-NN classifier; the value of $k = 3$ resulted to be the most effective in [21].

Figure 6.7 shows the results in the noise-free case on varying the moments order. Cross-markers refer to the characterisation accuracy, circles represent the identification accuracy, while straight lines are used to identify the percentage of unknowns. Figure 6.7a and b compares the performance of the Kr/pZ algorithm (red solid lines) with the original approach (dashed blue lines), in the case of classification with one sensor and the use of either 10 or 30 training images, respectively. No differences can be appreciated between characterisation and identification, while improvements are seen between the two approaches, pZ and Kr/pZ, both when the training is performed with 10 images per target and when it is carried out with 30 images.

Focussing on the moments orders equal to or greater than 8 and up to 23, these improvements can be quantified in terms of characterisation accuracy in 7.26 % and 4.51 %, when 10 and 30 images are used for the training, respectively, and in 10 % and 5.22 % in terms of percentage of unknowns. The maximum achieved accuracy is 60 % in the first case, and 87 % in the second case for the integrated Kr/pZ method, compared to 53.5 % and 82.3 % for the pZ algorithm.

Overall better performance is obtained when the classification is performed by using images acquired from two sensors, as shown in Figure 6.7c and d. Improvements due to the integrated Kr/pZ approach over the original algorithm are

¶Recognition accuracy is not taken into consideration, as it involves only two classes: car and industrial truck.

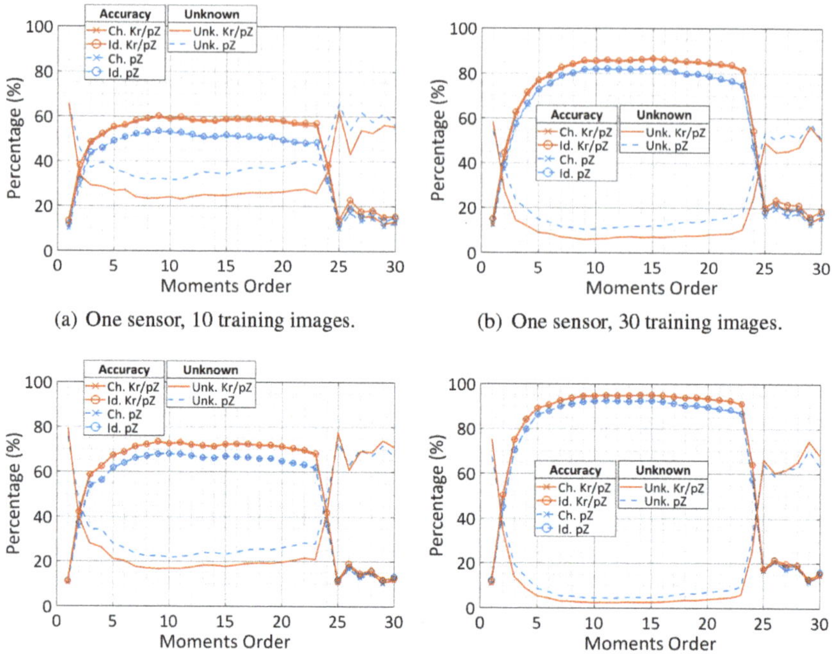

(a) One sensor, 10 training images.

(b) One sensor, 30 training images.

(c) Two sensors, 10 training images.

(d) Two sensors, 30 training images.

Figure 6.7 *Comparative analysis between Krogager pseudo-Zernike (Kr/pZ)- and pseudo-Zernike (pZ)-only approaches, noise-free case. Results in terms of characterisation (Ch.) and identification (Id.) accuracy, and percentage of unknowns (Unk.), on varying the number of training images (left-right), the number of sensors (top-bottom) and the moments order.*

also visible in this case: the improvement is of approximately 5.43 % in terms of characterisation accuracy and 5.73 % in terms of percentage of unknowns, when 10 images are used for the training, and of 2.90 % and 2.40 % for accuracy and unknowns, respectively, when the classifiers are trained with 30 images per target. In this scenario, the maximum accuracy is 73.5 % and 68 % for Kr/pZ and pZ, respectively, when the training is performed with 10 images, while it is 95.2 % and 92.6 %, respectively, when 30 images are used.

Interestingly, for moments orders greater than 23, both the pZ and the Kr/pZ algorithms fail in correctly classifying the targets, with the percentage of accuracy dropping below 20 % independently of both the number of sensors and the number of training images. This is caused by the numerical instability at the edge of the unit circle and the numerical approximation that is applied when computing the pZ polynomials, effects that are more evident as the order increases. Figure 6.8 shows the magnitude of some pZ polynomials computed on either a grid of 688 × 688 pixels, top row, or a grid of 69 × 69 pixels, bottom row, also used during the tests.

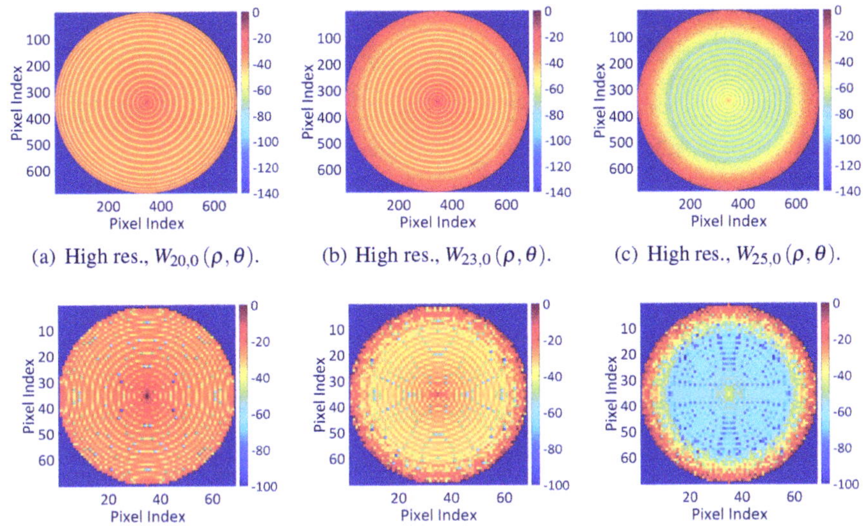

(a) High res., $W_{20,0}(\rho,\theta)$. (b) High res., $W_{23,0}(\rho,\theta)$. (c) High res., $W_{25,0}(\rho,\theta)$.

(d) Low res., $W_{20,0}(\rho,\theta)$. (e) Low res., $W_{23,0}(\rho,\theta)$. (f) Low res., $W_{25,0}(\rho,\theta)$.

Figure 6.8 Magnitude of the pseudo-Zernike polynomials (left) $W_{20,0}(\rho,\theta)$, *(middle)* $W_{23,0}(\rho,\theta)$ *and (right)* $W_{25,0}(\rho,\theta)$, *on a grid of either (top)* 688×688 *or (bottom)* 69×69 *pixels*

All the high-resolution polynomials $W_{20,0}(\rho,\theta)$, $W_{23,0}(\rho,\theta)$ and $W_{25,0}(\rho,\theta)$ depicted in Figure 6.8a–c, respectively, present a particular pattern with concentric circles, whose cardinality increases proportionally to the polynomial's degree. Moreover, especially for the polynomial of degree 25, undesired edge effects also appear. Comparing the high-resolution polynomials with the low-resolution ones, the pattern of concentric circles can be still distinguished for $W_{20,0}(\rho,\theta)$ and $W_{23,0}(\rho,\theta)$, in Figure 6.8d and e, while it does disappear for $W_{25,0}(\rho,\theta)$ in Figure 6.8(f). This behaviour makes the high-degree polynomials unable to extract distinctive features from different targets, which are then misclassified. For this reason, in the following, only polynomials up to degree 20 are considered.

Results similar to the noise-free case are obtained when compound Gaussian noise is added to the images, as observed from Figure 6.9, where characterisation accuracy, identification accuracy and percentage of unknowns are shown as functions of the SNR, for a fixed moments order equal to 10.

The accuracy increases as the SNR increases, while the percentage of unknown decreases. As expected, better performance is obtained when using more images for training the classifiers, as shown in Figure 6.9b and d; however, larger improvements of the integrated Kr/pZ algorithm over the pZ moments-based approach are visible when the training is performed with 10 images, as depicted in Figure 6.9a and c, confirming the effectiveness of the proposed algorithm especially when fewer images are available for training, as already demonstrated in the noise-free scenario. In the most likely scenario, that is, when 10 images per target are available

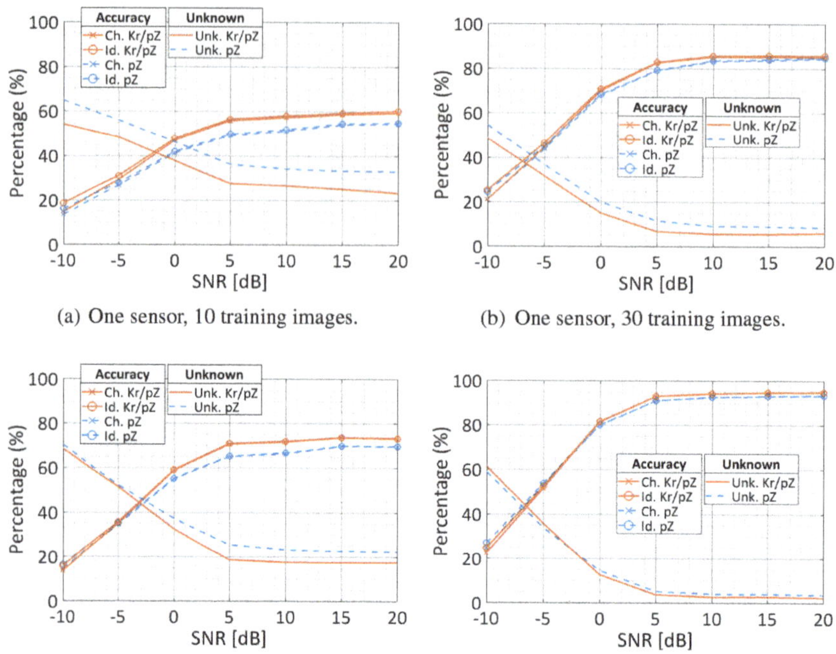

(a) One sensor, 10 training images.

(b) One sensor, 30 training images.

(c) Two sensors, 10 training images.

(d) Two sensors, 30 training images.

Figure 6.9 Comparative analysis between Krogager pseudo-Zernike (Kr/pZ)- and pseudo-Zernike (pZ)-only approaches, additive compound Gaussian noise case. Results in terms of characterisation (Ch.) and identification (Id.) accuracy, and percentage of unknowns (Unk.), on varying the number of training images (left-right), the number of sensors (top-bottom) and the SNR while fixing the moments order to 10.

for training and the classification is performed with only one sensor, for SNR above 0 dB, the average improvement is about 5.63 % in terms of accuracy and 8.48 % in terms of percentage of unknowns. Finally, for both 10 and 30 images cases, with an SNR of 5 dB, the Kr/Pz-based approach provides good results.

The final analysis evaluates the effect of multiplicative noise. Performance on varying the moments order is shown in Figure 6.10.

In agreement with the results obtained in the previous analyses, higher accuracy and lower percentage of unknowns are achieved when 30 images are used for the training and the classification is performed by using images acquired by two sensors. However, in all the analysed cases, the proposed combined Kr/pZ approach outperforms the original algorithm. Finally, it is worth highlighting that both the approaches are very sensitive to the multiplicative noise in particular when the moments order is high. It is clear that the optimum value of the order for this scenario is between 7 and 10.

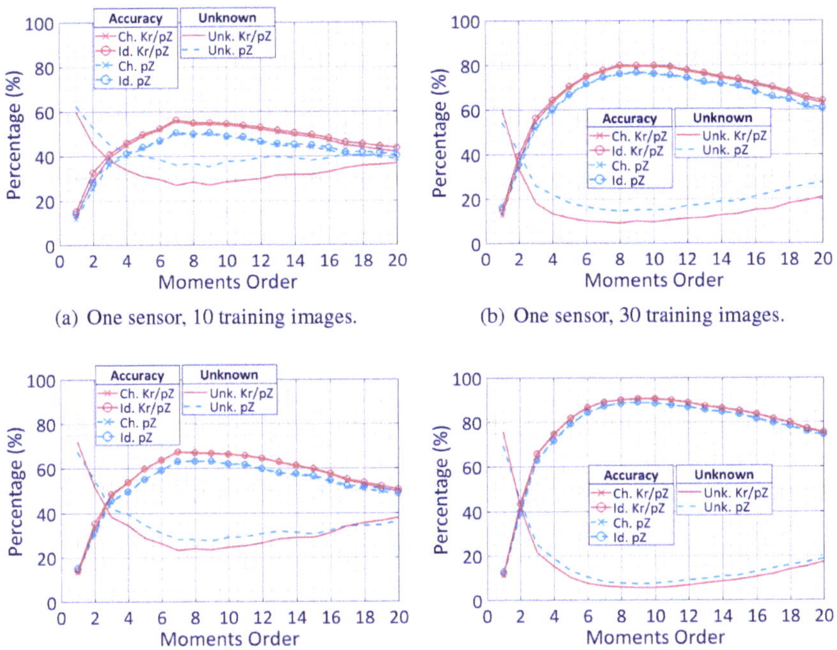

(a) One sensor, 10 training images.

(b) One sensor, 30 training images.

(c) Two sensors, 10 training images.

(d) Two sensors, 30 training images.

Figure 6.10 *Comparative analysis between Krogager pseudo-Zernike (Kr/pZ)- and pseudo-Zernike (pZ)-only approaches, multiplicative noise case. Results in terms of characterisation (Ch.) and identification (Id.) accuracy, and percentage of unknowns (Unk.), on varying the number of training images (left-right), the number of sensors (top-bottom) and the moments order.*

Hence, the proposed approach presents better performance for all the configurations in which it has been tested. In particular, the best improvements are achieved in the case in which 10 images for the training are used; this means that in order to reach the same performance, the Kr/pZ approach requires less a priori information compared to the pZ algorithm. This has also a positive impact on potential operational costs, since the acquisition of a targets database is often expensive and time-consuming. In addition to the general increase in accuracy, the decrease of the percentage of unknowns makes the Kr/pZ approach even more reliable than the pZ method. Moreover, the pZ moments properties of translation and rotation independence, combined with the roll-invariant characteristic of the Krogager decomposition, make the algorithm robust with respect to both the target orientation in the image plane and the acquisition elevation angle. Note that these improvements are obtained only at the cost of a small computational increase, as no additional information is required as an input to the algorithm with respect to the pZ approach.

6.5 Conclusions

This chapter has investigated the potential benefits of the use of polarimetry when dealing with ATR. CTD is particularly suitable to represent man-made targets and is physically interpretable. Moreover, it was shown how a particular decomposition belonging to this family of polarimetric decompositions can improve the performance of ATR frameworks. In particular, it was shown how the roll-invariant Krogager decomposition could be easy and effective to integrate in a feature-based ATR framework exploiting invariant image moments. The results presented demonstrate that the use of polarimetric information helps to enhance target-recognition capabilities as well as reduce data collection requirements for classifier training purposes.

However, a number of research questions still exist in this area, such as the selection of the best polarimetric decomposition for ATR purposes, or the possibility to design a custom ATR-oriented polarimetric decomposition with the aim to extract the most relevant information for the target-recognition task. Another topic to investigate is the necessity to differentiate approaches based on the level of target classification required. For example, it could be useful to assess which type of approach/decomposition would be more suited for different classification levels. As often only two polarimetric channels are available, it would be interesting to understand how much of the polarimetric information that supports the ATR task can still be extracted. A different perspective would be to investigate the latest advances in artificial intelligence with CNNs applied to the ATR task, and in this sense, recently some progress has been made [40, 41]. However, these techniques have not reached a sufficient level of maturity in the field and more research effort would be required in the future. The above are likely to be explored in the future by the research community, broadening the interest on this relevant application domain and strengthening the use case for polarimetric SAR sensors.

References

[1] NATO. *NATO AAP-06 [online].* 2013. Available from https://www.jcs.mil/Portals/36/Documents/Doctrine/Other_Pubs/aap6.pdf

[2] Alberga V., Krogager E., Chandra M. 'Potential of coherent decompositions in SAR polarimetry and interferometry'. *IEEE International Geoscience and Remote Sensing Symposium (IGARSS). vol. 3*; Anchorage, AK: IEEE, 2004. pp. 1792–95.

[3] Srikanth P., Ramana K.V., Deepika U., Kalyan Chakravarthi P., Sesha Sai M.V.R. 'Comparison of various polarimetric decomposition techniques for crop classification'. *Journal of the Indian Society of Remote Sensing*. 2016, vol. 44(4), pp. 635–42.

[4] Saville M.A., Jackson J.A., Fuller D.F. 'Rethinking vehicle classification with wide-angle polarimetric SAR'. *IEEE Aerospace and Electronic Systems Magazine*. 2014, vol. 29(1), pp. 41–9.

[5] Richards M.A., Scheer J.A., Holm W.A. *Principles of modern radar: basic principles*. IET Digital Library: SciTech Pub; 2010 Jan 1. Available from https://digital-library.theiet.org/content/books/ra/sbra021e
[6] Lee J.S., Pottier E. *Polarimetric radar imaging: from basics to applications*. Boca Raton, FL: CRC Press; 2009.
[7] Touzi R., Bhattacharya A., Mattar K. 'Multi-resolution target scattering decomposition for urban feature characterization using polarimetric SARs'. Presented at 2009 IEEE international geoscience and remote sensing symposium; Cape Town, South Africa, 2009.
[8] Turkar V., Rao Y. 'Applying coherent and incoherent target decomposition techniques to polarimetric SAR data'. *International Conference on Technology Systems and Management (ICTSM)*; Mumbai, India, 2011. pp. 23–29.
[9] Cameron W.L., Youssef N.N., Leung L.K. 'Simulated polarimetric signatures of primitive geometrical shapes'. *IEEE Trans Geosci Remote Sens*. 1996, vol. 34(3), pp. 793–803.
[10] Hellmann M., Krogager E. 'Comparison of decompositions for pol-SAR image interpretation'. *IEEE International Geoscience and Remote Sensing Symposium (IGARSS)*; Honolulu, HI: IEEE, 2000. pp. 1313–15.
[11] Krogager E. 'Properties of the sphere, diplane, helix (target scattering matrix) decomposition'. *Proceedings of JIPR*. 1995, vol. 3, p. 1995.
[12] Srinivas U., Monga V., Raj R.G. 'SAR automatic target recognition using discriminative graphical models'. *IEEE Transactions on Aerospace and Electronic Systems*. 2014, vol. 50(1), pp. 591–606.
[13] Chen S., Wang H., Xu F., Jin Y.-Q. 'Target classification using the deep convolutional networks for SAR images'. *IEEE Trans Geosci Remote Sens*. 2016, vol. 54(8), pp. 4806–17.
[14] Wagner S. 'Combination of convolutional feature extraction and support vector machines for radar ATR'. *17th International Conference on Information FUSION (FUSION)*; Salamanca, Spain, 2014. pp. 1–6.
[15] Yan F., Mei W., Chunqin Z. '2009 fifth international conference on information assurance and security'. 2009. xi'An china,
[16] Ming-KueiHVisual pattern recognition by moment invariants'. *IEEE Transactions on Information Theory*. 1962, vol. 8(2), pp. 179–87.
[17] Huang Z., Leng J. 'Analysis of hu's moment invariants on image scaling and rotation'. *2nd International Conference on Computer Engineering and Technology (ICCET)*. vol. 7; 2010. pp. V7–476.
[18] Chen W., Ji K., Xing X., Zou H., Sun H. 'Ship recognition in high resolution SAR imagery based on feature selection'. *2012 International Conference on Computer Vision in Remote Sensing (CVRS 2012)*; Xiamen, China, 2012. pp. 301–05. Available from https://ieeexplore.ieee.org/xpl/mostRecentIssue.jsp?punumber=6410226
[19] Amoon M., Rezai-rad G., ali Rezai-rad G. 'Automatic target recognition of synthetic aperture radar (SAR) images based on optimal selection of zernike moments features'. *IET Computer Vision*. 2014, vol. 8(2), pp. 77–85. Available from https://onlinelibrary.wiley.com/toc/17519640/8/2

[20] Clemente C., Pallotta L., Proudler I., De Maio A., Soraghan J.J., Farina A. 'Multi-sensor full-polarimetric SAR automatic target recognition using pseudo-zernike moments'. *2014 International Radar Conference (Radar)*; Lille, France, 2014. pp. 1–5.

[21] Clemente C., Pallotta L., Proudler I., De Maio A., Soraghan J.J., Farina A. 'Pseudo-Zernike-based multi-pass automatic target recognition from multi-channel synthetic aperture radar'. *IET Radar, Sonar & Navigation*. 2015, vol. 9(4), pp. 457–66. Available from https://onlinelibrary.wiley.com/toc/17518792/9/4

[22] Bhatia A.B., Wolf E. 'On the circle polynomials of zernike and related orthogonal sets'. *Math Proc Camb Philos Soc*. 1954, vol. 50(1), pp. 40–48.

[23] Teh C.-H., Chin R.T. 'On image analysis by the methods of moments'. *IEEE Transactions on Pattern Analysis and Machine Intelligence*. 1988, vol. 10(4), pp. 496–513.

[24] Ranade S.K., Kaur N. 'Empirical analysis of scale invariance in moment coefficients'. *International Journal of Computer Science and Technology*. 2015, vol. 6(2), pp. 237–42.

[25] Clemente C., Gaglione D., Pallotta L. *On the use of image moments for ATR from SAR images*. NATO STO; 2015.

[26] K.Ranade S., Anand S. 'Empirical analysis of rotation invariance in moment coefficients'. *International Journal of Computer Applications*. 2015, vol. 119(15), pp. 19–26.

[27] Kumar S. 'Comparative results of zernike moments and pseudo-zernike moments'. *3rd International Conference on Computing for Sustainable Global Development (INDIACom)*; New Delhi, India, 2016. pp. 1254–59.

[28] Nassery P., Faez K. 'Signature pattern recognition using pseudo zernike moments and a fuzzy logic classifier'. *3rd IEEE International Conference on Image Processing*; Lausanne, Switzerland, 1996. pp. 197–200.

[29] Nabatchian A., Abdel-Raheem E., Ahmadi M. 'Human face recognition using different moment invariants: A comparative study'. *Congress on Image and Signal Processing (CISPs)*; Sanya, China, 2008. pp. 661–66.

[30] Ahmady M., Ghasemi R., Rashidy kanan H. '*2013 13th Iranian conference on fuzzy systems (IFSC)*'. Qazvin, Iran, 2013. pp. 1–4.

[31] Sultana M., Gavrilova M., Expression Y.S. 'Pose, and illumination invariant face recognition using lower order pseudo zernike moments'. *International Conference on Computer Vision Theory and Applications (VISAPP)*; Lisbon, Portugal, 2014. pp. 216–21.

[32] Pallotta L., Clemente C., De Maio A., Soraghan J.J., Farina A. 'Pseudo-zernike moments based radar micro-doppler classification'. *2014 IEEE Radar Conference (RadarCon)*; Cincinnati, OH, USA, 2014.

[33] Shu H., Luo L., Caatrieux J. -l 'Moment-based approaches in imaging. 1. basic features [a look at ...]'. *IEEE Engineering in Medicine and Biology Magazine*. 2007, vol. 26(5), pp. 70–74.

[34] Fukunaga K. *Introduction to statistical pattern recognition. Computer science and scientific computing*. Elsevier Science; 1990.

[35] Zhao Q., Principe J.C. 'Support vector machines for SAR automatic target recognition'. *IEEE Transactions on Aerospace and Electronic Systems*. 2001, vol. 37(2), pp. 643–54.

[36] Gaglione D., Clemente C., Pallotta L. 'Krogager decomposition and pseudo-zernike moments for polarimetric distributed ATR'. *Sensor Signal Processing for Defence (SSPD)*; Edinburgh, UK, 2014. pp. 1–5.

[37] Ertin E., Austin C.D., Sharma S. 'GOTCHA experience report: three-dimensional SAR imaging with complete circular apertures'. *Proceedings Volume 6568, Algorithms for Synthetic Aperture Radar Imagery XIV*; Orlando, FL, 2007. pp. 656802–12.

[38] Conte E., Longo M. 'Characterisation of radar clutter as a spherically invariant random process'. *IEE Proceedings F Communications, Radar and Signal Processing*. 1987, vol. 134(2), pp. 191–7.

[39] Greco M.S., Gini F. 'Statistical analysis of high-resolution SAR ground Clutter data'. *IEEE Transactions on Geoscience and Remote Sensing*. 2007, vol. 45(3), pp. 566–75.

[40] Guo Y., Du L., Wei D., Li C. 'Robust SAR automatic target recognition via Adversarial learning'. *IEEE Journal of Selected Topics in Applied Earth Observations and Remote Sensing*. 2021, vol. 14, pp. 716–29.

[41] Zhou Y., Wang H., Xu F., Jin Y.-Q. 'Polarimetric SAR image classification using deep convolutional neural networks'. *IEEE Geoscience and Remote Sensing Letters*. 2016, vol. 13(12), pp. 1935–9.

Chapter 7
Polarimetric inverse synthetic aperture radar

Alberto Lupidi[1], *Elisa Giusti*[1], *and Marco Martorella*[1,2]

7.1 Introduction

Radar imaging has been a powerful tool for many applications in both civil and military scenarios. One of the main scopes of radar images is to provide information about an imaged area and allow for a number of tasks, such as target detection and recognition, parameter estimation, land classification, identification of natural phenomena as well as disaster assessment, maritime observation, snow/ice coverage and other applications in the area of earth observation. Benefits have been assessed when using multidimensional radar systems, e.g., multichannel, multi-frequency and polarimetric radar systems ([1]). Among several types and combination of multidimensional radar imaging systems, polarimetric radar has provided a way to enrich the information content that can be retrieved by a radar system. This information has proven to be powerful when analysing radar data in civil applications, such as environmental monitoring, and in military contexts, such as target detection and classification [2, 3]. Although polarimetric radar imaging has found a vast ground for applications, we will concentrate on recent advances in the field of polarimetric inverse synthetic aperture radar (pol-ISAR) imaging and its use for target classification. Inverse synthetic aperture radar (ISAR) imaging has been widely used to produce well-focused and high-resolution radar images of non-cooperative targets with the aim of classifying and/or recognizing them. Aspects of target recognition, with focus on Automatic Target Recognition (ATR), are of fundamental important in military- and homeland security-related applications. The interpretation of radar images, and therefore ISAR images, is generally more complex than that of optical images. This is to do with the fact that the human brain is not trained to recognize microwave images, which show characteristics that are fundamentally different from optical images, both in terms of their geometrical and radiometric features. Nevertheless, the amount of information, the complementarity with respect to optical images and the

[1]CNIT Radar and Surveillance System (RSS) National Laboratory, Pisa, Italy
[2]University of Pisa, Pisa PI, Italy

characteristics of radar systems to see independently of the illumination and weather conditions make radar images attractive. Moreover, the advent of ATR systems, also based on the use of neural networks (NN), overcomes the limitations posed by the human brain, which must be trained to interpret radar images. The inclusion of the polarimetric information enriches the content of radar images, which produces more accurate and effective ATR systems, with superior performances with respect to single polarization radar systems. This chapter will introduce the reader to polarimetric radar imaging and to the use of polarimetric radar images for ATR. The theoretical aspects will be complemented with practical examples through the use of real polarimetric radar data.

Section 7.2 will introduce the polarimetric framework and define how Pol-ISAR imaging is performed. Section 7.3 introduces and discusses the elements of target classification based on Pol-ISAR images, whereas section 7.4 focuses on ATR algorithms that are based on the use of Pol-ISAR images in combination with a series of standard and novel classifiers. The conclusions are drawn in section 7.5.

7.2 Polarimetric ISAR

Conventional image radar operates with a fixed polarization antenna both for transmission and for radio frequency signal reception. As a consequence, this system is not able to detect any changes in the polarization state of the received echo caused by the interaction with the target. Radar polarimetry is based on the concept that the polarization diversity gives more information about the target than the power of the received signal. This additional information cannot be used by conventional radar system. In fact, when an electromagnetic (EM) waveform hits a target, the polarization of the waveform may change. This change is due in addition to the polarization of the transmitted signal also to the structural and geometric characteristics of the target. The transformation of the EM wave polarization can be described from a mathematical point of view with a scattering matrix that can be considered a generalization of the radar cross section (RCS). The scattering matrix formulation by G. W. Sinclair is at the base of the first polarimetric techniques. In the 1950s, radar polarimetry was only a theoretical concept due to the lack of suitable hardware. Starting from the 1980s, radar polarimetry has become a consolidate technique, thanks to the growing number of polarimetric sensors. The polarimetric SAR (Pol-SAR) system has been used to improve performance in target detection and classification [4–6]. The advantages obtained by the exploitation of the SAR polarimetry techniques can also be extended to the ISAR geometry. The goal is to use all the polarimetric information to improve the quality of the ISAR image in term of image focalization. As explained in [1], one of the critical aspects in ISAR image reconstruction is the image autofocusing step that is the motion compensation of the target into the received signal. One of the main methods that are proposed for the image autofocusing is parametric method know as image contrast-based autofocusing (ICBA) algorithm [7]. This method is developed only for single polarization ISAR images. Therefore, the additional information provided by radar polarization can be used in the ISAR autofocusing framework. In past years, polarimetric radars have been used to maximize the signal noise ratio (SNR) in order to achieve a better performance in target detection. The concept of improving radar performance by looking for an optimal polarization can be extended to the ISAR autofocusing since it can be assumed that for a particular polarization the maximum focusing level of the ISAR image is achieved. In fact, the degree of focusing strongly depends both on the time invariance of the scatters and on the signal clutter ratio level. Both of the effects can be reduced by the use of polarization [2]. The aim of this chapter is to define a technique for the Pol-ISAR image reconstruction and to introduce an autofocusing algorithm for Pol-ISAR system. ISAR images prove useful also when used for classifying and recognizing targets [8]. For example, target shape and target dimension can be exploited in the classification framework. Nevertheless, the image data are often too large to implement real-time classifiers. For this reason, algorithms such as CLEAN technique can be used for reducing the data size significantly without losing useful information. CLEAN technique [9] iteratively estimates the point spread function (PSF) of the brightest scatter and removes it from the formed image. In this case, only the dominant scatters are used in target recognition, thus considerably reducing the size of the data. Polarimetric characteristics of the scatters can also be used as features in target classification. However, the problem of reducing the amount of data without losing useful information is even more critical when dealing with Pol-ISAR images since the data size is three or four times larger. An algorithm for centring extraction from polarimetric SAR images was proposed in [10]. In [10], the signal model was strongly based on the SAR geometry, which is *a priori* known. In the ISAR case, the non-cooperativity of the target does not allow using any such knowledge. So parametric model is introduced to account for unknown target motions. Therefore, the problem of estimating the model parameters and the scattering centre extraction must be solved jointly. In this report, a polarimetric CLEAN (Pol-CLEAN) technique is also proposed by extended the CLEAN technique in [9] in order to extract target features such as the position of scattering centres and their scattering matrix.

7.2.1 Pol-ISAR signal model

Target backscattering properties are described in the target reflectivity function $\varsigma(z)$, where z is the vector that defines the position of the target scattering centre in the $(z(1), z(2), z(3))$ reference system. Target can be considered as composed of some scattering points denoted as scattering centres. If the absence of EM iterations between scattering centres is assumed, it is possible to consider the target as composed of N scattering centres. Each scatterer introduces in the received signal a different time delay that is related to the scatter position. Under the previous hypothesis, the reflectivity function can be defined as:

$$\varsigma(z) = \sum_{n=1}^{N} k_n \delta(z - z_n) \tag{7.1}$$

where k_n is the complex amplitude describing the backscattering properties of the nth scatter. The received signal can be defined in the time-frequency domain as follow [5]:

$$S_R(f,t) = W(f,t)e^{-j\frac{4\pi f}{c}R_0(t)} \int_V \varsigma(z)e^{-j\frac{4\pi f}{c}[z^T i_{R_0}(t)]}dz + n(t) \tag{7.2}$$

where $W(f,t) = \text{rect}\left(\frac{t}{T_{obs}}\right)\text{rect}\left(\frac{f-f_0}{B}\right)$ defines the region in the Fourier where the signal is defined and V defines the spatial domain, respectively, while f_0 is the central frequency, B is the signal bandwidth, T_{obs} is the observation time and $i_{R_0(t)}$ is the $R_0(t)$ unit vector. $R_0(t)$ is the distance between radar and target. The additive Gaussian noise is represented as $n(t)$. In the case of Pol-ISAR system, the received signal can be extended as above:

$$\begin{aligned} S_R(f,t) &= \begin{bmatrix} S_R^{HH}(f,t) & S_R^{HV}(f,t) \\ S_R^{VH}(f,t) & S_R^{VV}(f,t) \end{bmatrix} \\ &= W(f,t)e^{-j\frac{4\pi f}{c}R_0(t)} \int_V \varsigma(z)e^{-j\frac{4\pi f}{c}[z^T i_{R_0}(t)]}dz + N(f,t) \end{aligned} \tag{7.3}$$

where

$$\varsigma(z) = \begin{bmatrix} S_R^{HH}(f,t) & S_R^{HV}(f,t) \\ S_R^{VH}(f,t) & S_R^{VV}(f,t) \end{bmatrix}$$

and

$$N(f,t) = \begin{bmatrix} N^{HH}(f,t) & N^{HV}(f,t) \\ N^{VH}(f,t) & N^{VV}(f,t) \end{bmatrix}$$

define the polarimetric matrix of the target and of the noise, i.e., the received signal from each polarimetric channel. It is quite clear that the complex elements of the scattering matrix $\varsigma(z)$ depend on the type of target, on the orientation between target and receiver, and on the central frequency. The information, included in the scattering matrix, can be represented in the same way by introducing the scattering vector as reported above:

$$k = \frac{1}{2}\text{tr}\left(\varsigma(z)\Psi\right) = \left[k(1), k(2), k(3), k(4)\right] \tag{7.4}$$

where Ψ is the 2×2 orthonormal base.

One of the common bases is the Pauli's decomposition:

$$\Psi_p = \left\{\sqrt{2}\begin{bmatrix} 1 & 0 \\ 0 & 1 \end{bmatrix}, \sqrt{2}\begin{bmatrix} 1 & 0 \\ 0 & -1 \end{bmatrix}, \sqrt{2}\begin{bmatrix} 0 & 1 \\ 1 & 0 \end{bmatrix}, \sqrt{2}\begin{bmatrix} 0 & -j \\ j & 0 \end{bmatrix}\right\} \tag{7.5}$$

where i is the imaginary part of the number. Applying (7.3) by using the Pauli's decomposition, the scattering vector becomes

$$k(z) = \frac{1}{\sqrt{2}}\left[\varsigma^{HH}(z) + \varsigma^{VV}(z), \varsigma^{VV}(z) - \varsigma^{HH}(z), \varsigma^{HV}(z) + \varsigma^{VH}(z), j(\varsigma^{VH}(z) - \varsigma^{HV}(z))\right] \tag{7.6}$$

In the isotropic media hypothesis, the scattering vector can be rewritten as follows:

$$\begin{aligned} k(z) &= \frac{1}{\sqrt{2}}\left[\varsigma^{HH}(z) + \varsigma^{VV}(z), \varsigma^{VV}(z) - \varsigma^{HH}(z), \varsigma^{HV}(z) + \varsigma^{VH}(z), 2\varsigma^{HV}(z)\right] \\ &= \frac{1}{\sqrt{2}}\left[k(1)(z), k(2)(z), k(3)(z)\right] \end{aligned} \tag{7.7}$$

The same decomposition defined in (7.7) can be applied to the polarization matrix of the received signal. Therefore, by using Pauli's decomposition jointly with the isotropic media hypothesis that is typically verified in ISAR applications, the received signal can be seen as

$$S_R(f,t) = \frac{1}{\sqrt{2}}\left[S_R^{HH} + S_R^{VV}, S_R^{VV} - S_R^{HH}, 2S_R^{HV}\right] \tag{7.8}$$

All possible projections can be obtained by means of an internal product between the received signal vector and a generic polarization vector p:

$$S_R^{(p)} = S_R P \tag{7.9}$$

where the dependence on (f, t) is omitted for notation simplicity.

7.2.2 Image formation

The definition of the steps that lead to the image formation will be carried out without including the noise contribution, that will be considered in subsequently analysis. This is a standard procedure in SAR/ISAR processing. The first step consists of removing the phase term due to the radial movement of the focusing point, i.e., $\exp\left[-j(4\pi f/c)R_0(t)\right]$. After motion compensation step, the received signal can be written as

$$S_R(f,t) = W(f,t)\int_V k(z)e^{-j\frac{4\pi f}{c}[z^T i_{R_0}(t)]}dz \tag{7.10}$$

where the scattering vector $k(z)$ is defined previously. For Pol-ISAR image reconstruction, the range Doppler (RD) processing can be considered [1]. Under the hypothesis to consider the target as composed of scattering centres, the reflectivity function defined for a particular polarimetric vector can be defined as:

$$\varsigma^P(z) = \sum_{n=1}^{N} k_n^{(p)}\delta\left(z - z_n^{(p)}\right) \tag{7.11}$$

where $\varsigma^p(z) = k(z) \cdot p$. It is worth pointing out that each scattering centre is defined by a specific scattering vector k_n and that $k_n^{(p)} = k_n \cdot p$ is the complex amplitude of the nth scattering centre in the polarimetric channel p. The position of the same scattering centre in the image plane referred to the same polarimetric channel is represented as $z_n^{(p)}$. Although the scattering centres are considered as point-like and independent their response changes according to the polarization. The RD image technique consists of a bidimensional Fourier transform (2D-FFT) along both the time and the frequency domain of the received signal after motion compensation [1]. The complex image will be reconstructed in the time-delay Doppler domain. For a generic polarization, the polarimetric-ISAR image can be expressed as

$$\begin{aligned} I^p(\tau, \nu) &= FT^2 \left\{ W(f,t) \sum_{n=1}^{N} k_n^{(p)} e^{-j\frac{4\pi f}{c}\left[\left(z - z_n^{(p)}\right)^T i_{R_0}^{(p)}(t)\right]} \right\} \\ &= T_{obs} B \sum_{n=1}^{N} k_n^{(p)} \mathrm{sinc}\left[T_{obs}\left(\nu - \nu_n^{(p)}\right)\right] \mathrm{sinc}\left[B\left(\tau - \tau_n^{(p)}\right)\right] e^{-j2\pi f_0\left(\tau - \tau_n^{(p)}\right)} \end{aligned} \tag{7.12}$$

where $\left(\tau_n^{(p)}, \nu_n^{(p)}\right)$ represents the scattering coordinates in the time-delay Doppler domain, FT^2 stands for 2D-FFT and sinc $(\cdot)$ is the cardinal sine function defined as $\mathrm{sinc}\,(x) = \sin(\pi x)/\pi x$.

7.2.3 Polarimetric autofocus

The aim of this section is to introduce an algorithm that jointly processes all polarimetric channels information obtaining highly focused ISAR images. The proposed method is derived from an extension of the ICBA algorithm defined for ISAR system with single polarization [7]. Since target is non-cooperative, the distance between radar and focusing point on the target, $R_0(t)$ in the single polarization case is not *a priori* known. As a consequence, this term must be estimated through autofocusing technique in order to compensate the term $\exp\left[-j(4\pi f/c)R_0(t)\right]$ in the received signal. Under the hypothesis of regular target motion, the radar-target distance can be well approximated with the second-order Taylor series:

$$R_0(t) \cong \alpha_0 + \alpha_1 t + \frac{\alpha_2}{2} t^2 \tag{7.13}$$

where $\alpha_0 = R_0(0)$, $\alpha_1 = \dot{R}_0(0)$ and $\alpha_2 = \ddot{R}_0(0)$. It is clear that the problem of the motion compensation can be reduced to the estimation of α parameters also defined as autofocusing parameters. It is worth pointing out that the estimation of α_0 is not strictly necessary since the residual phase term $\exp\left[-j(4\pi f/c)\alpha_0\right]$ varying linear with the frequency f and as a consequence introduce only a linear translation in the time-delay Doppler domain obtained after 2D-FTT. The ICBA technique is based on the maximization of the IC (Image Contrast) function and is consist of two steps: (1) a raw estimation of autofocusing parameters and (2) a fine estimation of autofocusing parameters obtained by maximizing the IC function. The IC function can be defined as

$$IC = \frac{\sqrt{\mathbb{E}\left\{\left[\left|I(\tau,\nu,\alpha_1,\alpha_2\right|^2 - E\left\{\left|I(\tau,\nu,\alpha_1,\alpha_2\right|^2\right\}\right]^2\right\}}}{\mathbb{E}\left\{\left|I(\tau,\nu,\alpha_1,\alpha_2\right|^2\right\}} \tag{7.14}$$

where $\mathbb{E}\left[\cdot\right]$ is the statistical expectation and I is the image before contrast optimization, It is clear that the denominator represents the mean power of the image and the normalization factor of IC function that must be taken into account when two images are compared in term of IC function. Moreover, the IC function is a measure of the focalization image level, i.e., when the image is highly focused, the IC function has a high value. As a consequence, the optimal value of the autofocusing parameters can be found with

$$\left(\alpha_1^{(opt)}, \alpha_2^{(opt)}\right) = \underset{(\alpha_1,\alpha_2)}{\arg\max}\left\{IC\left(\alpha_1,\alpha_2\right)\right\} \tag{7.15}$$

The convergence of the algorithm at the IC global maximum point depends on the initial values of (α_1, α_2) parameters [11]. Therefore, an accurate pre-estimation of the autofocusing parameters is a mandatory step in order to obtain a highly focused image. Applying the Fourier transform with respect to the frequency dimension on the received signal, the target range profiles will be obtained for each sweep. As a consequence, the time history distance $R_n(t)$ between radar and each scattering centre will be obtained. Under the hypothesis that the scattering centre radial velocity is constant during the observation time, the distance varies linearly with autofocus parameter α_1, i.e., $R_n(t) \cong R_0(t) + \alpha_1 t$. The $R_0(t)$ term represents the distance between the radar and the centre of the target and is assumed to have the same characteristic as of $R_n(t)$ distance with the same slope α_1. Under this assumption, the autofocus parameter α_1 can be estimated by the slope of the $R_n(t)$ terms:

$$\alpha_1 = \tan\left(\phi\right) \tag{7.16}$$

where ϕ is the angle between different terms and abscissa axis. This angle can be estimated with the Radon transform of the received signal as follows:

$$\hat{\phi} = \underset{(\alpha_1,\alpha_2)}{\arg\max}\left\{\left[RT_{S_R,\phi}(\phi)\right]\right\} - \frac{\pi}{2} \tag{7.17}$$

where $RT(\cdot)$ is the Radon transform. When an estimate of the autofocus parameter is obtained, an estimate of the autofocus parameter α_2 can be obtained by searching within a fixed interval $\left[\alpha_2^{(\min)}, \alpha_2^{(\max)}\right]$ for the maximum of the IC function $IC\left(\hat{\alpha}_1, \hat{\alpha}_2\right)$, i.e. $\hat{\alpha}_2 \underset{(\alpha_1,\alpha_2)}{\arg\max}\left\{\left[IC_{\hat{\alpha}_2}(\hat{\alpha}_1, \hat{\alpha}_2)\right]\right\}$.

In the case of Pol-ISAR systems, the optimization problem introduced previously can be defined in a larger domain that also includes polarimetric feature. An optimization problem can be introduced by defining a different IC function that also depends on the polarization vector $\boldsymbol{p}$. As a consequence, the optimization problem consist in the search for both the motion parameters and the polarization vector that maximizes the IC function, i.e.

$$\left(\alpha_1^{(opt)}, \alpha_2^{(opt)}, \boldsymbol{p}^{(opt)}\right) = \underset{(\alpha_1,\alpha_2,p)}{\arg\max}\, \{IC(\alpha_1, \alpha_2, \boldsymbol{p})\} \tag{7.18}$$

Also in this case, the algorithm requires a preliminary initialization of the parameters to be estimated. So an initial value of both the autofocus parameter and the polarization vector should be provided. The first step is the initialization of the polarimetric vector; the polarimetric vector initializing the autofocusing algorithm is the vector for which the projection of the received signal onto the polarimetric channel defined by the vector return the maximum SNR:

$$\hat{p}_M = \underset{p}{\arg\max} \left(\frac{\iint_{-\infty}^{+\infty} \left|S_R^{(p)}(f,t)\right|^2 dfdt}{\iint_{-\infty}^{+\infty} \left|N_R^{(p)}(f,t)\right|^2 dfdt} \right) \tag{7.19}$$

Since the noise level can be considered the same on the H channel and on the V channel, it can be observed that the SNR is maximum when the energy of the received signal is maximum:

$$\hat{p}_M = \underset{p}{\arg\max} \left(\iint_{-\infty}^{+\infty} \left|S_R^{(p)}(f,t)\right|^2 dfdt \right) \tag{7.20}$$

The second step is the initialization of the autofocus parameters: the initial values of (α_1, α_2) parameters are obtained by the same method described previously with the projection of the received signal onto the polarimetric channel defined by the polarimetric vector:

$$\left(\alpha_1^{(\hat{p}_M)}, \alpha_2^{(\hat{p}_M)}\right) = \underset{(\alpha_1,\alpha_2)}{\arg\max} \left\{IC^{\hat{p}_M}(\alpha_1, \alpha_2)\right\} \tag{7.21}$$

The final step is the search of the autofocus parameters and the polarization vector optimal values. Both the optimal autofocus parameters and the optimal polarization vector are jointly estimates by maximizing the IC function as follows:

$$(\hat{\alpha}_1, \hat{\alpha}_2, \hat{p}) = \underset{(\alpha_1,\alpha_2,p)}{\arg\max} \left\{IC^{(p)}(\alpha_1, \alpha_2)\right\} \tag{7.22}$$

where

$$IC^{(\hat{p})} = \frac{\sqrt{\mathbb{E}\left\{\left[\left|I^{(\hat{p})}(\tau, \nu, \alpha_1, \alpha_2\right|^2 - E\left\{\left|I^{(\hat{p})}(\tau, \nu, \alpha_1, \alpha_2\right|^2\right\}\right]^2\right\}}}{\mathbb{E}\left\{\left|I^{(\hat{p})}(\tau, \nu, \alpha_1, \alpha_2\right|^2\right\}} \tag{7.23}$$

In this way, not only the optimal estimates of α_1 and α_2 are found but also the optimal polarization where the image data can be projected.

The proposed algorithm provides an accurate estimation of the motion parameters that allow to remove the phase term due to the radial motion of the target. This algorithm also provides a more focused ISAR image since it gives the optimal polarization vector, and as a consequence, the resulting ISAR image obtained by processing the received signal onto the polarimetric channel $\hat{p}$ is the image with highest contrast. After motion compensation, ISAR image can be reconstructed by applying the RD algorithm. Through RD processing, a bidimensional image $I^{(p)}(\tau, \nu)$ can be obtained for each polarimetric channel. Therefore, for each nth range cell, $I^{(p)}(\tau_n, \nu_n)$ represents the reflectivity function of the target for a particular polarimetric channel p. By using orthonormal polarimetric vector, i.e., $p^{(1)} = [1, 0, 0]$, $p^{(2)} = [0, 1, 0]$, $p^{(3)} = [0, 0, 1]$, that defines the three-dimensional (3D) polarimetric space and applying the RD processing a $N \times M \times 3$ (where N is the number of sweep and M is the number of transmitted frequencies), datacube of two-dimensional (2D ISAR images can be obtained. Each ISAR image, i.e., $I(\tau, \nu) = [I^{(p,1)}(\tau, \nu), I^{(p,2)}(\tau, \nu), I^{(p,3)}(\tau, \nu)]$, represents an estimation of the reflectivity function for each range cell in each polarimetric channel. All the polarimetric information in the datacube can be synthesized in a single ISAR image through a color representation of the datacube. It is worth pointing out that in order to obtain a Pol-ISAR image, the autofocusing step processing must be applied using the same autofocus parameters for the signal received in each polarimetric channel.

In order to test the effectiveness of the polarimetric ICBA (Pol-ICBA), the proposed algorithm is applied to real data and compared with a single polarization ICBA. Real data are provided by the Defense Science & Technology Organization (DSTO) of Australia and are collected by DTSO's Ingara airborne radar. The target under test is a stationary commercial charter boat on the open ocean shown in Figure 7.1. The specifications are given in Table 7.1, while the experiment parameters are given in Table 7.2. Data set was collected over two days from circular fight paths roughly centred on the boat.

Figure 7.1 Target boat

Table 7.1 Target specification

Boat parameters	Dimension
Overall length	16.75 m
Maximum beam	5.80 m
Bulwark height	850 mm
Sill height	300 mm

The data sets were originally processed as spot-SAR images and had appropriate polarimetric channel calibration solution applied. In order to begin ISAR processing in the time frequency domain, an inverse Fourier transform must be applied. Results from Pol-ISAR image formation process are shown in Figure 7.2 and functional performance results over all data collections are given in Table 7.3. In Figure 7.2, the ISAR images were obtained by running the single-channel autofocusing approach across the four principal channels (HH, VV, HV and VH) and selection the motion compensation solution that gave the highest contrast in the ICBA case. This solution was then applied to the other three principal channels in order to obtain three co-registered ISAR images in order to correctly use the polarization that is associated with the local scattering mechanisms. In Figure 7.3, it is possible to see that the visual quality of the resultant images clearly improves when moves from single polarization autofocused image to the polarimetric autofocused image. This is expected as the amount of available information increases, the quality of the final focussed image increases too.

The improvement obtained from using pol-ICBA is presented in Table 7.3. In this table, the number of wins is the number of times a given approach achieves the best score for an IC measure. Finally, the Pol-ISAR image is shown in Figure 7.4. By using Pauli decomposition on the received signal and the scattering matrix, the red colour can be associated to the polarimetric channel $p^{(1)}$, the green to the polarimetric channel $p^{(2)}$ and the blue colour to $p^{(3)}$. In such way, all the information from the polarimetric channels can be visualized.

Table 7.2 Experimental collection parameters

Parameter name	Parameter value
Incidence angle	56°–57°
Slant range	6600 m
Centre frequency	10.1 GHz
Slant range resolution	0.35 m
Aircraft altitude	3600 m

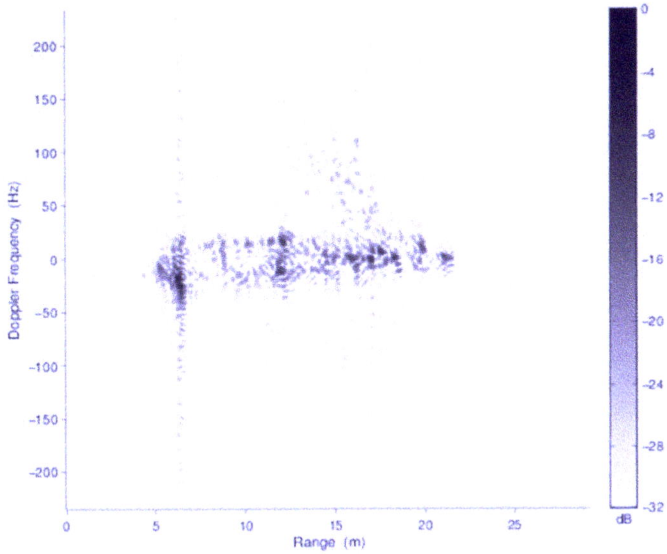

Figure 7.2 *ICBA – single polarization autofocusing – VV polarization*

7.3 Classification with Pol-ISAR

Polarimetric radar classification has gain the attention of the scientific community all over the world and since many years, as these systems may be of interest for many military and civilian applications. To be effective, a classification system must be able to extract a suitable set of features from the available data. The set of features must uniquely characterize the target to be recognized. Many different kinds of features can be extracted from a Pol-ISAR image, such as geometrical (target's size, target's shape, etc.) and polarimetric information. The geometrical interpretation of an ISAR image is strongly dependent on the target's own motions with respect to the radar that are usually unknown. A 2D ISAR image is in fact the projection of the 3D target's reflectivity function onto a 2D plane that is called image projection plane (IPP). Such plane is dependent on the radar-target geometry and kinematics. The use of the polarimetric information should, therefore, provide a more robust classification system with respect to the IPP orientation, because the IPP orientation does not depend on the target orientation and that the polarimetric features are robust with respect to the former and not to the latter.

Table 7.3 *ICBA comparison*

Approach	IC win
Single-channel autofocused ICBA	8
Polarimetric autofocused ICBA	15

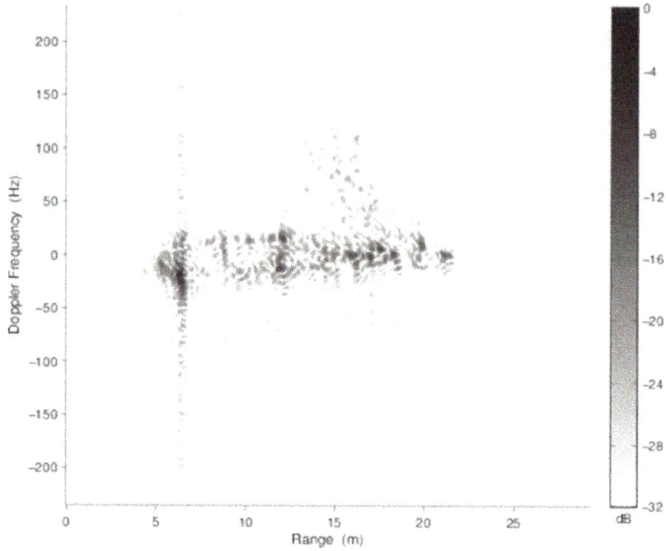

Figure 7.3 ICBA – polarimetric autofocusing – VV polarization

7.3.1 Feature extraction

ISAR images prove useful also when used for classifying and recognizing targets [12–16]. Nevertheless, the image data are often too large to implement real-time

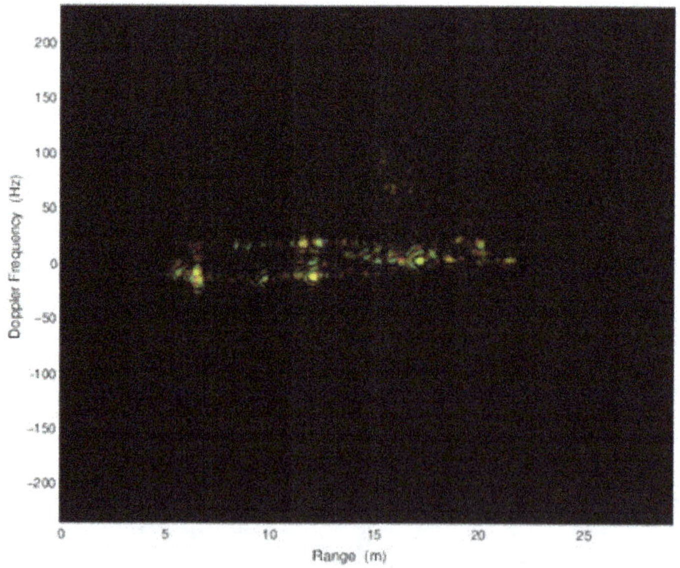

Figure 7.4 Polarimetric ISAR image

classifiers and additionally pose some constraints in terms of memory usage. To reduce data occupation, features can be used in place of the ISAR image to recognize the target if they unambiguously represent the targets. The problem of reducing the amount of data without losing useful information is even more critical when dealing with Pol-ISAR images since the data size is three or four times larger. However, polarimetry adds information on the target that, if well exploited, is useful for improving classification performance. Therefore, the role of the feature extraction algorithm is to extract the target main characteristics that make it distinguishable from others and discard useless information. Features can be drafted by a user or extracted using deep learning (DL) approaches typically implemented using convolutional neural network (CNN). In the first case, the classifier designer identifies the features based on their physical meaning, as for example, the power, the polarimetric behaviour of the target, shape [17–23]. Conversely, in the DL approach, CNNs automatically capture and extract feature by processing the input ISAR image [24]. The advantage of the latter over the former is that the feature extraction process does not require any human interaction and therefore it is not affected by the human interpretation. Moreover, CNNs are typically designed to mimic the human brain that extract hierarchical features. However, features extracted using CNNs do not have physical meaning and are therefore difficult to interpret. This makes the ATR system like a black box and the ATR system results difficult to explain. This, in turn, makes it difficult to evaluate the trustability of the recognition system. The dispute over which algorithm is the best is still open. Taking these considerations into account and being the focus of this chapter of the book on radar polarimetry and its use for ISAR imaging, this section focuses on hand-made polarimetric features. The polarimetric features presented in this section have been drafted taking into account the radar waveform and the physical mechanisms that regulate the interaction of the transmitted waveform with the target.

Section 7.3.2 presents the Pol-CLEAN algorithm. The Pol-CLEAN extracts the scattering matrix of each scatterer that composes the target. Some results on real data set are also shown to support the algorithm description. Section 7.3.3 focuses on the target decomposition (TD) algorithms. TD algorithms elaborate the polarimetric information of a Pol-ISAR image to extract to retrieve its dominant scattering mechanism. This may of further help for the target recognition. As for example, statistics on the target main scattering mechanism could be calculated and used to recognize the target. A set of TD methods has been developed for the interpretation of fully polarimetric data. A review of all the TD can be found in [25]. Section 7.3.3 recaps some of these algorithms because they have been used for ATR using Pol-ISAR images.

7.3.2 Pol-CLEAN

The Pol-CLEAN works iteratively by:

- locating the brightest scattering centre in the Pol-ISAR image;
- extracting its polarimetric vector by estimating its PSF;

- removing it from the image in order to extract the next brightest scattering centre.

To correctly eliminate a scatterer from an ISAR image, its PSF must be eliminated and subtracted from the image itself. Let $I^{(1)}(\tau, \nu)$, $I^{(2)}(\tau, \nu)$ and $I^{(3)}(\tau, \nu)$ be the three ISAR images in the three Pauli's channels. The brightest scatterer can be found by searching among the three ISAR images as follows:

$$\left(\tau_s, \nu_s, i_s\right) = \arg \max_{(\tau, \nu, i)} \quad |I^{(i)}(\tau, \nu)| \tag{7.24}$$

where $i \in 1, 2, 3$ indicates the Pauli channel and (τ_s, ν_s) are the range and Doppler coordinates of the scatterer.

The subsequent step aims at estimating the scatterer PSF and is based on the ISAR image signal model. Particularly, when the range-Doppler technique can be used to form the ISAR image, the received signal can be written as a product of a time and a frequency component:

$$s_R^{(i)}(t, f) = s_1^{(i)}(t)\, s_2^{(i)}(f) \tag{7.25}$$

where

$$s_1^{(i)} = B_i exp\left(j2\pi(\nu \cdot t + \frac{\mu}{2} \cdot t^2)\right) \cdot rect\left(\frac{t}{T_{obs}}\right) \tag{7.26}$$

$$s_2^{(i)} = A_i exp\left(2\pi f \tau\right) \cdot rect\left(\frac{f - f_0}{B}\right) \tag{7.27}$$

where the product $A_i \cdot B_i$ represents the complex amplitude in the i_{th} Pauli's channel, ν is the Doppler frequency, μ is the chirp rate and τ is the time delay associated with the scattering centre. It is worth pointing out that the parameter μ accounts for a quadratic radial motion, that is, it includes Doppler acceleration. Therefore, it should not be confused with the transmitted signal chirp rate if any is employed. It should be also noted that if the target can be considered as a rigid body each scatterer should undergo the same motion and hence the same μ.

Even though ISAR images are modelled as the superposition of independent scatterers, real ISAR images suffer from interference among scatterers, which may slightly modify the PSF of each scatterer. To tackle this aspect, the PSF parameters are estimated for each scatterer separately.

The algorithm aims at estimating $\left(B_i, \nu, \mu, A_i, \tau\right)$ for each scatterer. Being the range and Doppler profiles independent each other, when the hypothesis for the application of the RD algorithm hold, the PSF parameters are estimated in a two-step algorithm:

- first, using the Doppler profile of the selected scatterer the algorithm estimates $\left(B_i, \nu, \mu\right)$
- second, using the range profile of the selected scatterer the algorithm estimates the parameters $\left(A_i, \tau\right)$

Estimate of the PSF's parameters is performed by minimizing the image energy after scattering centre subtraction. The two-step algorithm described previously guarantees a better convergence of the optimization problem.

Following this approach, (7.24) is used to identified the polarimetry channel in which the scatterer PSF reaches its maximum value. Additionally, (7.24) provides a preliminary estimate of the scatterer range and Doppler coordinates that are used as initial parameters for the optimization algorithm described hereinafter. Then, the PSF's parameters are estimated as follows:

$$\{\hat{\nu}, \hat{\mu}, \hat{B}_{i_s}\} = \arg \min_{(\nu, \mu, B_{i_s})} \quad E_{D_{i_s}}\left(\nu, \mu, B_{i_s}\right) \tag{7.28}$$

$$\{\hat{\tau}, \hat{A}_{i_s}\} = \arg \min_{(\tau, A_{i_s})} \quad E_{R_{i_s}}\left(\nu, \mu, B_{i_s}\right) \tag{7.29}$$

where $E_{D_{i_s}}$ and $E_{R_{i_s}}$ are, respectively, the residual energy of the Doppler and range profile after the scatterer cancellation and the subscript i_s represents the polarimetric channel selected using (7.24). The optimization algorithm is initialized with $\hat{\tau} = \tau_s$, $\hat{\nu}_s$ and $\hat{\mu} = 0$. A precise estimation of (τ, ν, μ) is essential to eliminate in the best way possible the scatterer from the ISAR image, not only its peak but also its sidelobes that can mask weaker scatterer close to it. To this end, (τ_s, ν_s) are not precise enough and a better estimation of these parameters and of μ is required that can be obtained using (7.28).

It should be noted that the PSF shape parameters, namely (ν, μ, τ), do not change over polarimetric channel. Therefore, once these parameters have been estimated in the polarimetric channel i^*, the estimation of only (A_i, B_i) is repeated over the other two polarimetric channels, as follows:

$$\hat{B}_i = \underset{B_i}{\operatorname{argmin}} \quad E_{D_i}\left(\hat{\nu}, \hat{\mu}, B_i\right) \tag{7.30}$$

$$\hat{A}_i = \underset{A_i}{\operatorname{argmin}} \quad E_{R_i}\left(\hat{\tau}, A_i\right) \tag{7.31}$$

where $i \neq i_s$.

When the PSF's parameters have been estimated, the PSF of the selected scatterer is calculated applying the 2D-FFT ($2D - FT\cdot$), used to form the ISAR image, as follows:

$$I^{(i)}_{PSF}(\tau, \nu) = |2D - FT\{s^{(i)}_1(t|(\hat{\nu}, \hat{\mu}, \hat{B}_i)) \cdot s^{(i)}_2(f|(\hat{\tau}, \hat{A}_i))\}|e^{\angle I^{(i)}(\tau,\nu)} \tag{7.32}$$

The scattering vector of the selected scatterer is therefore composed of the peak of the three PSF centred in $\left(\tau_s, \nu_s\right)$. Finally the scatterer is removed and a new ISAR image is obtained as follows:

$$I^{(i)}_{k+1}(\tau, \nu) = I^{(i)}_k(\tau, \nu) - I^{(i)}_{PSF}(\tau, \nu) \tag{7.33}$$

The algorithm then repeats the previous steps and stops when a prefixed number of scatterer has been set or when the signal energy in the ISAR image becomes lower than a given threshold, λ:

$$\lambda = \Lambda E \frac{SNR}{SNR + 1} \tag{7.34}$$

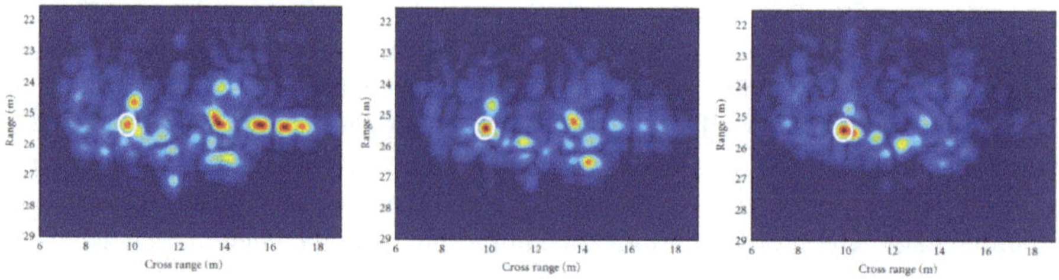

Figure 7.5 *ISAR images in the Pauli's channels, respectively, from left to right, HH+VV, VV−HH and 2HV [26]*

where E is the energy of the fully Pol-ISAR image. Therefore, the term $E\frac{SNR}{SNR+1}$ represents the energy of the signal component only (noiseless), namely P_S, and Λ is a number in the range $[0-1]$. As a consequence, the threshold is set to be equal to a certain percentage of the overall signal power, namely, $\lambda = \Lambda \cdot P_S$ and the algorithm stops when the residual energy after the $k-th$ scatterer removal is below this threshold.

As an example, Figure 7.5 shows the three ISAR images of a tank (T72) in the Pauli's polarimetric channels, namely, HH+VV, VV−HH and 2HV. Data are publicly available on the Sensor Data Management System website and released by the US Air Force [13, 27]. The white circle indicates the first selected scatterer. Then, the algorithm estimates the PSF by searching for the optimum PSF's parameters independently on the Doppler and range profile.

Figure 7.6 shows the range profiles in each polarimetric channel. Figure 7.7 shows the cross-range or Doppler profiles in each polarimetric channel. The red, blue and green lines are, respectively, the scatterer range profile, the final estimated range profile and the residual range profile.

Finally, Figure 7.8 shows the ISAR image when the selected scatterer is removed.

Then the algorithm is repeated. By setting $\Lambda \simeq 0.05$, 43 scatterers have been extracted. Figure 7.8 shows the original fully Pol-ISAR image using the RGB color coding in which red, green and blue are associated, respectively, to the first, second and third Pauli's channel. The coloured dots represent the

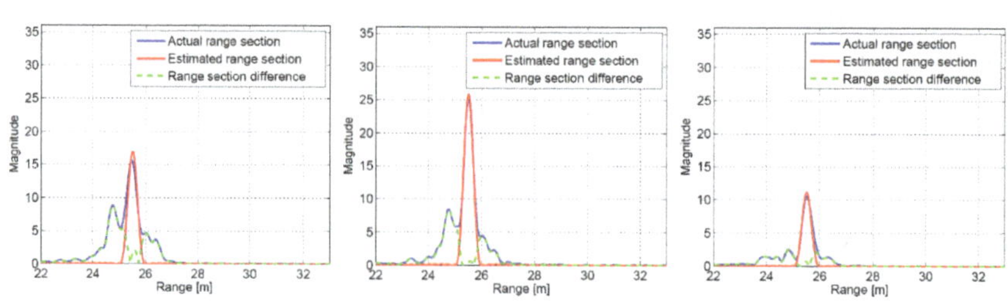

Figure 7.6 *Range profiles in the Pauli's channels, respectively, from left to right, HH+VV, VV−HH and 2HV [26]*

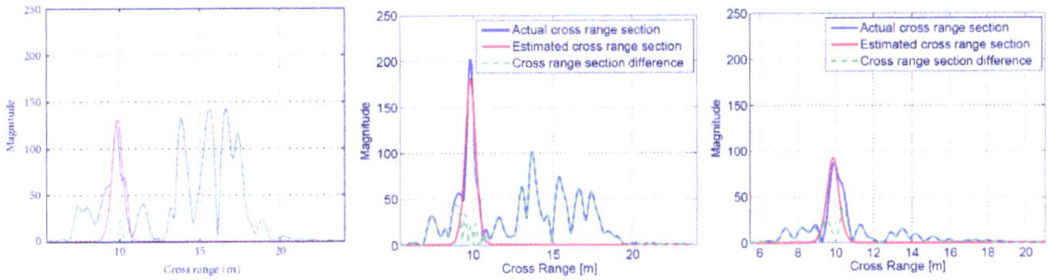

Figure 7.7 *Range profiles in the Pauli's channels, respectively, from left to right, HH+VV, VV−HH and 2HV [26]*

extracted scatterers. The colour of each dot represents the polarimetric signature of the extracted scattering centre in the Pauli basis. The coloured dots have been superimposed to the original Pol-ISAR image to visually assess the quality of the Pol-CLEAN algorithm. The colour inside the white circle represents the polarimetric PSF estimated by the Pol-CLEAN algorithm. It should be noted that the colour of the extracted dots is very close to the colour of the underlying ISAR image, especially for the brightest scatterers. Weaker scatterers do not match perfectly with underlying ISAR image because of two reasons:

- cancellation of each scatterers is not perfectly made and some residuals remain as it can be observed from Figure 7.8, that can alter the estimation process of further scatterers or even cause the extraction of false scatterers
- weaker scatterers are partially masked by stronger scatterer's sidelobes (it can be read as an interference problem).

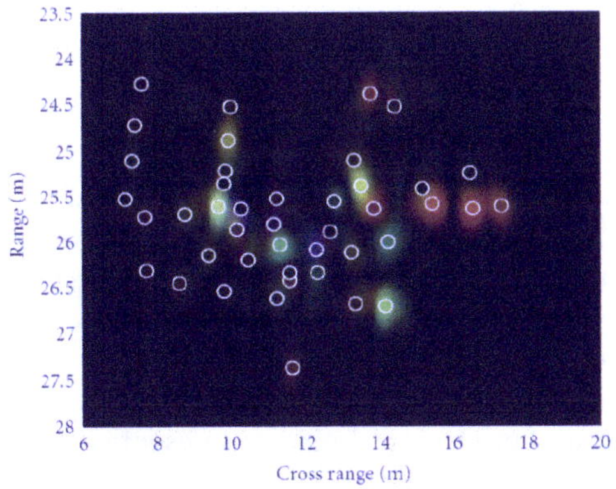

Figure 7.8 *RGB ISAR image. Red, green and blue are associated, respectively, the HH+VV, VV−HH and 2HV channels. Therefore, each colour represents a different polarimetric mechanism [26].*

7.3.3 Target decomposition

Several target decomposition (TD) theorems have been proposed in literature. These methods can be categorized into InCoherent TD (ICTD) methods and Coherent TD (CTD) methods. Both methods exhibit, however, pros and cons. Although the CTD is able to carefully characterize the polarimetric properties of each scattering centre composing the target, it can be effectively applied when the scatterer exhibits a high degree of coherency, which is, however, difficult to be assessed. Moreover, the number of scattering centres may change significantly among different ISAR images thus determining feature vectors of different size. On the contrary, ICTD methods allow the target to be characterized in a more compact way as they provide feature vectors of fixed size. As a drawback, they are able to characterize the target polarimetric behaviour from a global perspective only.

7.3.3.1 ICTD

The scattering matrix $[\boldsymbol{S}]$ is only able to characterize the so-called coherent or pure scatterers. On the contrary, this matrix cannot be employed to characterize, from a polarimetric point of view, the so-called distributed scatterers. Since noise must be reduced, only second-order polarimetric representations can be employed to analyse distributed scatterers. These second-order descriptors are the 3 × 3, Hermitian average covariance C and the coherency T matrices. These two representations of the polarimetric information are equivalent. To obtain these matrices, we proceed in this way. Given as polarimetric data set, $I_P\left(r_i, f_j\right)$, where r is the range and f the Doppler frequency, is possible to represent every pixel by a complex 3D scattering vector k. The coherency matrix T is obtained by

$$T\left(r_i, f_j\right) = k_P\left(r_i, f_j\right) k_P^{\dagger}\left(r_i, f_j\right) \tag{7.35}$$

The coherency matrix can be estimated averaging a number of coherency matrices of the single pixel, integrating the greatest part of the backscattered energy, obtaining

$$\bar{T} = \sum_{i=1}^{I} \sum_{j=1}^{J} T\left(r_i, f_j\right) \tag{7.36}$$

$\bar{T}$, matrix allows characterizing both deterministic and stochastic targets. If the rank is 1, the matrix represents a coherent target with 5 degrees of freedom, otherwise the target is (partially) incoherent and we have 9 degrees of freedom. The complexity of the scattering process makes extremely difficult the physical study of a given scatterer through the direct analysis of T. Hence, the objective of the incoherent decompositions is to separate the T matrix as the combination of second-order descriptors corresponding to simpler or canonical objects, presenting an easier physical interpretation. These decomposition theorems can be expressed as

$$\bar{T} = \sum_{i}^{L} p_i T_i \tag{7.37}$$

where the canonical response is represented by T_i, and p_i denotes the coefficients of these components. It is desirable that these components present some properties. First of all, these components should correspond to pure targets in order to simplify the physical study. Nevertheless, this is not absolutely necessary. In addition, the various components are independent, or in a more restrictive way, orthogonal. From now on, we will refer T as the averaged coherency matrix.

In the incoherent target decomposition proposed by Cloude and Pottier (ICTD) [28], the matrix T can be decomposed as

$$T = U \Lambda U^{-1} \tag{7.38}$$

with

$$\Lambda = \begin{bmatrix} \lambda_1 & 0 & 0 \\ 0 & \lambda_2 & 0 \\ 0 & 0 & \lambda_3 \end{bmatrix} \qquad U = \begin{bmatrix} u_1 & u_2 & u_3 \end{bmatrix} \tag{7.39}$$

where U is the eigenvector matrix and Λ is the eigenvalues matrix, where $\lambda_1 \geq \lambda_2 \geq \lambda_3$.

The eigenvectors are not unique, so we choose the ones that have a real first element

$$u_i = \begin{bmatrix} \cos(\alpha_i) \\ \sin(\alpha_i)\cos(\beta_i)\, e^{j\delta} \\ \sin(\alpha_i)\sin(\beta_i)\, e^{j\gamma} \end{bmatrix} \tag{7.40}$$

The eigenvalues and the eigenvectors are considered as the primary parameters of the eigen-decomposition of bT. In order to simplify the analysis of the physical information provided by this eigen-decomposition, three secondary parameters are defined as a function of the eigenvalues and the eigenvectors of T.

- Entropy:

$$H = \sum_{i=1}^{3} -p_i \log_3 p_i \qquad p_i = \frac{\lambda_i}{\sum_{k=1}^{3} \lambda_k} \tag{7.41}$$

where p_i, also called the probability of the eigenvalue λ_i, represents the relative importance of this eigenvalue with respect to the total scattered power, since

$$SPAN = \sum_{i=1}^{3} \lambda_i = |s_{hh}|^2 + 2|s_{vh}|^2 + |s_{vv}|^2 \tag{7.42}$$

- Mean alpha angle:

$$\alpha_{ML} = \alpha_1 p_1 + \alpha_2 p_2 + \alpha_3 p_3 \tag{7.43}$$

where α_i is the alpha angles of the first component of each eigenvector.

- Anisotropy:

$$A = \frac{\lambda_2 - \lambda_3}{\lambda_2 + \lambda_3} \tag{7.44}$$

The feature vector can be composed by nine features, but usually p_3 and α_3 are omitted because they are linearly dependent from the other value, so we obtain

$$(H, A, \alpha_{ML}, \alpha_1, p_1, \alpha_2, p_2) \tag{7.45}$$

The interpretation of the information provided by the eigen-decomposition of the coherency matrix must be performed in terms of the eigenvalues and eigenvectors of the decomposition or in terms of $H/A/\alpha$. Nevertheless, both interpretations have to be considered as complementary.

The study of the scattering mechanism is mainly performed through the interpretation of the mean alpha angle, since its values can be easily related with the physics behind the scattering process. The next list reports the interpretation of α:

- $\alpha = 0$: The scattering corresponds to single-bounce scattering produced by a rough surface.
- $\alpha = \pi/4$: The scattering mechanism corresponds to volume scattering.
- $\alpha = \pi/2$: The scattering mechanism is due to double-bounce scattering.

The second part in the interpretation of the eigen-decomposition is performed by studying the value of the eigenvalues of the decomposition. A given eigenvalue corresponds to the associated scattered power to the corresponding eigenvector. Consequently, the value of the eigenvalue gives the importance of the corresponding eigenvector or scattering mechanism. The entropy H determines the degree of randomness of the scattering process, which can be also interpreted as the degree of statistical disorder. In this way:

- $H = 0$:

$$\lambda_1 = SPAN \quad \lambda_2 = 0 \quad \lambda_3 = 0$$

Consequently, the coherence matrix $\boldsymbol{T}$ presents Rank 1 and the scattering process corresponds to a pure target.

- $H = 1$:

$$\lambda_1 = SPAN/3 \quad \lambda_2 = SPAN/3 \quad \lambda_3 = SPAN/3$$

In this situation, the coherence matrix $\boldsymbol{T}$ presents Rank 3, that is, the scattering process is due to the combination of three pure targets. Consequently, $\boldsymbol{T}$ corresponds to the response of a distributed target.

- $0 < H < 1$:

In this case, the final scattering mechanism given by T results from the combination of the three pure targets given by bu_i, but weighted by the corresponding eigenvalue.

7.3.3.2 CTD

CTD was developed to completely characterize coherent scattering, namely completely polarized scattered waves, for which the fully polarimetric information is contained in the scattering matrix. Well-known CTD methods are the Huynen [29], Krogager [30] and Cameron *et al.* [31] methods. CTD are therefore more suitable for high-resolution radar images of man-made targets for which it is likely that each spatial cell contains a dominant scatterers resulting in a coherent scatterer. Nonetheless, depending on the spatial resolution cell size with respect to a scatterer, a superposition of effects due to multiple scatterers within a resolution cell could happen anyway, which reduces the coherence of the signal reflected from that particular spatial cell. Furthermore, unbalances among polarimetric channels and clutter may further reduce the coherence degree. The coherent assessment of each scatterer is therefore useful prior the application of CTD methods, as proposed in Reference 32.

Differently from others CTD methods, Cameron's method permits to represent symmetric scattering and propose a metric to classify it. Specifically, it was shown that the diagonal form of symmetric scattering matrices can be represented with points in the unit disc of the complex plane. Moreover, the metric developed in Reference 31 provided a way for comparing symmetric scattering matrices and therefore classifying the type of elemental scattering mechanism. It was also shown that such a metric was insensitive to any rotation of the scattering centre about the radar line of sight (LoS). Hence, the insensitive of the results to any rotation around the LoS and the possibility to recognize the main scattering behaviour of a particular scatterer make Cameron's method attractive for ATR in general.

Cameron's method states that a scattering matrix, $S_t(\tau, \nu) \in \mathbb{C}^{2\times 2}$ of a scatterer located at range-Doppler cell (τ, ν), can be decomposed into three orthogonal components as follows:

$$S_t(\tau, \nu) = A\left[\cos(\theta_{rec})\left(\sin(\delta) S_{t\max} + \cos(\delta) S_{t\min}\right) + \sin(\theta_{rec}) S_{tnr}\right] \tag{7.46}$$

where $S_{tmax} \in \mathbb{C}^{2\times 2}$ and $S_{tmin} \in \mathbb{C}^{2\times 2}$ are the maximum symmetric components and the minimum symmetric components, while S_{tnr} is the non-reciprocal component. The angle θ_{rec} is a measure of how much $S_t \in \mathbb{C}^{2\times 2}$ is reciprocal. The reciprocity theorem states that a scattering matrix is reciprocal when the off diagonal element is equal. This is when the e.m. signals propagate in a reciprocal medium. However, some deviations can occur because of the imbalance among polarimetric channels or noise. The angle δ measures how much the reciprocal component of S_t is symmetric. A symmetric scattering has an axis of symmetry in the plane perpendicular to LoS and its scattering matrix can be diagonalized by means of a rigid rotation of the proper angle. Any symmetric scattering matrix can be written in terms of the Pauli's basis as follows:

$$S_{t_{sym}} = A\left(\alpha\Sigma_1 + \epsilon\left(\cos\left(\theta\right)\Sigma_2 + \sin\left(\theta\right)\Sigma_3\right)\right) \tag{7.47}$$

It can be also diagonalized by means of a rigid rotation of an angle, ϕ chosen within the set $\phi \in \{-\theta/2, -\theta/2 \pm \left(-\pi/2\right)\} \cap \left(-\pi/2\right]$, so that the upper diagonal element is maximum.

Furthermore, it can be proven that a symmetric matrix can be expressed in terms of a diagonal matrix D as follows:

$$S_{t_{sym}} = Ce^{j\rho}R\left(-\phi\right)D\left(z\right)R\left(\phi\right) \tag{7.48}$$

where ρ is a nuisance phase term that depends on the target to radar distance, $R(\phi)$ is a rotation matrix of the angle ϕ as defined above, z is a complex parameter and C represents the norm of $S_{t_{sym}}$.

The Cameron's algorithm first decomposes S_t into the three components $S_{t_{max}}$, $S_{t_{min}}$ and $S_{t_{nr}}$, as detailed in (7.47). Then it considers $S_{t_{max}}$ only, and diagonalises it using the estimated angle ϕ. Then, remembering that a symmetric scattering matrix becomes diagonal when rotated of a proper angle ϕ and that the angle ϕ is the one that both diagonalises $S_{t_{sym}}$ and makes the upper diagonal element maximum, it is possible to find

$$S_{t_{max}} = Ce^{j\rho}D\left(z\right) \tag{7.49}$$

where

$$D\left(z\right) = \frac{1}{\sqrt{\left(1 + |z|^2\right)}}\begin{pmatrix} 1 & 0 \\ 0 & z \end{pmatrix} \tag{7.50}$$

where $z \in \mathbb{C}$ and $|z| \leq 1$.

A similar procedure could be applied to $S_{t_{min}}$. However, the orthogonality between $S_{t_{min}}$ and $S_{t_{max}}$ limits the range of values that $S_{t_{min}}$ can assume for a fixed value of I_{max}. As a consequence it is possible to demonstrate that the first can be derived from the latter [31].

As a final remark it is possible to observe that any coherent scattering matrix, I can be uniquely represented by a set of parameters, namely, the $z \in \mathbb{C}$, the scattering matrix norm, C, and the angle ϕ. Cameron also derives the z parameter for some common symmetric scatterers and proposes a metric to measure the similarity between the scattering matrix under test and one of the selected common symmetric scatterers. This metric permits to identify if a scatter behaves as a trihedral, a dihedral, a cylinder, a dipole or a narrow Diplane.

With the aim to prove the effectiveness of the Cameron's method to distinguish among different scattering mechanism, Figure 7.9 shows some simulated results.

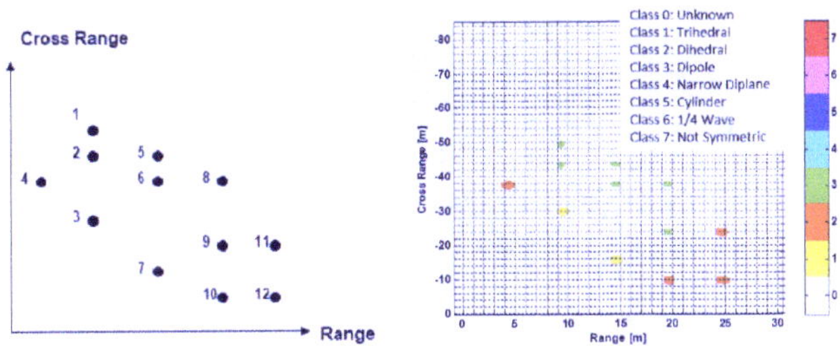

Figure 7.9 *Simulated target composed of ideal scatterer (left), and results of the Cameron's decomposition algorithm (right)*

An ISAR image has been simulated composed of scatterers with different polarimetric behaviour, as shown in Table 7.4. Then the scatterers are extracted and their scattering matrices decomposed using the Cameron's method. The classification results are shown in Figure 7.9, while Table 7.4 illustrates the estimation results in terms of ϕ.

Cameron's method has been also applied to an ISAR image of a tank, to show how it can be useful to characterize the target statistically. The data set is the one presented in section 7.3.2.

Figure 7.10 shows the optical image of the target taken approximatively at the same angles of the corresponding ISAR image and the results of the Cameron's algorithm.

The behaviour of each scatterer in a real environment is hard to understand due to the interactions among target structures and with the background. However, TD methods such as Cameron's method can be used to extract polarimetric statistical features useful for target recognition. For example, we could characterize the polarimetric behaviour of the target with the percentage of trihedral, dihedral, dipole, or using the Cameron's parameters for the brightest scatterers.

Table 7.4 *Cameron's results in terms of estimate of the rotation angle and classification of each scatterer polarimetric behaviour*

Scatterer	**Type**	ϕ	**Classified**	ϕ
3,7	Trihedral	0°	Trihedral	[0°, 0°]
4,12	Dihedral	90°	Dihedral	[90.4°, 89.9°]
10	Dihedral	45°	Dihedral	45.1°
11	Dihedral	70°	Dihedral	71.1°
2,5,6,9	Dipole	90°	Dipole	[88.9°, −89.6°, 88.2°, −89.9°]
8	Dipole	60°	Dipole	59°
1	Dipole	75°	Dipole	73.3°

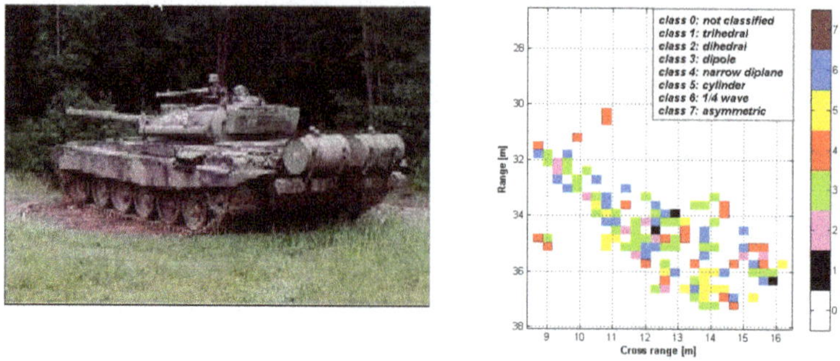

Figure 7.10 *Optical image of the target at nearly the same aspect angle of the corresponding ISAR image (left). Results of the Cameron's decomposition algorithm (right). Particularly, pixel correspond to the coherent scatterers forming the Pol-ISAR image.*

7.4 ATR using Pol-ISAR images

7.4.1 Feature-based classifier

For the analysis of the performances, two kinds of classifiers have been tested, namely, support vector machines and normal-based Bayesian classifier (BC).

7.4.1.1 Support vector machines

Support vector machines (SVM) were introduced initially by Cortes and Vapnik [33] and they represent an efficient technique for non-linear classification. Given an ensemble of input vectors in a L-dimensional space, first, a binary classification problem is analysed. The idea of SVMs is to estimate a separation hyperplane between two classes in the L-dimensional feature space. The training data set is composed of the couples $\{\boldsymbol{x}_i, y_i\}$, where the ith input vector $\boldsymbol{x}_i$ belongs to the features space and the output is $y_i \in \{-1, 1\}$, where 1 indicates the positives and −1 the falses. From the combination of various binary SVM, a k-class classification can be obtained. The one-against-one approach was chosen, where, after evaluating $k(k-1)/2$ binary classifiers (all possible combinations), we can obtain $k(k-1)/2$ hyperplanes separating the classes. Classification on the test set is made through a max-wins voting scheme.

7.4.1.2 Bayesian classifier

The BC [34] needs the knowledge of some *a-priori* information, namely, the class probability $P(c_i)$ and the probability of observing a specified feature vector x given the class, $P(\boldsymbol{x}|c_i)$. If we had such an information, the BC would be the optimum classifier with the minimum probability of error. In practice, we need to estimate these quantities from the training data set, having then a sub-optimal classifier. The method aims at maximizing the *a posteriori* probability, which according to Bayes theorem can be written as follows:

$$P(c_i|\boldsymbol{x}) = \frac{P(\boldsymbol{x}|c_i)\,P(c_i)}{P(\boldsymbol{x})} \tag{7.51}$$

The maximum a posteriori hypothesis is given by

$$P(c_i|\boldsymbol{x}) = \arg\max_{c\in C}\{P(c|\boldsymbol{x})\} = \arg\max_{c\in C}\{P(\boldsymbol{x}|c)\,P(c)\} \tag{7.52}$$

where $P(\boldsymbol{x})$ and $P(c_i)$ can be omitted as the first does not depend on c and the classes are assumed to be equiprobable. A decision hyper-surface can be easily computed in the case of normal-distributed features.

7.4.2 Features pre-processing

Feature pre-processing is an important step in any machine learning problems. Irrelevant information, such as, for example, feature value outside of the range, redundancy or noisy and unreliable data can damage the knowledge discovered during the training phase. Data pre-processing usually includes outliers removal, data normalization and missing values removal. In this section, we followed essentially the procedure illustrated in Reference 35. The feature analysis is carried out before the classifier in order to identify the features which allow for a better discrimination of the targets.

7.4.2.1 Fisher discriminant ratio

The first step is to look at each of the available features independently and test their discrimination capability. Although, to look at each feature separately is far from optimal, it helps us to discard easily features with a low discriminating power. The Fisher's discriminant ratio defined in Reference 35 is commonly adopted to evaluate the power of class-separability of the features

$$FDR = \sum_{i}^{M}\sum_{j\neq i}^{M}\frac{(\mu_i-\mu_j)^2}{\sigma_i^2+\sigma_j^2} \tag{7.53}$$

where μ_i and σ_i are, respectively, the mean and standard deviation of the ith feature estimated over the data set. All M features contribute to the ratio, which tends to be better when it is high (namely, the features distribution tends to be concentrated on the centroid).

7.4.2.2 Scatter matrices

The second step aims at assessing the discriminating power of different vectors of features. In particular, we adopted a simple criteria based on the so-called scatter matrices illustrated in Reference 35, which is built upon information related to the way feature vector samples are scattered in the L-dimensional feature space.

7.4.3 Application to real data set

In real ISAR applications usually it may difficult to a priori establish the best method to be used, as it depends on several factors, such as the spatial resolutions, the target's own motions, the autofocusing algorithm effectiveness, etc. Usually, when dealing with the classification of man-made target with Pol-ISAR data, it is customary the approach using CTD. Nevertheless, also the ICTD approach can provide very useful information in addition to the CTD. The idea that is presented is based on the jointly used of CTD and ICTD features with the aim of overcome the weaknesses of each method separately. To assess the performance of the proposed algorithm a data set collected in an anechoic chamber and composed of eight targets formed by elementary scatterers (dihedral, trihedral and flat plane) will be considered.

7.4.3.1 Setup description

The data have been collected in an anechoic chamber at the University of Adelaide, Australia. The transmitted signal was a stepped frequency pulsed waveform with centre frequency equal to $f_0 = 10$ *GHz* and a number of frequencies equal to $M = 401$ equally spaced by 10 *MHz*. The targets were on a turntable that was rotated of 0.1° after each radar sweep. A set of 101 radar sweep has been collected for each

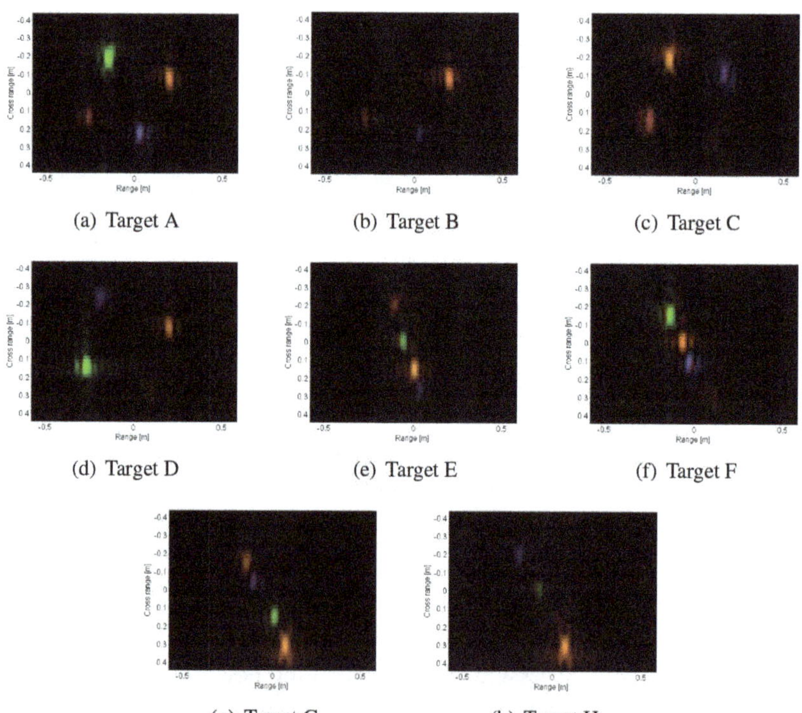

(a) Target A (b) Target B (c) Target C

(d) Target D (e) Target E (f) Target F

(g) Target G (h) Target H

Figure 7.11 Pol-ISAR images of the targets

radar acquisition. Each target was made of four simple scatterers, specifically trihedrals, dihedrals and flat planes of various size and orientation. More details and pictures of the targets can be found in Reference 3. Each target represents a class. Additive Gaussian white noise has been added to the raw data in order to assess the algorithm performance at different SNR values. The Pol-ISAR images are shown in Figure 7.11.

7.4.4 Results

Figure 7.12 shows the FDR (Fisher Discriminant Ratio) values for all the features extracted from the data set, both CTD and ICTD, against the SNR. As the targets are composed of four elementary scatterers, the number of classes is four. As a result, the size of x is equal to 19 (the real and the imaginary parts of z, the rotation angle about the LoS for each of the four scatterers and the 7 features extracted by the ICTD). By observing Figure 7.12, we can notice that as the SNR decreases, the FDR generally decreases as well. This is due to the fact that the different classes become more and more confused and therefore less separable. By focusing on $SNR = 5$ dB, we can see that the most discriminating CTD feature is $z_{R,1}$, while the most discriminating ICTD one is α_{ML}. Figures 7.13 and 7.14 show an estimate of the probability density function (PDF) relative to $z_{R,1}$ and α_{ML}, respectively, for all the classes.

From Figures 7.13 and 7.14, we can notice that in these two feature spaces the different classes are visibly well separated. Moreover, we can also qualitatively verify the hypothesis of Gaussian PDF.

Figures 7.15 and 7.16 show the results of classification with both classifiers in term of probability of correct classification (PCC). Both the classifiers (Bayesian and SVM) were trained with the data set at 0 dB. The comparison is made between CTD feature vector, ICTD feature vector and the fusion feature vector (where coherent and incoherent features were merged). For both classifiers, as the targets are composed of small geometrical objects, the CTD data set performs better than the ICTD data set. In both cases, we have PCC over 80% at SNR greater or equal to $-5dB$. Fusion technique shows performances equal or greater than both CTD or ICTD taken separately. Especially at SNR over $-5dB$, preliminary results show an

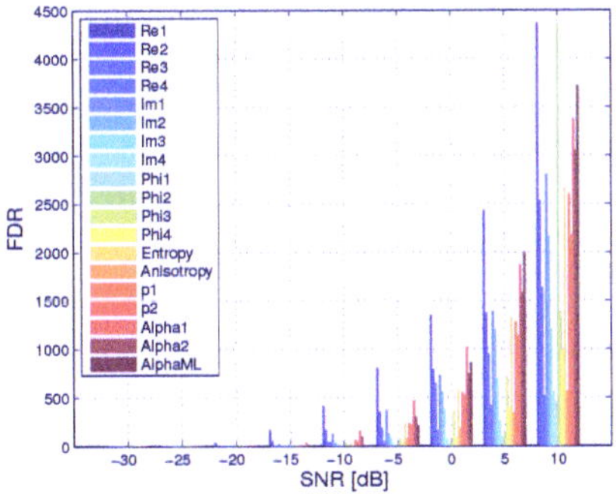

Figure 7.12 FDR for all the features against the SNR

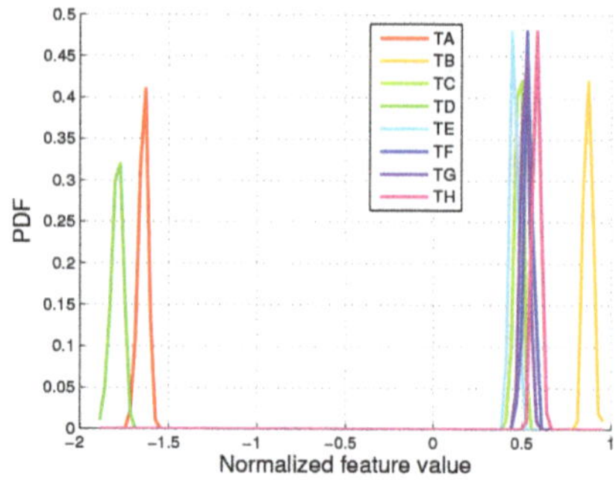

Figure 7.13 *Histogram of the normalized CTD $z_{R,1}$ for all the classes and SNR = 5 dB*

improvement between 3 and 5%. This is clear with SVM, while BC does not show appreciable improvements, while the PCC is slightly higher.

7.4.5 Neural network

Before the advent of DL and CNN, some experiments have been conducted using NN for NCTR purposes using 2D ISAR images. These networks are fed with feature vectors, and the most used is the multi-layer perceptron (MLP). Examples can be found in References 13, 36–38. An MLP structure is a fully connected multi-layered network composed by two or more layers. The first and the last layers represent the input and the output layers, whereas all the intermediate layers are defined as hidden layers. All neurons of the network are fully connected such that each neuron output of a given layer is fed to each neuron of the subsequent layer and so forth. The inputs of each neuron are weighted by a set of synaptic weights, which differ from one neuron to another. An example of MLP architecture is shown in Figure 7.17. The weights of the network, which are also called parameters, are evaluated during the training phase. The training algorithm can be selected among two different options: error back propagation and Marquardt back propagation. The latter is a faster method

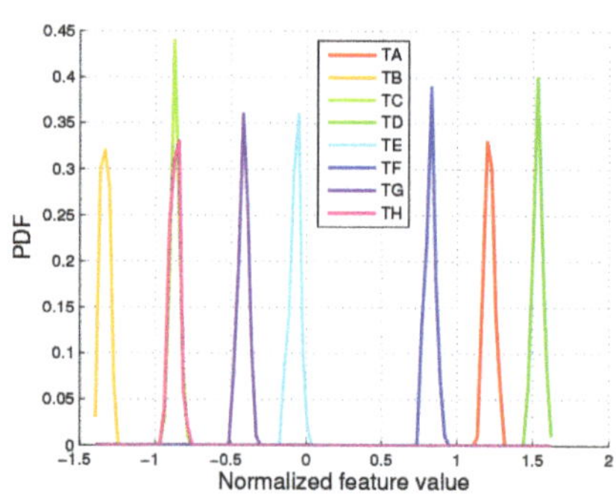

Figure 7.14 *Histogram of the normalized ICTD α_{ML} for all the classes and SNR = 5 dB*

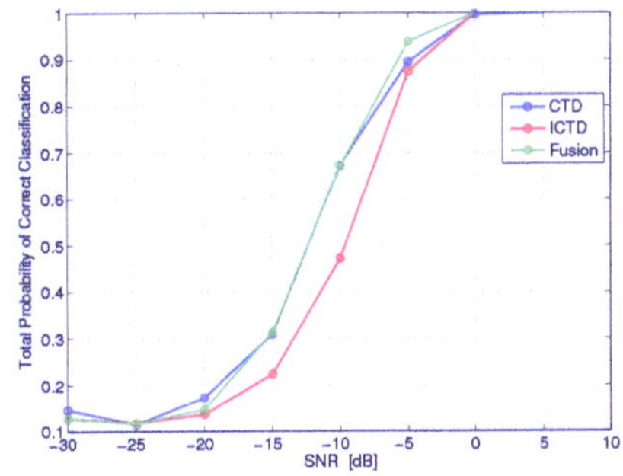

Figure 7.15 *PCC obtained with SVM classifier*

for feed-forward networks of moderate dimension, as proven in Reference 39. However, it is worth noting that both the training algorithms are gradient based, and that they differ only because of their software implementation. Parameter tuning is performed sequentially. For the hidden neurons, different transfer functions can be selected, as for example:

- Step function: the output is binary and depends on whether the input meets a threshold.
- Linear: the output is a linear combination of the inputs plus a bias term. This kind of transfer function is more used in the first layers of a network.
- Sigmoid function: is a non-linear function of the input. Sigmoid function is a mathematical function having a characteristic 'S'-shaped curve or sigmoid curve. The most common and used examples of a sigmoid function are the logistic and the hyperbolic tangent function.
- Rectified: the output is the positive part of the input. In modern CNN, this corresponds to the so-called ReLU (Rectified Linear activation Unit) function. It has been first introduced by Hahnloser *et al.* [39] with strong biological motivations and mathematical justifications. It has been demonstrated to enable better training of deeper networks compared to the widely used activation functions logistic sigmoid (which is inspired by probability theory) and the hyperbolic tangent.

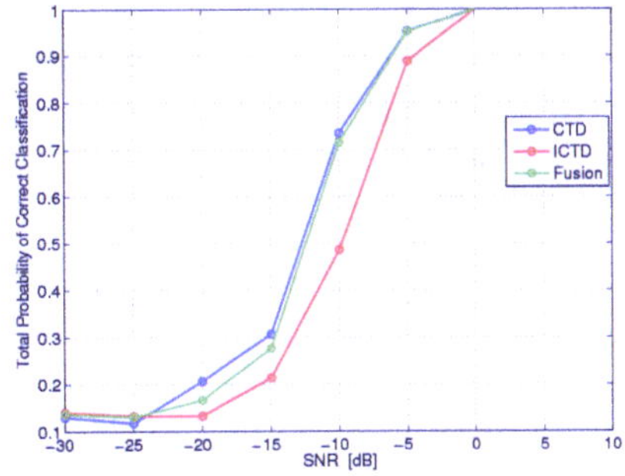

Figure 7.16 *PCC obtained with Bayesian classifier*

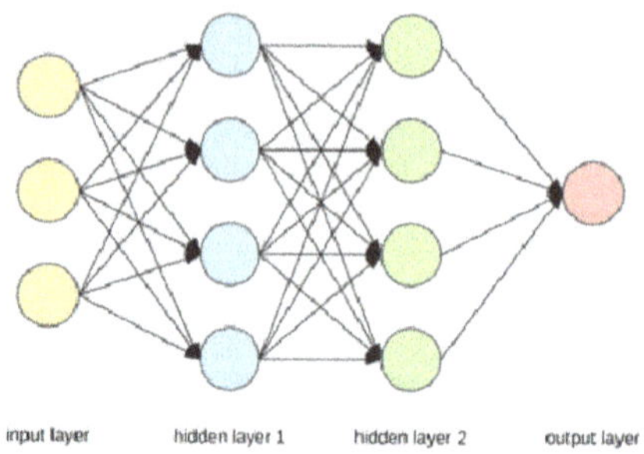

Figure 7.17 Architecture of the MLP NN

A rule of thumb, known as the Baum–Haussler rule [40] can be used to determine the number of hidden layer neurons. On the other hand, the universal approximation theorem recommends a number of three layers for an MLP network to approximate an arbitrary continuous multidimensional function up to any desired accuracy. An example of a typical MLP topology is shown in Figure 7.17. In this case the network is composed of an input layer, two hidden layers and an output layer. As it can be noted, each neuron is connected to all the neurons in the subsequent layer, and this makes the network fully connected. Each coloured circle represents a neuron with its proper transfer function.

Traditional implementations of ISAR produce 2D images that can be interpreted, under some hypotheses, as a filtered projection of the target's 3D reflectivity function. The orientation of the 2D imaging plane depends on the radar-target geometry and kinematics, which are typically unknown. This creates serious problems with the ISAR image interpretation and, consequently, with target recognition. To make the ATR algorithm less dependent on the radar-target geometry and target motions, only the polarimetric information of the most brightest scatterers is used as features to recognize the target. The polarimetric behaviour of a scatterer is in fact independent on its range/Doppler coordinates. As for example, the trihedral RCS has a main lobe quite large in both azimuth and vertical plane. But its range-Doppler coordinates may change rapidly if the target moves with respect to the radar. Hence, the geometrical information of that scatterer, namely, its position within the target's ISAR image is meaningless. Additionally, the only use of the polarimetric feature allows to better assess the advantages of using fully Pol-ISAR images.

In this section, two examples of application of NNs to Pol-ISAR images are shown, the former using a single Pol-ISAR image and the latter using a sequence of Pol-ISAR images. ISAR images are often formed out of long recorded data. Therefore, a sequence of ISAR images can be formed by properly processing the acquired data. Despite the example shown in the subsequent subsections, further ATR application of NNs can be found in References 41–43.

7.4.5.1 NN-based ATR using Pol-ISAR images

The adopted NN-based ATR algorithm is composed of the following main steps [13]:

- 2D ISAR image formation.
- Pol-CLEAN algorithm to extract the most brightest scatterers.
- Cameron's algorithm to extract the main polarimetric signature of each scatterer. The output of this block will be used to create the target's feature vector.
- NN-based classifier that produces a decision based on the target's feature vector at its input.

The adopted MLP topology is composed of a single hidden layer. Let N_i, N_n and N_c be, respectively, the feature vector size and the number of neurons in the input layer, the number of the hidden neuron and the number of classes. Hidden neurons are characterized by a sigmoidal transfer whereas the output layer is composed of linear neurons.

The feature vector represents the input of the MLP network, which is composed of $N_i = 3N$ elements, where N is the number of extracted scatterers. In fact, the Cameron's algorithm provides a three-element vector for each scatterer, composed of the real and imagery part of z and the angle ϕ. A test has been performed using the data set presented in section 7.3.2. This data set consists overall of 2465 ISAR images of the same target but acquired at different azimuth and elevation angles. The azimuth angles change in a turn with step of 4.25°. The elevation angles change in the range [27.99° – 31.93°] with a step of 0.145°. The transmitted signal is a stepped-frequency pulsed waveform with central frequency equal to $f_0 = 9.6GHz$. The number of transmitted frequencies is equal to 221. The frequencies are equally spaced by $3MHz$. The resultant bandwidth is $660MHz$, and the nominal range

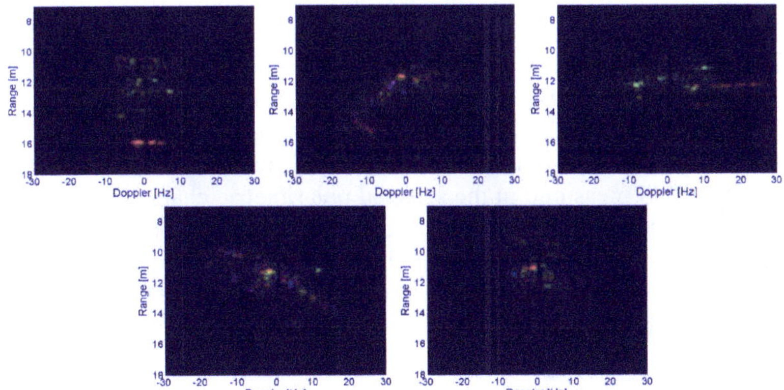

Figure 7.18 Pol-ISAR images of the T72 taken at different aspect angles (spaced roughly of 43°). Each Pol-ISAR image is red-green-blue coded where colours are associated to the three Pauli's channels.

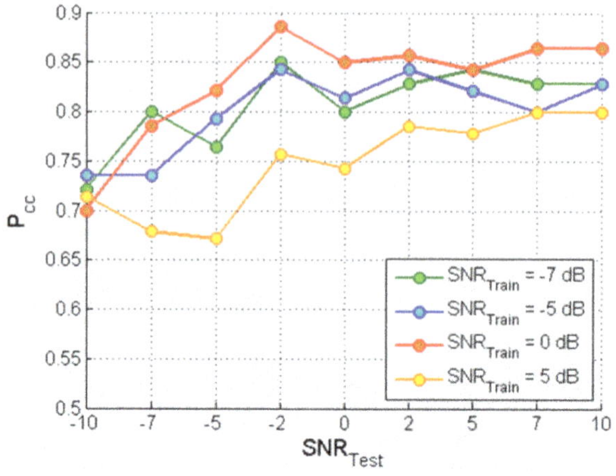

Figure 7.19 Performance measured in terms of correct classification rate

resolution is $0.3048m$. A number of 'virtual targets' can be created for significantly different aspect angles. Specifically, five virtual targets have been defined equally spaced of roughly 43° in the range [4.25° – 174.27°]. Figure 7.18 shows the Pol-ISAR images red-green-blue coded (red, green and blue represent, respectively, the three Pauli's channels).

For each class, a subset of elevation angles has been used to train the network and the complementary subset of elevation angles has been used to test the classifier. Additionally, zero mean Gaussian noise has been added to the raw data. The $N = 4$ brightest scatterers have been used to create the 12-element feature vector. The number of hidden neurons N_n and the training set size N_t are selected by means of an assisted process and by minimizing the MSE evaluated onto the training set. Figure 7.19 shows the P_{cc} measured onto the test set. The curves have been obtained using $N_n = 10$ and $N_t = 150$ and for different values of SNR_{train}. As it can be observed the SNR onto the test set changes from −10 to 10 dB.

It can be noted that the ATR performance is not seriously affected by the value of SNR_{train}, although, as expected, the best performance is obtained when the SNR_{test} and SNR_{train} are similar. The P_{cc} of the full pol classifier has been compared to the performance that would be obtained using a single polarimetric radar system. In this case, only the RCS measured in the selected polarimetric channel of each scatterer is considered instead of the Cameron's vector. The same network has been used and classification performance is shown in Figure 7.20.

Figure 7.20 shows the advantage of using fully polarimetric data over single polarimetric data for ATR.

7.4.5.2 NN-based ATR using Pol-ISAR images sequences

2D ISAR imaging has the disadvantage of concentrating a superposition of effects due to multiple scatterers within a resolution cell. This may happen either because different scatterer may have different height but same range/cross-range coordinates

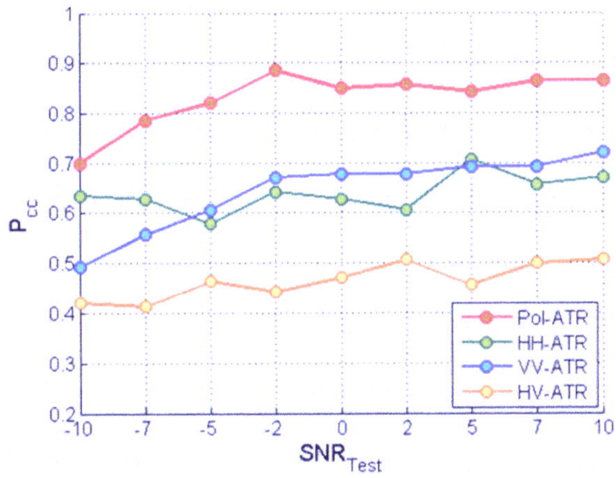

Figure 7.20 *Comparison between full-pol and single-pol acquisition mode*

or because spatial resolution cell is big if compared to the size of the scatterers. The problem associated with multiple scatterers is that the output signal strongly depends on the way that the contributing components are summed up, which may change rapidly when small rotations of the target occur or when the radar-target geometry change due to the target motions. With the aim of defining robust ATR systems, it is important to distinguish between multiple- and single-scatterer types of components in the received signal, as the latter type is less sensitive to small changes of the target's aspect angle. It is expected that single-scatterer-based ATR system is more robust with respect to an ATR system that does not discriminate between single- and multiple-scatterers. In Reference 15, an ATR system is proposed that makes the use of polarimetrically persistent scatterers (PPSs). A PPS is defined as a scatterer that does not change its polarimetric properties within a sequence of Pol-ISAR images. Two adjacent images extracted from a sequence of ISAR images of the same target are usually quite correlated when they are close in time because they are acquired by looking at the target from similar aspect angles. In such a condition, the concept of PPS becomes meaningful. Then a scatterer is a PPS if the following inequality is satisfied:

$$d_p < K \qquad \text{subject to} \qquad d_g < D \tag{7.54}$$

where d_p and d_g are the polarimetric and geometric Euclidean distances between two scatterers extracted from two Pol-ISAR images, and K and D are the thresholds. The geometric distance is calculated using the range and cross-range coordinates of the scatterers (unless 3D images are available in that case it would be a 3D vector). Conversely, the polarimetric distance is calculated using the polarimetric vector of each scatterer. The polarimetric vector size depends on the polarimetric TD algorithm adopted. In this case, we are using the Cameron's method and therefore the polarimetric vector is composed of three real values, namely the real and imagery part of z and the rotation angel ϕ.

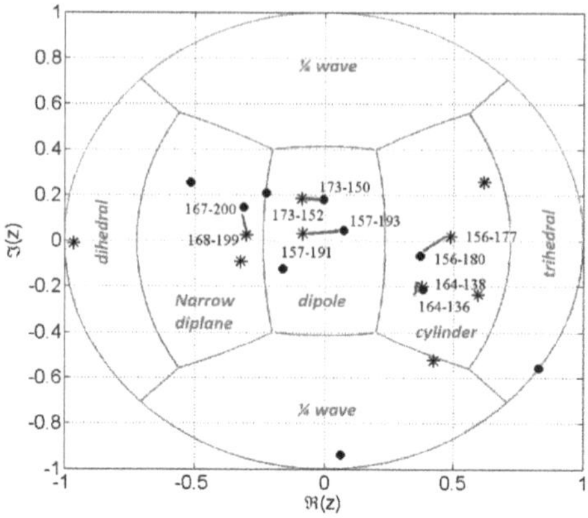

Figure 7.21 *Dots and stars shows the Cameron's z parameter of the ten brightest scatterers extracted from two Pol-ISAR images acquired using elevation angles that differ of roughly 0.28°. Numbers are the Doppler and range coordinates of the PPS. PPSs are those connected with a dark line, while the other scatterers exhibit different geometric and/or polarimetric features.*

Figure 7.21 shows the Cameron z parameter of the ten most brightest scatterers extracted from two Pol-ISAR images of the Tank T72 acquired with elevation angle difference of about 0.28°. Different marker are used to distinguish among the two Pol-ISAR images. As it can be noted, the Cameron's disc in Figure 7.21 is composed of separate areas representing classes of different types of scattering. For example, the scatterers represented by dots falling in the two areas at the top or at the bottom of the disk are classified as 1/4-wave scatterers. The scatterers represented by dots that fall in the central area of the disk are classified as dipoles. The scatterers that fall on the far right of the disc are classified as trihedrals, while those that fall on the far left of the disc are classified as dihedrals. Those that fall in between dipole and dihedral classes are classified as narrow Diplane. Those that fall in between dipole and trihedral classes are classified as cylinder. Although the two images are acquired with slightly different aspect angles, their polarimetric features look very different. Some of the scatterers exhibit similar polarimetric behaviour in both the Pol-ISAR images. The permanent scatterers are indicated by means of a connecting line, whereas the scatterers that appear in one image and not in the other one are not connected. The numbers in the figure indicate the Doppler and range bin. This confirms that the highlighted scatterers are permanent scatterers from a polarimetric and geometric point of view. It is clear that, among all the brightest scatterers, only a few are persistent, and therefore, they are suitable for being representative of the target.

This classifier performs an additional step if compared to the algorithm described in the previous subsection, namely the PPS tracking algorithm, which aims to select only those scattering centres that originate from a single scattering mechanism, based on the polarimetric stability of the scatterer. In other words, the classifier takes as inputs a sequence of Pol-ISAR images (instead of a single Pol-ISAR image) and the 'PPS tracking algorithm' tracks the scatterers in the sequence of Pol-ISAR images to identify the PPS. Then, the extracted PPSs are used to generate the feature vector that represents the polarimetric properties of a target when illuminated with aspect angles within a relatively wide angle range.

The PPS tracking algorithm assumes a parametric model for each scatterer. Then, the algorithm jointly finds the probability that each extracted scatterer belongs to a given model and the model parameters, which minimize a cost function. The adopted parametric models are polynomials. However, considering that the aim is to find PPS, the suggested model for the polarimetric features is a constant function. Conversely, for the geometric features, namely the range and Doppler coordinates, the models depend on the target movement whiting the observation time.

The estimation method adopted in [15] is based on probabilistic least square approach, and the cost function to minimize is defined as follows:

$$J(p,\omega) = \sum_{n=1}^{N}\sum_{i=1}^{M}\sum_{k=1}^{K} p_{i,k}(n)\, \Xi_{i,k}^{T}(n)\, R_{i,k}^{-1}(n)\, \Xi_{i,k}(n)\, p_{i,k}(n) \tag{7.55}$$

where P is a 3D matrix that contains the probabilities that the $k-th$ scatterer extracted from the $n-th$ Pol-ISAR image belongs to the $i-th$ model, namely $p_{i,k}(n)$, $\Xi_{i,k}(n) = w_k(n) - y_i(n)$ is the feature vector of the $k-th$ scatterer extracted from the $n-th$ Pol-ISAR image. It contains the scatterer Cameron's parameters $(\mathrm{Re}(z), \mathrm{Im}(z), \phi)$ and the scatterer spatial coordinates (range and Doppler) and, therefore, it is a column vector, $w_k(n) \in \mathbb{R}^5$. $y_i(n) \in \mathbb{R}^5$ is the $i-th$ parametric model vector valued at the $n-th$ time instant, $R_{i,k}(n)$ is the covariance matrix of $\Xi_{i,k}(n)$ is assumed here independent on both n and k, namely, $R_{i,k}(n) \simeq R_i$. ω is the vector that contains the parameter of the models used to model the spatial and geometric values of the scatterers over time. The optimization problem can be formalized as follows:

$$\left(\hat{P}, \hat{\omega}\right) = \underset{P,\omega}{\operatorname{argmin}}\ J(P,\omega) \tag{7.56}$$

A closed form solution of this optimization approach can be found in Reference 15 under the hypothesis of polynomial models.

The NN-based ATR algorithm has been tested using the same data set used in the previous subsection. As done previously five 'virtual targets' have been defined, which correspond to the same target but observed from different azimuth angles. In all, 29 Pol-ISAR images are available for each virtual class corresponding to different elevation angles. Half were used to train the network, while the complementary subset was used to test the algorithm. AWGN (Additive White Gaussian Noise) has been added to the raw data to build the training set. Differently from what done previously, however, 15 Pol-ISAR images are used to extract the PPS and hence a single feature vector. The following figures show the results of the PPS tracking algorithm obtained by setting $M = 9$ and $K = 10$. Moreover, a first-order polynomial function is assumed to model the range coordinates of the PPS scatterers, while a constant value model is used to model both the Doppler coordinates and the Cameron's parameters of each PPS. The estimated parametric models of both

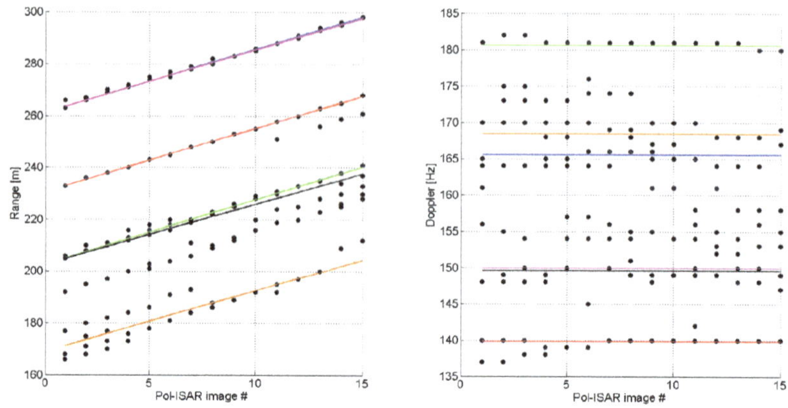

Figure 7.22 *Range (left) and Doppler (right) coordinates of the extracted scatterers. Coloured lines represents the selected scatterers models.*

the range and Doppler coordinates of the PPS are shown with different colour in Figure 7.22. The black dots represent the range and Doppler coordinates of the scatterers extracted in each Pol-ISAR image. To demonstrate that the selected scatterers correspond to PPSs, the Cameron's parameter z is plotted using the same color code in the Cameron's unit disk, as shown in Figure 7.23. The coloured stars represent the mean values of z. The standard deviation of z has been computed and used to generate a number of feature vectors with the aim to better generalize the training set.

As already said, training and test set have been created using complementary subset of Pol-ISAR images, such that, however, they are not completely interlaced. Three different cases have been tested as depicted in Figure 7.24.

Moreover, a rejection threshold has been applied onto the classification score provided by the NN, to measure P_{cc} but conditioned to a fixed P_{fa}. Conversely,

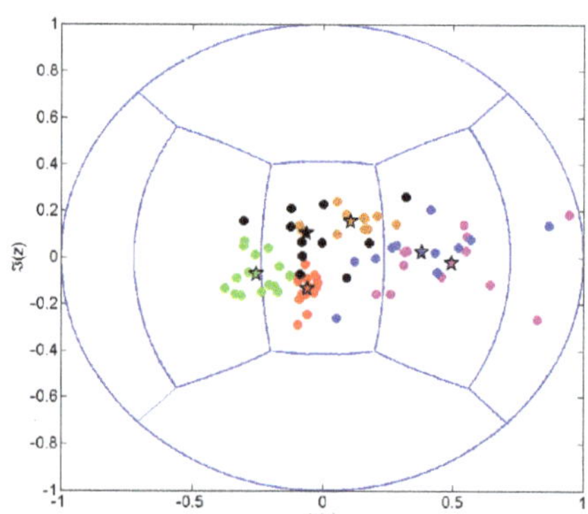

Figure 7.23 *Cameron's z parameter of the PPS plotted in the Cameron's unit disk. Colour code is the same of Figure 7.22.*

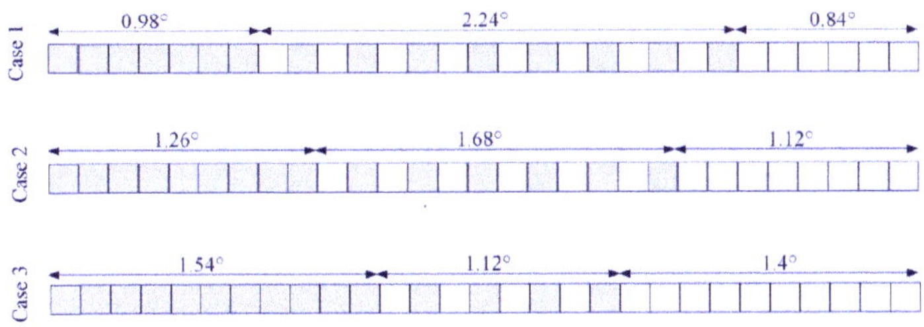

Figure 7.24 Training and test set strategy description

results in Figure 7.20 have been obtained without any threshold, which means a $P_{fa} = 1$.

To demonstrate the advantage of using PPS, two NN-based classifiers have been compared, the former that uses a feature vector composed of only the PPSs parameters, the latter that uses the features vector of the ten brightest scatterers. Differences between the two architecture are shown in Figure 7.25.

The curves in Figure 7.26 represent the P_{cc} as a function of P_{fa}. As expected, the performances of both algorithms get worse when the two data sets are less overlapped. Nevertheless, the results give some advice about the capability of both the algorithms to generalize even when the test set does not span the same angular sector of the training set.

The NN-ATR algorithm that uses the brightest scatterers recognizes the target as belonging to a target class when it is recognized as belonging to the same class at least N_{tj} times. In this experiment, such a threshold has been set equal to 8, being the sequence composed of 14 Pol-ISAR images. This is the reason why the BS-ATR

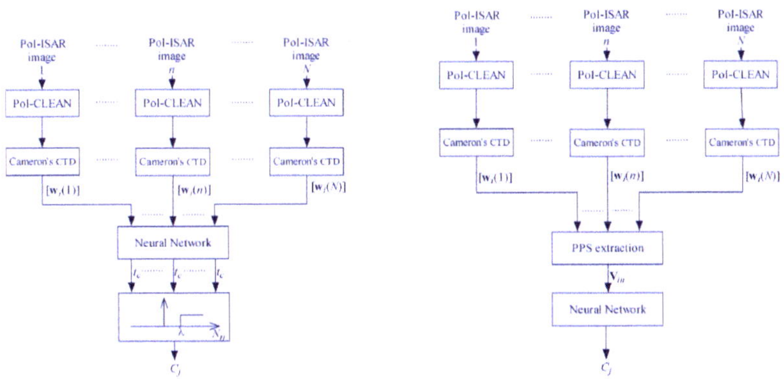

Figure 7.25 Block diagram of the NN-based ATR that uses the brightest scatterer. A decision is made for each one of the 15 Pol-ISAR images that compose the training set. The final decision is made when the classifier decides for the same target class at least $N_{tc} = 8$ times (left). Block diagram of the NN-based ATR that uses the PPS (right).

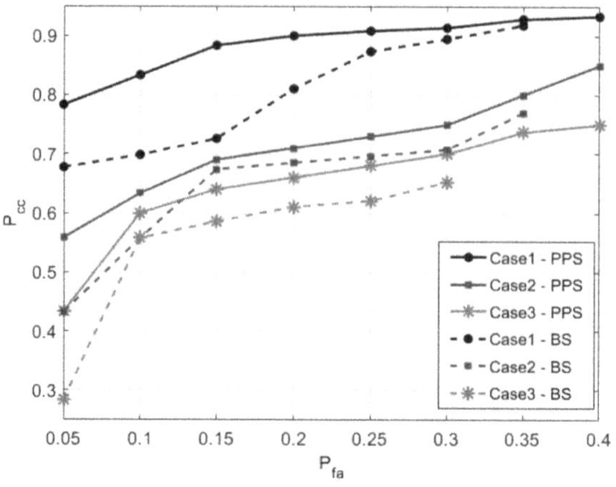

Figure 7.26 *Results in terms of P_{cc} and P_{fa} of both the NN-based classifier of Figure 7.25*

algorithm has a probability of false alarm lower than 0.4 in all the three examined cases. It should be also noted that the PPS-based ATR algorithm outperforms the BS-based algorithm, particularly for low probability of false alarm.

7.5 Conclusions

Pol-ISAR has proven to be more accurate and more effective than single-channel ISAR when used for target recognition. Given the limited amount of such systems and available data and literature, this is a field that can still benefit from additional study and experimentation to fully exploit its potential. The use of polarimetry involves additional receiving channels in the radar system and therefore it increases the complexity of cost of the overall radar system. This creates a trade-off that must be considered when designing ISAR systems for target recognition. Nevertheless, when size, weight, power and cost constraints are favourable, the benefit of using full polarization is significant both in terms of image quality, by producing better focused images, and in terms of the amount of information (features) that can be exploited to improve target classification performances.

References

[1] Chen M., Victor C. 'Inverse synthetic aperture radar imaging: principles, algorithms and applications' in *Radar, sonar and navigation*. Institution of Engineering and Technology; 2014.

[2] Novak L.M., Sechtin M.B., Cardullo M.J. 'Studies of target detection algorithms that use polarimetric radar data'. *IEEE Transactions on Aerospace and Electronic Systems*. 1989, vol. 25(2), pp. 150–65.

[3] Martorella M., Giusti E., Demi L., *et al.* 'Target recognition by means of polarimetric ISAR images'. *IEEE Transactions on Aerospace and Electronic Systems*. 2011, vol. 47(1), pp. 225–39.

[4] Cloude S.R., Pottier E. 'An entropy based classification scheme for land applications of polarimetric SAR'. *IEEE Trans Geosci Remote Sens*. 1997, vol. 35(1), pp. 68–78.

[5] Cloude S.R., Pottier E. 'A review of target decomposition theorems in radar polarimetry'. *IEEE Trans Geosci Remote Sens*. 1996, vol. 34(2), pp. 498–518.

[6] Migliaccio M., Gambardella A., Tranfaglia M. 'SAR polarimetry to observe oil spills'. *IEEE Trans Geosci Remote Sens*. 2007, vol. 45(2), pp. 506–11.

[7] Martorella M., Berizzi F., Haywood B. 'Contrast maximisation based technique for 2-D ISAR autofocusing'. *IEE Proceedings - Radar, Sonar and Navigation*. 2005, vol. 152(4), p. 253.

[8] Kyung-TaeK., Dong-KyuS., Hyo-TaeK.. 'Efficient classification of ISAR images'. *IEEE Trans Antennas Propag*. 2005, vol. 53(5), pp. 1611–21.

[9] Tsao J., Steinberg B.D. 'Reduction of sidelobe and speckle artifacts in microwave imaging: the clean technique'. *IEEE Trans Antennas Propag*. 1988, vol. 36(4), pp. 543–56.

[10] Vignaud L. 'Robust polarimetric scatterers extraction for SAR ATR'. Presented at Proceedings of the RTO-SET Symposium on Target Identification and Recognition Using RF System; Oslo.

[11] Lagarias J.C., Reeds J.A., Wright M.H., Wright P.E. 'Convergence properties of the nelder--mead simplex method in low dimensions'. *SIAM Journal on Optimization*. 1998, vol. 9(11), pp. 112–47.

[12] Kim K.T., Seo D.K., Kim H.T. 'Efficient classification of ISAR images'. *IEEE Transactions on Antennas and Propagation*. 2005, vol. 53(5), pp. 1611–21.

[13] Martorella M., Giusti E., Capria A., Berizzi F., Bates B. 'Automatic target recognition by means of polarimetric ISAR images and neural networks'. *IEEE Trans Geosci Remote Sens*. 2009, vol. 47(11), pp. 3786–94.

[14] Martorella M., Giusti E., Demi L., *et al.* 'Target recognition by means of polarimetric ISAR images'. *IEEE Transactions on Aerospace and Electronic Systems*. 2011, vol. 47(1), pp. 225–39.

[15] Giusti E., Martorella M., Capria A. 'Polarimetrically-persistent-scatterer-based automatic target recognition'. *IEEE Trans Geosci Remote Sens*. 2011, vol. 49(11), pp. 4588–99.

[16] Musman S., Kerr D., Bachmann C. 'Automatic recognition of ISAR SHIP images'. *IEEE Transactions on Aerospace and Electronic Systems*. 1996, vol. 32(4), pp. 1392–404.

[17] Saidi M.N., Toumi A., Khenchaf A. 'Pose estimation for ISAR image classification'. *IEEE International Geoscience and Remote Sensing Symposium*; 2010. pp. 4620–23.

[18] Saidi M.N., Hoeltzener B., Toumi A. 'Automatic recognition of ISAR images: target shapes features extraction'. *3rd International Conference on Information and Communication Technologies: From Theory to Applications*; 2008. pp. 1–6.

[19] Pastina D., Spina C. 'Multi-feature based automatic recognition of SHIP targets in ISAR images'. *IEEE Radar Conference*; 2008. pp. 1–6.
[20] Luo Y., Zhang Q., Cw Q, *et al.* 'Micro-doppler effect analysis and feature extraction in ISAR imaging with stepped-frequency chirp signals'. *IEEE Transactions on Geoscience and Remote Sensing*. 2010, vol. 48(4), pp. 2087–98.
[21] Yuan B., Chen Z., Xu S. 'Micro-doppler analysis and separation based on complex local mean decomposition for aircraft with fast-rotating parts in ISAR imaging'. *IEEE Transactions on Geoscience and Remote Sensing*. 2014, vol. 52(2), pp. 1285–98.
[22] Novak L.M., Halversen S.D., Owirka G., Hiett M. 'Effects of polarization and resolution on SAR ATR'. *IEEE Transactions on Aerospace and Electronic Systems*. 1997, vol. 33(1), pp. 102–16.
[23] Lischi S., Lupidi A., Giusti E. 'Enhanced ATR by jointly using coherent and incoherent target decomposition theorems on polarimetric ISAR images'. *11th European Radar Conference*; 2014. pp. 89–92.
[24] Malmgren-Hansen D., Kusk A., Dall J., Nielsen A.A., Engholm R., Skriver H. 'Improving SAR automatic target recognition models with transfer learning from simulated data'. *IEEE Geosci Remote Sens Lett*. 2017, vol. 14(9), pp. 1484–88.
[25] Cloude S.R., Pottier E. 'A review of target decomposition theorems in radar polarimetry'. *IEEE Trans Geosci Remote Sens*. 1996, vol. 34(2), pp. 498–518.
[26] Martorella M., Cacciamano A., Giusti E., Berizzi F., Haywood B., Bates B. 'CLEAN technique for polarimetric ISAR'. *International Journal of Navigation and Observation*. 2008, vol. 2008, pp. 1–12.
[27] database G. Available from https://www.sdms.afrl.af.mil/index.php?collection=gtri [Accessed Sep 15].
[28] Martorella M., Berizzi F., Haywood B. 'Contrast maximisation based technique for 2-D ISAR autofocusing'. *IEE Proceedings - Radar, Sonar and Navigation*. 2005, vol. 152(4), p. 253.
[29] Huynen J.R. 'Measurement of the target scattering matrix'. *Proceedings of the IEEE*. 1965, vol. 53(8), pp. 936–46.
[30] Krogager E. 'New decomposition of the radar target scattering matrix'. *Electron Lett*. 1990, vol. 26(18), p. 1525.
[31] Cameron W.L., Youssef N.N., Leung L.K. 'Simulated polarimetric signatures of primitive geometrical shapes'. *IEEE Trans Geosci Remote Sens*. 1996, vol. 34(3), pp. 793–803.
[32] Touzi R., Charbonneau F. 'Characterization of target symmetric scattering using polarimetric SARS'. *IEEE Trans Geosci Remote Sens*. 2002, vol. 40(11), pp. 2507–16.
[33] Cortes C., Vapnik V. 'Support-vector networks'. *Machine Learning*. 1995, vol. 20(3), pp. 273–97.
[34] Mitchell T.M. *Machine learning*. New York, NY: McGraw-Hill, Inc; 1997.
[35] Theodoridis S., Koutroumbas K. *Pattern recognition*. 3rd ed. San Diego, CA: Elsevier; 2006.

[36] Haykin S., Deng C. 'Classification of radar clutter using neural networks'. *IEEE Transactions on Neural Networks*. 1991, vol. 2(6), pp. 589–600.
[37] Chakrabarti S., Bindal N., Theagharajan K. 'Robust radar target classifier using artificial neural networks'. *IEEE Transactions on Neural Networks*. 1995, vol. 6(3), pp. 760–66.
[38] Jouny I., Garber F.D., Ahalt S.C. 'Classification of radar targets using synthetic neural networks'. *IEEE Transactions on Aerospace and Electronic Systems*. 1993, vol. 29(2), pp. 336–44.
[39] Hagan M.T., Menhaj M.B. 'Training feedforward networks with the marquardt algorithm'. *IEEE Transactions on Neural Networks*. 1994, vol. 5(6), pp. 989–93.
[40] Baum E.B., Haussler D. 'What size net gives valid generalization?'. *Neural Computation*. 1989, vol. 1(1), pp. 151–60.
[41] Principe J.C., Kim M., Fisher M. 'Target discrimination in synthetic aperture radar using artificial neural networks'. *IEEE Transactions on Image Processing*. 1998, vol. 7(8), pp. 1136–49.
[42] Yang S., Chang K.C. 'Modular neural net architecture for automatic target recognition'. *Proceedings of SPIE 2755, Signal Processing, Sensor Fusion, and Target Recognition*. 1996, vol. 2755, pp. 166–77.
[43] Ning W., Chen W., Zhang X. 'Automatic target recognition of ISAR object images based on neural network'. *Proceedings of the International Conference on Networks and Processing*. 2003, vol. 11, pp. 373–76.

Chapter 8

UWB short-range polarimetric imaging and its potential for target classification

Jianping Wang[1] and Alexander Yarovoy[1]

Ultra-wideband (UWB) short-range polarimetric imaging provides not only high spatial resolution but also polarimetric scattering features of targets, which has great application potentials in subsurface survey with ground penetrating radar, through-the-wall imaging, security check, and so on. In this chapter, two key aspects of the UWB short-range polarimetric imaging, i.e., full-polarimetric data acquisition and polarimetric image reconstruction, are discussed [1, 2].

8.1 Introduction

Despite the vectorial nature of electromagnetic (EM) waves, they are often used in the same way as acoustic (scalar) waves by the majority of radar systems and, in particular, for short-range imaging systems. Two principal exceptions are synthetic aperture radars (SAR) and the weather radars – in both types of systems, the polarimetric information is crucial for classification of the observed objects. Rarely are polarimetric properties of the reflected signal used in UWB ground penetrating radars (GPR). The reasons for this are the difficulties of coherent measurements of a full-polarimetric scattering matrix of a reflector, collocation of the antenna phased centers for different polarizations and measurements from a moving platform when simultaneous or quasi-simultaneous transmission of two orthogonally polarized EM fields and measurements of the reflected field at two orthogonal polarizations are required. Furthermore, implementation of full-polarimetric front-end doubles the radar system complexity and results in a very complicated antenna system.

The most advanced studies on UWB polarimetry have been conducted due to landmine detection. As the majority of landmines are rotationally symmetric objects, this can be exploited to classify detected reflectors as mine-like targets (rotationally symmetric) and friendly targets (asymmetric ones). Polarimetric features of the

[1]Delft University of Technology, Delft, The Netherlands

reflected wave in principle allow for classification of all detected objects as rotationally symmetric and asymmetric ones [3–7]. For a perfectly rotational symmetric target, the cross-polar components of this target signature are close to zero. However, despite their circular shape, some mines (e.g., PMN-2) show relatively strong cross-polar components $s_{xy}(t)$ and $s_{yx}(t)$ [8] in the reflected signal. These cross-polar components can most likely be attributed to the presence of a horizontal detonator in PMN-2 mines.

The need to classify objects detected by the short-range imaging systems has pushed system designers to develop polarimetric systems. Therefore, fully polarimetric measurements have indeed been implemented in a number of forward-looking systems [9, 10]. Down-looking systems typically do not have sufficient physical space for a fully polarimetric antenna system (two exceptions are DEMAND radar [11] and multi-wave form radar [12]). For that reason, some designers chose to use orthogonal polarizations for transmit and receive antennas [13]. In such systems, the surface reflection is considerably suppressed so as to improve the signal-to-clutter ratio.

In this chapter, a few efficient approaches to collect full polarimetric data by using conventional single-polarized antenna systems are presented. Furthermore, imaging algorithms that are using full-polarimetric measurement data are presented and advantages of full-polarimetric imaging are demonstrated.

8.2 Data acquisition for short-range polarimetric imaging

Short-range imaging techniques generally exploit UWB signals and large antenna arrays to probe the region of interest. After imaging operations, they provide high-resolution morphological characteristics of targets in space. However, due to the vector nature of the EM waves, spatial structures of targets could exhibit significant distinct sensitivities and scattering properties for incident EM waves with different polarizations. So, the obtained morphological features of targets are generally polarization dependent. To extract reliable morphological features of targets and improve the probability of target detection and recognition, differently polarized EM signals scattered from the region of interest are acquired for polarimetric imaging by using transmitting and receiving antennas with various polarizations. Specifically, for full polarimetry, both transmitter and receiver polarizations are, for instance, horizontal (H) and vertical (V) in a linear polarization basis [or left (L)- and right (R)-handed circular in a circular polarization basis]. Considering four combinations of transmitter and receiver polarizations in the linear polarization basis, HH-, HV-, VH-, and VV-polarized signals, where the two capital letters indicate the polarizations of transmitter and receiver in order, are collected (correspondingly, LL-, LR-, RL-, and RR-polarized signals are measured with the transmitter and receiver polarizations in the circular polarization basis).

For 3D high-resolution imaging, full-polarimetric measurements are generally acquired by using array techniques with three typical configurations:

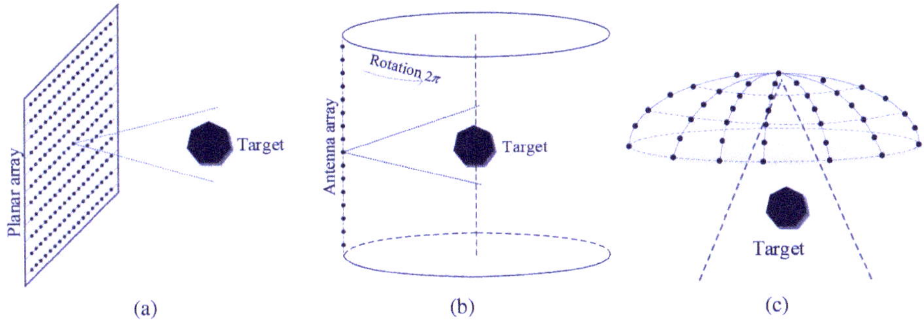

Figure 8.1 Geometrical configurations of data acquisition for short-range imaging with (a) planar array system, (b) cylindrical scanning system, and (c) hemispherical array system

planar array systems [14–17], cylindrical array systems [18, 19], and (hemi-) spherical array systems [20], as illustrated in Figure 8.1. Among them, the cylindrical and the (hemi-)spherical array systems are mainly popular in biomedical imaging (for instance, breast cancer detection and head imaging) and human security check. By contrast, the planar array systems are widely used in ground penetrating radar, through-the-wall imaging, security check, etc. Considering the fact that the planar array systems are the focus of this chapter in the following; nevertheless, some approaches to full-polarimetric imaging discussed below are also applicable to both cylindrical and (hemi-)spherical array imaging systems.

For high-resolution polarimetric imaging, full-polarimetric data are usually acquired by using a large polarimetric planar array or a linear polarimetric array combining SAR technique in the direction perpendicular to the array axis, where dual-polarized antenna elements are generally used. Such data acquisition schemes have been widely used for, for instance, GPR systems [13, 21, 22]. On the other hand, motivated by the GPR system used for TBM, we have proposed some novel and efficient full-polarimetric data acquisition schemes with single-polarized antenna systems [1, 2]. Before introducing the specific full-polarimetric data acquisition systems, some basic theories of the proposed approaches as well as the method for array system design are first discussed below.

8.2.1 Scattering formulation

Here, we consider the monostatic radar configuration for data collection. The antennas are deployed on the x_1-x_2 plane, and the x_3-axis points toward the observation scenario and forms a right-hand coordinate system. Based on the Born approximation, the scattering process can be represented by a linear expression [16]:

$$E^s_{\alpha\beta}\left(\boldsymbol{x}^r,\boldsymbol{x}^t,\omega\right) = \int_{V(x^c)} D_{\alpha\beta}\left(\boldsymbol{x}^r,\boldsymbol{x}^t\middle|\boldsymbol{x}^c,\omega\right)\chi\left(\boldsymbol{x}^c\right)J_\beta\left(\boldsymbol{x}^t,\omega\right)\mathrm{d}V \tag{8.1}$$

where $\omega = 2\pi f$ is the angular frequency and $\boldsymbol{x} = (x_1, x_2, x_3)$ collects the spatial coordinates. The superscripts r and t refer to the receiving and transmitting antennas while the subscripts α and β take values $\{1, 2\}$ and represent, respectively, the receiving and transmitting antennas' orientations along x_1 or x_2 direction. $\chi(\boldsymbol{x}^c)$ is the contrast function in the position $\boldsymbol{x}^c$, $J_\beta(\boldsymbol{x}', \omega)$ is the point source located at the position $\boldsymbol{x}'$, and $D_{\alpha\beta}(\boldsymbol{x}^r, \boldsymbol{x}' | \boldsymbol{x}^c, \omega)$ is the forward wavefield extrapolator from the transmitting antenna at $\boldsymbol{x}'$ to the scatterer at $\boldsymbol{x}^c$ and then to the receiving antenna at $\boldsymbol{x}^r$. The contrast function χ is defined as $\chi(\boldsymbol{x}^c) = \hat{\eta}^s - \hat{\eta}$, which is the difference of the physical properties of the scatterer $\hat{\eta}^s$ and the background $\hat{\eta}$. The physical property $\hat{\eta}$ is defined as $\hat{\eta} = \gamma + j\omega\varepsilon$, where γ is the conductivity and ε is the permittivity. The point source $J_\beta(\boldsymbol{x}', \omega)$ can be denoted as

$$J_\beta(\boldsymbol{x}', \omega) = S(\omega)\, b_\beta(\boldsymbol{x}') \tag{8.2}$$

where $S(\omega)$ is the source wavelet radiated by the source antenna and b_β indicates its orientation along x_1- or x_2-direction. The forward wavefield extrapolator $D_{\alpha\beta}(\boldsymbol{x}^r, \boldsymbol{x}', \omega)$ is defined by an inner product

$$D_{\alpha\beta}(\boldsymbol{x}^r, \boldsymbol{x}' | \boldsymbol{x}^c, \omega) = \sum_{\ell=1}^{3} G_{\alpha\ell}(\boldsymbol{x}^r | \boldsymbol{x}^c, \omega)\, G_{\ell\beta}(\boldsymbol{x}^c | \boldsymbol{x}', \omega) \tag{8.3}$$

where $\ell \in \{1, 2, 3\}$ denotes the electric field directions along the axes. Green's function $G_{\ell\beta}(\boldsymbol{x}^c | \boldsymbol{x}', \omega)$ describes the propagation of an EM wave from the source at $\boldsymbol{x}'$ to the scatterer at $\boldsymbol{x}^c$, and Green's function $G_{\alpha\ell}(\boldsymbol{x}^r | \boldsymbol{x}^c, \omega)$ expresses the propagation from the scatterer in the position $\boldsymbol{x}^c$ to the receiving antenna at $\boldsymbol{x}^r$. The forward wavefield extrapolator shown in (8.3) describes the scattering process with x_β-oriented transmitting antenna and x_α-oriented receiving antenna. Hence, (8.1) gives the scattered wave from an illuminated volume acquired with an x_α-oriented receiving antenna related to an x_β-oriented transmission. Accounting for a pair of orthogonal orientations of the receiving antennas on the acquisition plane, the observed waves in the two directions can be arranged as a vector

$$\begin{bmatrix} E_1^s(\boldsymbol{x}^r, \boldsymbol{x}', \omega) \\ E_2^s(\boldsymbol{x}^r, \boldsymbol{x}', \omega) \end{bmatrix} = S(\omega) \cdot \int_{V(x^c)} \boldsymbol{D}(\boldsymbol{x}^r, \boldsymbol{x}' | \boldsymbol{x}^c, \omega) \begin{bmatrix} b_1(\boldsymbol{x}') \\ b_2(\boldsymbol{x}') \end{bmatrix} \chi(\boldsymbol{x}^c)\, \mathrm{d}V \tag{8.4}$$

where $\left[E_1^s,\ E_2^s\right]^T$ is a vector measured by two receiving antennas with orthogonal orientations corresponding to two orthogonally polarized transmissions. It is given by

$$\begin{bmatrix} E_1^s(\boldsymbol{x}^r, \boldsymbol{x}', \omega) \\ E_2^s(\boldsymbol{x}^r, \boldsymbol{x}', \omega) \end{bmatrix} = \begin{bmatrix} E_{11}^s(\boldsymbol{x}^r, \boldsymbol{x}', \omega) + E_{12}^s(\boldsymbol{x}^r, \boldsymbol{x}', \omega) \\ E_{21}^s(\boldsymbol{x}^r, \boldsymbol{x}', \omega) + E_{22}^s(\boldsymbol{x}^r, \boldsymbol{x}', \omega) \end{bmatrix} \tag{8.5}$$

and $\boldsymbol{D}$ represents the forward wavefield extrapolator that is given by

$$\boldsymbol{D} = \begin{bmatrix} D_{11}\left(\boldsymbol{x}^r,\boldsymbol{x}^t \middle| \boldsymbol{x}^c,\omega\right) & D_{12}\left(\boldsymbol{x}^r,\boldsymbol{x}^t \middle| \boldsymbol{x}^c,\omega\right) \\ D_{21}\left(\boldsymbol{x}^r,\boldsymbol{x}^t \middle| \boldsymbol{x}^c,\omega\right) & D_{22}\left(\boldsymbol{x}^r,\boldsymbol{x}^t \middle| \boldsymbol{x}^c,\omega\right) \end{bmatrix} = \begin{bmatrix} G^r_{11} & G^r_{21} & G^r_{31} \\ G^r_{12} & G^r_{22} & G^r_{32} \end{bmatrix} \begin{bmatrix} G^t_{11} & G^t_{12} \\ G^t_{21} & G^t_{22} \\ G^t_{31} & G^t_{32} \end{bmatrix} \tag{8.6}$$

where G^r is short for $G^r\left(\boldsymbol{x}^r|\boldsymbol{x}^c,\omega\right)$ and G^t for $G^t\left(\boldsymbol{x}^c|\boldsymbol{x}^t,\omega\right)$. We focus on the monostatic configuration here, so transmitting and receiving antennas are located at the same position $\boldsymbol{x}^a$, i.e., $\boldsymbol{x}^t = \boldsymbol{x}^r = \boldsymbol{x}^a$ for each observation. Consequently, Green's functions of transmitting and receiving antennas are equal in the corresponding electric field directions. Meanwhile, using the reciprocity properties of propagation, the elements of $\boldsymbol{D}$ can be explicitly written as

$$\begin{cases} D_{11} = G^2_{11} + G^2_{21} + G^2_{31} \\ D_{12} = G_{11}G_{12} + G_{21}G_{22} + G_{31}G_{32} \\ D_{21} = G_{11}G_{12} + G_{21}G_{22} + G_{31}G_{32} \\ D_{22} = G^2_{12} + G^2_{22} + G^2_{32} \end{cases} \tag{8.7}$$

where Green's functions G are functions of $\boldsymbol{x}^c$, $\boldsymbol{x}^a$, and ω. From (8.7), it can be observed that D_{12} is equal to D_{21} in the monostatic configuration, which is the result of the reciprocity theorem.

8.2.2 Wavefield extrapolator for rotated antennas

The variation of the orientations of transmitting/receiving antennas changes the polarizations of the radiated/received EM fields. In the monostatic configuration, simultaneously rotating the orientations of the transmitting and receiving antennas equivalently rotates the polarization coordinate system of measurement. Assume the new polarization coordinate system is rotated with an angle θ in a clockwise direction with respect to the original one [e.g., $\left(b_1, b_2\right)$ base], then the received signal in the new polarization bases can be related to the measurements before rotation at the same position through the rotation matrix

$$\begin{bmatrix} E^s_1\left(\boldsymbol{x}^r,\boldsymbol{x}^t,\omega\right) \\ E^s_2\left(\boldsymbol{x}^r,\boldsymbol{x}^t,\omega\right) \end{bmatrix} = R \begin{bmatrix} E^s_\theta\left(\boldsymbol{x}^r,\boldsymbol{x}^t,\omega\right) \\ E^s_{\theta\perp}\left(\boldsymbol{x}^r,\boldsymbol{x}^t,\omega\right) \end{bmatrix}$$
$$\begin{bmatrix} b_1\left(\boldsymbol{x}^t\right) \\ b_1\left(\boldsymbol{x}^t\right) \end{bmatrix} = R \begin{bmatrix} b_\theta\left(\boldsymbol{x}^t\right) \\ b_{\theta\perp}\left(\boldsymbol{x}^t\right) \end{bmatrix} \tag{8.8}$$

where R is the rotation matrix and is expressed as

$$R = \begin{bmatrix} \cos\theta & \sin\theta \\ -\sin\theta & \cos\theta \end{bmatrix} \tag{8.9}$$

Inserting (8.8) into (8.4) and performing a simple algebraic manipulation result in

$$\begin{bmatrix} E^s_\theta(\boldsymbol{x}^r,\boldsymbol{x}^t,\omega) \\ E^s_{\theta_\perp}(\boldsymbol{x}^r,\boldsymbol{x}^t,\omega) \end{bmatrix} = S(\omega)\int_{V(\boldsymbol{x}^c)} R^{-1}\boldsymbol{D}\left(\boldsymbol{x}^r,\boldsymbol{x}^t \middle| \boldsymbol{x}^c,\omega\right) R \begin{bmatrix} b_\theta(\boldsymbol{x}^t) \\ b_{\theta_\perp}(\boldsymbol{x}^t) \end{bmatrix} \chi(\boldsymbol{x}^c)\,\mathrm{d}V \tag{8.10}$$

Equation (8.10) formulates the scattering process in the polarization bases $(\theta, \theta_\perp)$. Compared to (8.4), in the polarization bases $(\theta, \theta_\perp)$ the forward wavefield extrapolator, denoted by $\hat{\boldsymbol{D}}$, can be defined as

$$\hat{\boldsymbol{D}} = R^{-1}\boldsymbol{D}\left(\boldsymbol{x}^r,\boldsymbol{x}^t \middle| \boldsymbol{x}^c,\omega\right) R \tag{8.11}$$

Equation (8.11) describes the relationship between the forward wavefield extrapolators in two different polarization bases (b_1, b_2) and $(\theta, \theta_\perp)$. Substituting (8.9) for R, $\hat{\boldsymbol{D}}$ can be explicitly written as

$$\hat{\boldsymbol{D}} = \begin{bmatrix} \hat{D}_{11} & \hat{D}_{12} \\ \hat{D}_{21} & \hat{D}_{22} \end{bmatrix} \tag{8.12}$$

where

$$\begin{cases} \hat{D}_{11} = \cos^2\theta \cdot D_{11} - \sin\theta\cos\theta \cdot D_{21} - \sin\theta\cos\theta \cdot D_{12} + \sin^2\theta \cdot D_{22} \\ \hat{D}_{12} = \sin\theta\cos\theta \cdot D_{11} - \sin^2\theta \cdot D_{21} + \cos^2\theta \cdot D_{12} - \sin\theta\cos\theta \cdot D_{22} \\ \hat{D}_{21} = \sin\theta\cos\theta \cdot D_{11} + \cos^2\theta \cdot D_{21} - \sin^2\theta \cdot D_{12} - \sin\theta\cos\theta \cdot D_{22} \\ \hat{D}_{22} = \sin^2\theta \cdot D_{11} + \sin\theta\cos\theta \cdot D_{21} + \sin\theta\cos\theta \cdot D_{12} + \cos^2\theta \cdot D_{22} \end{cases} \tag{8.13}$$

In practice, quad-polarimetric radar systems radiate two orthogonally polarized electric fields separately in time or with orthogonal waveforms for identification; then for each polarized transmission, two orthogonally polarized components are collected. This operation can be represented by setting the vector $[b_\theta(\boldsymbol{x}^t),\ b_{\theta_\perp}(\boldsymbol{x}^t)]^T$ as $[1,\ 0]^T$ or $[0,\ 1]^T$ for the two orthogonally polarized transmissions. Utilizing the four combinations of the polarizations of transmitting and receiving antennas, the acquired polarimetric measurements are

$$\begin{cases} E^s_{\theta\theta}(\boldsymbol{x}^r,\boldsymbol{x}^t,\omega) = S(\omega)\int_{V(\boldsymbol{x}^c)} \hat{D}_{11}\chi(\boldsymbol{x}^c)\,\mathrm{d}V \\ E^s_{\theta_\perp\theta}(\boldsymbol{x}^r,\boldsymbol{x}^t,\omega) = S(\omega)\int_{V(\boldsymbol{x}^c)} \hat{D}_{12}\chi(\boldsymbol{x}^c)\,\mathrm{d}V \\ E^s_{\theta\theta_\perp}(\boldsymbol{x}^r,\boldsymbol{x}^t,\omega) = S(\omega)\int_{V(\boldsymbol{x}^c)} \hat{D}_{21}\chi(\boldsymbol{x}^c)\,\mathrm{d}V \\ E^s_{\theta_\perp\theta_\perp}(\boldsymbol{x}^r,\boldsymbol{x}^t,\omega) = S(\omega)\int_{V(\boldsymbol{x}^c)} \hat{D}_{22}\chi(\boldsymbol{x}^c)\,\mathrm{d}V \end{cases} \tag{8.14}$$

Applying (8.1) and (8.13) to (8.14), we can get

$$\begin{cases} E^s_{\theta\theta}\left(\mathbf{x}^r,\mathbf{x}^t,\omega\right) = \cos^2\theta\cdot E_{11} - \sin\theta\cos\theta\cdot E_{21} \\ \qquad - \sin\theta\cos\theta\cdot E_{12} + \sin^2\theta\cdot E_{22} \\ E^s_{\theta_\perp\theta}\left(\mathbf{x}^r,\mathbf{x}^t,\omega\right) = \sin\theta\cos\theta\cdot E_{11} - \sin^2\theta\cdot E_{21} \\ \qquad +\cos^2\theta\cdot E_{12} - \sin\theta\cos\theta\cdot E_{22} \\ E^s_{\theta\theta_\perp}\left(\mathbf{x}^r,\mathbf{x}^t,\omega\right) = \sin\theta\cos\theta\cdot E_{11} + \cos^2\theta\cdot E_{21} \\ \qquad -\sin^2\theta\cdot E_{12} - \sin\theta\cos\theta\cdot E_{22} \\ E^s_{\theta_\perp\theta_\perp}\left(\mathbf{x}^r,\mathbf{x}^t,\omega\right) = \sin^2\theta\cdot E_{11} + \sin\theta\cos\theta\cdot E_{21} \\ \qquad + \sin\theta\cos\theta\cdot E_{12} + \cos^2\theta\cdot E_{22} \end{cases} \tag{8.15}$$

As $E_{21} = E_{12}$ in the monostatic configuration, (8.15) can be further simplified as

$$\begin{cases} E^s_{\theta\theta}\left(\mathbf{x}^r,\mathbf{x}^t,\omega\right) = E_{11}\cos^2\theta - E_{12}\sin 2\theta + E_{22}\sin^2\theta \\ E^s_{\theta_\perp\theta}\left(\mathbf{x}^r,\mathbf{x}^t,\omega\right) = \dfrac{\left(E_{11}-E_{22}\right)}{2}\sin 2\theta + E_{12}\cos 2\theta \\ E^s_{\theta\theta_\perp}\left(\mathbf{x}^r,\mathbf{x}^t,\omega\right) = \dfrac{\left(E_{11}-E_{22}\right)}{2}\sin 2\theta + E_{12}\cos 2\theta \\ E^s_{\theta_\perp\theta_\perp}\left(\mathbf{x}^r,\mathbf{x}^t,\omega\right) = E_{11}\sin^2\theta + E_{12}\sin 2\theta + E_{22}\cos^2\theta \end{cases} \tag{8.16}$$

In (8.16), $E^s_{\theta_\perp\theta}$ is equal to $E^s_{\theta\theta_\perp}$, which agrees with the reciprocity theorem. Another fact one can observe is that the measurement $E^s_{\theta\theta}$ should be obtained by turning 90° (or 270°) clockwise the orientations of the transmitting-receiving antennas for $E^s_{\theta_\perp\theta_\perp}$. This can be demonstrated by replacing θ in the last line of (8.16) with $\theta + 90°$ (or $\theta + 270°$) such that the first line in (8.16) arrives. It shows the self-consistency of the derivation.

The equations in (8.16) give the relationships among the polarimetric measurements in two sets of polarization bases, i.e., $(\theta, \theta_\perp)$ and (b_1, b_2). So one can see that as long as we get the fully polarimetric measurements with respect to a rotated polarization coordinate system, the fully polarimetric signals in terms of the aligned polarization bases, for example, (b_1, b_2), can be reconstructed via (8.16). This provides the first method to design a rotated array for fully polarimetric imaging.

Moreover, the equations in (8.16) reveal that each polarimetric measurement in the $(\theta, \theta_\perp)$ bases is a function of the fully polarimetric signals in the (b_1, b_2) bases at the same position. Taking the first equation in (8.16) as an example, it shows that the θ co-polarized scattered waves, which are acquired with transmitting and receiving antennas with orientations of θ from the x_1-axis, contain both the co-polarized (i.e., E_{11} and E_{22}) and cross-polarized (i.e., E_{12}) information that are measured by b_1- or b_2-polarized antennas. Therefore, to extract or reconstruct the fully polarimetric information of targets with rotated antennas, a second approach is to take three co-polarized measurements with antennas of three different orientations. Then, the application of the relation in (8.16) helps to reconstruct the full-polarized scattered signals that could be observed by b_1- or b_2-polarized antennas. More specifically, assume the three antennas for co-polarized measurements are oriented with angles of θ_1, θ_2, and θ_3 from the x_1-axis, then this approach can be expressed as

$$\begin{cases} E^s_{\theta_1\theta_1}\left(\boldsymbol{x}^r,\boldsymbol{x}^t,\omega\right) = \cos^2\theta_1 \cdot E_{11} - \sin 2\theta_1 \cdot E_{12} + \sin^2\theta_1 \cdot E_{22} \\ E^s_{\theta_2\theta_2}\left(\boldsymbol{x}^r,\boldsymbol{x}^t,\omega\right) = \cos^2\theta_2 \cdot E_{11} - \sin 2\theta_2 \cdot E_{12} + \sin^2\theta_2 \cdot E_{22} \\ E^s_{\theta_3\theta_3}\left(\boldsymbol{x}^r,\boldsymbol{x}^t,\omega\right) = \cos^2\theta_3 \cdot E_{11} - \sin 2\theta_3 \cdot E_{12} + \sin^2\theta_3 \cdot E_{22} \end{cases} \tag{8.17}$$

which gives the relations of three co-polarized measurements with orientations of θ_1, θ_2, and θ_3 with the fully polarimetric measurements E_{11}, E_{12}, and E_{22} in the linear polarization bases (b_1, b_2). Solving the system of linear equations in (8.17) reconstructs the observables E_{11}, E_{12}, and E_{22} in the linear polarization bases (b_1, b_2) required by the conventional fully polarimetric imaging approaches. Then, the reconstructed measurements E_{11}, E_{12}, and E_{22} can be processed by employing either the scalar-wave based imaging algorithms or the matrix-based inversion algorithm.

8.2.3 Rotated antenna arrays design

8.2.3.1 Array topologies

In this section, the proposed approaches to fully polarized signal acquisition above are applied to design rotated antenna arrays that could be used in, for example, the GPR systems for the tunnel boring machine (TBM) applications.

First, one rotated array topology can be proposed based on the first approach, which measures two co-polarized and one cross-polarized signals at each position with respect to the "local" polarization coordinate system. Assume the antenna elements are distributed along three radial directions over several concentric circles. The antennas on two of the radial directions are placed along the "local" polarization bases to collect the two co-polarized components while the transmitting and receiving antennas on the third radial directions are placed orthogonally at each position to take the cross-polarized measurements [2]. Figure 8.2(a) illustrates such a topology, where the "local" polarization bases are defined as the vectors parallel and perpendicular to the local radius. For convenience, this rotated array is referred to as rotated array I in the following. Its operation scheme is shown in Figure 8.3. With its rotation, the rotated array I collects two "local" co-polarized and one cross-polarized EM signals when the antennas pass by the same sampling positions. As an example, the polarizations of the three measurements acquired by the rotated array I at a position on the y-axis are indicated in Figure 8.3.

Next, utilizing the second approach proposed above, another rotated array topology can be designed for full-polarized imaging by just recording three co-polarized measurements at each position [1]. Similar to the rotated array I, the antennas are placed in three radial directions over a series of concentric circles. The antennas in the same radial direction are set with the same orientation, but in different radial directions the antenna orientation angles formed with respect to the corresponding radius are different. Figure 8.2(b) shows an example of a

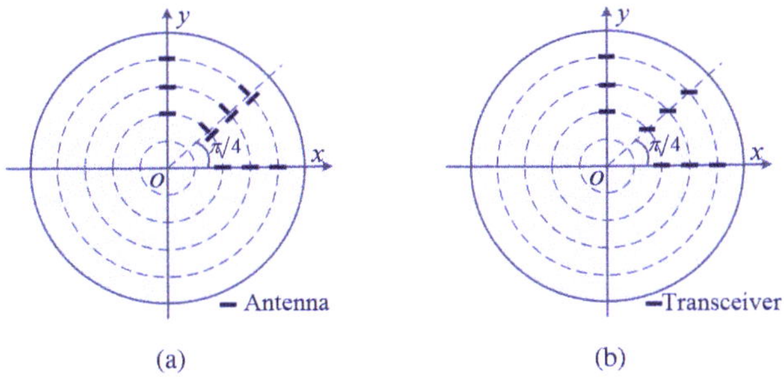

Figure 8.2 Topologies of the rotated arrays for fully polarimetric imaging designed with (a) the first approach based on (8.16) (referred to as rotated array I), and (b) the second approach based on (8.17) (referred to as rotated array II)

rotated array when the three angles formed between antenna axis and the corresponding radius are 0°, 45°, and 90°, which is referred to as rotated array II. With the rotation of the circular disc, three co-polarized measurements are acquired when the three antennas on the same circle sequentially pass by one particular sampling position. This operation mechanism for the signal acquisition at a point on the y-axis is shown in Figure 8.4. One may note that polarizations of the three measurements with both rotated array I and II are still distinct at different spatial points as the antenna array rotates circularly. But employing (8.16) and (8.17), the full-polarized signals in an aligned polarization coordinate system can be reconstructed from the measurements of rotated arrays I and II, respectively. Furthermore, to get sufficient spatial observations within the aperture, the radii of concentric circles and the azimuthal sampling intervals with the rotation are determined by the sampling criteria and will be discussed in the next section.

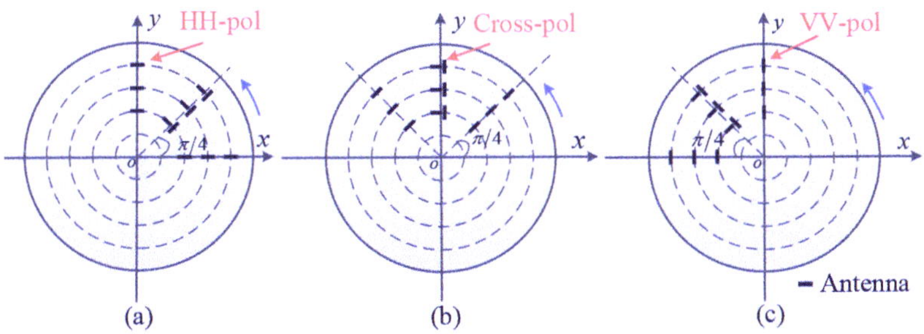

Figure 8.3 Operation scheme of the rotated array I to acquire two "local" co-polarized and one cross-polarized measurements. (a), (b), and (c) illustrate the two co-polarized and one cross-polarized measurements at a position on the y-axis acquired sequentially by three antennas on the same circle.

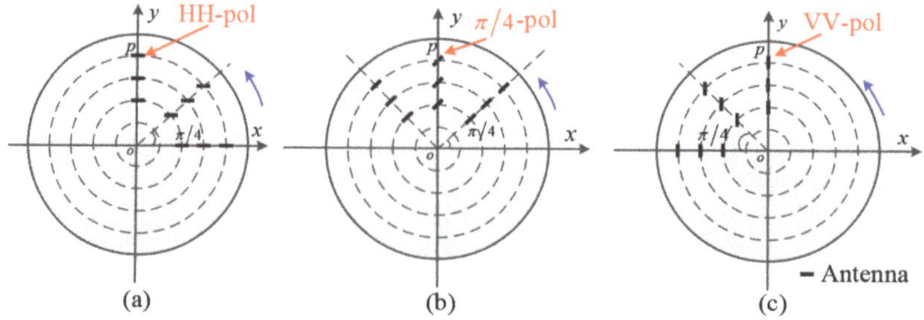

Figure 8.4 Operation scheme of the rotated array II to acquire three co-polarized measurements. (a), (b), and (c) illustrate the three different co-polarized measurements at a sampling position on the y-axis acquired sequentially by three antennas on the same circle.

8.2.3.2 Sampling criteria

The rotated antenna array synthesizes a planar circular aperture (i.e., RadSAR) for the 3D imaging and takes spatial samples over a polar grid, as shown in Figure 8.5. The numbers of antennas along a radial direction and azimuthal samples can be determined based on the polar sampling analysis. To avoid aliasing, it is derived for narrowband (or monochromatic) systems in Reference [23] that regular sampling is performed along the radial direction and equi-arc-length sampling in azimuth. The azimuthal sampling interval is

$$\Delta\varphi = 2\pi/(2N+1), \quad N = \lceil K_a r \rceil \tag{8.18}$$

where K_a is the maximum wavenumber related to the scenario, r denotes the radius, and N is the smallest integer larger than or equal to $K_a r$. The radial sampling constraint can be obtained in a similar way to the linear arrays. Based on the Nyquist criterion, it should satisfy

$$\Delta r \leqslant \lambda \sqrt{(R_a + a)^2 + R_0^2} \,/\, [4(R_a + a)] \tag{8.19}$$

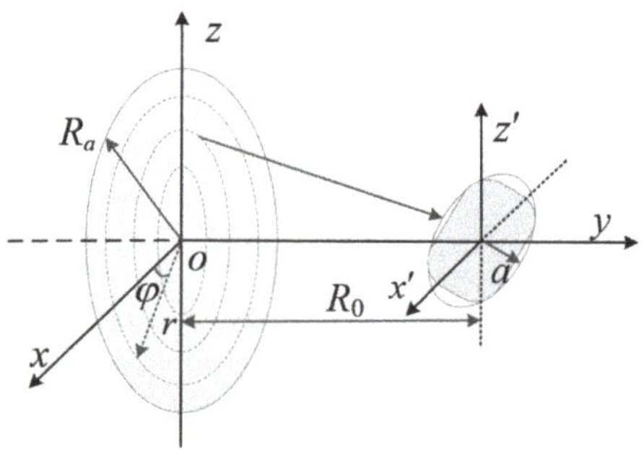

Figure 8.5 Geometry configuration for rotated array imaging

where λ is the wavelength of the highest frequency of the signal, R_a is the radius of the antenna aperture, R_0 is the distance from the target to antenna aperture, and a is the radius of the smallest sphere circumscribing the object. For typical imaging systems, the values of the radius of the antenna aperture, the extension of the object, and the distance between antenna aperture and the object are comparable; thus, the radial sampling interval is on the order of $\lambda/2$.

In addition, we have to mention that combining UWB techniques these sampling constraints could be relaxed to take sparse measurements without causing aliasing, especially when the fractional bandwidth is larger than 100%. This is due to the limited interference region of UWB pulses in the space domain [24]. Design of a particular sparse sampling strategy for UWB systems involves the spatial sampling (or antenna array) optimization [25]. For simplicity, the equal-angular sampling strategy based on the Nyquist criterion is used for the examples and analyses in this chapter.

8.3 Image reconstruction

As mentioned above, after acquiring the full-polarimetric data in an aligned polarization basis, they can be processed in two ways: (i) four polarimetric components are focused separately by using the radar-based (or scalar-wave based) approaches to reconstruct four polarimetric images of targets; (ii) all polarimetric components are jointly processed by using inverse-based approaches to get an integrated image [16, 26]. For the full-polarimetric imaging with the radar-based approaches, four reconstructed polarimetric images are fused through polarimetric decomposition to extract the polarimetric scattering features of targets, which is widely used in the SAR community to characterize different scattering mechanism of targets in the imaging scene. Considering its simplicity and high efficiency, we also use this kind of approach to manipulate the full-polarimetric measurements for short-range imaging in this chapter. Specifically, Kirchhoff migration is used as the scalar-wave-based approach to reconstruct the image of each polarimetric component.

Kirchhoff migration utilizes the Kirchhoff integral to backpropagate the acquired signals to the time zero to reconstruct the image of scatterers [27]. The reconstructed image with the Kirchhoff migration algorithm can be expressed in

$$f(\boldsymbol{r}) = \frac{1}{2\pi} \int \frac{\partial d_r}{\partial \boldsymbol{n}} \left[\frac{\partial U(\boldsymbol{r}_0, t - d_r/v)}{\partial t} \frac{1}{d_r v} + \frac{U(\boldsymbol{r}_0, t - d_r/v)}{d_r^2} \right] dS_0|_{t=0} \tag{8.20}$$

where $\boldsymbol{r}$ is the vector from the origin to a wavefield point. U is the wave field and $\boldsymbol{r}_0$ is the vector from the origin to a point on the surface of integration $S_0 \cdot \boldsymbol{n}$ is the unit vector normal to the surface S_0. $d_r = \|\boldsymbol{r} - \boldsymbol{r}_0\|$ is the distance between the field points $\boldsymbol{r}$ and $\boldsymbol{r}_0$. For imaging systems with the monostatic configuration, the exploding reflector model, which assumes that the reflected signals can be considered as the radiated signals from a virtual source placed at the scatterer's position but propagating with half of the real speed in the practical medium, is exploited by the Kirchhoff migration algorithm for wave field extrapolation. So in practice v is half of the wave speed in the medium in (8.20).

8.4 Numerical simulations of UWB full-polarimetric imaging

In this section, numerical EM simulations were performed to demonstrate the effectiveness of the proposed approaches to fully polarimetric information retrieval and imaging. The simulation models were implemented with the applied EM simulation software FEKO in which the Method of Moments solver is utilized to solve the integral equations. In these models, Hertz dipoles were used as transmitting and receiving antennas. The operational signal bandwidth was from 2 to 10 GHz. For comparison, simulations were carried out for both the proposed rotated arrays [Figure 8.2(a) and (b)] and their traditional counterpart (Figure 8.6) for fully polarimetric imaging. Due to the feature of the Method of Moments solver, the EM simulations were implemented in the frequency domain. The synthetic data were converted to the time domain by using the fast fourier transform (FFT) after applying a Hanning window. Then, the time-domain-polarized data were focused using Kirchhoff migration (8.20) to obtain the fully polarimetric images of objects.

Both the traditional fully polarimetric and the rotated arrays performed the same spatial sampling. In the rotated arrays, the antennas were placed on a series of concentric circles with the topologies in Figure 8.2. The radii of these concentric circles varied from 0.02 to 0.5 m with intervals of 2 cm (i.e., $0.4\lambda_c$, where λ_c is the wavelength of the center frequency). In azimuth, the samples were taken every 4° over each circle. Then, it resulted in a circular antenna aperture of radius 0.5 m (i.e., $10\lambda_c$). With the same spatial sampling intervals, the simulations were also conducted for the traditional fully polarimetric arrays (Figure 8.6). For convenience, the simulation parameters are summarized in Table 8.1.

To evaluate the imaging performance, the point spread functions (PSF) of the traditional and rotated arrays for differently polarized signals are obtained by focusing the scattered signals from a point-like target (i.e., a small sphere) and shown

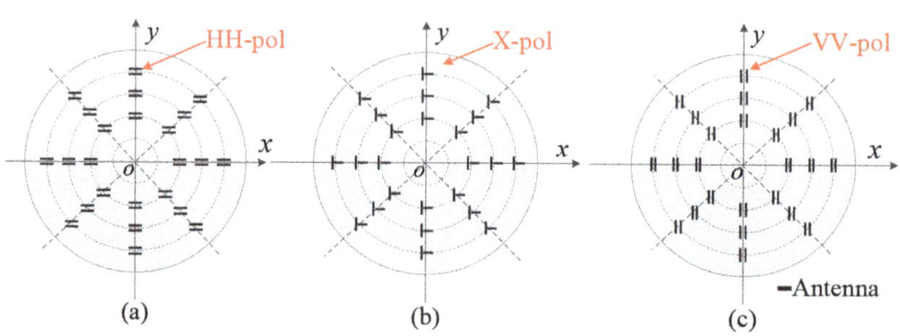

Figure 8.6 Topologies of traditional fully polarimetric imaging arrays: (a) HH co-polarized array, (b) HV cross-polarized array, and (c) VV co-polarized array

Table 8.1 Parameters for numerical simulations

Parameters	Value
Radius of antenna aperture	0.5 m
Sampling interval in radial direction	2 cm
Sampling interval in azimuth	4°
Operational frequencies	2 – 10 GHz
Frequency steps	100 MHz
Distance between target and antenna array	0.5 m

in Figure 8.7. The polarization effects on the focused patterns are noticeable. For all HH-, HV-, and VV-polarized signals, the nearly same PSFs were obtained for the three arrays. The sidelobes of the PSFs for differently polarized signals are all lower than −25 dB. However, slightly stronger sidelobes around the focal point can be seen in the PSFs of HV-polarized signals [Figure 8.7(e), (k), and (q)]. As the cross-polarized radiation pattern does not have main lobes but sidelobes, the cross-polarization antenna picks up scattered energy via sidelobes when scatterers are off the broadside direction of the antennas. Thus, it results in relatively stronger sidelobes of the cross-polarization PSF compared to that of the co-polarization components. In addition, we have to mention that the equal-angle sampling of rotated arrays causes a non-uniform distribution of the samples in the synthetic aperture where the sampling distance depends on the radius. The non-uniform distribution of samples inherently introduces a space-tapering and may influence the resolutions of a target. To tackle this effect, the samples were weighted by the effective area (e.g., the areas of Voronoi cells) surrounding them within the aperture in the imaging process. That is to say, smaller weighting factors were imposed on the densely sampled region while larger weighting factors were used for relatively sparse samples within the aperture. This technique has been utilized for image formation in all experiments in this chapter.

Next, a numerical simulation was performed for a complex "E"-shaped perfect electric conductor object that was placed in front of the antenna array at a distance of 0.5 m (i.e., $10\lambda_c$) in free space. The "E"-shaped object is illustrated in Figure 8.8, which contains a vertical column of the length 15 cm, a horizontal bar of the length 10 cm in the middle, and two inclined bars joined with the vertical column at the two ends. The two inclined bars were 10 cm in length and rotated 30° away from the horizontal direction. The width and thickness of all the parts of the "E"-shaped object were 3 cm. The synthetic data with rotated arrays and traditional arrays at 2 GHz are shown in Figure 8.9. For simplicity of notation, the polarizations of the measurements of rotated arrays at each spatial position are defined with respect to the local radius: (i) PP, where the antenna axis is parallel to the radius; (ii) NN, where the antenna axis is perpendicular (normal) to the radius; (iii) DD, where the antenna axis forms an angle of 45° with the radius; (iv) LX, i.e., the local cross-polarization measurements that are only acquired by rotated array I.

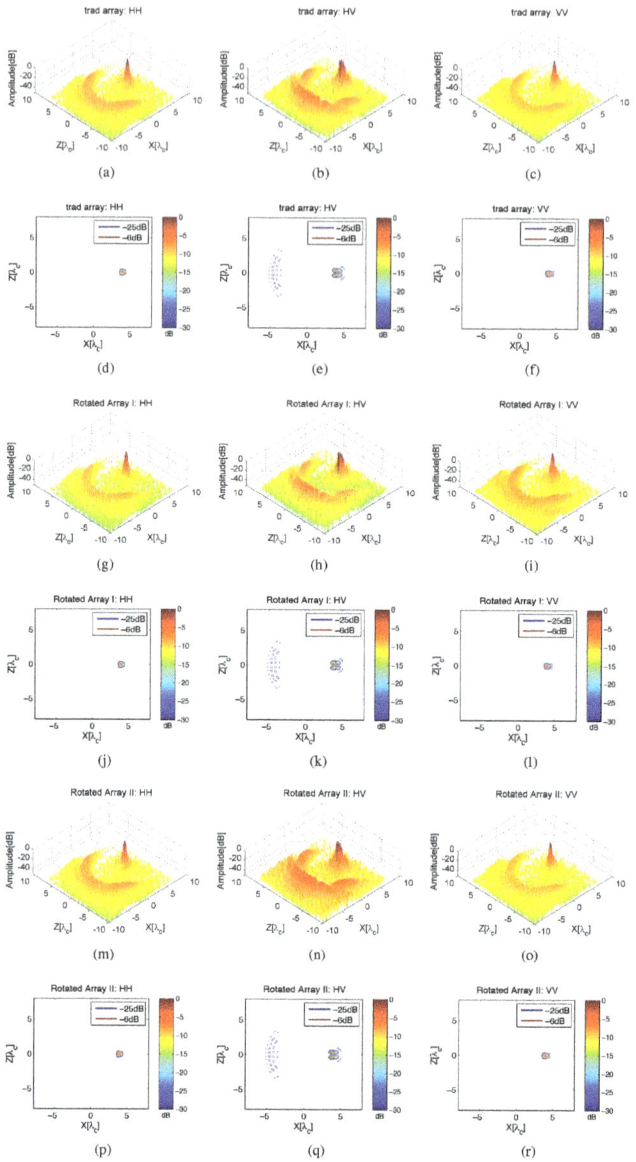

Figure 8.7 *The normalized focused patterns of traditional and rotated arrays with focal point at* $(4\lambda_c,\ 10\lambda_c,\ 0)$ *for differently polarized signals. The focusing patterns of traditional array for (a) HH polarization, (b) HV polarization, and (c) VV polarization. (d), (e), and (f) Corresponding contour plots at −25 dB beamwidths. The focused patterns of rotated array I for (g) HH polarization, (h) HV polarization, and (i) VV polarization. (j), (k), and (l) Corresponding contour plots at − 25 dB beamwidths.*

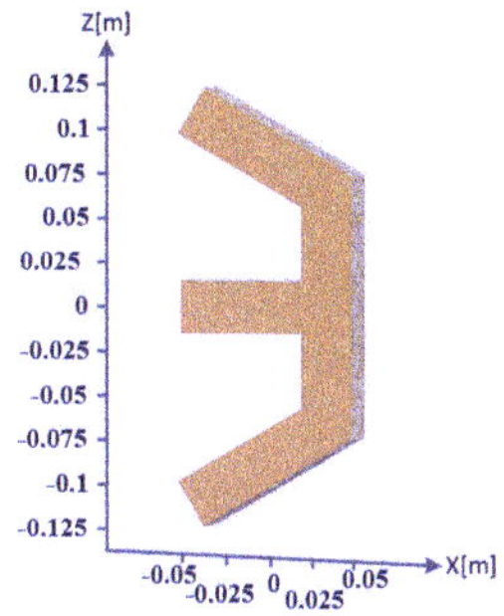

Figure 8.8 The "E"-shaped object used in the simulation

Using (8.16) and (8.17), the polarimetric [i.e., HH-, HV(VH)-, and VV-polarized] signals were retrieved from the synthetic signals collected by rotated arrays and denoted as E_{HH}^{rti}, E_{HV}^{rti}, and E_{VV}^{rti}, where the superscript rt refers to "retrieved" signals and $i \in \{1, 2\}$ indicates rotated array I or II. Arbitrarily choosing a spatial sample position within the aperture, the retrieved time-domain signals are shown in Figure 8.10. Here, the selected spatial position is $(0.3\,\text{m}, 236°)$. In addition, the polarimetric signals E_{HH}^{m}, E_{HV}^{m}, and E_{VV}^{m} acquired with traditional arrays are displayed as references. The differences between the retrieved signals and their corresponding references are presented in Figure 8.10(d), (e), and (f).

First, from Figure 8.10(a)–(c), one can see that the retrieved polarimetric signals with both rotated arrays I and II agree very well with their references at all three positions. It is confirmed by the small differences between the retrieved and reference signals in Figure 8.10(d)–(f) where the relative differences observed are generally smaller than 0.5%. So, both rotated arrays I and II provide the capability to estimate the full-polarized signals from their measurements with sufficient accuracy compared to the traditional full-polarized arrays.

However, some subtle differences between the two rotated arrays are also perceived in terms of the accuracies of the retrieved signals. Based on Figure 8.10(d) and (f), one can see that the co-polarized (i.e., HH- and VV-) signals retrieved with rotated array I exhibit larger differences than those with rotated array II relative to the reference signals. On the other hand, the differences between the cross-polarized signals retrieved with rotated array I and the reference are slightly smaller than that between the cross-polarized signal retrieved with rotated array II and the reference one at the position $(0.3\,\text{m}, 236°)$. So, rotated array I provides slightly better or comparable estimation for the HV-polarized signals in contrast to rotated array II. Moreover, it can be seen that the differences between the polarized signals retrieved with rotated array I and the corresponding references are relatively comparable. By contrast, with the measurements of rotated array II, the co-polarized signals are more accurately estimated than the cross-polarized one.

To quantitatively analyze the accuracy of the retrieved signals, we introduce the L_2 relative error as a metric that is defined as the energy of the differential signal divided by the energy of the reference signal acquired with traditional arrays

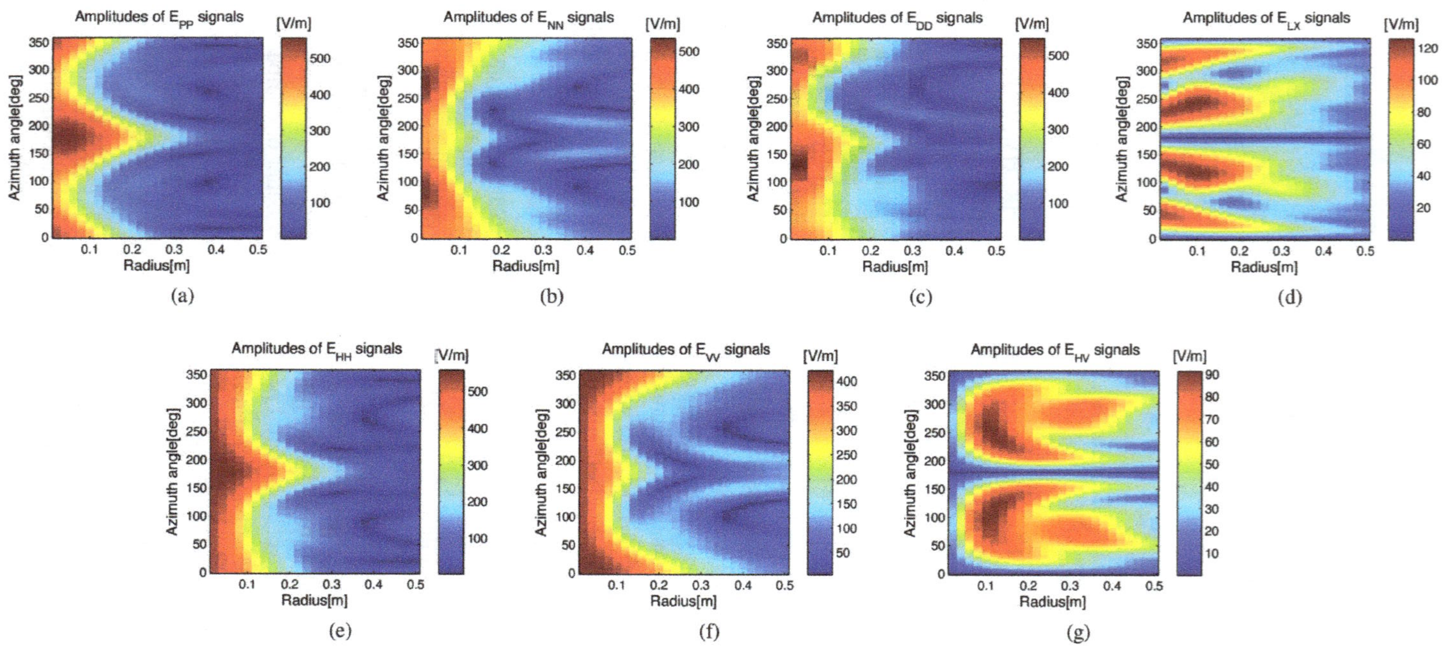

Figure 8.9 *Amplitudes of the signals acquired with rotated arrays and traditional polarimetric arrays at* f = 2 GHz*: (a), (b), and (d) show the amplitudes of PP-, NN-, and LX-polarized measurements obtained with rotated array I; (a), (b), and (c) show the amplitudes of PP-, NN-, and DD-polarized signals acquired with rotated array II; and (e), (f), and (g) are the amplitudes of HH, VV, and HV signals obtained with traditional polarimetric arrays*

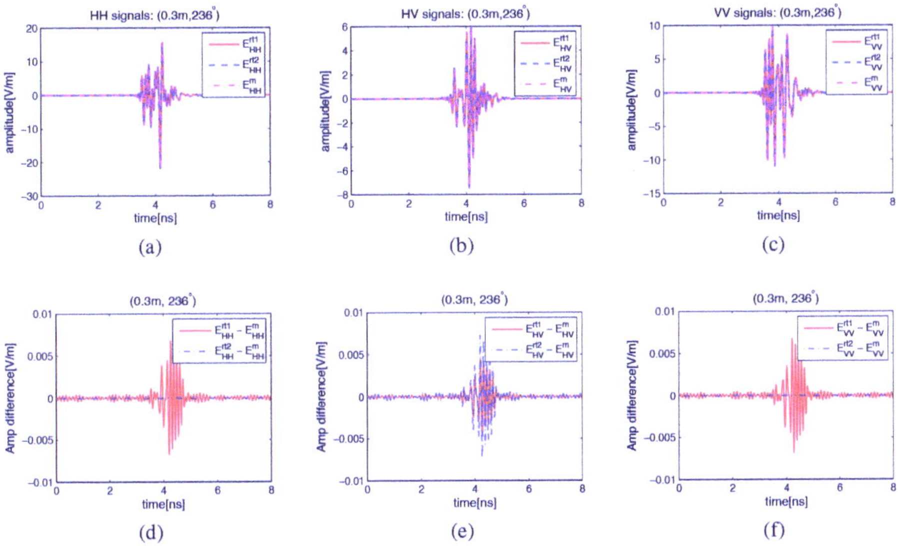

Figure 8.10 *Comparison of the retrieved signals with the rotated arrays and the measured signals with traditional array at the position* $(0.3\,\text{m}, 236^\circ)$. *(a), (b), and (c) show the measured and retrieved signals while their corresponding differences are presented in (d), (e), and (f).*

$$\alpha = \frac{\sum_{k=0}^{N} \left| E^{rt}(t_k) - E^{m}(t_k) \right|^2}{\sum_{i=0}^{N} E^{m}(t_i)^2} \tag{8.21}$$

where N is the number of discretized samples of the signal. One can see that the smaller the differences between the retrieved signal $E^{rt}(t)$ and the reference signal $E^{m}(t)$, the closer α to zero, thus, the more accurate the retrieved signal.

Figure 8.11 illustrates the relative errors of the full-polarized signals estimated from the measurements of rotated arrays I and II on three circles (i.e., $R = 0.1$, 0.2, and 0.3 m) within the synthetic aperture. In general, considerably small relative errors are observed for the polarized signals retrieved with both rotated arrays (i.e., less than 10^{-4}). However, some differences are also found between the relative errors of the signals retrieved with the two rotated arrays as well as among the differently polarized signals. Comparing Figure 8.11(a), (e) with (b), (f), respectively, one can see that the relative errors of the co-polarized signals estimated with rotated array II are much (i.e., about two or three orders of magnitude) smaller than that of the co-polarized ones retrieved with rotated array I. On the other hand, the accuracy of the HV signals estimated with rotated array I is slightly better or comparable to that obtained with rotated array II.

Moreover, for differently polarized signals obtained with rotated array I, their relative errors are more or less at the same level. However, with rotated array II, the relative errors of the estimated HV signals are much larger than that of the co-polarized ones [see Figure 8.11(b), (d), and (f)], which has been noticed in Figure 8.10 as well. This can

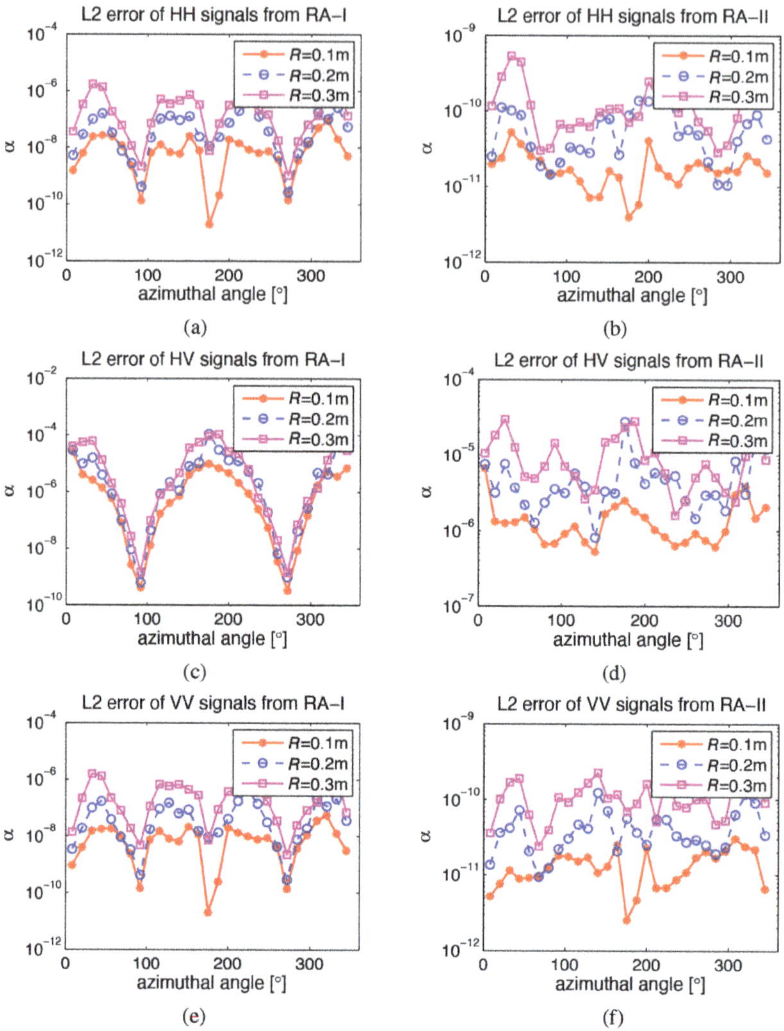

Figure 8.11 Relative errors of the retrieved polarimetric signals on three circles (i.e., R = *0.1, 0.2* and *0.3 m) within the synthetic aperture. (a), (c), and (e) show the* L_2 *relative errors of the HH-, HV-, and VV-signals retrieved with rotated array I; (b), (d), and (f) show the* L_2 *relative errors of the corresponding signals estimated with rotated array II.*

be explained as follows. As the relations (8.16) and (8.17) used for full-polarized signal retrieval were derived based on the Born approximation, so they are accurate for weak and point-like scatterers. But for distributed targets and strong scatterers, these relations are only approximate. In our simulation, an extended "E"-shaped object was used, which is a strong scatterer. Rotated array I acquires two co-polarized and one cross-polarized signal in terms of the "local" polarization basis. Then using (8.16) and algebraic theory, relatively uniform differences result for all retrieved polarized signals. On the other hand, rotated array II measures three co-polarized signals scattered from the targets. Since two of the co-polarized measurements are the same as the co-polarized ones measured by rotated array I, the third co-polarized (i.e., DD) measurement provides some redundancy about the co-polarized components scattered from the target. So, applying (8.17) to the measurements of rotated array II, relatively higher accuracies are achieved for the co-polarized signal estimation than for the cross-polarized ones. Furthermore, rotated array I takes a local cross-polarized measurement instead of a third co-polarized one, which improves the accuracies of the HV signal estimations, for instance, at the positions from 60° to 120° and from 240° to 310° in azimuth compared to rotated array II [see Figure 8.11(c)]. But at the position from 0 to 30° and from 160° to 190°, the accuracies of the signals obtained with rotated array I are slightly worse than those acquired with rotated array II. So, rotated array I is not definitely superior to rotated array II in terms of the accuracy of the estimated HV signals, but the accuracies of the signals retrieved with rotated array I have a relatively larger variation. Nevertheless, with both rotated arrays the relative errors of the estimated signals are sufficiently small [about 10^{-4} in Figure 8.11(c)] so that they are practically negligible for object imaging.

After imaging operation with the Kirchhoff migration [27], the accurately retrieved polarimetric signals lead to almost identical images as those generated by the reference polarimetric signals (Figure 8.12). The images formed with the two rotated arrays and the traditional arrays reveal the similar polarization dependence of different parts of the target. For example, the horizontal bar shows higher amplitudes in the HH-polarized images while the vertical column is highlighted in the VV-polarized images. In the HV-pol images, the inclined bars are well reconstructed and exhibit stronger scattering properties than other parts. The similarities of the corresponding polarimetric images obtained with the three arrays can also be quantitatively examined via the L_2 and L_∞ relative errors. The L_2 relative error is defined as in (8.21), but the time-domain samples of signals are replaced by the voxels of images. The L_∞ relative error is defined as the maximum difference between the voxel values of a polarimetric image and its reference divided by the maximum voxel value of the reference image. The relative errors for the images of the "E"-shaped object are listed in Table 8.2. According to Table 8.2, one can see that the L_2 and L_∞ errors of the images obtained with the two rotated arrays are smaller than 5×10^{-5} and 5×10^{-3}, respectively. Therefore, both rotated arrays I and II provide sufficient accuracy for fully polarimetric imaging. Moreover, the relative errors of the HV-polarized images acquired with both rotated arrays are on the same order of magnitude. For different polarimetric images acquired with rotated array I, both L_2 and L_∞ errors are relatively uniform while relative errors with different orders of magnitude are obtained for co-polarized and cross-polarized images with rotated array II, which are consistent with the observations about the differences between the retrieved polarimetric signals and their references.

In addition, comparing the co-polarized measurements of rotated array II and the signals recorded by traditional arrays in Figure 8.9, one can see that the amplitudes of co-polarized signals measured with rotated array II are generally larger than

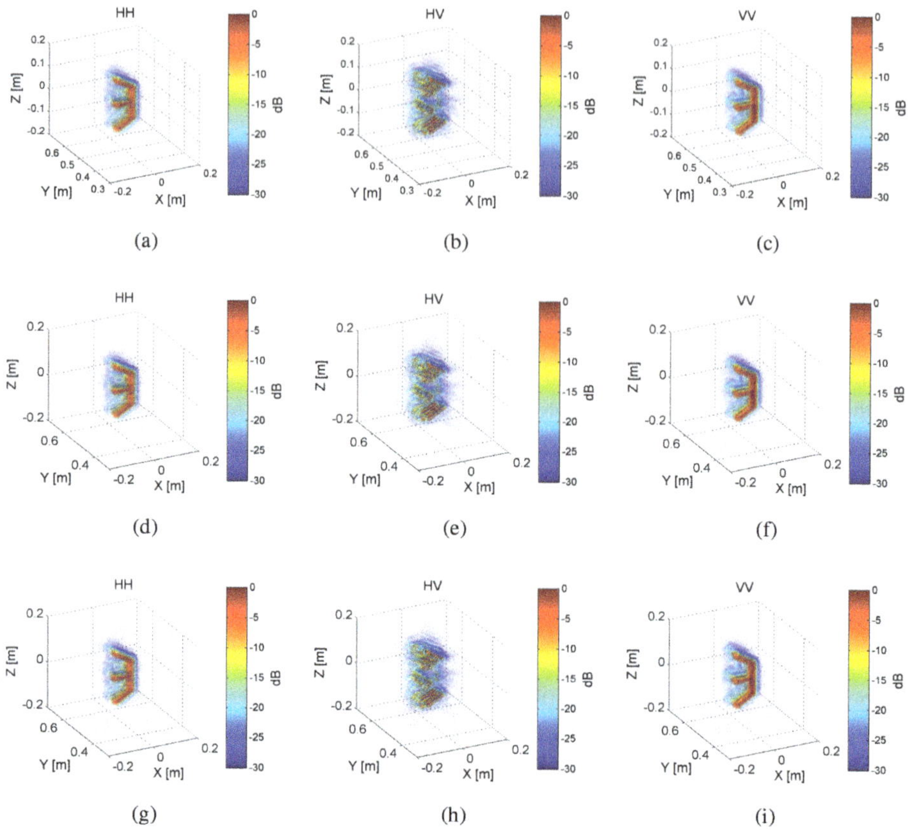

Figure 8.12 Polarimetric images reconstructed from the measurements of the three antenna arrays. (a), (b), and (c) are obtained with rotated array I; (d), (e), and (f) are obtained with rotated array II; and (g), (h), and (i) are obtained with traditional antenna arrays.

Table 8.2 The accuracy of the polarimetric images of "E"-shaped object obtained with the rotated arrays and the traditional arrays

		HH	**HV**	**VV**
Rotated array I	L_2 error	5.8479e-8	7.5763e-6	6.0008e-8
	L_∞ error	2.5080e-4	2.9873e-3	2.7552e-4
Rotated array II	L_2 error	1.0163e-12	5.3444e-6	9.8370e-13
	L_∞ error	2.8061e-7	2.8859e-3	3.1629e-7

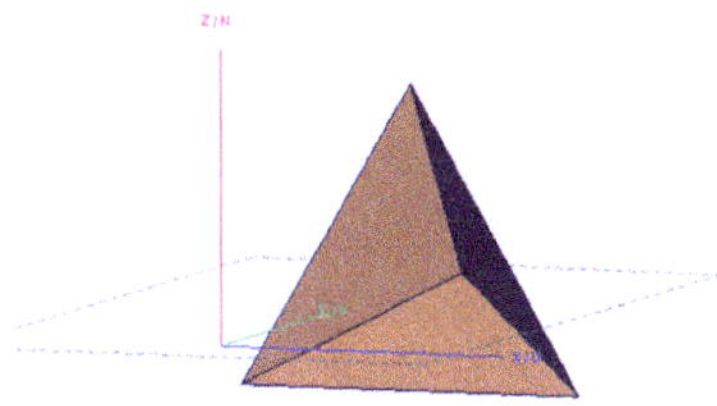

Figure 8.13 Geometrical configuration and corner reflector used for simulation

that of cross-polarized signals recorded with the traditional array. So considering the same noise level, rotated array II could acquire signals with low susceptibility to noise.

To further examine the accuracy of the full-polarized imaging for strong scattering objects with the rotated arrays proposed, one more simulation was carried out with a trihedral corner reflector (TCR) in free space. We took the parameters listed in Table 8.1 for the antenna arrays except that the operational bandwidth was 2–12 GHz. The antenna arrays were located on the xoz plane and centered at the origin. A TCR was placed at a distance of 0.4 m from the corner point to the antenna array. The TCR contains three surfaces that are mutually perpendicular, and their intersecting sides are 15 cm in length. Figure 8.13 shows the geometrical configuration. Again, the EM data were synthesized with the EM software FEKO for the three antenna arrays. Then, the signals acquired with rotated arrays I and II were used to estimate the fully polarimetric signals via (8.16) and (8.17). Comparing the estimated polarized signals with the references acquired by the traditional full-polarized arrays, the similar phenomenon was observed as for the "E"-shaped object: the errors of the co-polarized signals estimated with rotated array I are larger than those of their counterparts estimated with rotated array II; by contrast, the cross-polarized signals estimated with rotated array I are slightly more accurate than that with rotated array II relative to the references. To make the text compact, these figures are omitted here.

Focusing the polarimetric signals estimated from the measurements of the two rotated antenna arrays as well as collected by traditional full-polarized arrays, the polarimetric images of the TCR were reconstructed, as shown in Figure 8.14. First, one can see that in Figure 8.14 all the images are well focused, and different scattering characteristics of the TCR are revealed in the HH-, HV-, and VV-polarized images. Specifically, the edge of the bottom surface is well reconstructed in the HH-polarized images while the edges of the two top surfaces are clearly presented in the VV-polarized images. Meanwhile, the HH- and VV-polarized images distinctly show the three intersecting sides of the surfaces and the corner point appears as the brightest spot. In the HV-polarized images, the two bottom intersecting sides demonstrate the strongest cross-polarization effects with the orientation of the TCR placed in the simulation. As its projection on the antenna aperture is a vertical line, the top intersecting side induces negligible cross-polarized signals. Thus, it is missing in the focused HV-polarized images [i.e., the middle gaps in Figure 8.14(b), (e), and (h)]. Comparing the polarized images in each column, the HH-, HV- and

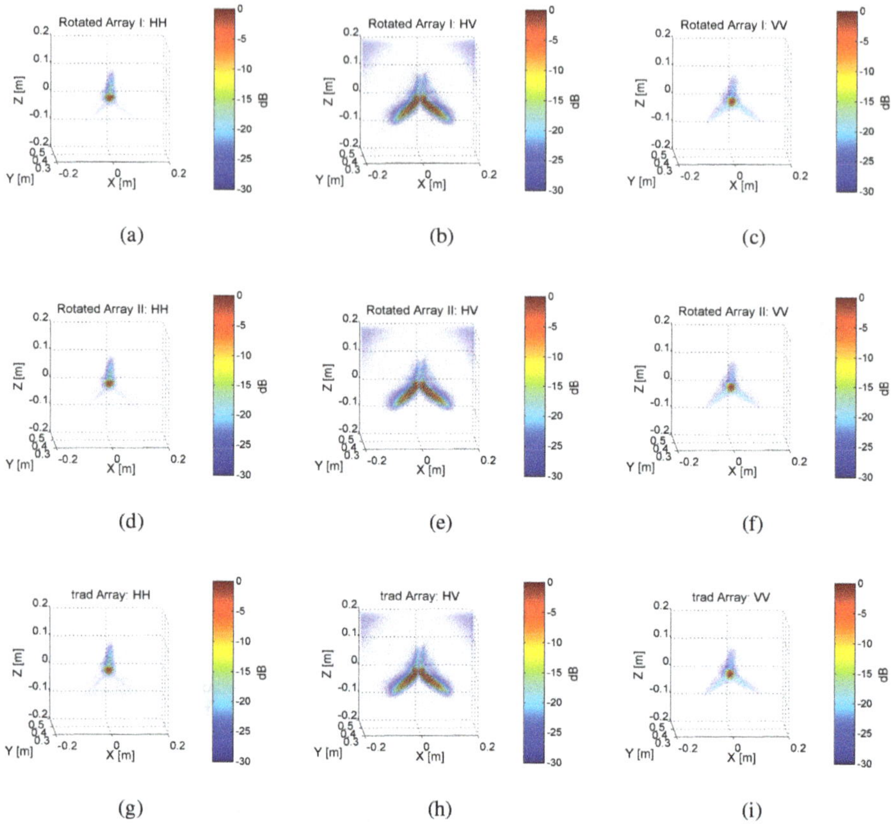

Figure 8.14 Full-polarimetric images reconstructed from the measurements of the three antenna arrays. (a), (b), and (c) are obtained with rotated array I; (d), (e), and (f) are obtained with rotated array II; and (g), (h), and (i) are obtained with traditional antenna arrays.

VV-polarized images obtained with the two rotated antenna arrays and the traditional arrays are visually equivalent. The relative errors of the reconstructed images were also analyzed with respect to the reference images, and the results are listed in Table 8.3. Similar to the "E"-shaped object case, rotated array I achieves roughly similar L_2 and L_∞ errors for different polarimetric images while the relative errors of HH- and VV-polarized images obtained with rotated array II are much smaller than that of the HV-polarized image. But the accuracies of the HV-polarized images acquired with rotated arrays I and II are comparable (i.e., about 10^{-5} for L_2 error and 10^{-3} for L_∞ error). Therefore, both rotated arrays I and II achieve sufficient accuracies, even for strong scattering objects, for full-polarized imaging although some subtle differences are perceived in the relative errors of their full-polarized images.

Table 8.3 The accuracy of the polarimetric images of the TCR obtained with the rotated arrays and the traditional arrays

		HH	HV	VV
Rotated array I	L_2 error	1.6374e-6	2.6249e-5	1.6333e-6
	L_∞ error	5.5975e-4	3.89e-3	5.4874e-4
Rotated array II	L_2 error	2.1335e-13	1.733e-5	2.1401e-13
	L_∞ error	3.0898e-8	3.578e-3	3.2136e-8

8.5 Experimental studies of UWB polarimetric imaging

Experimental measurements in free space were also performed to further demonstrate the effectiveness and accuracy of the proposed arrays for fully polarimetric imaging. The experimental setup for the rotated arrays (i.e., rotated arrays I and II) is shown in Figure 8.15(a). To simulate the rotated arrays, a stepper motor was used to drive a vertical column on top of which a polyethylene plastic panel was mounted to support

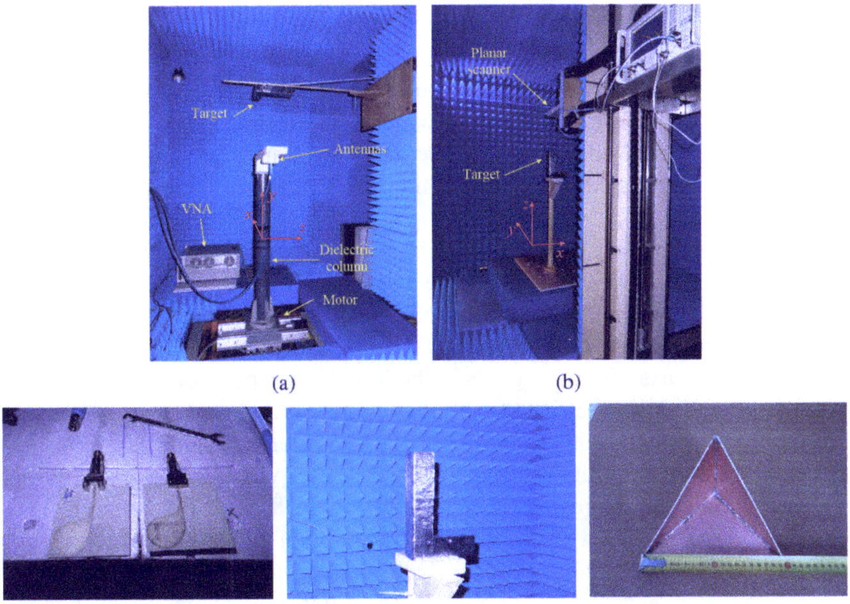

(a) (b)

(c) (d) (e)

Figure 8.15 The experimental setups for rotated and traditional planar arrays. (a) is the setup for the rotated arrays and (b) for the traditional planar array. (c) shows the employed antipodal Vivaldi antennas. (d) and (e) display the "L"-shaped object and the TCR used for the experiments, respectively.

antennas. The stepper motor was precisely controlled by a computer for positioning and rotating the column. Antipodal Vivaldi antennas [28] were used for transmission and reception and connected to a vector network analyzer for signal acquisition in the frequency domain. As the monostatic radar configuration was considered, at each spatial sampling position S_{11} was measured to get the co-polarized components for all three antenna arrays while S_{21} was measured to get the cross-polarized components with a quasi-monostatic configuration where a transmitting antenna and a receiving antenna were placed with a separation of 6 cm. The operational frequencies swept from 3 to 15 GHz with steps of 20 MHz. Through the translation and rotation of the vertical column, the antennas measured the data over 15 circles of the radii ranging from 11 to 53 cm with steps of 3 cm (i.e., $0.9\lambda_c$) and the azimuth sampling interval $d\theta$ along each circle was 1.2°. So, a planar circular array of the radius 0.53 m (i.e., $15.9\lambda_c$) was synthesized. An "L"-shaped distributed target which was coated with aluminum foil was placed at a distance of 0.5 m (i.e., $15\lambda_c$) in front of the center of the "equivalent" circular array for the test [Figure 8.15(d)]. The two arms of the "L"-shaped object are about 20 and 30 cm in length, respectively. Their width and thickness are 6 and 5.5 cm. Arranging antennas with orientations as required by the two rotated arrays and repeating the measurements over the same sampling grid, then the corresponding polarized signals, i.e., PP-, NN-, DD-, and LX-polarized signals were acquired at each sampling position.

For comparison, the reference HH-, HV-, and VV-polarized signals were also measured with the traditional polarimetric arrays over the same sampling grid, which was implemented with the planar scanner [Figure 8.15 (b)]. The signals scattered from the background were also measured in the absence of the target with all three arrays. Applying the Hanning window to all the signals measured in the frequency domain and taking the inverse FFT, the time-domain scattered signals were obtained. After background subtraction, the EM signals scattered from targets were extracted.

Taking advantage of (8.16) and (8.17), the HH-, HV- and VV-signals were retrieved from the measurements of the two rotated arrays. As an example, the retrieved signals at $(R, \theta) = (0.29\,\text{m}, -163.2°)$, as well as the corresponding reference signals, are shown in Figure 8.16. To suppress the strong antenna reflections, time clipping was used for the signals in the first three nanoseconds. In Figure 8.16, the full-polarized signals retrieved with the two rotated arrays have relatively good agreements with the HH-, HV-, and VV-polarized reference signals in terms of the wavelets. Focusing both the retrieved and reference polarized signals, the reconstructed polarimetric images are presented in Figure 8.17. It can be seen that the major features of the target are well reconstructed with all the three antenna arrays. The corresponding polarimetric images are in good agreement. Specifically, the polarization dependence of the horizontal and vertical part of the "L"-shaped target is clearly visible in the HH- and VV-polarized images, respectively. Meanwhile, the edges of the target can be perceived in the HV images.

However, some discrepancies are observed between the amplitudes of the retrieved and reference signals in Figure 8.16 as well as between the images in Figure 8.17. In particular, relatively larger discrepancies between the cross-polarized images [i.e., Figure 8.17(b), (e), and (h)] obtained with the traditional arrays and the

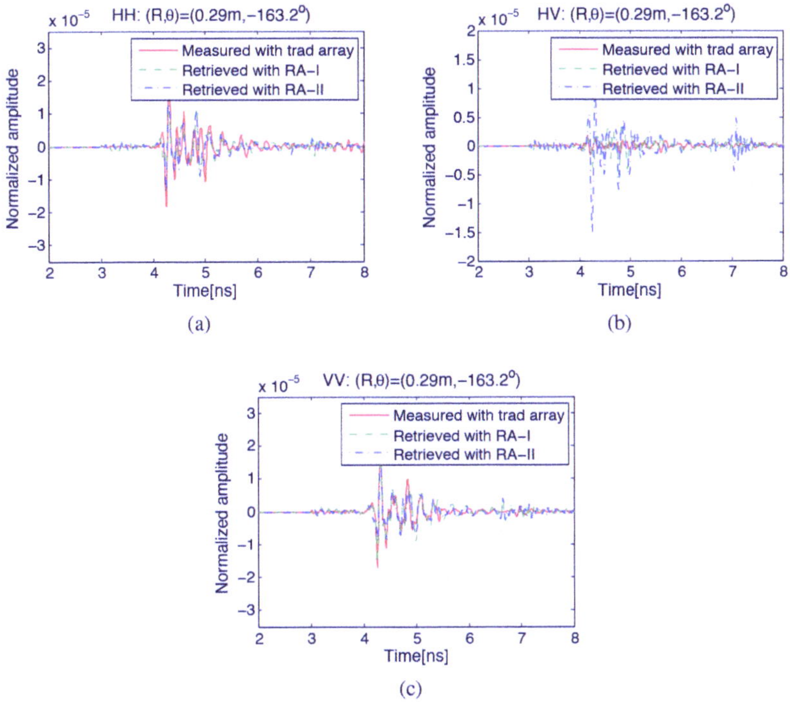

Figure 8.16 Comparison of the retrieved polarimetric signals and the reference ones at the position $(\mathrm{R}, \theta) = (0.29\,\mathrm{m}, -163.2^\circ)$*. (a) HH-polarized signals, (b) HV-polarized signals, and (c) VV-polarized signals. The scattered signal spectrum was normalized with respect to the transmitted one.*

rotated arrays are seen compared to that between the co-polarized images. As the cross-polarized signals acquired with traditional full-polarized arrays are measured with slightly separated transmitting and receiving antenna pairs (i.e., bistatic configuration), then not only the edges and corners of the target but also the flat surfaces exhibit high intensity in the reference HV image [Figure 8.17(b)]. By contrast, the HV images obtained with the two rotated arrays demonstrate the cross-polarization features of the target as theoretically expected, where the edges and corners are mainly reconstructed. So, the two rotated arrays enable to effectively reconstruct both the co- and cross-polarized images of the target. Furthermore, small differences between the HV images acquired with the two rotated arrays [Figure 8.17(e) and (h)] are also noticed. These differences might be induced by the non-purity of the linear polarization of the antipodal Vivaldi antenna, especially for high frequencies. Due to this non-purity of the antenna polarization, the HV-polarized signal retrieved with the rotated array II may contain some co-polarized components, which leads to its larger amplitude in Figure 8.16(b) (compared to that retrieved with

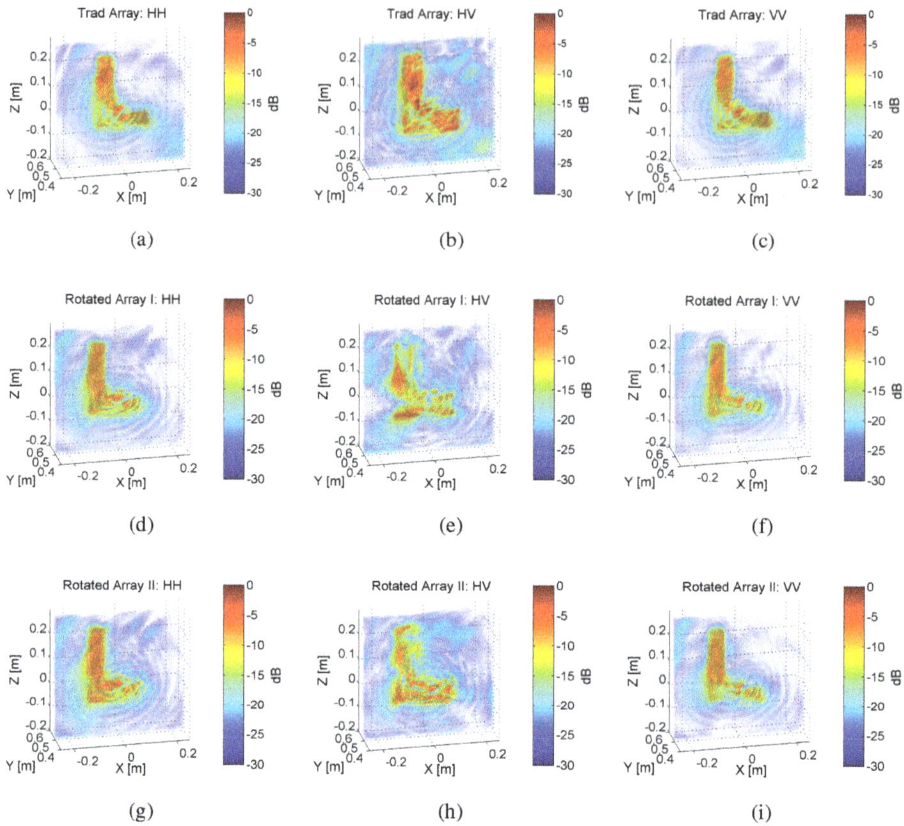

Figure 8.17 The HH, HV, and VV polarimetric images obtained with (a)–(c) traditional array, (d)–(f) rotated array I, and (g)–(i) rotated array II

rotated array I) and partially reconstructed flat surface in Figure 8.17(h) [in contrast to Figure 8.17(e)].

To clearly illustrate different polarization features of images in Figure 8.17, the color-coded slice images at the target position are shown in Figure 8.18. The horizontal and vertical bars are displayed in red and pink while the edges of the targets are represented in green where the cross-polarization effect induces the HV polarized signals. According to Figure 8.18, the pseudocolor images obtained with the rotated arrays reveal similar polarimetric features of the targets as that acquired with the traditional full-polarized imaging arrays. Hence, the two rotated arrays provide comparable imaging performance as the traditional polarimetric arrays.

Using the same experimental setups for both the rotated and traditional polarimetric arrays, a TCR was used as a strong scattering object for a second experimental measurement. The edges of the TCR are about 19.7 cm in length, as shown in Figure 8.15(e). The TCR was placed in front of the antenna aperture at a distance of

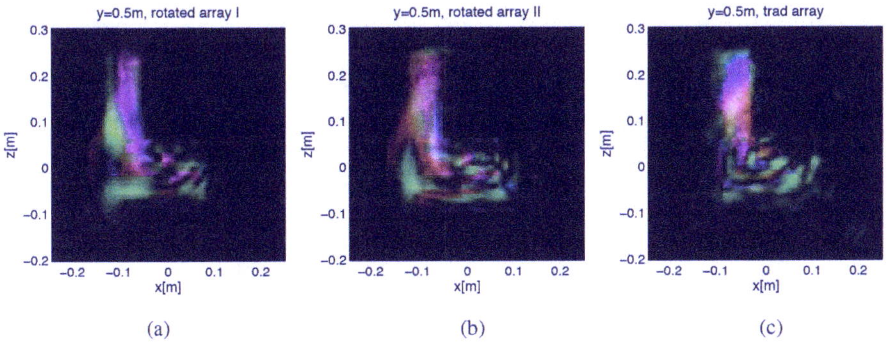

Figure 8.18 Pseudocolor slice images at the target position obtained with (a) rotated array I, (b) rotated array II, and (c) traditional array with dynamic range of 15 dB. Lexicographic color coding is used for visualization (HH: red; HV: green, and VV: blue).

0.5 m. After performing the same experimental measurement operations and image formation as for the "L"-shaped object, the fully polarimetric images of the TCR with the three antenna arrays are obtained, which are shown in Figure 8.19.

It can be seen that both rotated antenna arrays reconstruct the comparable polarimetric features of the TCR in the corresponding images in contrast to the traditional arrays. However, some differences are still noticeable between the polarimetric images formed with the two rotated arrays. Visually, rotated array I achieves slightly better fully polarimetric images in terms of the similarity to those obtained with the traditional arrays. Specifically, in both HH- and VV-polarized images obtained with rotated array II the edges along the z-axis direction are much weaker than that in the corresponding images of rotated array I. Moreover, the cross-polarized image focused with rotated array II [Figure 8.19(e)] is slightly asymmetric with respect to the xoy plane compared to the one obtained with rotated array I [Figure 8.19(b)]. Meanwhile, the sidelobes/artifacts surrounding the images of the TCR seem slightly rotated. These may be caused by non-purity of the linear polarization of the antipodal Vivaldi antenna used for the measurement, as mentioned above. As a consequence, the cross-polarized signals were inaccurately estimated with the measurements of rotated array II. However, in rotated array I, the cross-polarized signals were measured with two orthogonally oriented Vivaldi antennas at each spatial position, which to some extent mitigates the effect of the non-purity of their linear polarization on the HV-polarized signal estimation.

Furthermore, although a well-reconstructed HV-polarized image is obtained with the traditional array through a quasi-monostatic measurement configuration, the HV-polarized image obtained with rotated array I [Figure 8.19(b)] reveals the slit induced by the intersecting side formed by the two top surfaces in a full agreement with the HV-polarized images in Figure 8.14 obtained in the numerical simulations. This demonstrates further the effectiveness and accuracy of rotated array I for fully polarimetric imaging and indicates the

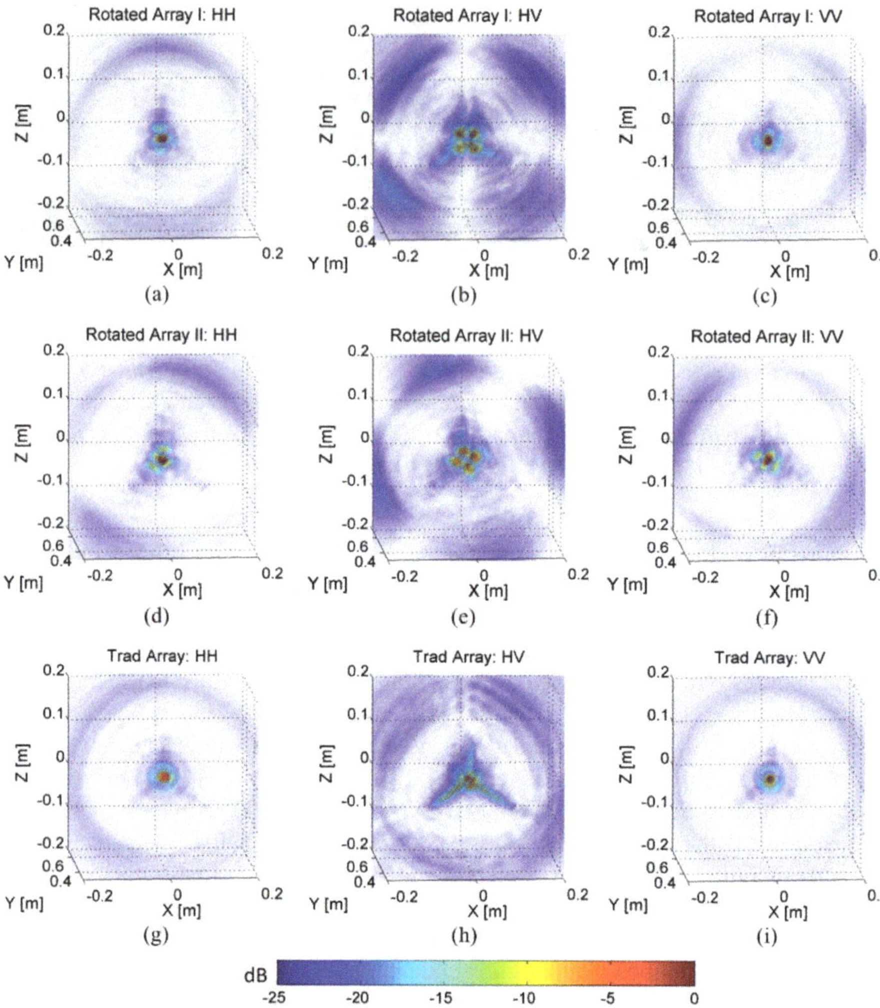

Figure 8.19 3D polarimetric images of a TCR obtained with the experimental measurements of the three antenna arrays. (a)–(c) are the HH-, HV-, and VV-polarized images obtained with rotated array I, (d)–(f) are obtained with rotated array II, and (g)–(i) are obtained with the traditional antenna arrays.

relatively higher tolerance of rotated array I to the quasi-monostatic measurement configuration for full-polarized imaging compared to the traditional fully polarimetric imaging arrays and rotated array II. This advantage could be very attractive for the implementation of practical full-polarized imaging systems, especially in the circumstances where antennas with relatively lower cross-polarized isolation are used.

8.6 Conclusions

While being found useful for target classification, UWB polarimetry is rarely used in imaging systems due to the complexity of full-polarimetric front-ends. In this chapter, a few efficient approaches to collect full polarimetric data by using conventional single-polarized antenna systems are presented. Accuracy of these approaches to reconstruct polarimetric scattering matrices have been investigated both in simulations and experiments. Furthermore, imaging algorithms that are using full-polarimetric measurement data are presented and advantages of full-polarimetric imaging are demonstrated.

References

[1] Wang J., Aubry P., Yarovoy A. 'A novel approach to full-polarimetric short-range imaging with copolarized data'. *IEEE Trans Antennas Propag.* 2016, vol. 64(11), pp. 4733–44.

[2] Wang J., Aubry P., Yarovoy A. 'A novel rotated antenna array topology for near-field 3-D fully polarimetric imaging'. *IEEE Trans Antennas Propag.* 2018, vol. 66(3), pp. 1584–89.

[3] Carin L., Geng N., McClure M., Sichina J., Lam Nguyen. 'Ultra-wide-band synthetic-aperture radar for mine-field detection'. *IEEE Antennas and Propagation Magazine.* 41 Feb.; 1999. pp. 18–33.

[4] Chi-ChihC.Higgins M.B., O'Neill K., Detsch R. 'Ultrawide-bandwidth fully-polarimetric ground penetrating radar classification of subsurface unexploded ordnance'. *IEEE Trans Geosci Remote Sens.* 2001, vol. 39(6), pp. 1221–30.

[5] Sabath F., Mokole E.L., Schenk U., Nitsch D. *Ultra-wideband, short-pulse electromagnetics 7*. New York, NY: Springer New York; 2007. pp. 655–64. Available from http://www.springerlink.com/index/10.1007/978-0-387-37731-5

[6] Dubey A.C., Stiles J.M., Parra-Bocaranda P, *et al.* 'Detection of object symmetry using bistatic and polarimetric GPR observations' in Dubey AC., Broach JT. (eds.). *AeroSense '99*. Bellingham, Washington: International Society for Optics and Photonics. SPIE; 1999. pp. 992–1002.

[7] Sagues L., Lopez-Sanchez J.M., Fortuny J., Fabregas X., Broquetas A., Sieber A.J. 'Polarimetric radar interferometry for improved mine detection and surface clutter rejection'. *IEEE Trans Geosci Remote Sens.* 2001, vol. 39(6), pp. 1271–78.

[8] Kovalenko V., Harmon R.S., Holloway J.H, *et al.* 'Fully polarimetric measurements over a minefieldlike test site with video impulse ground-penetrating radar' in Harmon RS., Holloway JH., Broach JT. (eds.). *AeroSense 2003*. Bellingham, Washington: Society of Photo-Optical Instrumentation Engineers (SPIE) Conference Series; 2003. pp. 448–56.

[9] Witten T.R., Dubey A.C., Harvey J.F., Broach J.T. 'Present state of the art in ground-penetrating radars for mine detection' in Dubey AC., Harvey JF.,

Broach JT. (eds.). *Aerospace/defense sensing and controls*. Bellingham, Washington: International Society for Optics and Photonics. SPIE; 1998. pp. 576–85.

[10] Dubey A.C., Kositsky J., Milanfar P., Harvey J.F., Broach J.T., Dugan R.E. 'Forward-looking high-resolution GPR system' in Dubey AC., Harvey JF., Broach JT. (eds.). *AeroSense '99*. Bellingham, Washington: International Society for Optics and Photonics. SPIE; 1999. pp. 1052–62.

[11] Sachs J. 'M-sequence ultra-wideband-radar: state of development and applications'. *Proceedings of the International Conference on Radar (IEEE Cat. No.03EX695)*; Adelaide, Australia, 2003. pp. 224–29.

[12] Yarovoy A.G., Kovalenko V., Bloemenkamp R.F. 'Multiwaveform full-polarimetric GPR sensor for landmine detection: first experimental results'. *EUDEM2-SCOT – 2003, International Conference on Requirements and Technologies for the Detection, Removal and Neutralization of Landmines and UXO*; Brussels, Belgium, 2003. pp. 554–560.

[13] Chignell R.J., Dabis H., Frost N. 'Radar requirements for detecting anti-personnel mines'. *Eighth International Conference on Ground Penetrating Radar*; Gold Coast, Australia, 2000. pp. 861–66.

[14] Sheen D.M., McMakin D.L., Hall T.E. 'Three-dimensional millimeter-wave imaging for concealed weapon detection'. *IEEE Transactions on Microwave Theory and Techniques*. 2001, vol. 49(9), pp. 1581–92.

[15] Zhuge X., Yarovoy A.G. 'A sparse aperture MIMO-SAR-based UWB imaging system for concealed weapon detection'. *IEEE Trans Geosci Remote Sens*. 2011, vol. 49(1), pp. 509–18.

[16] Kruk J., Wapenaar C.P.A., Fokkema J.T., van den Berg P.M. 'Three-dimensional imaging of multicomponent ground-penetrating radar data'. *GEOPHYSICS*. 2003, vol. 68(4), pp. 1241–54.

[17] Zhang W., Hoorfar A. 'Three-dimensional real-time through-the-wall radar imaging with diffraction tomographic algorithm'. *IEEE Trans Geosci Remote Sens*. 2013, vol. 51(7), pp. 4155–63.

[18] Fear E.C., Li X., Hagness S.C., Stuchly M.A. 'Confocal microwave imaging for breast cancer detection: localization of tumors in three dimensions'. *IEEE Transactions on Bio-Medical Engineering*. 2002, vol. 49(8), pp. 812–22.

[19] Sheen D., McMakin D., Hall T. 'Near-field three-dimensional radar imaging techniques and applications'. *Applied Optics*. 2010, vol. 49(19), pp. E83–93.

[20] Klemm M., Craddock I.J., Leendertz J.A., Preece A., Benjamin R. 'Radar-based breast cancer detection using a hemispherical antenna array—experimental results'. *IEEE Transactions on Antennas and Propagation*. 2009, vol. 57(6), pp. 1692–704.

[21] Sato M., Feng X., Kobayashi T., Zhou Z.-S., Savelyev T.G., Fujiwara J. 'Development of an array-antenna GPR system (SAR-GPR)'. *Proc. SPIE 5794, Detection and Remediation Technologies for Mines and Minelike Targets X, (10 June)*; 2005.

[22] Simi A., Manacorda G., Miniati M., Bracciali S., Buonaccorsi A. 'Underground asset mapping with dual-frequency dual-polarized GPR massive array'. *13th*

International Conference on Ground Penetrating Radar; Lecce, Italy, 2010. pp. 1–5.

[23] Wang J., Cetinkaya H., Yarovoy A. 'On polar sampling of GPR for tunneling boring machine'. *Proceedings of the 15th International Conference on Ground Penetrating Radar*; Brussels, Belgium, 30 Jun-04 Jul; 2014. pp. 330–3.

[24] Schwartz J.L., Steinberg B.D. 'Ultrasparse, ultrawideband arrays'. *IEEE Transactions on Ultrasonics, Ferroelectrics and Frequency Control*. 1998, vol. 45(2), pp. 376–93.

[25] Wang J., Yarovoy A. 'Sampling design of synthetic volume arrays for three-dimensional microwave imaging'. *IEEE Transactions on Computational Imaging*. 2018, vol. 4(4), pp. 648–60.

[26] Wang J., Aubry P., Yarovoy A. 'Efficient implementation of GPR data inversion in case of spatially varying antenna polarizations'. *IEEE Transactions on Geoscience and Remote Sensing*. 2018, vol. 56(4), pp. 2387–96.

[27] Wiggins J.W. 'Kirchhoff integral extrapolation and migration of nonplanar data'. *GEOPHYSICS*. 1984, vol. 49(8), pp. 1239–48.

[28] Zhuge X., Yarovoy A. 'Design of low profile antipodal Vivaldi antenna for ultra-wideband near-field imaging'. *Proceedings of the Fourth European Conference on Antennas and Propagation*; Barcelona, Spain, Apr; 2010. pp. 1–5.

Chapter 9

Robust transceiver design for polarimetric radars

Xu Cheng[1], Augusto Aubry[2], Domenico Ciuonzo[2], Antonio De Maio[2], and Alfonso Farina[3]

9.1 Introduction

Polarization, together with the amplitude, time, frequency, phase, and bearing descriptors of radar signals, completes the information that can be obtained on target returns in modern radars [1]. Radar polarization exploits the vector nature of electromagnetic fields and has become an indispensable tool in radar systems. Since its birth in the 1950s, polarimetric radar has played an important role in military and civilian areas including air defense, remote sensing, weather surveillance, etc. [2–5]. Not surprisingly, in the last decades, the exploitation of information on the echo polarization state to improve radar performance has represented a research topic of remarkable interest [6–15].

On the other hand, waveform diversity represents another valuable degree of freedom to boost sensing systems. It refers to the radar transmitter adaptation in several domains, such as spatial, temporal, spectral, and also recently polarimetric, aimed at dynamically optimizing the radar performance to specific scenarios and tasks. This advanced and powerful feature is enabled by the availability of new computing architectures, high-speed and off-the-shelf processors, arbitrary digital waveform generators, solid-state transmitters [16], and provides unique capabilities to reliably accomplish some critical tasks, such as target detection, tracking, and classification. Not surprisingly, several studies concerning radar transmit waveform design (jointly with bespoke receive processing architectures) appeared over the past decades [17–23], mainly with reference to single polarization radars.

According to the above considerations, it appears evident that the joint exploitation of polarization and waveform diversity may provide a viable means to improve

[1]School of Electronics and Communication Engineering, Sun Yat-Sen University, China
[2]Dipartimento di Ingegneria Elettrica e Tecnologie dell'Informazione, Università degli Studi di Napoli "Federico II" Napoli, Italy
[3]Selex ES (retired), Italy

radar performance, especially in some challenging scenarios such as the detection of stealthy targets, as well as the identification of static or slow-moving targets buried in heavy clutter. In this context, the work in [24] represents a milestone, addressing for the first time the joint optimization of transmit waveform and receive filter of polarimetric radars to optimize the signal-to-interference plus noise ratio (SINR). Therein, it is assumed the exact knowledge of the full-polarization scattering behavior of the (extended) target, characterized in terms of target impulse response matrix (TIRM), whose formal definition is reported in main body of the chapter. As a consequence, the analysis conducted in [24] provides insights on the achievable SINR benefits resulting from the exploitation of full-polarization matched-illumination. However, the developed iterative technique is not guaranteed to converge to an optimal solution. Therefore, in [25], with reference to a more general sensing system, an improved algorithm, which ensures a non-decreasing SINR along with the iterations and the convergence to local maximum SINR value, is proposed. In particular, although the technique in [25] is devised for a multiple-input multiple-output radar, it can be readily applied to the full-polarimetric radar model in [24] by setting both the numbers of transmit and receive channels equal to two and suitably specializing the system parameters. However, the works in [24, 25] are based on the following *restrictive assumptions:*

- the TIRM is assumed to be exactly known;
- only an energy constraint is accounted for at the design of the radar illumination;
- the clutter is modeled as range-bin independent with just possible polarimetric correlation at each cell.

Indeed, due to the high sensitivity of the TIRM to the actual target aspect angle (TAA), supposing its exact knowledge at the synthesis stage is somewhat overoptimistic, and a reliable design should endow robustness against unavoidable uncertainties. Additionally, it is practically appealing to enforce constraints on the shape of the transmit waveform, in addition to an upper bound on the available energy, in order to control some relevant hallmarks of the signal, including range-Doppler resolution and peak-to-sidelobe level (PSL) [20]. Finally, with a high range resolution (HRR) radar, statistical independence of the clutter returns among the range cells may be no longer a reasonable assumption.

Aimed at overcoming the above shortcomings, in this chapter, a new framework to jointly optimize the transmit waveform and the receive filter of a fully polarimetric radar system is detailed. Specifically, building upon the extended target model in signal-dependent interference [26], the focus is on a less restrictive formulation than [24] and [25]. Indeed, although the exact TAA is unknown, it is assumed to lie in a given uncertainty set. A filter bank*, with each branch tuned to a specific TAA (viz. to a specific TIRM), as opposed to a single filter, is

*A similar situation occurs with reference to spatial processing applications making use of stacked beams to handle angle of arrival uncertainties.

employed and the worst case SINR at the output of the filter bank, is considered as performance metric. This choice naturally ensures performance robustness with respect to TAA uncertainty. Second, along with a restriction on the available transmit energy, a similarity constraint is forced to bestow some desirable attributes (e.g. Doppler tolerance, integrated sidelobe level, PSL) to the radar probing waveform. Finally, clutter correlation in both polarization and range domains is assumed.

The resulting design problem belongs to the challenging class of non-convex max–min optimization problems quite hard to solve. Thus, an equivalent formulation of the original design problem is provided, which paves the way to the development of a novel iterative optimization procedure monotonically improving the worst-case SINR. Each iteration of the algorithm, whose convergence to a stationary point is analytically proved, requires the solution of both a convex and a max–min fractional problem. The proposed approach is also able (by properly selecting the number of iterations according to a prescribed computational time limit) to gracefully trade off SINR improvement for complexity, with the starting solution coinciding with traditional code/radar filter bank processing. At the analysis stage, the effectiveness of the proposed full-polarization transmit code and receive filter bank technique is assessed by testing its performance on an experimental data set of a T-72 tank [27] under (signal-dependent) clutter and (signal-independent) noise background. The comparison between this novel algorithm and its single-polarization counterpart as well as the (full-polarimetric) methods in [24, 25] is then carried out. The reported case studies illustrate the advantages of the proposed polarimetric robust design.

The remainder of the chapter is organized as follows. In section 9.2, the system model as well as the description of the full-polarization target response and the statistical characterization of the interference are introduced. In section 9.3, the constrained optimization problem for the joint robust design of the radar waveform and the receive filter bank is formulated. Then, in section 9.4 the iterative optimization procedure able to synthesize optimized waveform and filter bank structure is developed, whereas experimental results are reported and discussed in section 9.5. Finally, concluding remarks and open issues are provided in section 9.6.

9.2 System model

In the following, it is assumed that a full-polarimetric HRR radar is employed to perform environment sensing activities. Nowadays, with the development of broadband microwave devices, HRR systems are widely applied in civil and military contexts. Thus, the overall scattering behavior of a prospective target is completely specified by its TIRM. Without the loss of generality, a monostatic configuration is considered in this chapter.

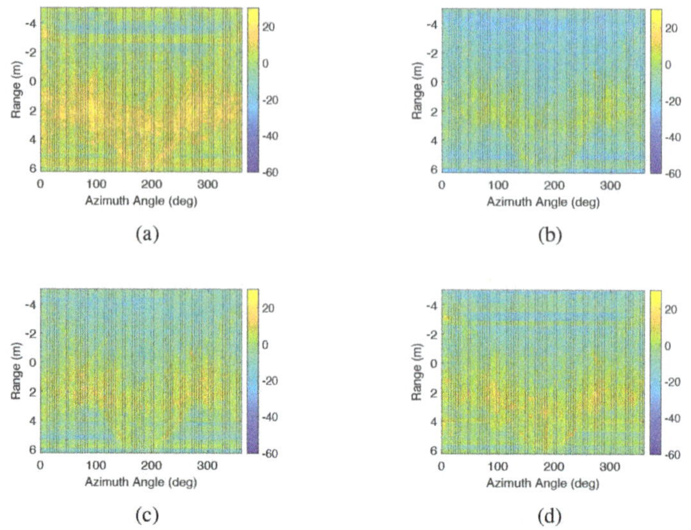

Figure 9.1 *The target impulse response (dB scale) of (a) HH, (b) HV, (c) VH and (d) VV components for a T-72 tank in [27], covering the full range of azimuth [0°, 360°] and an elevation equal to 29.28°, where "XY" channel refers to the signal received on the "X" polarization channel due to transmission on the "Y" polarization channel*

9.2.1 Target model

It is well known that the TIRM is strongly dependent on the TAA [1, 28], namely, the target scattering behavior is sensitive to even a small change of the TAA†. As an example, Figure 9.1 shows the TIRM of a T-72 tank with the azimuth angle ranging within [0°, 360°], at a fixed elevation angle of 29.28°. The subplots in Figure 9.1a,b,c,d depict HH, HV, VH, and VV components of the target response, respectively. Clearly, these subplots emphasize that the target response of all the polarization channels considerably changes with the azimuth angle. Therefore, accounting for TIRM variability with respect to the TAA represents the key point for an effective transmit-receive design. As a consequence, the dependence of the TIRM on the TAA is explicitly reported hereafter.

9.2.2 Signal model

Let us consider a radar system that transmits a fully polarimetric waveform, and is denoted in

†For notational simplicity, the focus is on a two-dimensional sensing scenario. Nonetheless, the following framework can be readily generalized to a three-dimensional scenario, with the TIRM being dependent on both elevation and azimuth angles.

$$\mathbf{s} \triangleq \left[\left(s_{H,0}, s_{V,0}\right), \ldots, \left(s_{H,N-1}, s_{V,N-1}\right)\right]^T \in \mathbb{C}^{2N} \tag{9.1}$$

the fast time transmit code in an interleaved form, where the subscripts H and V indicate the horizontal and vertical polarization components, respectively‡. The deinterleaved counterpart to (9.1) is given as

$$\mathbf{s}_d \triangleq \left[s_{H,0}, \ldots, s_{H,N-1}, s_{V,0}, \ldots, s_{V,N-1}\right]^T \tag{9.2}$$

Evidently, by means of a simple interleaving/deinterleaving operation, $\mathbf{s}$ and $\mathbf{s}_d$ can be easily obtained one from the other.

Note that the higher the time-bandwidth product of the transmitted waveform, the higher the N. Let Q be an upper bound to the size support of the TIRM (with respect to the TAA θ) and the observations be collected over a discrete interval M, where $M = Q + N - 1$ so that the most of the extended target echoes is received. For a certain TAA θ and range bin n, the full-polarization target scattering matrix $\mathbf{T}_n(\theta) \in \mathbb{C}^{2\times 2}$, corresponding to a specific slice of the TIRM $\mathbf{T}(\theta)$, is expressed as [5]:

$$\mathbf{T}_n(\theta) \triangleq \begin{bmatrix} T_{HH,n}(\theta) & T_{HV,n}(\theta) \\ T_{VH,n}(\theta) & T_{VV,n}(\theta) \end{bmatrix}, \quad \theta \in [0, 2\pi), n \in \{0, 1, \ldots, Q-1\} \tag{9.3}$$

where $T_{XY,n}(\theta)$ denotes the scattering coefficient measured (at the nth range bin) on Y receive polarization channel due to transmission on X polarization, when the target exhibits an aspect angle θ. In this chapter, it is supposed that the parameter Q is known, as it can be determined from the maximum target length and radar range resolution, i.e., $\mathbf{T}_n(\theta) \approx \mathbf{0}$ unless $n \in \{0, 1, \ldots, Q-1\}$. Similarly, the clutter scattering matrix $\mathbf{C}_n \in \mathbb{C}^{2\times 2}$ associated with the nth range bin, corresponding to a specific slice of the clutter impulse response matrix (CIRM) $\mathbf{C}$, is defined as

$$\mathbf{C}_n \triangleq \begin{bmatrix} C_{HH,n} & C_{HV,n} \\ C_{VH,n} & C_{VV,n} \end{bmatrix}, \quad n \in \{-N+1, -N+2, \ldots, M-1\} \tag{9.4}$$

After stacking all the observations into a single vector:

$$\mathbf{r} \triangleq \left[\left(r_{H,0}, r_{V,0}\right), \ldots, \left(r_{H,M-1}, r_{V,M-1}\right)\right]^T \in \mathbb{C}^{2M} \tag{9.5}$$

the signal model in compact vectorial form can be cast as

$$\mathbf{r} = \alpha_T \mathbf{T}(\theta)\mathbf{s} + \mathbf{c} + \mathbf{v} = \alpha_T \mathbf{T}(\theta)\mathbf{s} + \mathbf{C}_s + \mathbf{v} \tag{9.6}$$

where α_T is a complex parameter that accounts for radar equation and propagation effects, while $\mathbf{v}$ is the $2M$-dimensional additive interference:

‡This representation is fully general, since any circularly polarized or elliptically polarized waveform can be expressed as a linear combination of two orthogonal linear polarizations [1].

$$\boldsymbol{v} \triangleq \left[\left(v_{H,0}, v_{V,0}\right), \ldots, \left(v_{H,M-1}, v_{V,M-1}\right)\right]^T \in \mathbb{C}^{2M} \tag{9.7}$$

The matrices:

$$\boldsymbol{T}(\theta) \triangleq \sum_{n=0}^{Q-1} \boldsymbol{J}_n \otimes \boldsymbol{T}_n(\theta) \tag{9.8}$$

and

$$\boldsymbol{C} \triangleq \sum_{n=-N+1}^{M-1} \boldsymbol{J}_n \otimes \boldsymbol{C}_n \tag{9.9}$$

are the $2M \times 2N$-dimensional TIRM and CIRM, respectively, with $\boldsymbol{J}_n$ being the $M \times N$-dimensional shift matrix defined as

$$\boldsymbol{J}_n(\ell_1, \ell_2) \triangleq \begin{cases} 1 & \text{if } \ell_1 - \ell_2 = n \\ 0 & \text{if } \ell_1 - \ell_2 \neq n \end{cases}, \quad \ell_1 \in \{1, \ldots, M\}, \quad \ell_2 \in \{1, \ldots, N\} \tag{9.10}$$

Note that due to the convolution operation involved in the clutter return, the first clutter patch affecting the useful return is located at the $(N-1)$th bin before to the first target scatterer, while the last clutter patch is positioned at the $(M-N)$th bin and then the last target scatter is included, which motivates range indices of $\boldsymbol{C}_n$ in (9.9) start from $-N+1$ and end at $M-1$.

Compared with the **TIR** of a single-polarization measurement in [29], the full-polarization **TIRM** has a similar form except for the fact that the scalar scattering coefficients are replaced by 2×2 matrices and the presence of the Kronecker product operation. Furthermore, $\boldsymbol{T}(\theta)$ can also be written in an extended (viz. explicit) form as the block-Toeplitz matrix (assuming for simplicity $Q \geq N$):

$$\boldsymbol{T}(\theta) = \begin{bmatrix} \boldsymbol{T}_0(\theta) & \mathbf{0} & \cdots & \mathbf{0} \\ \boldsymbol{T}_1(\theta) & \boldsymbol{T}_0(\theta) & \cdots & \mathbf{0} \\ \vdots & \vdots & \ddots & \vdots \\ \boldsymbol{T}_{N-1}(\theta) & \boldsymbol{T}_{N-2}(\theta) & \cdots & \boldsymbol{T}_0(\theta) \\ \vdots & \vdots & \ddots & \vdots \\ \boldsymbol{T}_{Q-1}(\theta) & \boldsymbol{T}_{Q-2}(\theta) & \cdots & \boldsymbol{T}_{Q-N}(\theta) \\ \mathbf{0} & \boldsymbol{T}_{Q-1}(\theta) & \cdots & \boldsymbol{T}_{Q-N+1}(\theta) \\ \vdots & \vdots & \ddots & \vdots \\ \mathbf{0} & \mathbf{0} & \cdots & \boldsymbol{T}_{Q-1}(\theta) \end{bmatrix} \tag{9.11}$$

Similarly, the CIRM $\boldsymbol{C}$ is expressed as the block-Toeplitz matrix:

$$C = \begin{bmatrix} C_0 & C_{-1} & \dots & C_{-N+1} \\ C_1 & C_0 & \dots & C_{-N+2} \\ \vdots & \vdots & \ddots & \vdots \\ C_{N-1} & C_{N-2} & \dots & C_0 \\ C_N & C_{N-1} & \dots & C_1 \\ \vdots & \vdots & \ddots & \vdots \\ C_{M-1} & C_{M-2} & \dots & C_{M-N} \end{bmatrix} \tag{9.12}$$

As to the statistical characterization of the noise vector $\boldsymbol{v}$, it is assumed that it is a zero-mean (i.e. $\mathbb{E}[\boldsymbol{v}] = \boldsymbol{0}$), circularly symmetric random vector with covariance matrix $\boldsymbol{\Sigma}_v = \mathbb{E}[\boldsymbol{v}\boldsymbol{v}^\dagger] = \sigma_v^2 \boldsymbol{I}$, with σ_v^2 being the mean square value of each filtered noise sample.

As to the clutter statistical characterization, due to the reciprocity property involved in the monostatic radar configuration, it is supposed that $C_{HV,n} = C_{VH,n}$, $n \in \{-N+1, \dots, M-1\}$. In the following, no (restrictive) assumptions on the multivariate statistical characterization of c are made. In particular, it is reasonable to assume that the clutter scatterer in the nth range bin provides a zero-mean contribution, i.e., $\mathbb{E}[\boldsymbol{C}_n] = \boldsymbol{0}$. This straightforwardly implies that $\mathbb{E}[\boldsymbol{C}] = \boldsymbol{0}$. Unlike [24] and [25], the clutter correlation in both polarization and range domains is taken into account. In this respect, the normalized correlation parameter between the clutter scatterers at nth and n'th range bin with reference to the j polarization channel and the k channel, with $r_{jk}(n, n')$ ($j, k \in \{HH, HV, VV\}$) given in

$$r_{jk}(n, n') \triangleq \frac{\mathbb{E}[C_{j,n} C^*_{k,n'}]}{\sqrt{\mathbb{E}[|C_{j,n}|^2]}\sqrt{\mathbb{E}[|C_{k,n'}|^2]}} \tag{9.13}$$

is introduced. As to the polarization correlation of the clutter, the polarimetric clutter model in [30] is exploited. Also, note that matrix $\boldsymbol{C}$ can be cast in the following form:

$$\boldsymbol{C} = \sum_{n=-N+1}^{M-1} \boldsymbol{J}_n \otimes \boldsymbol{C}_n = \sum_{n=-N+1}^{M-1} \left[C_{HH,n}\boldsymbol{J}_n \otimes \boldsymbol{A}_1 + C_{HV,n}\boldsymbol{J}_n \otimes \boldsymbol{A}_2 + C_{VV,n}\boldsymbol{J}_n \otimes \boldsymbol{A}_3 \right] \tag{9.14}$$

where $\boldsymbol{A}_1 \triangleq \begin{bmatrix} 1 & 0 \\ 0 & 0 \end{bmatrix}$, $\boldsymbol{A}_2 \triangleq \begin{bmatrix} 0 & 1 \\ 1 & 0 \end{bmatrix}$, and $\boldsymbol{A}_3 \triangleq \begin{bmatrix} 0 & 0 \\ 0 & 1 \end{bmatrix}$. Let us also define the compact index notation $\{1, 2, 3\} = \{HH, HV, VV\}$. Hence, after some algebraic manipulations, the $2M \times 2M$ clutter covariance matrix can be expressed as

$$
\begin{aligned}
\boldsymbol{\Sigma}_c(\boldsymbol{s}) &= \mathbb{E}[\boldsymbol{c}\boldsymbol{c}^\dagger] = \mathbb{E}[\boldsymbol{C}\boldsymbol{s}\boldsymbol{s}^\dagger\boldsymbol{C}^\dagger] \\
&= \sum_{n=-N+1}^{M-1}\sum_{n'=-N+1}^{M-1} \Big[\sqrt{\sigma_n}\sqrt{\sigma_{n'}}r_{11}(n,n')\left(\boldsymbol{J}_n \otimes \boldsymbol{A}_1\right)\boldsymbol{s}\boldsymbol{s}^\dagger\left(\boldsymbol{J}_{n'}^T \otimes \boldsymbol{A}_1\right) \\
&+\sqrt{\varepsilon_n\sigma_n}\sqrt{\varepsilon_{n'}\sigma_{n'}}r_{22}(n,n')\left(\boldsymbol{J}_n \otimes \boldsymbol{A}_2\right)\boldsymbol{s}\boldsymbol{s}^\dagger\left(\boldsymbol{J}_{n'}^T \otimes \boldsymbol{A}_2\right) \\
&+\sqrt{\sigma_n}\sqrt{\sigma_{n'}\chi_{n'}}r_{13}(n,n')\left(\boldsymbol{J}_n \otimes \boldsymbol{A}_1\right)\boldsymbol{s}\boldsymbol{s}^\dagger\left(\boldsymbol{J}_{n'}^T \otimes \boldsymbol{A}_3\right) \\
&+\sqrt{\sigma_n\chi_n}\sqrt{\sigma_{n'}}r_{13}^*(n',n)\left(\boldsymbol{J}_n \otimes \boldsymbol{A}_3\right)\boldsymbol{s}\boldsymbol{s}^\dagger\left(\boldsymbol{J}_{n'}^T \otimes \boldsymbol{A}_1\right) \\
&+\sqrt{\sigma_n\chi_n}\sqrt{\sigma_{n'}\chi_{n'}}r_{33}(n,n')\left(\boldsymbol{J}_n \otimes \boldsymbol{A}_3\right)\boldsymbol{s}\boldsymbol{s}^\dagger\left(\boldsymbol{J}_{n'}^T \otimes \boldsymbol{A}_3\right)\Big]
\end{aligned}
\tag{9.15}
$$

where

$$
\sigma_n \triangleq \mathbb{E}[|C_{HH,n}|^2],\ \varepsilon_n \triangleq \frac{\mathbb{E}[|C_{HV,n}|^2]}{\mathbb{E}[|C_{HH,n}|^2]},\ \chi_n \triangleq \frac{\mathbb{E}[|C_{VV,n}|^2]}{\mathbb{E}[|C_{HH,n}|^2]}
\tag{9.16}
$$

It is worth pointing out that the expression (9.15) relies on the assumption that the copolarized components are independent of the cross-polarized (physically grounded) components, namely:

$$
\mathbb{E}\left[C_{HH,n}C_{HV,n'}^*\right] = \mathbb{E}\left[C_{HV,n}C_{HH,n'}^*\right] = \mathbb{E}\left[C_{HV,n}C_{VV,n'}^*\right] = \mathbb{E}\left[C_{VV,n}C_{HV,n'}^*\right] = 0
$$

Indeed, as stated in [31], this is reasonable for clutter, especially arising from rough surfaces or a random medium model. To characterize the correlation among the *HH* and *VV* channels, it is also introduced the coefficient ρ_n defined as

$$
\rho_n \triangleq \frac{\mathbb{E}[C_{HH,n}C_{VV,n}^*]}{\left[\mathbb{E}[|C_{HH,n}|^2]\mathbb{E}[|C_{VV,n}|^2]\right]^{\frac{1}{2}}}
\tag{9.17}
$$

which allows to specify $r_{13}(\cdot,\cdot)$ according to the parametric model reported in [32]. As a consequence, in line with [32], the clutter covariance is specified by parameters[§] $\{(\sigma_n, \epsilon_n, \chi_n, \rho_n),\ n = -N+1, \ldots, M-1\}$ and normalized correlations $r_{jk}(\cdot,\cdot)$.

Finally, clutter and noise are modeled as uncorrelated random quantities.

9.3 Problem formulation

Note that the received vector $\boldsymbol{r}$ of length $2M$ includes both the target echo term $\alpha_T\boldsymbol{T}(\theta)\boldsymbol{s}$ and the undesirable contributions due to the system noise v and (signal dependent) clutter return $\boldsymbol{C}\boldsymbol{s}$. Therefore, let the observation vector be filtered through $\boldsymbol{w} \triangleq [w(0), \ldots, w(2M-1)]^T \in \mathcal{C}^{2M}$, then the system output can be written as

$$
y = \boldsymbol{w}^\dagger\boldsymbol{r} = \alpha_T\boldsymbol{w}^\dagger\boldsymbol{T}(\theta)\boldsymbol{s} + \boldsymbol{w}^\dagger\boldsymbol{C}\boldsymbol{s} + \boldsymbol{w}^\dagger\boldsymbol{v}
\tag{9.18}
$$

[§]In [32], a set of typical values of σ, ε, χ, and ρ for trees, grass, mixed scrub, and shadows clutter have been reported.

with the corresponding output SINR:

$$\mathrm{SINR}_\theta(\boldsymbol{s}, \boldsymbol{w}) \triangleq \frac{|\alpha_T|^2 |\boldsymbol{w}^\dagger \boldsymbol{T}(\theta)\boldsymbol{s}|^2}{\mathbb{E}[|\boldsymbol{w}^\dagger \boldsymbol{C}\boldsymbol{s}|^2] + \mathbb{E}[|\boldsymbol{w}^\dagger \boldsymbol{v}|^2]} \tag{9.19}$$

Before proceeding further, an equivalent expression for the SINR in (9.19) is introduced, as summarized in the following proposition.

Proposition 9.3.1. The output SINR in (9.19) can be recast as

$$\mathrm{SINR}_\theta(\boldsymbol{s}, \boldsymbol{w}) = \frac{|\alpha_T|^2 |\boldsymbol{w}^\dagger \boldsymbol{T}(\theta)\boldsymbol{s}|^2}{\boldsymbol{w}^\dagger \left(\boldsymbol{\Sigma}_{\boldsymbol{C}}(\boldsymbol{s}) + \sigma_{\boldsymbol{V}}^2 \boldsymbol{I}\right) \boldsymbol{w}} \tag{9.20}$$

Proof. See Appendix A of [26] (for completeness it is also reported in Appendix 9.7.1 of this chapter).

Note that it is customary to assume that some relevant features of the illuminated target of interest can be obtained or roughly estimated by pre-scan procedures, or capitalizing on environmental cognition according to the recently emerging and powerful cognitive processing architecture [16, 20, 22]. For example, in the case of an already-tracked target, available features may include an accurate confidence on the type of target (e.g. obtained from a target classifier) and the potential range of TAA (e.g. obtained from a target tracker) [29]. This implies that the exact TAA is not known a priori, resulting in the lack of perfect knowledge of the TIRM. Therefore, in order to circumvent such an uncertainty on target scattering behavior, at the design stage, a complete data set of the actual TIRM (encompassing a set of relevant TAAs) is assumed available, and a bank of filters, each tuned to a specific TAA (viz. to a specific TIRM), is jointly optimized with polarimetric transmit code.

Precisely, let the vector r be processed through the array of filters¶ $\{\boldsymbol{w}_i\}_{i=1}^K$ each associated with a specific TAA θ_i, $i = 1, 2, \ldots, K$. Figure 9.2 illustrates a pictorial representation of the filter bank processing.

Then, the output SINR at the ith branch (SINR_i) $i = 1, 2, .., K$ can be expressed as

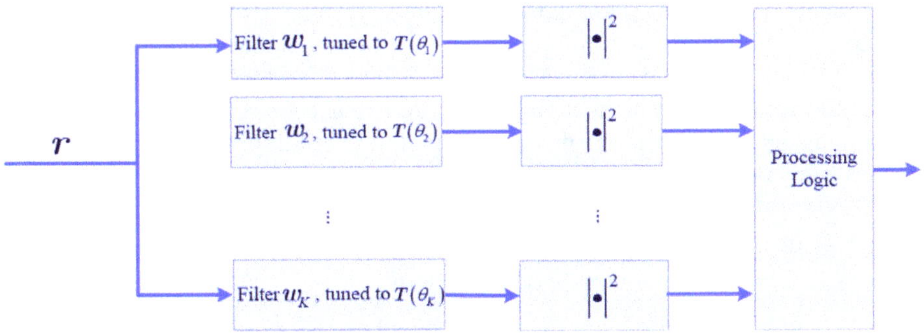

Figure 9.2 Notional representation of the filter bank processing

¶Obviously, it is assumed that $\boldsymbol{w}_i \neq \boldsymbol{0}$, $i = 1, 2, ..., K$.

$$\mathrm{SINR}_i(\boldsymbol{s}, \boldsymbol{w}_i) = \frac{|\alpha_T|^2 |\boldsymbol{w}_i^\dagger \boldsymbol{T}(\theta_i)\boldsymbol{s}|^2}{\boldsymbol{w}_i^\dagger \boldsymbol{\Sigma}_{\boldsymbol{C}}(\boldsymbol{s})\boldsymbol{w}_i + \sigma_V^2 \boldsymbol{w}_i^\dagger \boldsymbol{w}_i} \tag{9.21}$$

where the numerator is the useful energy at the output of ith filter with reference to the matched TAA θ_i, while the denominator denotes the signal-independent noise and the signal-dependent clutter energies at the output of the filter $\boldsymbol{w}_i$, respectively. Now, considering the worst-case SINR at the output of the filter array as the relevant performance metric, i.e.:

$$\mathrm{SINR} \triangleq \min_{i=1,2,\ldots,K} \mathrm{SINR}_i \tag{9.22}$$

a robust design with respect to the TAA uncertainty can be conceived. A few important considerations are now in order in (9.22). First, the considered filter-bank approach endows robustness to the receive end, since each receive filter may be optimized to its presumed TAA. Second, considering (9.22) as the relevant figure of merit allows for robustness also at the transmit stage, since the optimized waveform can avoid scenarios where non-satisfactory SINR values are observed (even though after matched processing), corresponding to less favorable TAA cases.

To make the proposed design practically appealing, some constraints on the probing signal need to be enforced. More specifically, the design is conducted under a finite transmit energy budget, i.e., $||\boldsymbol{s}||^2 = 1$. Moreover, to control some relevant features of the probing waveform, a similarity constraint is forced on the transmit sequence, namely:

$$||\boldsymbol{s} - \boldsymbol{s}_0||^2 \leq \gamma \tag{9.23}$$

where the parameter $0 \leq \gamma < 2$ rules the size of the trust hypervolume and $\boldsymbol{s}_0$ is a prefixed code ($||\boldsymbol{s}_0||^2 = 1$). The constraint (9.23) is motivated by the following reasons. An unconstrained optimization can lead to signals with significant modulus variations, poor range resolutions, high PSL, and more generally to a non-desirable ambiguity function response. These drawbacks can be partially circumvented by forcing the solution to be similar to a known code s_0, which shares some excellent properties such as constant modulus, reasonable range resolution, and PSL [20]. According to the previous considerations and guidelines, under such energy and similarity constraints, the optimization problem for the joint design of full-polarization radar transmit code and the receive filter bank can be formulated as

$$\mathcal{P} \begin{cases} \displaystyle\max_{\boldsymbol{s},\{\boldsymbol{w}_i\}_{i=1}^K} \; \min_{i=1,\ldots,K} \frac{|\boldsymbol{w}_i^\dagger \boldsymbol{T}(\theta_i)\boldsymbol{s}|^2}{\boldsymbol{w}_i^\dagger \boldsymbol{\Sigma}_c(\boldsymbol{s})\boldsymbol{w}_i + \sigma_v^2 \boldsymbol{w}_i^\dagger \boldsymbol{w}_i} \\ \text{s.t.} \quad ||\boldsymbol{s}||^2 = 1, \\ \qquad ||\boldsymbol{s} - \boldsymbol{s}_0||^2 \leq \gamma \end{cases} \tag{9.24}$$

where since the term $|\alpha_T|^2$ does not affect the optimization process to be handled, it is not included.

Inspection of (9.24) leads to an interesting consideration. In standard radar processing, when a fixed transmit waveform $\bar{s}$ is employed, the conventional matched filters are given by

$$w_{i,\mathrm{MF}} = \alpha T(\theta_i)\bar{s}, \quad i = 1, 2, \ldots, K \tag{9.25}$$

with α being an arbitrary constant. Substituting (9.25) into (9.21) and (9.22) yields the worst-case matched-filter SINR:

$$\mathrm{SINR}_{\mathrm{MF}} \triangleq \min_{i=1,\ldots,K} \mathrm{SINR}_i^{\mathrm{MF}} = \min_{i=1,\ldots,K} \frac{|\alpha_T|^2 |\bar{s}^\dagger T(\theta_i)^\dagger T(\theta_i)\bar{s}|^2}{\bar{s}^\dagger T(\theta_i)^\dagger \Sigma_c(\bar{s}) T(\theta_i)\bar{s} + \sigma_v^2 \| T(\theta_i)\bar{s}\|^2} \tag{9.26}$$

On the other hand, (9.21) can be upper bounded by

$$\mathrm{SINR}_{\mathrm{UB}} \triangleq \frac{|\alpha_T|^2 \max\limits_{i=1,\ldots,K} \|T(\theta_i)\|_2^2}{\min\limits_{s,\|s\|=1} \lambda_{\min(\Sigma_c(s))} + \sigma_v^2} \tag{9.27}$$

Thus, the following order relationship holds true:

$$\mathrm{SINR}_{\mathrm{UB}} \geq \max_{s,\{w_i\}_{i=1}^K} \mathrm{SINR} \geq \mathrm{SINR}_{\mathrm{MF}} \tag{9.28}$$

Therefore, Equation (9.28) defines the region of the potential SINR improvement achieved by the joint transmitter/receiver-optimized design with respect to the classic processing. Evidently, the width of the interval $[\mathrm{SINR}_{\mathrm{MF}}, \mathrm{SINR}_{\mathrm{UB}}]$ depends on the specific clutter environment (through the matrix $(\Sigma_c(s))$).

Before concluding this section, an alternative expression of the SINR, as summarized in the following proposition, is introduced. This objective function reformulation represents one of the ingredient for the development of an effective SINR optimization procedure.

Proposition 2. The objective function of $\mathscr{P}$ can be recast as

$$\max_{s,\{w_i\}_{i=1}^K} \min_{i=1,\ldots,K} \frac{|w_i^\dagger T(\theta_i)s|^2}{s^\dagger \left(\bar{\Sigma}_c(w_i) + \sigma_v^2 w_i^\dagger w_i I\right) s} \tag{9.29}$$

where

$$\begin{aligned}
\bar{\Sigma}_c(w_i) &= \mathbb{E}[cc^\dagger] = \mathbb{E}[C^\dagger w_i w_i^\dagger C] \\
&= \sum_{n=-N+1}^{M-1} \sum_{n'=-N+1}^{M-1} \Big[\sqrt{\sigma_n}\sqrt{\sigma_{n'}} r_{11}^*(n,n') \left(J_n^T \otimes A_1\right) \cdot \\
&\quad w_i w_i^\dagger \left(J_{n'} \otimes A_1\right) + \sqrt{\varepsilon_n \sigma_n}\sqrt{\varepsilon_{n'}\sigma_{n'}} r_{22}^*(n,n') \left(J_n^T \otimes A_2\right) \cdot \\
&\quad w_i w_i^\dagger \left(J_{n'} \otimes A_2\right) + \sqrt{\sigma_n}\sqrt{\sigma_{n'}\chi_{n'}} r_{13}^*(n,n') \left(J_n^T \otimes A_1\right) \cdot \\
&\quad w_i w_i^\dagger \left(J_{n'} \otimes A_3\right) + \sqrt{\sigma_n \chi_n}\sqrt{\sigma_{n'}} r_{13}(n',n) \left(J_n^T \otimes A_3\right) \cdot \\
&\quad w_i w_i^\dagger \left(J_{n'} \otimes A_1\right) + \sqrt{\sigma_n \chi_n}\sqrt{\sigma_{n'}\chi_{n'}} r_{33}^*(n,n') \left(J_n^T \otimes A_3\right) \cdot \\
&\quad w_i w_i^\dagger \left(J_{n'} \otimes A_3\right)\Big]
\end{aligned}$$

Proof. See Appendix B of [26] (for completeness it is also reported in Appendix 9.7.2 of this chapter).

9.4 Code and filter bank design

In this section, a solution technique, with some quality guarantees, is proposed to handle the non-convex optimization $\mathcal{P}$ (the objective is non-convex and $\|s\|^2 = 1$ defines a non-convex set). A first step toward such goal is represented by the following proposition that represents the main technical result of this chapter, from an optimization theory point of view. More specifically, the design $\mathcal{P}$ is cast as an equivalent optimization problem where, with respect to either the code or the filter vectors, the feasible set is convex and each objective function is the ratio between a bilinear function and a marginally convex function.

Proposition 3. Problem $\mathcal{P}$ is solvable** and, for any $0 \leq \gamma < 2$, can be equivalently recast†† as

$$\mathcal{P}_1 \begin{cases} \max\limits_{\boldsymbol{s},\{\boldsymbol{w}_i\}_{i=1}^K} & \min\limits_{i=1,\ldots,K} \dfrac{\mathrm{Re}\left(\boldsymbol{w}_i^\dagger \boldsymbol{T}(\theta_i)\boldsymbol{s}\right)}{\sqrt{\boldsymbol{w}_i^\dagger \boldsymbol{\Sigma}_c(\boldsymbol{s})\, \boldsymbol{w}_i + \sigma_v^2 \boldsymbol{w}_i^\dagger \boldsymbol{w}_i \|\boldsymbol{s}\|^2}} \\ s.t. & \|\boldsymbol{s}\|^2 \leq 1, \\ & \mathrm{Re}\left(\boldsymbol{s}_0^\dagger \boldsymbol{s}\right) \geq \gamma_1, \\ & \mathrm{Re}\left(\boldsymbol{w}_i^\dagger \boldsymbol{T}(\theta_i)\boldsymbol{s}\right) \geq 0, \quad i = 1,2,\ldots,K \end{cases}, \tag{9.30}$$

where $\gamma_1 = 1 - \frac{\gamma}{2} > 0$. Indeed, given an optimal solution to $\mathcal{P}$, it is possible to construct an optimal solution to $\mathcal{P}_1$, and vice versa.

Proof. The proof can be obtained similarly as that given in [23] (for completeness it is also reported in Appendix 9.7.3 of this chapter).

Evidently, $\mathcal{P}_1$ is still a non-convex optimization problem because the objective function is still a non-convex function. Nevertheless, as opposed to $\mathcal{P}$, the equivalent formulation provided by Proposition 3 shares a convex feasible set with a quasi-concave objective with respect to either the code or the filter bank. Hereinafter, aimed at finding an optimized solution to the aforementioned problem, an optimization procedure that iteratively optimizes

$$g(\boldsymbol{s}, \{\boldsymbol{w}_i\}_{i=1}^K) \triangleq \min_{i=1,\ldots,K} \frac{\mathrm{Re}\left(\boldsymbol{w}_i^\dagger \boldsymbol{T}(\theta_i)\boldsymbol{s}\right)}{\sqrt{\boldsymbol{w}_i^\dagger \boldsymbol{\Sigma}_c(\boldsymbol{s})\, \boldsymbol{w}_i + \sigma_v^2 \boldsymbol{w}_i^\dagger \boldsymbol{w}_i \|\boldsymbol{s}\|^2}} \tag{9.31}$$

is developed. More specifically, starting from an array of receive filters $\{\boldsymbol{w}_i^{(m-1)}\}_{i=1}^K$, a feasible radar code $\boldsymbol{s}^{(m)}$ at step m maximizing $g(\boldsymbol{s}, \{\boldsymbol{w}_i^{(m-1)}\}_{i=1}^K)$ is searched. Whenever $\boldsymbol{s}^{(m)}$ is found, the filter bank $\{\boldsymbol{w}_i^{(m)}\}_{i=1}^K$ that maximizes the objective function $g(\boldsymbol{s}^{(m)}, \{\boldsymbol{w}_i\}_{i=1}^K)$ is determined. Otherwise stated, $\{\boldsymbol{w}_i^{(m)}\}_{i=1}^K$ is used as the starting point at step $m+1$. To trigger the procedure, the optimal receive filter $\{\boldsymbol{w}_i^{(0)}\}_{i=1}^K$, for the admissible code $\boldsymbol{s}^{(0)}$, is considered. The algorithm can be initialized with $\boldsymbol{s}_0$.

**"Solvable" means that the problem is feasible and bounded above, and the optimal value is attained (see [33, p. 13]).

††The case $\gamma = 2$ is not of great practical relevance since it refers to the absence of the similarity constraint. In this specific case, the optimal value can be approximated through (1.30), with arbitrary precision, just setting $\gamma = 2 - \frac{1}{n}$ with n large enough.

From an analytical point of view, $s^{(m)}$ and $\{w_i^{(m)}\}_{i=1}^K$ are the optimal solutions to the optimization problems $\mathscr{P}_s^{(m)}$ and $\mathscr{P}_w^{(m)}$, respectively, defined as

$$\mathscr{P}_{s^{(m)}} \begin{cases} \max_{s} \min_{i=1,\dots,K} \dfrac{\mathrm{Re}\left(w_i^{(m-1)\dagger} T(\theta_i)\, s\right)}{\sqrt{w_i^{(m-1)\dagger}\left(\Sigma_c(s) + \sigma_v^2 \|s\|^2 I\right) w_i^{(m-1)}}} \\ \text{s.t.} \quad \|s\|^2 \le 1, \\ \qquad \mathrm{Re}\left(s_0^\dagger s\right) \ge \gamma_1, \\ \qquad \mathrm{Re}\left(w_i^{(m-1)\dagger} T(\theta_i)\, s\right) \ge 0, \quad i = 1, 2, \dots, K \end{cases} \tag{9.32}$$

and

$$\mathscr{P}_{w^m} \begin{cases} \max_{\{w_i\}_{i=1}^K} \min_{i=1,\dots,K} \dfrac{\mathrm{Re}\left(w_i^\dagger T(\theta_i)\, s^{(m)}\right)}{\sqrt{w_i^\dagger\left(\Sigma_c\left(s^{(m)}\right) + \sigma_v^2 \|s^{(m)}\|^2 I\right) w_i}} \\ \text{s.t.} \quad \mathrm{Re}\left(w_i^\dagger T(\theta_i)\, s^{(m)}\right) \ge 0, \quad i = 1, 2, \dots, K \end{cases} \tag{9.33}$$

The proposed procedure exhibits some interesting properties summarized in the following proposition.

Proposition 4. Assume that problems $\mathscr{P}_w^{(m)}$ and $\mathscr{P}_s^{(m)}$ are solvable‡‡. Let $\left\{\left(s^{(m)}, \{w_i^{(m)}\}_{i=1}^K\right)\right\}$ be the sequence of points obtained through the proposed iterative optimization procedure; let also $\mathrm{SINR}^{(m)}$ be the worst-case filter-bank SINR value corresponding to the point $\left(s^{(m)}, \{w_i^{(m)}\}_{i=1}^K\right)$ at the mth iteration. Then:

- the sequence $\mathit{SINR}^{(m)}$ is monotonically increasing with m and converges to a finite value $\mathit{SINR}^\star$;
- starting from the sequence $\left\{\left(s^{(m)}, \{w_i^{(m)}\}_{i=1}^K\right)\right\}$, it is possible to construct a subsequence $\left\{\left(\tilde{s}^{(m')}, \{\tilde{w}_i^{(m')}\}_{i=1}^K\right)\right\}$, which converges to a feasible point of $\mathscr{P}_1$ (i.e. $(\tilde{s}^\star, \{\tilde{w}_i^\star\}_{i=1}^K)$) and such that the worst SINR evaluated at $\left(\tilde{s}^\star, \{\tilde{w}_i^\star\}_{i=1}^K\right)$ is equal to $\mathit{SINR}^\star$;
- any cluster point $\left(\tilde{s}^\star, \{\tilde{w}_i^\star\}_{i=1}^K\right)$ to $\left\{\left(s^{(m)}, \{w_i^{(m)}\}_{i=1}^K\right)\right\}$, with $\tilde{w}_i^\star \neq \mathbf{0}$, $i = 1, \dots, K$, is a stationary point for $\mathscr{P}_1$.

Proof. The proof is similar to that found in [23] (for completeness it is also reported in Appendix 9.7.4 of this chapter).

Proposition 4 highlights that the devised algorithm monotonically converges to the limit value associated with the employed starting point and that any cluster point is a stationary point. The former ensures the algorithm stability (in terms of convergence) while the latter shows the quality of the obtained solution since it satisfies necessary optimality conditions.

‡‡In subsections 1.4.1 and 1.4.2, optimization procedures to globally solve both $\mathscr{P}_w^{(m)}$ and $\mathscr{P}_s^{(m)}$, respectively, are developed.

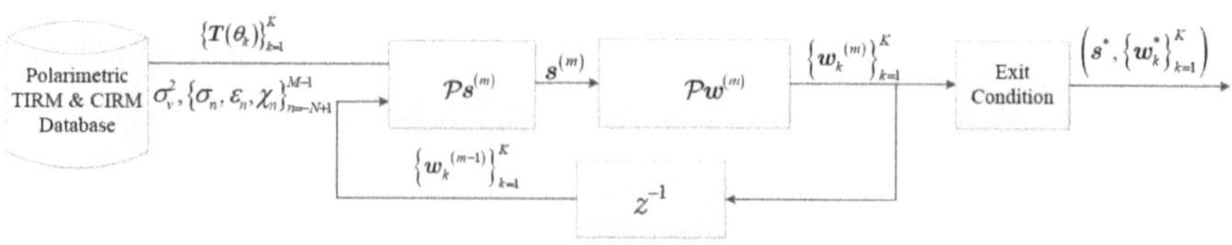

Figure 9.3 Block diagram of the proposed polarimetric transmit code and receive filter bank optimization procedure

Remarks: the proposed iterative optimization procedure requires an exit condition. Some natural choices consist in considering the maximum number of tolerable iterations, or the difference between two consecutive values of the partially optimized SINR (i.e. forcing an iteration gain constraint), or mixing them. Figure 9.3 presents a block scheme representation of the joint optimization process of the radar code and receive filter bank, based on the available information of the target and clutter environment. Precisely, the joint optimization process of the polarimetric radar code and receive filter bank, based on the available information of the target and clutter environment, can be described as follow. Through the use of a TIRM and CIRM database, the required knowledge about target and clutter, in terms of $\boldsymbol{T}(\theta_i)$, $\{\sigma_n, \epsilon_n, \chi_n, \rho_n\}$ and $r_{jk}(\cdot,\cdot)$, $\{j,k\} \in \{HH, HV, VV\}$, respectively, is obtained. Triggering the optimization procedure with the initial code s_0, the filter bank $\{\boldsymbol{w}_i^{(0)}\}_{i=1}^K$ is designed. Then, the SINR is iteratively optimized by solving $\mathcal{P}_s^{(m)}$ and $\mathcal{P}_w^{(m)}$ and at the step m, $\left(\boldsymbol{s}^{(m)}, \{\boldsymbol{w}_i^{(m)}\}_{i=1}^K\right)$ is synthesized. The process continues until the exit condition is satisfied, returning (optimized) radar code $\boldsymbol{s}^\star$ and the receive filters $\{\boldsymbol{w}_i^\star\}_{i=1}^K$. The next subsections will be devoted to the study of the optimization problems $\mathcal{P}_w^{(m)}$ and $\mathcal{P}_s^{(m)}$ required to implement the proposed iterative optimization algorithm.

9.4.1 Receive filter optimization solution to problem $\mathcal{P}_w^{(m)}$

The following Lemma provides a convex formulation to $\mathcal{P}_w^{(m)}$ that paves the way to the design of the optimal receive filter bank structure.

Lemma 1. An optimal solution to $\mathcal{P}_w^{(m)}$ is obtained by solving the set of the decoupled problems, $\mathcal{P}^i$, $i = 1, \ldots, K$, defined as

$$\mathcal{P}^i \quad \begin{cases} \min\limits_{\boldsymbol{w}} & \boldsymbol{w}^\dagger \boldsymbol{\Sigma}_c\left(\boldsymbol{s}^{(m)}\right)\boldsymbol{w} + \sigma_v^2 \boldsymbol{w}^\dagger \boldsymbol{w} \|\boldsymbol{s}^{(m)}\|^2 \\ \text{s.t.} & \mathrm{Re}\left(\boldsymbol{w}^\dagger \boldsymbol{T}(\theta_i)\, \boldsymbol{s}^{(m)}\right) = 1, \quad i = 1, \ldots, K \end{cases} \tag{9.34}$$

i.e., given optimal solutions $\boldsymbol{w}_i^\star$ respectively to $\mathcal{P}^i$, $i = 1, \ldots, K$, $\{\boldsymbol{w}_i^\star\}_{i=1}^K$ is an optimal solution to $\mathcal{P}_w^{(m)}$.

Proof. The interested reader may refer to [26].

Leveraging Lemma 1 and [34], an optimal solution to $\mathscr{P}_w^{(n)}$ is given in

$$w_i^{(m)} = \frac{\left(\boldsymbol{\Sigma}_c\left(\boldsymbol{s}^{(m)}\right) + \sigma_v^2 \|\boldsymbol{s}^{(m)}\|^2 \boldsymbol{I}\right)^{-1} \boldsymbol{T}(\theta_i)\, \boldsymbol{s}^{(m)}}{\left\| \left(\boldsymbol{\Sigma}_c\left(\boldsymbol{s}^{(m)}\right) + \sigma_v^2 \|\boldsymbol{s}^{(m)}\|^2 \boldsymbol{I}\right)^{-1/2} \boldsymbol{T}(\theta_i)\, \boldsymbol{s}^{(m)} \right\|^2}, \quad i = 1, \ldots, K \tag{9.35}$$

(9.35) implies that the solution for ith receive filter in the bank at mth iteration is the well-known Capon filter (also known as minimum-variance distortionless response in the literature) [35]. It also highlights the influence of $\boldsymbol{s}^{(m)}$ as well as the nominal TIRM on the optimized responses of the filters.

9.4.2 Radar code optimization: solution to problem $\mathscr{P}_s^{(m)}$

In this subsection, the main properties of $\mathscr{P}_s^{(m)}$ are analyzed. Specifically, it is shown that the problem is solvable and an algorithm that finds an optimal solution is described. To this end, results from the generalized fractional programming (GFP) theory [36, 37] are exploited, which are summarized hereinafter in the form of a lemma.

Lemma 2. Let $\mathscr{X} \subseteq \mathbb{C}^N$ be a convex compact set, $\{f_i(\boldsymbol{x})\}_{i=1}^I$ be nonnegative concave functions over $\mathscr{X}$, and $\{g_i(\boldsymbol{x})\}_{i=1}^I$ positive convex functions over $\mathscr{X}$. Hence, the GFP problem

$$P_{GFP} \quad \begin{cases} \max_x \quad \min_{i=1,\ldots,I} \dfrac{f_i(\boldsymbol{x})}{g_i(\boldsymbol{x})} \\ \text{s.t.} \quad \boldsymbol{x} \in \mathscr{X} \end{cases} \tag{9.36}$$

is solvable and an optimal solution can be obtained through **Algorithm 1**.

Algorithm 1 : Generalized Dinkelbach's Algorithm

Input: $\mathscr{X} \subseteq \mathbb{C}^N$, $\{f_i(\boldsymbol{x})\}_{i=1}^I$, and $\{g_i(\boldsymbol{x})\}_{i=1}^I$.
Output: A solution for the optimal $\boldsymbol{x}$.

1: set $m = 0$, $\lambda_m = 0$.
2: **do**
3: find $\boldsymbol{x}_m^\star = \arg\max_{\boldsymbol{x}\in\mathscr{X}} \min_{1\le i\le I} \{f_i(\boldsymbol{x}) - \lambda_m g_i(\boldsymbol{x})\}$;
4: $F_\lambda = \min_{1\le i\le I} \{f_i(\boldsymbol{x}_m^\star) - \lambda_m g_i(\boldsymbol{x}_m^\star)\}$;
5: $m = m + 1$;
6: $\lambda_m = \min_{1\le i\le I} \dfrac{f_i(\boldsymbol{x}_m^\star)}{g_i(\boldsymbol{x}_m^\star)}$;
7: **until** $F_\lambda = 0$
8: output $\boldsymbol{x}^\star = \boldsymbol{x}_m^\star$.

Algorithm 1 is characterized by a linear convergence rate [37] and, at each iteration, it only requires the solution of a convex problem, which can be obtained in polynomial time using many convex programming algorithms [38, 39]. Additionally, the objective function of P_{GFP} monotonically converges to the optimal value of P_{GFP} and the exit condition $F_\lambda = 0$, in practice, is replaced by $F_\lambda \le \varsigma$, with ς being a prescribed accuracy. Interestingly, the bisection method [39] could also be used to solve (9.36) even if it usually involves a higher number of iterations than the counterpart to convergence to the optimal solution.

Let us now observe that $\mathscr{P}_s^{(m)}$ belongs to the class of GFP problems. In fact, based on Proposition 2, $\mathscr{P}_s^{(m)}$ in (9.32) can be equivalently recast as

$$\begin{cases} \max\limits_{s} & \min\limits_{i=1,\ldots,K} \dfrac{\operatorname{Re}\left(\boldsymbol{w}_i^{(m-1)^\dagger} \boldsymbol{T}(\theta_i)\, \boldsymbol{s} \right)}{\sqrt{\boldsymbol{s}^\dagger \left(\bar{\boldsymbol{\Sigma}}_c \left(\boldsymbol{w}_i^{(m-1)} \right) + \sigma_v^2 \|\boldsymbol{w}_i^{(m-1)}\|^2 \boldsymbol{I} \right) \boldsymbol{s}}} \\ \text{s.t.} & \|\boldsymbol{s}\|^2 \leq 1, \\ & \operatorname{Re}\left(\boldsymbol{s}_0^\dagger \boldsymbol{s} \right) \geq \gamma_1, \\ & \operatorname{Re}\left(\boldsymbol{w}_i^{(m-1)^\dagger} \boldsymbol{T}(\theta_i)\, \boldsymbol{s} \right) \geq 0, \quad i = 1, 2, \ldots, K \end{cases} \tag{9.37}$$

Note that, since the numerator of the objective function of $\mathscr{P}_s^{(m)}$ is an affine function with respect to s, it is concave. As to the denominator, the following matrix, i.e.,

$$\left(\bar{\boldsymbol{\Sigma}}_c(\boldsymbol{w}_i^{(m-1)}) + \sigma_v^2 \boldsymbol{w}_i^{(m-1)\dagger} \boldsymbol{w}_i^{(m-1)} \boldsymbol{I} \right) \tag{9.38}$$

is Hermitian and positive definite for any $\boldsymbol{w}_i^{(m-1)} \neq \boldsymbol{0}$, $i = 1, \ldots, K$, namely, for any feasible $\boldsymbol{s}$,$\boldsymbol{s}\boldsymbol{s}^\dagger \left(\bar{\boldsymbol{\Sigma}}_c(\boldsymbol{w}_i^{(m-1)}) + \sigma_v^2 \boldsymbol{w}_i^{(m-1)\dagger} \boldsymbol{w}_i^{(m-1)} \boldsymbol{I} \right) \boldsymbol{s} > 0$. Thus, introducing

$$\left(\bar{\boldsymbol{\Sigma}}_c(\boldsymbol{w}_i^{(m-1)}) + \sigma_v^2 \boldsymbol{w}_i^{(m-1)\dagger} \boldsymbol{w}_i^{(m-1)} \boldsymbol{I} \right) \triangleq \boldsymbol{B}^\dagger \boldsymbol{B} \tag{9.39}$$

it follows that $\boldsymbol{s}^\dagger \left(\bar{\boldsymbol{\Sigma}}_c(\boldsymbol{w}_i^{(m-1)}) + \sigma_v^2 \boldsymbol{w}_i^{(m-1)\dagger} \boldsymbol{w}_i^{(m-1)} \boldsymbol{I} \right) \boldsymbol{s} = (\boldsymbol{B}\boldsymbol{s})^\dagger \boldsymbol{B}\boldsymbol{s}$. As a consequence, the denominator is the norm of $2N$-dimensional vector $\boldsymbol{B}\boldsymbol{s}$, which ensures its convexity. It can be seen that the feasible set is a convex compact set and the functions in the objective comply with the assumptions of Lemma 2. As a result, Algorithm 1 can be applied to get an optimal solution. Interestingly, the optimization problem involved at each step of Algorithm 1 for the specific instance at hand can be cast as a second-order cone programming (SOCP) problem exploiting its epigraph form [39].

9.4.3 Transmit-receive system design: optimization procedure

In this subsection, the devised iterative optimization procedure is summarized as Algorithm 2. To trigger the recursion, an initial radar code $\boldsymbol{s}^{(0)}$ from which obtaining the optimal filter bank $\{\boldsymbol{w}_i^{(0)}\}_{i=1}^K$ is required; a natural choice is $\boldsymbol{s}^{(0)} = \boldsymbol{s}_0$, as already stated.

Algorithm 2 : Polarization Transmit-Receive System Design

Input: $\{\sigma_v^2\}, \{\sigma_n, \varepsilon_n, \chi_n, \rho_n\}, r_{jk}(\cdot,\cdot), \{\boldsymbol{T}(\theta_i)\}, \boldsymbol{s}_0, \gamma$.

Output: An optimized solution $(\boldsymbol{s}^\star, \{\boldsymbol{w}_i^\star\}_{i=1}^K)$ to $\mathscr{P}$.

1: set $m := 0$, $\boldsymbol{s}^{(m)} = \boldsymbol{s}_0$, then the receive filter optimizer $\boldsymbol{w}_i^{(m)}$ is given in Eq. (1.35), and $\mathrm{SINR}^{(m)} = \left(g(\boldsymbol{s}^{(m)}, \{\boldsymbol{w}_i^{(m)}\}_{i=1}^K)\right)^2$;

2: **do**

3: $m := m + 1$;

4: construct $\left(\bar{\boldsymbol{\Sigma}}_{\boldsymbol{C}}(\boldsymbol{w}_i^{(m)}) + \sigma_v^2 \|\boldsymbol{w}_i^{(m)}\|^2 \boldsymbol{I}\right)$, $i = 1, \ldots, K$;

5: solve $\mathscr{P}_{\boldsymbol{s}}{}^{(m)}$ finding an optimal radar code $\boldsymbol{s}^{(m)}$, via the use of **Algorithm 1**;

6: construct the matrix $\boldsymbol{\Sigma}_{\boldsymbol{C}}\left(\boldsymbol{s}^{(m)}\right)$;

7: solve $\mathscr{P}_{\boldsymbol{w}}{}^{(m)}$ finding an optimal receive filter bank using Eq. (1.35).

8: let $\mathrm{SINR}^{(m)} = \left(g(\boldsymbol{s}^{(m)}, \{\boldsymbol{w}_i^{(m)}\}_{i=1}^K)\right)^2$;

9: **until** $|\mathrm{SINR}^{(m)} - \mathrm{SINR}^{(m-1)}| \leq \zeta$

10: output $\boldsymbol{s}^\star = \boldsymbol{s}^{(m)}$ and $\boldsymbol{w}_i^\star = \boldsymbol{w}_i^{(m)}$, $i = 1, \ldots, K$.

First, it is important to remark that the devised procedure is able to improve the performance of any feasible transmit-receive pair, due to its monotonically increasing property. Second, it is worth noticing that the computational complexity connected with the implementation of Algorithm 2 depends on the number of outer iterations $\overline{N}$, as well as on the complexity burden required at each iteration. Precisely, the overall complexity is linear with respect to $\overline{N}$, while the computational complexity in each iteration is listed in Table 9.1. Interestingly, even if Algorithm 1 is stopped after a maximum number of allowed iterations, the overall procedure still preserves the monotonic increasing property if at $(n+1)$th iteration the starting value of the parameter λ is set equal to $\lambda_0 = \sqrt{\mathrm{SINR}^{(n)}}$. Finally, it is worth pointing out that the computational complexity issue for a real-time transceiver design can be mitigated by the fact that for each set of TIRM and clutter parameters a corresponding routine (based on Algorithm 2) could be run offline to obtain the optimized waveform/receive filter bank, which can be thus stored to form a corresponding library, as proposed by Howard *et al.* [40].

Table 9.1 Computational complexity of each iteration of Algorithm 2, where η is a prescribed accuracy

Term	**Complexity**
$(\boldsymbol{\Sigma}_c(\boldsymbol{s}) + \sigma_v^2 \boldsymbol{I})^{-1}$	$O(K(2N)^3)$
Algorithm 1	$O\left((2M)^{3.5}\log(1/\eta)\right)$

Table 9.2 Basic information of the target data set

Target	T-72 tank
Central frequency	9.6 GHz
Bandwidth	660 MHz
Elevation angle interval (see Figure 9.4(a))	[28°, 32°]
Azimuth angle interval (see Figure 9.4(b))	[0°, 360°]

9.5 Numerical results

In this section, the performance analysis of the proposed algorithm is presented. The Georgia Technology Research Institute publicly released data set [27] is utilized to gather foreknowledge of the target. The basic information of the data set is listed in Table 9.2. Throughout the simulations, the elevation angle is fixed to 29.28°. Each file in the data set contains 79 radar sweeps for a fixed elevation angle and corresponds to a specific central azimuth angle. After a radar sweep, there is a turntable rotation of 0.05° in azimuth; therefore, a total azimuth variation of 3.90° is spanned with respect to the corresponding central azimuth angle. Based on the length extent of the considered target [42] and the radar range resolution, the TIRMs support size is set as $Q = 37$. Besides, an unitary energy target response, see (9.3), $\boldsymbol{T}_n(\theta)$ is considered, performing:

$$\boldsymbol{T}_n(\theta) = \frac{\bar{T}_n(\theta)}{\sqrt{\sum_{n=0}^{Q-1} \|\bar{T}_n(\theta)\|_F^2}} \tag{9.40}$$

where $\bar{\boldsymbol{T}}_n(\theta)$ and $\boldsymbol{T}_n(\theta)$ denote the unnormalized and normalized target scattering matrix as a function of n, respectively.

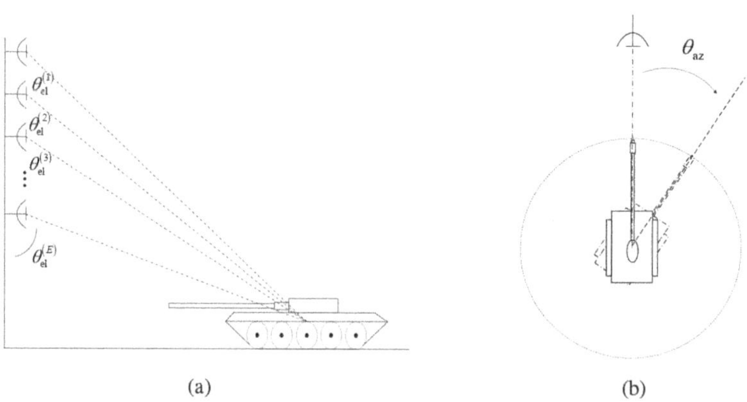

Figure 9.4 Notional representation of the possible measurement setups for the turntable object in 41: (a) elevation θ_{el} belongs to [28°, 32°]; (b) azimuth θ_{az} belongs to [0°, 360°]

Throughout the simulations, unless otherwise stated, $|\alpha_T|^2 = 1$ is assumed. Moreover, a code length $2N$=80 is considered, and the reference code $\boldsymbol{s}_0$ is defined using the same reference chirp waveform for horizontal and vertical components, i.e.,

$$\boldsymbol{s}_{H0}(n) = \boldsymbol{s}_{V0}(n) = \frac{1}{\sqrt{2N}} e^{j\pi \frac{n^2}{N}}, \quad n = 0, \ldots, N-1 \tag{9.41}$$

which corresponds to the samples of a chirp signal whose chirp rate is $\frac{f_s^2}{N}$. Furthermore, with reference to the similarity constraint, unless otherwise stated, $\gamma = 0.5$ is forced. Regarding to the exit condition of the devised methods, $\varsigma = 10^{-6}$ and $\zeta = 10^{-3}$ are set for Algorithms 1 and 2, respectively. All the convex optimization problems are solved via CVX toolbox [43].

Additionally, a uniformly distributed clutter is assumed with the parameters $\varepsilon_n = \varepsilon = 0.16$, $\chi_n = \chi = 0.89$, and $\rho_n = \rho = 0.65$ by employing the clutter second-order polarization statistics of trees reported in [32]. As to the noise and clutter power parameters σ_n and σ_v^2, respectively, they depend on the target-to-noise ratio (TNR) and clutter-to-noise ratio (CNR), with their definitions given, respectively, as

$$\text{TNR} \triangleq \frac{\sum_{n=0}^{Q-1} \|\boldsymbol{T}_n(\theta)\|_F^2}{2 \times M \times \sigma_v^2} \tag{9.42}$$

and

$$\text{CNR} \triangleq \frac{\sigma_n^2 \left(\sum_{n=0}^{M-1} \text{tr}\,[\mathbb{E}[\boldsymbol{C}_n \boldsymbol{C}_n^\dagger]]|_{\sigma_n=1}\right)}{2M\sigma_v^2} \tag{9.43}$$

where $\text{tr}\,[\mathbb{E}[\boldsymbol{C}_n \boldsymbol{C}_n^\dagger]]|_{\sigma_n=1}$ means that the expectation is evaluated by setting $\sigma_n = 1$. Throughout the simulation, TNR = 5 dB and CNR = 10 dB are considered. Furthermore, the clutter returns at different range bins exhibit an exponentially shaped autocorrelation function:

$$r_{jj}(n, n') = a_{jj}^{|n-n'|}, \; j \in \{HH, HV, VV\}, \; n, n' \in \{-N+1, \ldots, M-1\} \tag{9.44}$$

While for the special case $j = HH$, $k = VV$, the following exponentially shaped cross-correlation function, i.e.,

$$r_{jk}(n, n') = \rho a_{jk}^{|n-n'|}, \; n, n' \in \{-N+1, \ldots, M-1\} \tag{9.45}$$

is employed. Clearly, such model encompasses the special case of mutually independent clutter returns from different range bins. This simplified setup can be obtained setting $a_{jk} = a = 0$ in (9.44) and (9.45). Throughout the simulations, unless otherwise explicitly stated, range-independent clutter ($a = 0$) is assumed. As to the range-correlated clutter case, $a = 0.5$ is set, for this specific case studies.

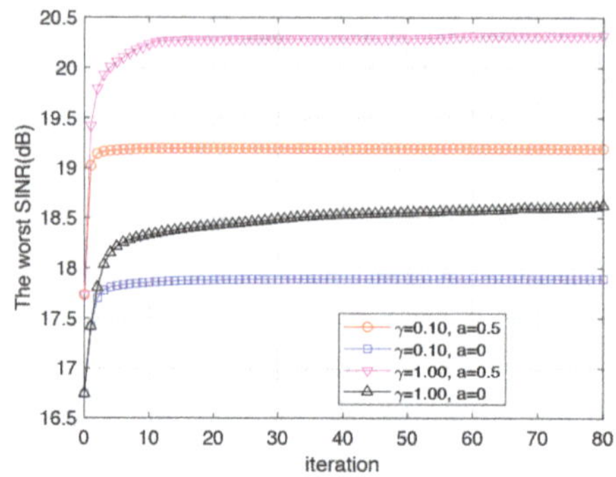

Figure 9.5 Optimized worst-case SINR behavior versus the number of outer iterations, for different values of the similarity parameter $\gamma \in \{0.10, 1.00\}$ and clutter correlation parameter $a \in \{0, 0.5\}$ corresponding to a filter bank size of $K = 4$ with the TAA belonging to the interval $\mathscr{I}_T = [115.40°, 116.20°]$

9.5.1 *Monotonic property of the proposed method and the impact of the similarity constraint*

Let the TAA belong to the azimuth interval $\mathscr{I}_T = [115.40°, 116.20°]$[§§]. Figure 9.5 plots the worst-case SINR versus the number of outer iterations, for different values of the similarity $\gamma \in \{0.10, 1.00\}$ and clutter range-correlation $a \in \{0, 0.5\}$, assuming a filter bank with parameter $K = 4$ branches. Precisely, the discrete TAAs θ_i, $i = 1, \ldots, K$, are obtained uniformly sampling $\mathscr{I}_T$ with a step size 0.20°, namely, $\theta_i = 115.50° + 0.20° \times (i - 1)$, $i = 1, \ldots, 4$. As expected, the devised method monotonically increases the lowest SINR at the output of the filter array. Moreover, as γ increases, the optimized worst-case SINR value improves since the feasible set of the optimization problem is larger and larger. On the other hand, it can be observed that, for the case of range-correlative clutter, larger values of the worst SINR are obtained with respect to the range-independent case. This is because, the higher the clutter correlation the better the achievable interference suppression. The results also highlight the effectiveness of the proposed procedure to provide a significant performance gain with respect to conventional receiver design, corresponding to $m = 0$. Finally, the lower the similarity parameter the faster the convergence to the steady state.

[§§]It is worth pointing out that similar results to those reported have been observed when other central azimuth angles are analyzed.

9.5.2 The effects of the TAA uncertainty sets and receive filter bank size

The devised iterative procedure relies on several design parameters including the TAA uncertainty set, and the number of filters in the receiver bank. In this subsection, the impact of the aforementioned parameters is analyzed.

Let us start with the influence of the TAA uncertainty interval on the transceiver performance. To this end, the SINR values associated with all the filters are shown in Figure 9.6(a)–(d) for a clutter correlation parameter $a \in \{0, 0.5\}$ and four different sizes of the filter bank, i.e., $K \in \{4, 8, 16, 32\}$, whose discrete TAAs θ_i, $i = 1, 2, \ldots, K$, are obtained uniformly sampling the TAA uncertainty interval with a step of 0.20°. Specifically, the possible TAA uncertainty intervals considered are

$$\mathscr{I}_T^{(1)} = [115.40°, 116.20°]$$
$$\mathscr{I}_T^{(2)} = [115.40°, 117.00°]$$
$$\mathscr{I}_T^{(3)} = [115.40°, 118.60°]$$
$$\mathscr{I}_T^{(4)} = [115.40°, 119.30°] \cup [119.65°, 122.10°]$$

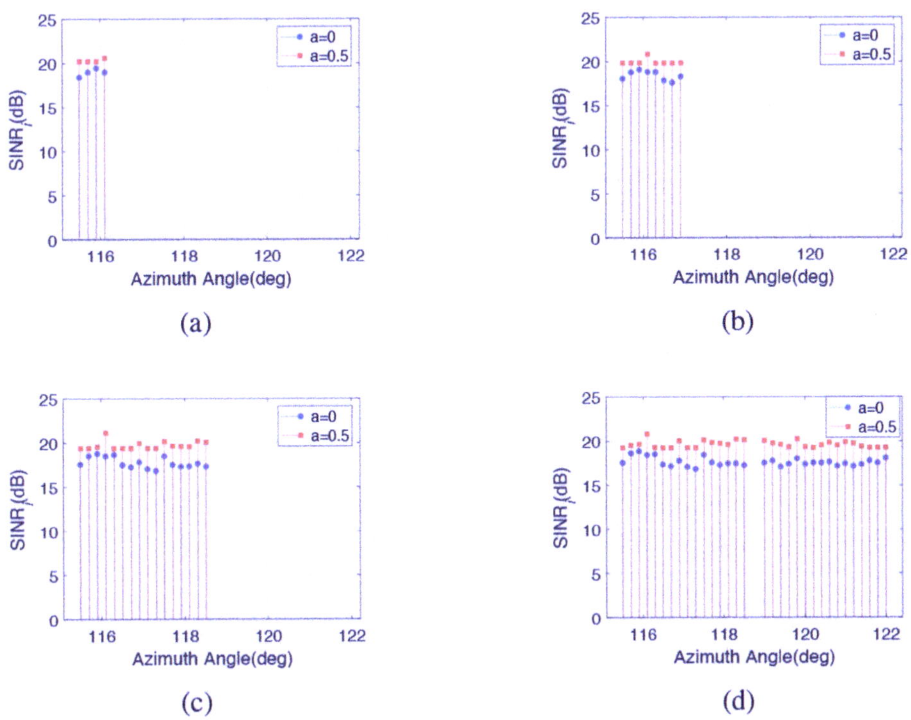

Figure 9.6 SINR values associated with filters of the devised bank for both range-correlated (red "□" markers) and range-independent (blue "○" markers) clutter scenarios. TAA uncertainty belongs to the interval $\mathscr{I}_T^{(1)}$ in (a), $\mathscr{I}_T^{(2)}$ in (b), $\mathscr{I}_T^{(3)}$ in (c), and $\mathscr{I}_T^{(4)}$ in (d).

The inspection of the figures clearly shows that the wider the uncertainty set of the TAA the smaller the worst-case SINR. This is reasonable as a wider $\mathcal{I}_T$ implies a higher uncertainty on the TIRM. In addition, the results of Figure 9.6 reveal that in the presence of range-correlated clutter larger values of the worst SINR than range-independent counterpart are obtained. This is not surprising and complies with analogous findings of Figure 9.5.

Let us now assess the effects of the filter bank size K on the system performance. As in [23], the quantity $\mathrm{SINR}_i(\theta)$, $1 \le i \le K$, is defined as the output of the ith filter due to a target with the TAA θ, i.e.:

$$\mathrm{SINR}_i(\theta) = \frac{|\alpha_T|^2 |\boldsymbol{w}_i^\dagger \boldsymbol{T}(\theta)\boldsymbol{s}|^2}{\boldsymbol{w}_i^\dagger \boldsymbol{\Sigma}_c(\boldsymbol{s})\boldsymbol{w}_i + \sigma_v^2 \boldsymbol{w}_i^\dagger \boldsymbol{w}_i}, \quad i = 1, \ldots, K \tag{9.46}$$

In the following case study, the TAA uncertainty interval is $\mathcal{I}_T = [115.40°, 117.00°]$ and the discrete TAAs are chosen according to the following rule:

$$\theta_i = 115.50° + (i-1)\Delta\theta, \quad i = 1, \ldots, K \tag{9.47}$$

where $\Delta\theta$ represents the azimuth sampling step. In this subsection, three different filter bank sizes $K = 4, 8, 16$ are considered, corresponding to $\Delta\theta = 0.4°$, $0.2°$, and $0.1°$, respectively. Figure 9.7(a) displays $\mathrm{SINR}_i(\theta)$, $i = 1,2,3,4$, versus θ, for the case study $K = 4$. Additionally, the curve corresponding to the metric $\mathrm{SINR}_{\max}(\theta) \triangleq \max_{1\le i\le K} \mathrm{SINR}_i(\theta)$ is also reported in the figure. It can be observed that mismatches between the nominal and the actual TIRM lead to an SINR loss. This is due to the finite number of filters K within the bank and it is usually referred to as "straddle loss." Hence, in Figure 9.7(b), the behavior of $\mathrm{SINR}_{\max}(\theta)$ versus θ is illustrated for $K = 8$ and $K = 16$. As expected, the larger the K, the smaller the straddle loss. To provide a quantitative analysis of the straddle loss L, defined as

$$L \triangleq \frac{\mathrm{SINR}}{\min\limits_{\theta\in\mathcal{I}_T} \mathrm{SINR}_{\max}(\theta)} = \frac{\min\limits_{1\le i\le K} \mathrm{SINR}_i}{\min\limits_{\theta\in\mathcal{I}_T} \max\limits_{1\le i\le K} \mathrm{SINR}_i(\theta)} \tag{9.48}$$

Figure 9.7(c) reports L versus K. The curve clearly shows that L monotonically decreases with K.

Finally, let us assess the influence of the straddle loss on detection performance. The detection probability of the equivalent coherent detector sharing a signal-to-noise ratio equal to the worst SINR of the devised filter bank is considered. The mentioned detection probability is referred to as $P_{\mathrm{d,worst}}$ and, assuming a non-fluctuating target embedded in Gaussian background, it is analytically given by Naghsh *et al.* [44]:

$$P_{\mathrm{d,worst}} = Q\left(\sqrt{2\mathrm{SINR}^\star}, \sqrt{-2\log(P_{fa})}\right) \tag{9.49}$$

where $\mathrm{SINR}^\star = |\alpha_T|^2 \min_{\theta\in\mathcal{I}_T} \mathrm{SINR}_{\max}(\theta)$ with $|\alpha_T|^2$ selected to control the extended target strength, P_{fa} denotes the probability of false alarm, and $Q(\cdot)$ is the generalized

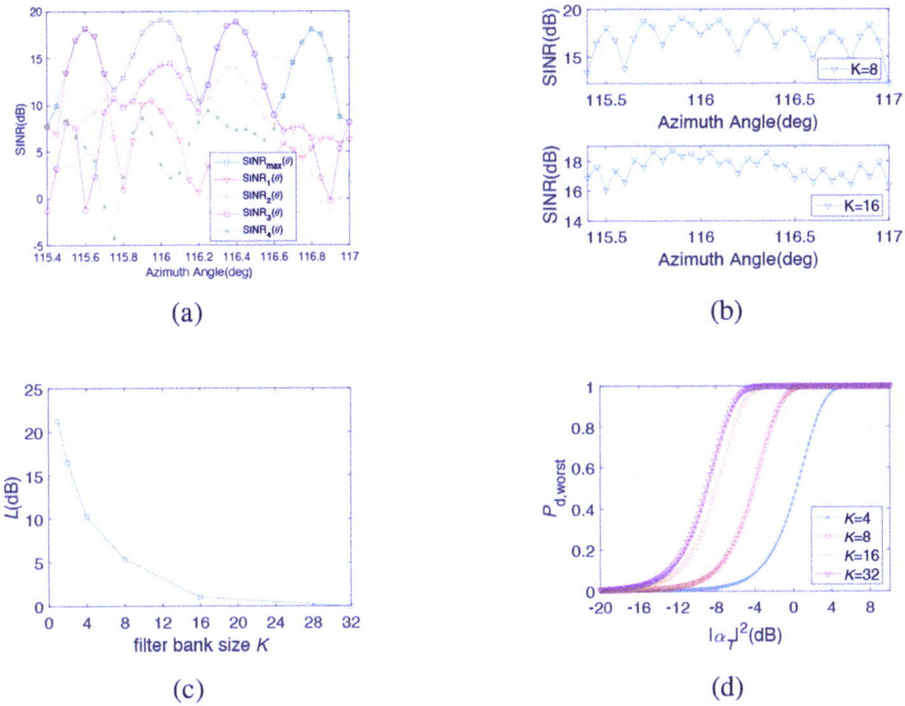

Figure 9.7 *Impact of the filter bank size K on system performance: (a) $SINR_i$ versus θ for $K = 4$, (b) $SINR_{max}$ versus θ for $K = 8$ and $K = 16$, (c) straddle loss L versus K, (d) $P_{d,worst}$ versus $|\alpha_T|^2$ for different values of KK ($P_{fa} = 10^{-3}$)*

Marcum Q function [45]. Figure 9.7 shows $P_{\text{d,worst}}$ versus $|\alpha_T|^2$, with $P_{fa} = 10^{-3}$. The curves confirm the intuition that, for a given TAA uncertainty set, the larger the filter bank size K, the better the detection performance.

9.5.3 *Performance comparison with other methods*

Several methods have been proposed in the open literature to optimize the SINR at the output of the receiver, exploiting fully polarimetric transmit code and receive filter for the extended target under (signal dependent) clutter background. In this subsection, the algorithm is compared with the proposed in [24, 25]. It is worth pointing out that, differently from the filter bank design depicted in this chapter, the strategies in [24, 25] employ one filter and the optimized (matched) polarimetric waveform-filter pair is computed based on a nominal design, namely, the TIRM is assumed known (i.e. absence of mismatches). Moreover, the techniques developed in [24, 25] do not assume a similarity constraint, so ambiguity function properties

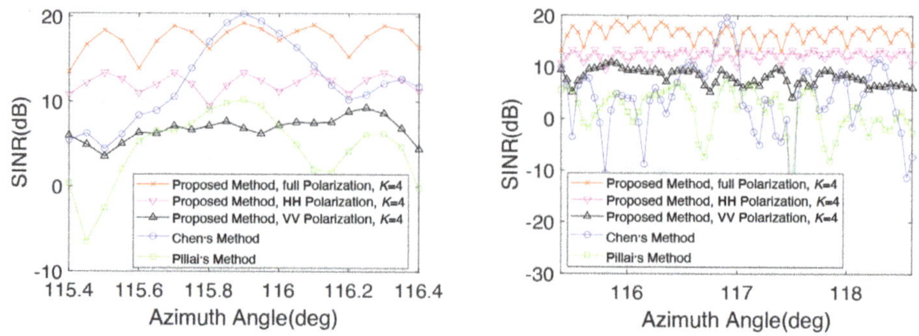

(a)

(b)

Figure 9.8 Comparison of the proposed method with single polarization robust designs (including only HH polarization and only VV polarization cases) and the methods in [24, 25], in term of $SINR_{max}$ *versus* θ*: (a)* $\mathscr{I}_T = [115.40°, 116.40°]$ $(K = 4)$*, (b)* $\mathscr{I}_T = [115.40°, 118.60°]$ $(K = 16)$

cannot be controlled in both cases. Finally, the robust design method illustrated in this chapter is also specialized to the single-polarimetric radar counterpart (using only HH or VV channel), to ascertain the potential improvement in terms of SINR performance of a fully polarimetric radar with respect to the single-channel case.

Let the TAA uncertainty interval $\mathscr{I}_T = [115.40°, 116.40°]$. Regarding to the proposed method and its single-polarimetric (only HH or VV channel) counterparts, the same experimental setup as in section 9.5.1 is utilized, namely, the TAA uncertainty interval is sampled uniformly with a step size 0.20°, thus resulting in a filter-bank size of $K = 5$. With reference to the fully polarimetric methods [24, 25], the nominal TAA is set to 115.90°.

Figure 9.8(a) plots $\mathrm{SINR}_{\max}(\theta)$ versus θ corresponding to the aforementioned five design strategies. Inspection of the figure highlights that, although the robust method suffers of a slight SINR loss (less than 1.12 dB) with respect to the matched SINR (in correspondence of $\theta = 115.90°$) achieved by the nominal design [25]), it exhibits higher and flatter SINR values over the entire TAA uncertainty set with respect to the counterparts (with a gain on the minimum SINR of 13.38 dB over [25]). This behavior was expected, due to the basic idea behind the robust design methodology. It is worth pointing out that, the method in [25] achieves slightly better performance than that in [24], as already highlighted in [25]. The reason is that the algorithm in [24] is not guaranteed to converge to a stationary point. Moreover, the results reveal that, the new proposed approach outperforms both HH and VV channel single-polarimetric robust design in terms of worst-case SINR. The same comparison is also shown in Figure 9.8 for a wider TAA uncertainty interval $\mathscr{I}_T = [115.40°, 118.60°]$ where the same step size as in Figure 9.8 is employed for filter bank methods (thus resulting in $K = 16$) and the nominal TAA for methods in [24, 25] has been differently set to 117.00°. Looking over the curves, similar conclusions as those in Figure 9.8) hold true, the same conclusion can be obtained, thus confirming the effectiveness of the proposed method.

9.6 Conclusion

Owing to a suitable reformulation of the considered non-convex design problem, an iterative optimization procedure over the transmit signal and the receive filter bank has been devised. It monotonically improves the worst-case SINR and converges to a stationary point. Each iteration of the algorithm involves both a convex and a GFP problem, which can be globally solved via the generalized Dinkelbach's procedure with a polynomial computational complexity. The complexity of the proposed method is linear with the number of outer iterations whereas at each iteration, it mainly requires the implementation effort of the Dinkelbach's procedure, that demands the solution of convex SOCP problems.

Several case studies have been provided to assess the effectiveness of the described robust design method, considering challenging experimental scenarios. The results have shown the capability of the procedure to ensure high SINR values regardless of the actual TAA, with significant SINR gains with respect to the traditional approach adapting the receiver side only. Moreover, the conducted analysis has also highlighted the capability of the proposed framework to outperform both some approaches available in the open literature in terms of robustness, and also single-polarimetric robust designs.

Possible future research avenues might concern the robust transceiver design under separable energy and similarity constraints. Moreover, it might be of interest to robustify the developed method to account for possible inaccuracies on clutter/interference statistics.

9.7 Appendix

9.7.1 Proof of proposition 1

Considering the code $\boldsymbol{s}$ as a fixed parameter, the denominator of (9.19) can be rewritten as

$$\begin{aligned}
&\mathbb{E}\left[|\boldsymbol{w}^\dagger \boldsymbol{C}\boldsymbol{s}|^2\right] + \mathbb{E}\left[|\boldsymbol{w}^\dagger \boldsymbol{v}|^2\right] \\
&= \boldsymbol{w}^\dagger \mathbb{E}\left[\boldsymbol{C}\boldsymbol{s}\boldsymbol{s}^\dagger \boldsymbol{C}^\dagger\right] \boldsymbol{w} + \boldsymbol{w}^\dagger \mathbb{E}\left[\boldsymbol{v}^\dagger \boldsymbol{v}\right] \boldsymbol{w} \\
&= \boldsymbol{w}^\dagger \mathbb{E}\left[\boldsymbol{C}\boldsymbol{s}\boldsymbol{s}^\dagger \boldsymbol{C}^\dagger\right] \boldsymbol{w} + \sigma_v^2 \boldsymbol{w}^\dagger \boldsymbol{w} \\
&\triangleq \boldsymbol{w}^\dagger \left(\boldsymbol{\Sigma}_{c(s)+\sigma_v^2 I}\right) \boldsymbol{w}
\end{aligned} \tag{9.50}$$

9.7.2 Proof of proposition 2

Since the energy constraint $\|\boldsymbol{s}\|^2 = 1$ holds for $\mathscr{P}$, the following equation

$$\begin{aligned} & \boldsymbol{w}_i^\dagger \left(\boldsymbol{\Sigma}_{c(s)+\sigma_v^2 I} \right) \boldsymbol{w}_i \\ & = \mathbb{E}\left[|\boldsymbol{w}_i^\dagger \boldsymbol{C}\boldsymbol{s}|^2 \right] + \sigma_v^2 \boldsymbol{w}_i^\dagger \boldsymbol{w}_i \|\boldsymbol{s}\|^2 \\ & = \boldsymbol{s}^\dagger \mathbb{E}\left[\boldsymbol{C}^\dagger \boldsymbol{w}_i \boldsymbol{w}_i^\dagger \boldsymbol{C} \right] \boldsymbol{s} + \sigma_v^2 \boldsymbol{w}_i^\dagger \boldsymbol{w}_i \boldsymbol{s}^\dagger \boldsymbol{s} \\ & \triangleq \boldsymbol{s}^\dagger \left(\bar{\boldsymbol{\Sigma}}_{c(w_i)+\sigma_v^2 \|w_i\|^2 I} \right) \boldsymbol{s} \end{aligned} \tag{9.51}$$

is obtained. Then the explicit form of $\bar{\boldsymbol{\Sigma}}_C(\boldsymbol{w}_i)$ can be directly derived based on (9.9).

9.7.3 Proof of proposition 3

The proof is organized in the following two steps:
P1. The first step is to show that $\mathscr{P}$ is solvable and shares the same optimal solutions and optimal value of:

$$\mathscr{P}_2 \quad \begin{cases} \max\limits_{s,\{w_i\}_{i=1}^K} & \min\limits_{i=1,\ldots,K} \dfrac{|\boldsymbol{w}_i^\dagger \boldsymbol{T}(\theta_i)\boldsymbol{s}|^2}{\boldsymbol{w}_i^\dagger \boldsymbol{\Sigma}_c(\boldsymbol{s})\boldsymbol{w}_i + \sigma_v^2 \boldsymbol{w}_i^\dagger \boldsymbol{w}_i \|\boldsymbol{s}\|^2} \\ \text{s.t.} & \|\boldsymbol{s}\|^2 \le 1 \\ & \Re\left(\boldsymbol{s}_0^\dagger \boldsymbol{s}\right) \ge \gamma_1, \\ & \Re\left(\boldsymbol{s}_0^\dagger \boldsymbol{s}\right) \ge \gamma_1 \|\boldsymbol{s}\| \end{cases} \tag{9.52}$$

where $\gamma_1 = 1 - \frac{\gamma}{2}$;
P2. The second step is to prove that $\mathscr{P}_2$ is equivalent to $\mathscr{P}_1$.

As to P1, let us prove that $\mathscr{P}$ is solvable. To this end, notice that solving $\mathscr{P}$ is equivalent to solve:

$$\bar{\mathscr{P}} \quad \begin{cases} \max\limits_{s,\{w_i\}_{i=1}^K} & \min\limits_{i=1,\ldots,K} \dfrac{|\boldsymbol{w}_i^\dagger \boldsymbol{T}(\theta_i)\boldsymbol{s}|^2}{\boldsymbol{w}_i^\dagger \boldsymbol{\Sigma}_c(\boldsymbol{s})\boldsymbol{w}_i + \sigma_v^2 \boldsymbol{w}_i^\dagger \boldsymbol{w}_i} \\ \text{s.t.} & \|\boldsymbol{s}\|^2 = 1 \\ & \|\boldsymbol{w}_i\|^2 = 1, \ i = 1, 2, \ldots, K \\ & \|\boldsymbol{s} - \boldsymbol{s}_0\|^2 \le \gamma \end{cases} \tag{9.53}$$

In fact, the two problems share the same objective function. Additionally, any feasible solution to $\bar{\mathscr{P}}$ is a feasible solution to $\mathscr{P}$; moreover, for any nontrivial feasible solution $(\boldsymbol{s}, \boldsymbol{w}_1, \ldots, \boldsymbol{w}_K)$ to $\mathscr{P}$, $\left(\boldsymbol{s}, \dfrac{\boldsymbol{w}_1}{\|\boldsymbol{w}_1\|}, \ldots, \dfrac{\boldsymbol{w}_K}{\|\boldsymbol{w}_K\|}\right)$ is a feasible solution to $\bar{\mathscr{P}}$ achieving the same objective value as$(\boldsymbol{s}, \boldsymbol{w}_1, \ldots, \boldsymbol{w}_K)$.

Now observe that the objective function of $\bar{\mathscr{P}}$ is a continuous function (the minimum of continuous functions is still a continuous function), and the constraint set defines a compact set. Therefore, from Weierstrass theorem, there exists a feasible point $(\boldsymbol{s}^\star, \boldsymbol{w}_1^\star, \ldots, \boldsymbol{w}_K^\star)$ for $\bar{\mathscr{P}}$ such that:

$$v(\bar{\mathscr{P}}) = \max_{s^\star,\{w_i\}_{i=1}^K} \min_{i=1,\ldots,K} \frac{|\boldsymbol{w}_i^\dagger \boldsymbol{T}(\theta_i)\boldsymbol{s}^\star|^2}{\boldsymbol{w}_i^\dagger \boldsymbol{\Sigma}_c(\boldsymbol{s}^\star)\boldsymbol{w}_i + \sigma_v^2 \boldsymbol{w}_i^\dagger \boldsymbol{w}_i}$$

Hence, both $\bar{\mathscr{P}}$ and $\mathscr{P}$ are solvable.

Next, let us focus on the equivalence between $\mathscr{P}$ and $\mathscr{P}_2$. Since

$$\|s - s_0\|^2 = 2 - 2\Re\left(s_0^\dagger s\right)$$

$\mathscr{P}$ shares the same optimal solutions and the same optimal value as

$$\bar{\mathscr{P}}_2 \quad \begin{cases} \max\limits_{s,\{w_i\}_{i=1}^K} & \min\limits_{i=1,\dots,K} \dfrac{|w_i^\dagger T(\theta_i)s|^2}{w_i^\dagger \Sigma_c(s) w_i + \sigma_v^2 w_i^\dagger w_i \|s\|^2} \\ \text{s.t.} & \|s\|^2 = 1 \\ & \Re\left(s_0^\dagger s\right) \geq \gamma_1, \\ & \Re\left(s_0^\dagger s\right) \geq \gamma_1 \|s\| \end{cases} \tag{9.54}$$

Let us show that the non-convex constraint $\|s\|^2 = 1$ in $\bar{\mathscr{P}}_2$ can be relaxed into the convex constraint $\|s\|^2 \leq 1$, namely, $\bar{\mathscr{P}}_2$ is equivalent to $\mathscr{P}_2$. Clearly, $v(\bar{\mathscr{P}}_2) \leq v(\mathscr{P}_2)$, because the feasible set of $\bar{\mathscr{P}}_2$ is contained in the feasible set of $\mathscr{P}_2$. Moreover, given a feasible point $\hat{s}$ to $\mathscr{P}_2$, $s_1 = \dfrac{\hat{s}}{\|\hat{s}\|}$ is feasible to $\bar{\mathscr{P}}_2$. In fact,

- $\|s_1\|^2 = 1$ since $\hat{s} \neq \mathbf{0}$;
- $\Re\left(s_0^\dagger s_1\right) \geq \gamma_1 = \gamma_1 \|s_1\|$, due to $\frac{1}{\|\hat{s}\|}\Re\left(s_0^\dagger \hat{s}\right) \geq \gamma_1$;
- $\Re\left(s_0^\dagger s_1\right) \geq \frac{\gamma_1}{\|\hat{s}\|} \geq \gamma_1$, because $\frac{1}{\|\hat{s}\|} \geq 1$.

Owing to the scale invariance of the objective function of $\mathscr{P}_2$ with respect to $\|s\|$, the proof of step 1 is completed.

As to P2, it is observed that $\mathscr{P}_2$ shares the same optimal solutions as

$$\bar{\mathscr{P}}_1 \quad \begin{cases} \max\limits_{s,\{w_i\}_{i=1}^K} & \min\limits_{i=1,\dots,K} \dfrac{|w_i^\dagger T(\theta_i)s|}{\sqrt{w_i^\dagger \Sigma_c\left(s\right) w_i + \sigma_v^2 w_i^\dagger w_i \|s\|^2}} \\ \text{s.t.} & \|s\|^2 \leq 1 \\ & \Re\left(s_0^\dagger s\right) \geq \gamma_1, \\ & \Re\left(s_0^\dagger s\right) \geq \gamma_1 \|s\| \end{cases} \tag{9.55}$$

and $v(\bar{\mathscr{P}}_1) = v(\sqrt{\mathscr{P}_2}) = v(\sqrt{\mathscr{P}})$. Specifically, the two optimization problems share the same feasibility set and their objective functions are related through a monotonic increasing one-to-one mapping; thus, $\mathscr{P}_2$ and $\bar{\mathscr{P}}_1$ exhibit the same set of optimal solutions. Indeed, for any $x_1, x_2, \dots, x_K \geq 0$, $\sqrt{\min(x_1, x_2, \dots, x_K)} = \min\left(\sqrt{x_1}, \sqrt{x_2}, \dots, \sqrt{x_K}\right)$.

Now, let $(s^\star, w_1^\star, \dots, w_K^\star)$ be an optimal solution to $\bar{\mathscr{P}}_1$, then

$$\left(s^\star, w_1^\star \exp\left(j\arg\left(w_1^{\star\dagger}\left[T(\theta_1)s^\star\right]\right)\right), \dots, w_K^\star \exp\left(j\arg\left(w_K^{\star\dagger}\left[T(\theta_K)s^\star\right]\right)\right)\right)$$

is another optimal solution to $\bar{\mathscr{P}}_1$ that is also feasible to $\mathscr{P}_1$; hence

$$v(\mathscr{P}_1) \geq v(\bar{\mathscr{P}}_1)$$

Furthermore, any feasible point $(\boldsymbol{s}, \boldsymbol{w}_1, \ldots, \boldsymbol{w}_K)$ to $\mathscr{P}_1$ is also feasible to $\bar{\mathscr{P}}_1$, and the objective function of $\mathscr{P}_1$ in $\left(\boldsymbol{s}^\star, \boldsymbol{w}_1^\star, \ldots, \boldsymbol{w}_K^\star\right)$ is lower than or equal to the objective function of $\bar{\mathscr{P}}_1$, evaluated at the same point. This implies that

$$v(\mathscr{P}_1) \le v(\bar{\mathscr{P}}_1)$$

As a consequence,

$$v(\mathscr{P}_1) = v(\bar{\mathscr{P}}_1)$$

and any optimal solution to $\mathscr{P}_1$ is also optimal to $\bar{\mathscr{P}}_1$. As a final remark, note that given an optimal solution to $\bar{\mathscr{P}}_1$, an optimal solution to $\mathscr{P}_1$ can be immediately constructed.

9.7.4 Proof of proposition 4

The proof of the first two items of the proposition mainly follows from [20, Proposition 2.1] and the fact that $\mathrm{SINR}^{(n)} = \left(g\left(\boldsymbol{s}^{(n)}, \boldsymbol{w}_1^{(n)}, \ldots, \boldsymbol{w}_K^{(n)}\right)\right)^2$. Let us show that any cluster point to the generated sequence is a stationary point for the optimization problem. To this end, let $\left\{\left(\tilde{\boldsymbol{s}}^{(n')}, \tilde{\boldsymbol{w}}_1^{(n')}, \ldots, \tilde{\boldsymbol{w}}_K^{(n')}\right)\right\}$ be a subsequence of $\left\{\left(\boldsymbol{s}^{(n)}, \boldsymbol{w}_1^{(n)}, \ldots, \boldsymbol{w}_K^{(n)}\right)\right\}$ converging to the cluster point $\left(\tilde{\boldsymbol{s}}^\star, \tilde{\boldsymbol{w}}_1^\star, \ldots, \tilde{\boldsymbol{w}}_K^\star\right)$. Hence, it will be proved that:

$$\begin{aligned} g(\boldsymbol{s}, \tilde{\boldsymbol{w}}_1^\star, \tilde{\boldsymbol{w}}_2^\star, \ldots, \tilde{\boldsymbol{w}}_K^\star) &\le g(\tilde{\boldsymbol{s}}^\star, \tilde{\boldsymbol{w}}_1^\star, \tilde{\boldsymbol{w}}_2^\star, \ldots, \tilde{\boldsymbol{w}}_K^\star) \\ & \quad \text{for any feasible } \boldsymbol{s} \\ g(\tilde{\boldsymbol{s}}^\star, \boldsymbol{w}_1, \boldsymbol{w}_2, \ldots, \boldsymbol{w}_K) &\le g(\tilde{\boldsymbol{s}}^\star, \tilde{\boldsymbol{w}}_1^\star, \tilde{\boldsymbol{w}}_2^\star, \ldots, \tilde{\boldsymbol{w}}_K^\star) \\ & \quad \text{for any feasible } \left(\boldsymbol{w}_1, \boldsymbol{w}_2, \ldots, \boldsymbol{w}_K\right) \end{aligned}$$

Let $\bar{\boldsymbol{s}}^\star$ be an optimal solution to:

$$\begin{cases} \max\limits_{\boldsymbol{s}} & \min\limits_{i=1,\ldots,K} \dfrac{\Re\left(\tilde{\boldsymbol{w}}_i^{\star\dagger}\left(\boldsymbol{T}(\theta_i)\boldsymbol{s}\right)\right)}{\sqrt{\left(\tilde{\boldsymbol{w}}_i^{\star\dagger}\boldsymbol{\Sigma}_{\boldsymbol{C}}\left(\boldsymbol{s}\right)\tilde{\boldsymbol{w}}_i^\star + \sigma_v^2\tilde{\boldsymbol{w}}_i^{\star\dagger}\tilde{\boldsymbol{w}}_i^\star\|\boldsymbol{s}\|^2\right)}} \\ \text{s.t.} & \|\boldsymbol{s}\|^2 \le 1 \\ & \Re\left(\boldsymbol{s}_0^\dagger\boldsymbol{s}\right) \ge \gamma_1, \\ & \Re\left(\tilde{\boldsymbol{w}}_i^{\star\dagger}\left(\boldsymbol{T}(\theta_i)\boldsymbol{s}\right)\right) \ge 0, \quad i = 1, 2, \ldots, K \end{cases} \tag{9.56}$$

Then, the following inequality chain holds true:

$$\begin{aligned} & g(\bar{\boldsymbol{s}}^\star, \tilde{\boldsymbol{w}}_1^{(n')}, \tilde{\boldsymbol{w}}_2^{(n')}, \ldots, \tilde{\boldsymbol{w}}_K^{(n')}) \le \\ & \quad g(\tilde{\boldsymbol{s}}^{(n'+1)}, \tilde{\boldsymbol{w}}_1^{(n')}, \tilde{\boldsymbol{w}}_2^{(n')}, \ldots, \tilde{\boldsymbol{w}}_K^{(n')}) \le \\ & \quad g(\tilde{\boldsymbol{s}}^{(n'+1)}, \tilde{\boldsymbol{w}}_1^{(n'+1)}, \tilde{\boldsymbol{w}}_2^{(n'+1)}, \ldots, \tilde{\boldsymbol{w}}_K^{(n'+1)}) \le \\ & \quad g(\tilde{\boldsymbol{s}}^{(\{n+1\}')}, \tilde{\boldsymbol{w}}_1^{(\{n+1\}')}, \tilde{\boldsymbol{w}}_2^{(\{n+1\}')}, \ldots, \tilde{\boldsymbol{w}}_K^{(\{n+1\}')}) \end{aligned} \tag{9.57}$$

where the first inequality is straightforward if $\bar{\boldsymbol{s}}^\star$ is a feasible point to $\mathscr{P}_s^{(n'+1)}$, otherwise it follows from the fact that in such case $g(\bar{\boldsymbol{s}}^\star, \tilde{\boldsymbol{w}}_1^{(n')}, \tilde{\boldsymbol{w}}_2^{(n')}, \ldots, \tilde{\boldsymbol{w}}_K^{(n')}) < 0$, while

$g(\tilde{s}^{(n'+1)}, \tilde{w}_1^{(n')}, \tilde{w}_2^{(n')}, \ldots, \tilde{w}_K^{(n')}) \geq 0$. The second inequality is due to the optimality of the point $(\tilde{w}_1^{(n'+1)}, \tilde{w}_2^{(n'+1)}, \ldots, \tilde{w}_K^{(n'+1)})$ to $\mathscr{P}_w^{(n'+1)}$. Finally, the last inequality stems from the fact that $g(\tilde{s}^{(n)}, \tilde{w}_1^{(n)}, \tilde{w}_2^{(n)}, \ldots, \tilde{w}_K^{(n)})$ is an increasing sequence of values and $n'+1 \leq \{n+1\}'$. Therefore, due to the continuity of the function $g(\cdot)$, as $n \to \infty$, it follows that:

$$g(\tilde{s}^\star, \tilde{w}_1^\star, \tilde{w}_2^\star, \ldots, \tilde{w}_K^\star) \;\leq\; g(\bar{s}^\star, \tilde{w}_1^\star, \tilde{w}_2^\star, \ldots, \tilde{w}_K^\star) \tag{9.58}$$

Similarly, let $\left(\bar{w}_1^\star, \bar{w}_2^\star, \ldots, \bar{w}_K^\star\right)$ be an optimal solution to:

$$\begin{cases} \max\limits_{w_1,\ldots,w_K} & \min\limits_{i=1,\ldots,K} \dfrac{\Re\left(w_i^\dagger \left(T(\theta_i)\tilde{s}^\star\right)\right)}{\sqrt{\left(w_i^\dagger \Sigma_c\left(\tilde{s}^\star\right) w_i + \sigma_v^2 w_i^\dagger w_i \|\tilde{s}^\star\|^2\right)}} \\ \text{s.t.} & \Re\left(w_i^\dagger\left(T(\theta_i)\tilde{s}^\star\right)\right) \geq 0, \quad i = 1, 2, \ldots, K \end{cases} . \tag{9.59}$$

Hence:

$$g(\tilde{s}^{(n')}, \bar{w}_1^\star, \bar{w}_2^\star, \ldots, \bar{w}_K^\star) \;\leq\; g(\tilde{s}^{(n')}, \tilde{w}_1^{(n')}, \tilde{w}_2^{(n')}, \ldots, \tilde{w}_K^{(n')})$$

where the inequality is straightforward if $(\bar{w}_1^\star, \bar{w}_2^\star, \ldots, \bar{w}_K^\star)$ is feasible to $\mathscr{P}_w^{(n')}$, otherwise, as before, it is due to the fact that $g(\tilde{s}^{(n')}, \bar{w}_1^\star, \bar{w}_2^\star, \ldots, \bar{w}_K^\star) < 0$, while $g(\tilde{s}^{(n')}, \tilde{w}_1^{(n')}, \tilde{w}_2^{(n')}, \ldots, \tilde{w}_K^{(n')}) \geq 0$. Thus, as $n \to \infty$, due to the continuity of the function $g(\cdot)$, the following inequality holds true

$$g(\tilde{s}^\star, \bar{w}_1^\star, \bar{w}_2^\star, \ldots, \bar{w}_K^\star) \;\leq\; g(\tilde{s}^\star, \tilde{w}_1^\star, \tilde{w}_2^\star, \ldots, \tilde{w}_K^\star)$$

This concludes the proof.

References

[1] Giuli D. 'Polarization diversity in radars'. *Proceedings of the IEEE*. 1986, vol. 74(2), pp. 245–69.

[2] Stiglitz M.R., Callahan J. 'Antennas for radar and communications: a polarimetric approach'. *Microwave Journal*. 1994, vol. 37(1), p. 169.

[3] Zrnic D.S., Ryzhkov A.V. 'Polarimetry for weather surveillance radars'. *Bulletin of the American Meteorological Society*. 1999, vol. 80(3), pp. 389–406.

[4] Lee J.S., Pottier E. 'Polarimetric radar imaging: from basics to applications radar imaging: from basics to applications' in Thompson B.J. (ed.). *Francis T*. Boca Raton, FL: CRC Press; 2009.

[5] Melvin W.L., Scheer J. 'Principles of modern radar: advanced techniques' in 2012 Jan 1.

[6] Pastina D., Lombardo P., Bucciarelli T. 'Adaptive polarimetric target detection with coherent radar. I. detection against gaussian background'. *IEEE Transactions on Aerospace and Electronic Systems*. 2001, vol. 37(4), pp. 1194–206.

[7] Lombardo P., Pastina D., Bucciarelli T. 'Adaptive polarimetric target detection with coherent radar. II. detection against non-gaussian background'. *IEEE Transactions on Aerospace and Electronic Systems*. 2001, vol. 37(4), pp. 1207–20.

[8] Park H.-R., Li J., Wang H. 'Polarization-space-time domain generalized likelihood ratio detection of radar targets'. *Signal Processing*. 1995, vol. 41(2), pp. 153–64.

[9] Park H.-R., Kwak Y.-G., Wang H. 'Efficient joint polarisation-space-time processor for nonhomogeneous clutter environments'. *Electron Lett*. 2002, vol. 38(25), p. 1714.

[10] Park H.-R., Wang H. 'Adaptive polarisation-space–time domain radar target detection in inhomogeneous clutter environments'. *IEE Proceedings - Radar, Sonar and Navigation*. 2006, vol. 153(1), pp. 35–43.

[11] De Maio A. 'Polarimetric adaptive detection of range-distributed targets'. *IEEE Transactions on Signal Processing*. 2002, vol. 50(9), pp. 2152–59.

[12] De Maio A., Alfano G. 'Polarimetric adaptive detection in non-gaussian noise'. *Signal Processing*. 2003, vol. 83(2), pp. 297–306.

[13] De Maio A., Alfano G., Conte E. 'Polarization diversity detection in compound-gaussian clutter'. *IEEE Transactions on Aerospace and Electronic Systems*. 2004, vol. 40(1), pp. 114–31.

[14] Hurtado M., Zhao T., Nehorai A. 'Adaptive polarized waveform design for target tracking based on sequential Bayesian inference'. *IEEE Transactions on Signal Processing*. 2008, vol. 56(3), pp. 1120–33.

[15] Cheng X., Shi L., Chang Y., Li Y., Wang X. 'Target scattering estimation in clutter with polarization optimization'. *EURASIP Journal on Advances in Signal Processing*. 2015, vol. 2015(1), pp. 1–12.

[16] Gini F., De Maio A., Patton L. *Waveform design and diversity for advanced radar systems*. London: Institution of Engineering and Technology; 2012 Jan 1. Available from https://digital-library.theiet.org/content/books/ra/pbra022e

[17] Li J., Guerci J.R., Xu L. 'Signal waveform's optimal under restriction design for active sensing'. *Fourth IEEE Workshop on Sensor Array and Multichannel Processing*; Waltham, MA, USA, 2006. pp. 382–86.

[18] De Maio A., De Nicola S., YongweiH., Shuzhong Z., Farina A. 'Code design to optimize radar detection performance under accuracy and similarity constraints'. *IEEE Transactions on Signal Processing*. 2008, vol. 56(11), pp. 5618–29.

[19] Stoica P., He H., Li J. 'Optimization of the receive filter and transmit sequence for active sensing'. *IEEE Transactions on Signal Processing*. 2012, vol. 60(4), pp. 1730–40.

[20] Aubry A., DeMaio A., Farina A., Wicks M. 'Knowledge-aided (potentially cognitive) transmit signal and receive filter design in signal-dependent clutter'. *IEEE Transactions on Aerospace and Electronic Systems*. 2013, vol. 49(1), pp. 93–117.

[21] Kay S. 'Waveform design for multistatic radar detection'. *IEEE Transactions on Aerospace and Electronic Systems*. 2009, vol. 45(3), pp. 1153–66.

[22] Aubry A., Maio A.D., Iommelli S., Piezzo M., Farina A., Wicks M. '2012 Tyrrhenian workshop on advances in radar and remote sensing (Tywrrs 2012)'. Naples, Italy, 2012.

[23] Aubry A., De Maio A., Naghsh M.M. 'Optimizing radar waveform and doppler filter bank via generalized fractional programming'. *IEEE Journal of Selected Topics in Signal Processing*. 2015, vol. 9(8), pp. 1387–99.

[24] Garren D.A., Odom A.C., Osborn M.K., Goldstein J.S., Pillai S.U., Guerci J.R. 'Full-polarization matched-illumination for target detection and identification'. *IEEE Transactions on Aerospace and Electronic Systems*. 2002, vol. 38(3), pp. 824–37.

[25] Chun-YangC., Vaidyanathan P.P. 'MIMO radar waveform optimization with prior information of the extended target and clutter'. *IEEE Transactions on Signal Processing*. 2009, vol. 57(9), pp. 3533–44.

[26] Cheng X., Aubry A., Ciuonzo D., De Maio A., Wang X. 'Robust waveform and filter bank design of polarimetric radar'. *IEEE Transactions on Aerospace and Electronic Systems*. 2017, vol. 53(1), pp. 370–84.

[27] GTRI *GTRI public dataset [online]*. 2016. Available from https://www.sdms.afrl.af.mil/index.php?collection=gtri

[28] Skolnik M.I. *Radar Handbook*. New York: McGraw-Hill; 1991.

[29] Karbasi S.M., Aubry A., De Maio A., Bastani M.H. 'Robust transmit code and receive filter design for extended targets in clutter'. *IEEE Transactions on Signal Processing*. 2015, vol. 63(8), pp. 1965–76.

[30] Novak L.M., Sechtin M.B., Cardullo M.J. 'Studies of target detection algorithms that use polarimetric radar data'. *IEEE Transactions on Aerospace and Electronic Systems*. 1989, vol. 25(2), pp. 150–65.

[31] Zebker H.A., Van Zyl J.J. 'Imaging radar polarimetry: a review'. *Proceedings of the IEEE*. 1991, vol. 79(11), pp. 1583–606.

[32] Novak L.M., Burl M.C., Irving W.W. 'Optimal polarimetric processing for enhanced target detection'. *IEEE Transactions on Aerospace and Electronic Systems*. 1993, vol. 29(1), pp. 234–44.

[33] Ben-Tal A., Nemirovski A. *Lectures on modern convex optimization [online]*. 2012. Available from http://www.isye.gatech.edu/faculty-staff/profile.php?entry=an63

[34] Trees V L.H. 'Detection of signals-estimation of signal parameters' in *Detection, estimation, and modulation theory, part I*. John Wiley & Sons; 2001. pp. 239–422.

[35] Li J., Stoica P., Zhisong W. 'On robust capon beamforming and diagonal loading'. *IEEE Transactions on Signal Processing*. 2003, vol. 51(7), pp. 1702–15.

[36] Crouzeix J.-P., Ferland J.A. 'Algorithms for generalized fractional programming'. *Math Program*. 1991, vol. 52(1–3), pp. 191–207.

[37] Barros A.I., Frenk J.B.G., Schaible S., Zhang S. 'A new algorithm for generalized fractional programs'. *Math Program*. 1996, vol. 72(2), pp. 147–75.

[38] Grant M., Boyd S. *CVX package [online]*. 2012. Available from http://www.cvxr.com/cvx

[39] Boyd S., Vanderberghe L. (eds.) *Convex optimization*. Cambridge: Cambridge University Press; 2004.
[40] Cochran D., Suvorova S., Howard S., Moran B. 'Waveform libraries'. *IEEE Signal Processing Magazine*. 2009, vol. 26(1), pp. 12–21.
[41] Giusti E., Martorella M., Capria A. 'Polarimetrically-persistent-scatterer-based automatic target recognition'. *IEEE Trans Geosci Remote Sens*. 2011, vol. 49(11), pp. 4588–99.
[42] *Basic knowledge of T-72 main battle tank [online]*. 2016. Available from http://en.wikipedia.org/wiki/T-72
[43] CVX Research I *CVX: matlab software for disciplined convex programming, version 2.0 [online]*. 2011. Available from http://cvxr.com/cvx
[44] Naghsh M.M., Soltanalian M., Stoica P., Modarres-Hashemi M. 'Radar code design for detection of moving targets'. *IEEE Transactions on Aerospace and Electronic Systems*. 2014, vol. 50(4), pp. 2762–78.
[45] Cantrell P., Ojha A. 'Comparison of generalized Q- function algorithms (corresp.)'. *IEEE Transactions on Information Theory*. 1987, vol. 33(4), pp. 591–96.

Chapter 10

An ambiguity suppression scheme for quad-pol SAR based on quasi-orthogonal waveforms

Guodong Jin[1], Augusto Aubry[2], Antonio De Maio[2], Robert Wang[3], and Wei Wang[4]

10.1 Introduction

Radar waveform design has received considerable attention along the years [1–3] and represents a key factor to boost radar performance. Quasi-orthogonal waveforms represent a promising methodology to suppress the unwanted ambiguity energy in synthetic aperture radar (SAR) systems [4, 5]. However, unlike communication systems, the orthogonality constraint for SAR demands more challenges, due to the need of achieving desired imaging performance in the presence of a distributed target scene. The cross-correlation energy (CCE) between the transmitted pulses, resulting from range-focusing process, introduces artifacts at different unambiguous range-azimuth positions, which can be either smeared or concentrated [6, 7]. Therefore, to mitigate the effects of ambiguities, quasi-orthogonal waveforms with a low CCE delineate a convenient option. However, according to the Parseval's theorem, Krieger [8] pointed out that waveforms with flat spectra (e.g. up-down chirp waveforms) present a fixed CCE value regardless of the specific pair, thus cueing the adoption of non-flat spectra to grant CCE reduction. Last but not least, SAR is an imaging sensor which calls for the transmitted waveforms to exhibit low peak to side lobe ratios (PSLRs) and integrated side lobe ratios (ISLRs) in order to ensure satisfactory imaging performance.

Before proceeding further, it is worth providing an overview on SAR systems that capitalize on polarimetric information to boost imaging capabilities. Quadrature-polarimetric (quad-pol) SAR aims at acquiring environmental scattering features

[1]Nanjing University of Aeronautics and Astronautics, College of Electronic and Information Engineering, Nanjing, China
[2]Universit degli Studi di Napoli "Federico II" Dipartimento di Ingegneria Elettrica e delle Tecnologie dell'Informazione, Napoli, Italy
[3]Chinese Academy of Sciences, Institute of Aerospace Information Research, Science and Technology on Microwave Imaging Laboratory, Beijing, China
[4]Department of Space Microwave Remote Sensing System, Chinese Academy of Sciences, Institute of Aerospace Information Research, Beijing, China

from all polarimetric components (HH, HV, VH, VV) and represents an important remote sensing instrument [9–11]. Existing quad-pol SAR systems are mainly realized according to the following transmission modes: (1) conventional quad-pol SAR that alternately transmits H- and V-polarized pulses [12], (2) hybrid quad-pol SAR that alternately transmits left- and right-circular polarizations [13], (3) $\pm\pi/4$ quad-pol SAR that alternately transmits (H − V) and (H + V) polarizations [14]. At the receiver end, the echoes from the foregoing transmission modes are gathered on both H and V channels. For a conventional quad-pol SAR, the full scattering matrix can be directly built up and this mode can suffer from severe range ambiguities in the cross-polarized (HV and VH) channels [12]. Hybrid and $\pm\pi/4$ quad-pol SAR systems employ a simple mathematical transformation to obtain the full scattering matrix (HH, HV, VH, VV components) starting from the collected measurements. Furthermore, hybrid and $\pm\pi/4$ quad-pol SAR systems guarantee a much better balance between the desired and the range ambiguous signals, ensuring reduced range ambiguities in the cross-polarized links [12]. On the other hand, it has been shown that hybrid and $\pm\pi/4$ quad-pol SARs suffer from severe azimuth ambiguities in the cross-polarized channels [15].

Azimuth phase coding (APC) [16] entails the use of an azimuth phase modulation on the transmitted pulses together with a matched demodulation on the receive side. It induces alternating 0 and π phases on the odd-numbered range ambiguous signals [16], corresponding to a shift in the Doppler frequency domain. Therefore, it can convert range ambiguities to azimuth ambiguities through azimuth phase modulation on transmission, and the range ambiguity energy can be filtered out in the azimuth dimension. Resorting to this technique, it has been shown that the range ambiguities in both conventional SAR and conventional quad-pol SAR systems can be suitably mitigated [17]. Another strategy with limited extra costs for range ambiguities suppression is the diverse transmitting scheme [4, 5, 18]. Unlike the APC method, the ambiguity suppression principle behind the alternating transmission scheme leverages the low CCE of the used signals. While the implementation of quasi-orthogonal waveforms in the conventional SAR system for range ambiguity suppression has been widely discussed [4, 5], its applicability to the quad-pol SAR has not yet been completely explored.

Considering many design requirements connected with the use of quasi-orthogonal waveforms for SAR, non-linear frequency modulation (NLFM) signals represent promising candidates. Along with the existing research that focused on NLFM signals' generation [19, 20] and optimization [21, 22], the implementation of the NLFM waveform with low sidelobes has been discussed in [23–25]. Nevertheless, all these methods only pay attention to the autocorrelation function behaviour. Moreover, a high-precision NLFM signal generator with the ability of predistortion compensation is developed and employed in the LuTan-1 (LT-1)* [26]. The successful use of NLFM signal transmissions in SAR is thus a fundamental reason as to why this chapter is focused on quasi-orthogonal NLFM signals in the context of polarimetric SAR systems.

*The LuTan-1 (LT-1) mission (approved in 2016 and planned to be launched in 2020) is an important part of the medium- and long-term development plan for China's civil space infrastructure. LT-1 has the remote-sensing capabilities of spaceborne L-band interferometry, differential interferometry and quad-pol interferometry.

10.1.1 Main contribution and organization of the chapter

The present study addresses the design and the implementation of quasi-orthogonal NLFM waveforms for ambiguity suppression in the quad-pol SAR system. The main contributions are:

- Design of quasi-orthogonal NLFM waveforms with a low CCE and PSLRs. The autocorrelation features of the synthesized signals are compatible with the imaging performance without windowing and can thus avoid the SNR loss.
- The range and azimuth ambiguity-to-signal ratios (RASR and AASR, respectively) for quad-pol SARs have been studied in [15]. However, since the effect of different transmitted waveforms is not considered therein, this chapter tries to fill this gap providing more general AASR and RASR evaluation formulas accounting for waveform diversity [27].
- Starting from the designed quasi-orthogonal NLFM waveforms, a new intertwined polarimetric transmission scheme, combining fast-time waveform diversity and APC, is proposed to improve ambiguity suppression performance.

The rest of the chapter is organized as follows. In section 10.2, the signal model and the optimization method to get quasi-orthogonal NLFM signals are discussed. Joint exploitation of quasi-orthogonal NLFM waveforms and APC in the conventional quad-pol SAR system is handled in section 10.3. Section 10.4 provides detailed simulation experiments to corroborate the theoretical developments. Finally, conclusions are drawn in section 10.5.

Notation: '$\leq$' is to be intended elementwise.

10.2 The design of quasi-orthogonal NLFM waveforms

10.2.1 Signal model

This section is focused on the design of NLFM waveforms constructed stacking together several LFM chirp sub-pulses, as depicted in Figure 10.1. Compared with the orthogonal frequency-division multiplexing (OFDM)-chirp waveforms [28], the instantaneous frequency is a continuous piecewise linear function.

Denoting the bandwidth and pulse length of the NLFM waveform by B and T, the signal can be constructed using $N-1$ chirp sub-pulses, as

$$\bar{s}(t) = \sum_{n=1}^{N-1} s_n(t-(n-1)T_s) \tag{10.1}$$

where

$$s_n(t) = \mathrm{rect}\left(\frac{t-T_s/2}{T_s}\right)\exp(j\pi\alpha_n t^2)\exp(j2\pi b_n t) \tag{10.2}$$

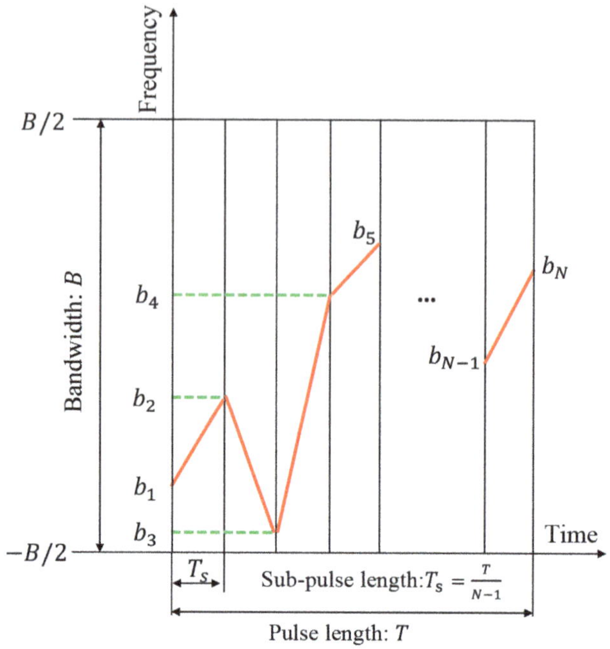

Figure 10.1 Illustration of the instantaneous frequency history

with

$$\text{rect}\,(t) = \begin{cases} 1, & -0.5 \leq t \leq 0.5 \\ 0, & \text{elsewhere} \end{cases} \tag{10.3}$$

where $T_s = \frac{T}{N-1}$ denotes the sub-pulse length, α_n and b_n are the frequency modulation slope and offset of the nth sub-pulse, respectively. In particular, to force instantaneous frequency continuity:

$$\alpha_n = \frac{b_{n+1} - b_n}{T_s} \tag{10.4}$$

where b_n and b_{n+1} are the start and final instantaneous frequencies of the nth sub-pulse, $n = 1, \cdots, N-1$. Obviously, the constructed NLFM waveform depends on the control point vector $\boldsymbol{b} = (b(1), b(2), \cdots, b(N))$ (where $b(n) = b_n$), which will be synthetically denoted by $\bar{s}(t; \boldsymbol{b})$.

10.2.2 Correlation function

The cross-correlation function between two NLFM waveforms $\bar{s}_1(t; \boldsymbol{b}_1)$ and $\bar{s}_2(t; \boldsymbol{b}_2)$ can be represented as

$$
\begin{aligned}
g(\tau;\boldsymbol{b}_1;\boldsymbol{b}_2) &= \int_{-\infty}^{\infty} \bar{s}_1(t;\boldsymbol{b}_1)\bar{s}_2^*(t-\tau;\boldsymbol{b}_2)dt \\
&= \int_{-\infty}^{\infty} \sum_{n=1}^{N-1} s_{1,n}\left(t-(n-1)T_s\right) \sum_{m=1}^{N-1} s_{2,m}^*\left(t-(m-1)T_s-\tau\right)dt \\
&= \sum_{n=1}^{N-1}\sum_{m=1}^{N-1} \int_{-\infty}^{\infty} s_{1,n}\left(t-(n-1)T_s\right) s_{2,m}^*\left(t-(m-1)T_s-\tau\right)dt \\
&= \sum_{n=1}^{N-1}\sum_{m=1}^{N-1} f_{n,m}^{1,2}(\tau)
\end{aligned}
\tag{10.5}
$$

where $s_{i,n}\left(t-(n-1)T_s\right)$ denotes the nth sub-pulse of $\bar{s}_i(t;\boldsymbol{b}_i)$ with $n \in \{1,\cdots,N-1\}$ and $i \in \{1,2\}$. In addition, $f_{n,m}^{1,2}(\tau)$ is the sub-pulse cross-correlation function, given by

$$
f_{n,m}^{1,2}(\tau) = \int_{-\infty}^{\infty} s_{1,n}\left(t-(n-1)T_s\right) s_{2,m}^*\left(t-(m-1)T_s-\tau\right)dt. \tag{10.6}
$$

Now, letting $t' = t-(n-1)T_s$ and denoting by $b_{i,n}$ and $\alpha_{i,n}$ the instantaneous frequency offset and slope of the nth sub-pulse of the ith signal $\bar{s}_i(t;\boldsymbol{b}_i)$, respectively (1.6) can be recast as

$$
\begin{aligned}
f_{n,m}^{1,2}(\tau) &= \int_{-\infty}^{\infty} s_{1,n}(t')\, s_{2,m}^*\left(t'+(n-m)T_s-\tau\right)dt' \\
&= \exp\left(-j\pi\alpha_{2,m}\tau^2\right)\exp\left(j2\pi\left(b_{2,m}+\alpha_{2,m}(n-m)T_s\right)\tau\right)\times \\
&\quad \exp\left(-j2\pi b_{2,m}(n-m)T_s\right)\exp\left(-j\pi\alpha_{2,m}(n-m)^2T_s^2\right)\times \\
&\quad \int_{-\infty}^{\infty} \mathrm{rect}\left(\frac{t'-T_s/2}{T_s}\right)\mathrm{rect}\left(\frac{t'+(n-m-1/2)T_s-\tau}{T_s}\right) \\
&\quad \times\exp\left(j\pi\left(\alpha_{1,n}-\alpha_{2,m}\right)t'^2\right)\times \\
&\quad \exp\left(j2\pi\left(\alpha_{2,m}\tau+b_{1,n}-b_{2,m}-\alpha_{2,m}(n-m)T_s\right)t'\right)dt' \\
&= \exp\left(-j\pi\alpha_{2,m}\tau^2\right)\exp\left(j2\pi\left(b_{1,n}-\Delta b_{n,m}^{1,2}\right)\tau\right)\Phi_{n,m}^{1,2}\times \\
&\quad \mathrm{rect}\left(\frac{\tau-(n-m)T_s}{2T_s}\right)\times \\
&\quad \int_{t_1'}^{t_2'} \exp\left(j\pi\Delta\alpha_{n,m}^{1,2}t'^2\right)\exp\left(j2\pi\left(\alpha_{2,m}\tau+\Delta b_{n,m}^{1,2}\right)t'\right)dt'
\end{aligned}
\tag{10.7}
$$

where

$$
\begin{aligned}
\Delta\alpha_{n,m}^{1,2} &= \alpha_{1,n}-\alpha_{2,m} \\
\Delta b_{n,m}^{1,2} &= b_{1,n}-b_{2,m}-\alpha_{2,m}(n-m)T_s \\
\Phi_{n,m}^{1,2} &= \exp\left(-j2\pi b_{2,m}(n-m)T_s\right)\exp\left(-j\pi\alpha_{2,m}(n-m)^2T_s^2\right)
\end{aligned}
\tag{10.8}
$$

and

$$
\begin{cases}
t_1' = 0 \quad \text{and} \quad t_2' = (1+m-n)T_s+\tau, \\
\qquad \text{if } (n-m-1)T_s < \tau \le (n-m)T_s \\
\\
t_1' = \tau+(m-n)T_s \quad \text{and} \quad t_2' = T_s, \\
\qquad \text{if } (n-m)T_s \le \tau \le (1+n-m)T_s
\end{cases}
\tag{10.9}
$$

To proceed further, different cases depending on the sign of $\Delta\alpha_{n,m}^{1,2}$ must be analysed. If $\Delta\alpha_{n,m}^{1,2} > 0$, (10.7) can be expressed as

$$\begin{aligned} f_{n,m}^{1,2}(\tau) &= \exp\left(-j\pi\left(\alpha_{2,m} + \frac{(\alpha_{2,m})^2}{\Delta\alpha_{n,m}^{1,2}}\right)\tau^2\right) \times \\ &\exp\left(j2\pi\left(b_{1,n} - \Delta b_{n,m}^{1,2} - \frac{\alpha_{2,m}\Delta b_{n,m}^{1,2}}{\Delta\alpha_{n,m}^{1,2}}\right)\tau\right)\Phi_{n,m}^{1,2} \times \\ &\exp\left(-j\pi\frac{(\Delta b_{n,m}^{1,2})^2}{\Delta\alpha_{n,m}^{1,2}}\right)\mathrm{rect}\left(\frac{\tau-(n-m)T_S}{2T_S}\right) \times \\ &\int_{t_1'}^{t_2'} \exp\left(j\pi\left(\sqrt{\Delta\alpha_{n,m}^{1,2}}t' + \frac{\alpha_{2,m}\tau + \Delta b_{n,m}^{1,2}}{\sqrt{\Delta\alpha_{n,m}^{1,2}}}\right)^2\right)dt'. \end{aligned} \tag{10.10}$$

Hence, indicating with:

$$\xi(t') = \sqrt{\Delta\alpha_{n,m}^{1,2}}t' + \frac{\alpha_{2,m} + \Delta b_{n,m}^{1,2}}{\sqrt{\Delta\alpha_{n,m}^{1,2}}} \tag{10.11}$$

and performing the change of variable:

$$\xi = \sqrt{\Delta\alpha_{n,m}^{1,2}}t' + \frac{\alpha_{2,m} + \Delta b_{n,m}^{1,2}}{\sqrt{\Delta\alpha_{n,m}^{1,2}}} \tag{10.12}$$

(10.10) can be further represented as

$$\begin{aligned} f_{n,m}^{1,2}(\tau) &= \exp\left(-j\pi\left(\alpha_{2,m} + \frac{(\alpha_{2,m})^2}{\Delta\alpha_{n,m}^{1,2}}\right)\tau^2\right) \times \\ &\exp\left(j2\pi\left(b_{1,n} - \Delta b_{n,m}^{1,2} - \frac{\alpha_{2,m}\Delta b_{n,m}^{1,2}}{\Delta\alpha_{n,m}^{1,2}}\right)\tau\right) \times \\ &\frac{\Phi_{n,m}^{1,2}\exp\left(-j\pi\frac{(\Delta b_{n,m}^{1,2})^2}{\Delta\alpha_{n,m}^{1,2}}\right)}{\sqrt{\Delta\alpha_{n,m}^{1,2}}}\mathrm{rect}\left(\frac{\tau-(n-m)T_s}{2T_s}\right)\int_{\xi(t_1')}^{\xi(t_2')}\exp\left(j\pi\xi^2\right)d\xi \\ &= \exp\left(-j\pi\left(\alpha_{2,m} + \frac{(\alpha_{2,m})^2}{\Delta\alpha_{n,m}^{1,2}}\right)\tau^2\right) \times \\ &\exp\left(j2\pi\left(b_{1,n} - \Delta b_{n,m}^{1,2} - \frac{\alpha_{2,m}\Delta b_{n,m}^{1,2}}{\Delta\alpha_{n,m}^{1,2}}\right)\tau\right)\frac{\Phi_{n,m}^{1,2}\exp\left(-j\pi\frac{(\Delta b_{n,m}^{1,2})^2}{\Delta\alpha_{n,m}^{1,2}}\right)}{\sqrt{\Delta\alpha_{n,m}^{1,2}}} \times \\ &\mathrm{rect}\left(\frac{\tau-(n-m)T_S}{2T_S}\right)\left(R\left(\xi\left(t_2'\right)\right) + jI\left(\xi\left(t_2'\right)\right) - R\left(\xi\left(t_1'\right)\right) - jI\left(\xi\left(t_1'\right)\right)\right) \end{aligned} \tag{10.13}$$

In (10.13), $R(\xi)$ and $I(\xi)$ are the Fresnel integrals [29] given in

$$\begin{aligned} R(\xi) &= \int_0^{\xi}\cos\left(\pi\xi^2\right)d\xi \\ I(\xi) &= \int_0^{\xi}\sin\left(\pi\xi^2\right)d\xi \end{aligned} \tag{10.14}$$

respectively. If $\Delta\alpha_{n,m}^{1,2} < 0$, (10.7) can be cast as

$$\begin{aligned}f_{n,m}^{1,2}(\tau) = &\exp\left(-j\pi\left(\alpha_{2,m}+\frac{(\alpha_{2,m})^2}{\Delta\alpha_{n,m}^{1,2}}\right)\tau^2\right)\times\\ &\exp\left(j2\pi\left(b_{1,n}-\Delta b_{n,m}^{1,2}-\frac{\alpha_{2,m}\Delta b_{n,m}^{1,2}}{\Delta\alpha_{n,m}^{1,2}}\right)\tau\right)\Phi_{n,m}^{1,2}\times\\ &\exp\left(-j\pi\frac{(\Delta b_{n,m}^{1,2})^2}{\Delta\alpha_{n,m}^{1,2}}\right)\operatorname{rect}\left(\frac{\tau-(n-m)T_S}{2T_S}\right)\times\\ &\int_{t_1'}^{t_2'}\exp\left(-j\pi\left(\sqrt{-\Delta\alpha_{n,m}^{1,2}}t'-\frac{\alpha_{2,m}\tau+\Delta b_{n,m}^{1,2}}{\sqrt{-\Delta\alpha_{n,m}^{1,2}}}\right)^2\right)dt'\end{aligned} \tag{10.15}$$

Thus, letting:

$$\xi'(t') = \sqrt{-\Delta\alpha_{n,m}^{1,2}}t' - \frac{\alpha_{2,m}+\Delta b_{n,m}^{1,2}}{\sqrt{-\Delta\alpha_{n,m}^{1,2}}} \tag{10.16}$$

and performing the change of variable:

$$\xi' = \sqrt{-\Delta\alpha_{nm}^{1,2}}t' - \frac{\alpha_{2,m}+\Delta b_{n,m}^{1,2}}{\sqrt{-\Delta\alpha_{n,m}^{1,2}}} \tag{10.17}$$

(10.15) can be further written as

$$\begin{aligned}f_{n,m}^{1,2}(\tau) = &\exp\left(-j\pi\left(\alpha_{2,m}+\frac{(\alpha_{2,m})^2}{\Delta\alpha_{n,m}^{1,2}}\right)\tau^2\right)\times\\ &\exp\left(j2\pi\left(b_{1,n}-\Delta b_{n,m}^{1,2}-\frac{\alpha_{2,m}\Delta b_{n,m}^{1,2}}{\sqrt{\Delta\alpha_{n,m}^{1,2}}}\right)\tau\right)\times\\ &\frac{\Phi_{n,m}^{1,2}\exp\left(-j\pi\frac{(\Delta b_{n,m}^{1,2})^2}{\Delta\alpha_{n,m}^{1,2}}\right)}{\sqrt{-\Delta\alpha_{n,m}^{1,2}}}\operatorname{rect}\left(\frac{\tau-(n-m)T_s}{2T_s}\right)\times\\ &\left(R\left(\xi'\left(t_2'\right)\right)-jI\left(\xi'\left(t_2'\right)\right)-R\left(\xi'\left(t_1'\right)\right)+jI\left(\xi'\left(t_1'\right)\right)\right)\end{aligned} \tag{10.18}$$

Finally, if $\Delta\alpha_{n,m}^{1,2}=0$, (10.7) can be reformulated as (see Appendix I of [27]):

$$\begin{aligned}f_{n,m}^{1,2}(\tau) = &\exp\left(-j\pi\alpha_{1,n}\tau^2\right)\exp\left(j2\pi\left(b_{1,n}-\Delta b_{n,m}^{1,2}\right)\tau\right)\times\\ &\Phi_{n,m}^{1,2}\exp\left(j\pi\left(\alpha_{1,n}\tau+\Delta b_{n,m}^{1,2}\right)\left(T_s+(m-n)T_s+\tau\right)\right)\times\\ &\left(T_s-\left|\tau+(m-n)T_s\right|\right)\operatorname{rect}\left(\frac{\tau-(n-m)T_S}{2T_S}\right)\times\\ &\operatorname{sinc}\left(\left(\alpha_{1,n}\tau+\Delta b_{n,m}^{1,2}\right)\left(T_s-\left|\tau+(m-n)T_s\right|\right)\right)\end{aligned} \tag{10.19}$$

Following the same line of reasoning as in (10.5), the autocorrelation function of $\bar{s}_i(t;\boldsymbol{b}_i)$ can be represented as

$$h(\tau;\boldsymbol{b}_i) = \sum_{n=1}^{N-1}\sum_{m=1}^{N-1} f_{n,m}^{i,i}(\tau) = \sum_{n=1}^{N-1} f_{n,n}^{i,i}(\tau) + \sum_{n=1}^{N-1}\sum_{m=1,m\neq n}^{N-1} f_{n,m}^{i,i}(\tau) \tag{10.20}$$

where $f_{n,n}^{i,i}(\tau)$ is the autocorrelation of the nth sub-pulse ($f_{n,m}^{i,i}(\tau)$ is the cross-correlation function between nth sub-pulse and mth sub-pulse) of $\bar{s}_i(t;\boldsymbol{b}_i)$, which stems from (1.9) (setting $n=m$, $\Delta b_{n,m}^{1,2}=0$ and $\Phi_{n,m}^{1,2}=1$), i.e.:

$$f_{n,n}^{i,i}(\tau) = (T_s - |\tau|)\exp\left(j\pi(\alpha_{i,n}T_s + 2b_{i,n})\tau\right)\operatorname{rect}\left(\frac{\tau}{2T_s}\right)\operatorname{sinc}\left(\alpha_{i,n}\tau\left(T_s - |\tau|\right)\right). \tag{10.21}$$

10.2.3 Optimization method

Both PSLR and ISLR metrics are commonly used to evaluate the imaging performance of a given waveform. They are formally defined, respectively, as

$$f_{\text{PSLR}}(\boldsymbol{b}) = \max_{\tau}\left|\frac{h(\tau;\boldsymbol{b})}{h(0;\boldsymbol{b})}\right|, \tau \in [\tau_m, T] \tag{10.22}$$

$$f_{\text{ISLR}}(\boldsymbol{b}) = \frac{\int_{\tau_m}^{T}|h(\tau;\boldsymbol{b})|^2 d\tau}{\int_0^{\tau_m}|h(\tau;\boldsymbol{b})|^2 d\tau} \tag{10.23}$$

where the interval $[-\tau_m, \tau_m]$ corresponds to the main-lobe temporal support with τ_m depending, in general, on $\boldsymbol{b}$. Besides, the CCE can be used as performance metric to assess the degree of orthogonality between waveforms and is defined as

$$f_{\text{CCE}}(\boldsymbol{b}_1;\boldsymbol{b}_2) = \int_{-T}^{T}|g(\tau;\boldsymbol{b}_1;\boldsymbol{b}_2)|^2 d\tau \tag{10.24}$$

Generally, the PSLR should be lower than −20 dB. Furthermore, considering the time-varying system error, the PSLR of an ideal waveform should be around −35 dB [25]. In the sequel, both the CCE and PSLR are considered as performance indicators to address the problem of designing NLFM signals with piecewise linear frequency functions. The aforementioned problem can be formulated as the following constrained optimization:

$$\mathscr{P}(\boldsymbol{b}_1;\boldsymbol{b}_2)\begin{cases}\min\limits_{\boldsymbol{b}_1;\boldsymbol{b}_2} f_{\text{CCE}}(\boldsymbol{b}_1;\boldsymbol{b}_2)\\ \text{s.t. } f_{\text{PSLR}}(\boldsymbol{b}_1) \le PSLR_1, f_{\text{PSLR}}(\boldsymbol{b}_2) \le PSLR_2,\\ \boldsymbol{b}_1, \boldsymbol{b}_2 \in \Omega\end{cases} \tag{10.25}$$

where both $PSLR_1$ and $PSLR_2$ are design thresholds for PSLRs, while Ω identifies the feasible set of the vector $\boldsymbol{b}$, i.e.:

$$\Omega = \left\{\boldsymbol{b} \in \mathbb{R}^N : b(1) = -\frac{B}{2}, b(N) = \frac{B}{2}, -\frac{B}{2} \le \boldsymbol{b} \le \frac{B}{2}\right\}. \tag{10.26}$$

As to the non-convex problem $\mathscr{P}(\boldsymbol{b}_1;\boldsymbol{b}_2)$, the coordinate descent (CD) method is employed to approximatively solve it and the corresponding design scheme for quasi-orthogonal NLFM waveforms is summarized in Table 10.1, where both $\boldsymbol{b}_1^{(l)}$ and $\boldsymbol{b}_2^{(l)}$ denote the optimized vectors at the step l. Moreover, the efficient search method proposed in [30] (the corresponding ©MATLAB function is "*GlobalSearch*") is used to solve $\mathscr{P}\left(\boldsymbol{b}_1^{(l)};\boldsymbol{b}_2\right)$ and $\mathscr{P}\left(\boldsymbol{b}_1;\boldsymbol{b}_2^{(l+1)}\right)$.

Table 10.1 Generation of quasi-orthogonal NLFM waveforms

Input: Signal parameters B and T, initial feasible frequency control vector $\boldsymbol{b}_{\text{Initial}}$ and minimum required improvement δ;

Output: CD solution for the first ($\boldsymbol{b}_1^{\text{CD}}$) and the second ($\boldsymbol{b}_2^{\text{CD}}$) signal:

(1) Initialization.

- Set $l = 0$ and construct the initial frequency control vectors for the first and the second signal as $\boldsymbol{b}_1^{(l)} = \boldsymbol{b}_{\text{Initial}}$ and $\boldsymbol{b}_2^{(l)} = -\boldsymbol{b}_{\text{Initial}}$.

(2) Improvement.

- Fix on $\boldsymbol{b}_1^{(l)}$ and optimize $\boldsymbol{b}_2$, i.e., solve $\mathscr{P}\left(\boldsymbol{b}_1^{(l)}; \boldsymbol{b}_2\right)$ with the initial value $\boldsymbol{b}_2 = -\boldsymbol{b}_1^{(l)}$ to obtain $\boldsymbol{b}_2^{(l+1)}$.
- Fix on $\boldsymbol{b}_2^{(l+1)}$ and optimize $\boldsymbol{b}_1$, i.e., solve $\mathscr{P}\left(\boldsymbol{b}_1; \boldsymbol{b}_2^{(l+1)}\right)$ with the initial value $\boldsymbol{b}_1 = -\boldsymbol{b}_2^{(l+1)}$ to obtain $\boldsymbol{b}_1^{(l+1)}$.

(3) Stopping Criterion.

- If $\left| f_{\text{CCE}}\left(\boldsymbol{b}_1^{(l+1)}; \boldsymbol{b}_2^{(l+1)}\right) - f_{\text{CCE}}\left(\boldsymbol{b}_1^{(l)}; \boldsymbol{b}_2^{(l)}\right) \right| < \delta$, stop. Otherwise, $l = l + 1$ and go to step (2).

(4) Output.

- Set $\boldsymbol{b}_1^{\text{CD}} = \boldsymbol{b}_1^{(l+1)}$ and $\boldsymbol{b}_2^{\text{CD}} = \boldsymbol{b}_2^{(l+1)}$.

10.3 Conventional quad-pol SAR

In this section, the exploitation of quasi-orthogonal NLFM waveforms (possibly jointly combined with APC) in the conventional quad-pol SAR system is addressed. Furthermore, the corresponding AASR and RASR evaluation formulas are reported.

10.3.1 Ambiguity performance using quasi-orthogonal NLFM waveforms

A conventional quad-pol system is realized by alternately transmitting H and V pulses and simultaneously receiving both H-polarized and V-polarized echoes. The received pulses in the range frequency (f_t) and azimuth time (η) domains can be represented as [12]:

$$\begin{pmatrix} \hat{S}_{\text{HH}}(f_t,\eta) & \hat{S}_{\text{HV}}(f_t,\eta) \\ \hat{S}_{\text{VH}}(f_t,\eta) & \hat{S}_{\text{VV}}(f_t,\eta) \end{pmatrix} = \begin{pmatrix} \tilde{S}_{\text{HH}}(f_t,\eta) & \tilde{S}_{\text{HV}}(f_t,\eta) \\ \tilde{S}_{\text{VH}}(f_t,\eta) & \tilde{S}_{\text{VV}}(f_t,\eta) \end{pmatrix}$$
$$+ \sum_{k=\text{odd}} \begin{pmatrix} \check{S}_{\text{HV}_k}(f_t,\eta) & \check{S}_{\text{HH}_k}(f_t,\eta) \\ \check{S}_{\text{VV}_k}(f_t,\eta) & \check{S}_{\text{VH}_k}(f_t,\eta) \end{pmatrix} \tag{10.27}$$
$$+ \sum_{k=\text{even}} \begin{pmatrix} \check{S}_{\text{HV}_k}(f_t,\eta) & \check{S}_{\text{HH}_k}(f_t,\eta) \\ \check{S}_{\text{VV}_k}(f_t,\eta) & \check{S}_{\text{VH}_k}(f_t,\eta) \end{pmatrix}$$

where $\hat{S}_{pq}(f_t, \eta)$ $(p, q \in \{H, V\})$ indicates the observed signals at the pq polarimetric channel (p in Rx and q in Tx), $\tilde{S}_{pq}(f_t, \eta)$ and $\check{S}_{pq_k}(f_t, \eta)$ denote the desired and range ambiguous scattered signals, respectively, and the subscript $k \in \mathbb{N}, k \neq 0$ refers to the kth range ambiguity. The RASR is defined as the ratio between the range ambiguity and the desired signal power [31–33]. Here, considering the amplitude weighting of the Doppler spectrum $W(f_\eta)$, the RASR after range-matched filtering for a single polarization SAR system without waveform diversity can be computed as [15]:

$$\mathrm{RASR} \cong \frac{\sum_{k=1}^{N_A} \frac{\sigma^0(\phi_k) \int_{-Ba/2}^{Ba/2} G_r^2(\theta_k, f_\eta) W^2(f_\eta) df_\eta}{R_k^3 \sin(\phi_k)}}{\frac{\sigma^0(\phi_0) \int_{-Ba/2}^{Ba/2} G_r^2(\theta_0, f_\eta) W^2(f_\eta) df_\eta}{R_0^3 \sin(\phi_0)}} \tag{10.28}$$

where $\sigma^0(\phi)$ is the backscatter reflectivity versus the incidence angle ϕ, $G_r^2(\theta, f_\eta)$ is the two-way antenna power pattern in elevation as a function of the look angle θ and the azimuth frequency f_η, whereas B_a is the processed Doppler bandwidth; θ_k and ϕ_k (θ_0 and ϕ_0) are the look angle and incidence angle for the kth range ambiguity (desired scattered signal), respectively, and N_A indicates the total number of range ambiguities. Besides, R_0 and R_k are the range for the desired echo and for the kth range ambiguity signal, respectively, i.e.:

$$R_k = R_0 + k\frac{\mathrm{PRI}}{2}c, \tag{10.29}$$

where PRI denotes the pulse repetition interval and c is the speed of light.

Unlike [12] and [15], in this chapter, the effect of the transmitted waveforms is accounted for using diverse pulses $s_1(t)$ and $s_2(t)$ for the H and V polarization,

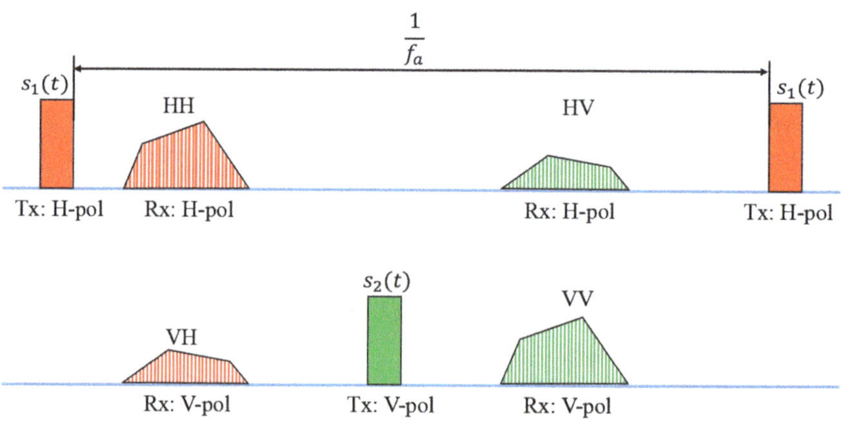

Figure 10.2 Timing diagram for a conventional quad-pol SAR system, where f_a denotes the PRF for the transmitted pulses sharing the same polarization

respectively. The corresponding timing diagram is illustrated in Figure 10.2 and the range compression processing in each channel is shown in Table 10.2, where $S_i(f_t)$ is the Fourier transform of $s_i(t)$. Precisely, the range compressed signal in the range frequency and azimuth time domains can be cast as

$$\begin{aligned}
&\begin{pmatrix} \hat{S}^1_{\mathrm{HH}}(f_t,\eta) & \hat{S}^2_{\mathrm{HV}}(f_t,\eta) \\ \hat{S}^1_{\mathrm{VH}}(f_t,\eta) & \hat{S}^2_{\mathrm{VV}}(f_t,\eta) \end{pmatrix} \odot \begin{pmatrix} S_1^*(f_t) & S_2^*(f_t) \\ S_1^*(f_t) & S_2^*(f_t) \end{pmatrix} = \\
&\left\{ \begin{pmatrix} \tilde{S}^1_{\mathrm{HH}}(f_t,\eta) & \tilde{S}^2_{\mathrm{HV}}(f_t,\eta) \\ \tilde{S}^1_{\mathrm{VH}}(f_t,\eta) & \tilde{S}^2_{\mathrm{VV}}(f_t,\eta) \end{pmatrix} + \sum_{k=\mathrm{odd}} \begin{pmatrix} \check{S}^2_{\mathrm{HV}_k}(f_t,\eta) & \check{S}^1_{\mathrm{HH}_k}(f_t,\eta) \\ \check{S}^2_{\mathrm{VV}_k}(f_t,\eta) & \check{S}^1_{\mathrm{VH}_k}(f_t,\eta) \end{pmatrix} \right. \\
&\left. + \sum_{k=\mathrm{even}} \begin{pmatrix} \check{S}^1_{\mathrm{HH}_k}(f_t,\eta) & \check{S}^2_{\mathrm{HV}_k}(f_t,\eta) \\ \check{S}^1_{\mathrm{VH}_k}(f_t,\eta) & \check{S}^2_{\mathrm{VV}_k}(f_t,\eta) \end{pmatrix} \right\} \odot \begin{pmatrix} S_1^*(f_t) & S_2^*(f_t) \\ S_1^*(f_t) & S_2^*(f_t) \end{pmatrix} \\
&= \begin{pmatrix} \tilde{S}^{1,1}_{\mathrm{HH}}(f_t,\eta) & \tilde{S}^{2,2}_{\mathrm{HV}}(f_t,\eta) \\ \tilde{S}^{1,1}_{\mathrm{VH}}(f_t,\eta) & \tilde{S}^{2,2}_{\mathrm{VV}}(f_t,\eta) \end{pmatrix} + \sum_{k=\mathrm{odd}} \begin{pmatrix} \check{S}^{2,1}_{\mathrm{HV}_k}(f_t,\eta) & \check{S}^{1,2}_{\mathrm{HH}_k}(f_t,\eta) \\ \check{S}^{2,1}_{\mathrm{VV}_k}(f_t,\eta) & \check{S}^{1,2}_{\mathrm{VH}_k}(f_t,\eta) \end{pmatrix} \\
&+ \sum_{k=\mathrm{even}} \begin{pmatrix} \check{S}^{1,1}_{\mathrm{HH}_k}(f_t,\eta) & \check{S}^{2,2}_{\mathrm{HV}_k}(f_t,\eta) \\ \check{S}^{1,1}_{\mathrm{VH}_k}(f_t,\eta) & \check{S}^{2,2}_{\mathrm{VV}_k}(f_t,\eta) \end{pmatrix}
\end{aligned} \tag{10.30}$$

where$\hat{S}^i_{pq}(f_t,\eta)(p,q\in\{\mathrm{H},\mathrm{V}\},i\in\{1,2\})$indicatesthesignalatthe$pq$polarimetricchannel generated via the waveform $S_i(f_t)$.$S\sim^{i,j}_{pq}(f_t,\eta)$ and $\check{S}^{i,j}_{pq_k}(f_t,\eta)\,(p,q\in\{\mathrm{H},\mathrm{V}\},i,j\in\{1,2\})$ denotes the outputs of the range-matched filter $S_j^*(f_t)$ applied to the signal $\tilde{S}^i_{pq}(f_t,\eta)$ (desired return generated by $S_i(f_t)$) and $\check{S}^i_{pq_k}(f_t,\eta)$ (kth range ambiguous contribution produced by $S_i(f_t)$), respectively.

Moreover, using the specific value of the backscatter reflectivity for the contribution of each channel, the RASR can be represented as (10.31), where the antenna patterns are assumed to be equal for all polarizations. Cross-polarized returns (viz. $\sigma^0_{\mathrm{HV}}(\phi_k),\sigma^0_{\mathrm{VH}}(\phi_k)$) are generally 6–10 dB weaker than their co-polarized counterparts (viz. $\sigma^0_{\mathrm{HH}}(\phi_k),\sigma^0_{\mathrm{VV}}(\phi_k)$). Moreover, the odd-numbered range ambiguities in cross-polarized (HV and VH) channels are due to co-polarized (viz. HH and VV) contributions. Therefore, the range ambiguities in cross-polarized channel can be very strong.

Table 10.2 Range compression processing for the desired and range ambiguous signals

Channel	**Desired signal**	**Odd ambiguities**	**Even ambiguities**
HH	$S_1(f_t)S_1^*(f_t)$	$S_2(f_t)S_1^*(f_t)$	$S_1(f_t)S_1^*(f_t)$
HV	$S_2(f_t)S_2^*(f_t)$	$S_1(f_t)S_2^*(f_t)$	$S_2(f_t)S_2^*(f_t)$
VH	$S_1(f_t)S_1^*(f_t)$	$S_2(f_t)S_1^*(f_t)$	$S_1(f_t)S_1^*(f_t)$
VV	$S_2(f_t)S_2^*(f_t)$	$S_1(f_t)S_2^*(f_t)$	$S_2(f_t)S_2^*(f_t)$

In general, the introduction of diverse quasi-orthogonal NLFM waveforms leads to a decrease in CCE, i.e., $\int_{-T}^{T}|g(\tau,\boldsymbol{b}_1,\boldsymbol{b}_2)|^2 d\tau$, with the final goal of reducing significantly the energy of the ambiguities:

$$
\begin{aligned}
\mathrm{RASR}_{\mathrm{HH}} &\cong \frac{\sum_{k=\mathrm{odd}}^{N_A} \dfrac{\sigma_{\mathrm{HV}}^{0}(\phi_k)\int_{-\tau m}^{\tau m}|g(\tau;\boldsymbol{b}_2;\boldsymbol{b}_1)|^2 d\tau \int_{-Ba/2}^{Ba/2} G_r^2(\theta_k,f_n)\,W^2(f_n)df_n}{R_k^3\sin(\phi_k)}}{\dfrac{\sigma_{\mathrm{HH}}^{0}(\phi_0)\int_{-\tau m}^{\tau m}|h(\tau;\boldsymbol{b}_1)|^2 d\tau \int_{-Ba/2}^{Ba/2} G_r^2(\theta_0,f_n)\,W^2(f_n)df_n}{R_0^3\sin(\phi_0)}} \\
&+ \frac{\sum_{k=\mathrm{even},\neq 0}^{N_A} \dfrac{\sigma_{\mathrm{HH}}^{0}(\phi_k)\int_{-Ba/2}^{Ba/2} G_r^2(\theta_k,f_n)\,W^2(f_n)df_n}{R_k^3\sin(\phi_k)}}{\dfrac{\sigma_{\mathrm{HH}}^{0}(\phi_0)\int_{-Ba/2}^{Ba/2} G_r^2(\theta_0,f_n)\,W^2(f_n)df_n}{R_0^3\sin(\phi_0)}} \\
\mathrm{RASR}_{\mathrm{HV}} &\cong \frac{\sum_{k=\mathrm{odd}}^{N_A} \dfrac{\sigma_{\mathrm{HH}}^{0}(\phi_k)\int_{-\tau m}^{\tau m}|g(\tau;\boldsymbol{b}_1;\boldsymbol{b}_2)|^2 d\tau \int_{-Ba/2}^{Ba/2} G_r^2(\theta_k,f_n)\,W^2(f_n)df_n}{R_k^3\sin(\phi_k)}}{\dfrac{\sigma_{\mathrm{HV}}^{0}(\phi_0)\int_{-\tau m}^{\tau m}|h(\tau;\boldsymbol{b}_2)|^2 d\tau \int_{-Ba/2}^{Ba/2} G_r^2(\theta_0,f_n)\,W^2(f_n)df_n}{R_0^3\sin(\phi_0)}} \\
&+ \frac{\sum_{k=\mathrm{even},\neq 0}^{N_A} \dfrac{\sigma_{\mathrm{HV}}^{0}(\phi_k)\int_{-Ba/2}^{Ba/2} G_r^2(\theta_k,f_n)\,W^2(f_n)df_n}{R_k^3\sin(\phi_k)}}{\dfrac{\sigma_{\mathrm{HV}}^{0}(\phi_0)\int_{-Ba/2}^{Ba/2} G_r^2(\theta_0,f_n)\,W^2(f_n)df_n}{R_0^3\sin(\phi_0)}} \\
\mathrm{RASR}_{\mathrm{VH}} &\cong \frac{\sum_{k=\mathrm{odd}}^{N_A} \dfrac{\sigma_{\mathrm{VV}}^{0}(\phi_k)\int_{-\tau m}^{\tau m}|g(\tau;\boldsymbol{b}_2;\boldsymbol{b}_1)|^2 d\tau \int_{-Ba/2}^{Ba/2} G_r^2(\theta_k,f_n)\,W^2(f_n)df_n}{R_k^3\sin(\phi_k)}}{\dfrac{\sigma_{\mathrm{VH}}^{0}(\phi_0)\int_{-\tau m}^{\tau m}|h(\tau;\boldsymbol{b}_1)|^2 d\tau \int_{-Ba/2}^{Ba/2} G_r^2(\theta_0,f_n)\,W^2(f_n)df_n}{R_0^3\sin(\phi_0)}} \\
&+ \frac{\sum_{k=\mathrm{even},\neq 0}^{N_A} \dfrac{\sigma_{\mathrm{VH}}^{0}(\phi_k)\int_{-Ba/2}^{Ba/2} G_r^2(\theta_k,f_n)\,W^2(f_n)df_n}{R_k^3\sin(\phi_k)}}{\dfrac{\sigma_{\mathrm{VH}}^{0}(\phi_0)\int_{-Ba/2}^{Ba/2} G_r^2(\theta_0,f_n)\,W^2(f_n)df_n}{R_0^3\sin(\phi_0)}}
\end{aligned}
\tag{10.31}
$$

$$\mathrm{RASR_{VV}} \cong \frac{\sum_{k=\mathrm{odd}}^{N_A} \dfrac{\sigma_{\mathrm{VH}}^0(\phi_k) \int_{-\tau_m}^{\tau_m} |g(\tau; \boldsymbol{b}_1; \boldsymbol{b}_2)|^2 d\tau \int_{-B_a/2}^{B_a/2} G_r^2(\theta_k, f_\eta) W^2(f_\eta) df_\eta}{R_k^3 \sin(\phi_k)}}{\dfrac{\sigma_{\mathrm{VV}}^0(\phi_0) \int_{-\tau_m}^{\tau_m} |h(\tau; \boldsymbol{b}_2)|^2 d\tau \int_{-B_a/2}^{B_a/2} G_r^2(\theta_0, f_\eta) W^2(f_\eta) df_\eta}{R_0^3 \sin(\phi_0)}} + \frac{\sum_{k=\mathrm{even},\neq 0}^{N_A} \dfrac{\sigma_{\mathrm{VV}}^0(\phi_k) \int_{-B_a/2}^{B_a/2} G_r^2(\theta_k, f_\eta) W^2(f_\eta) df_\eta}{R_k^3 \sin(\phi_k)}}{\dfrac{\sigma_{\mathrm{VV}}^0(\phi_0) \int_{-B_a/2}^{B_a/2} G_r^2(\theta_0, f_\eta) W^2(f_\eta) df_\eta}{R_0^3 \sin(\phi_0)}}$$

The AASR is defined as the ratio between the azimuth ambiguous signal power and the main signal power for a uniform backscatter scene [34]. In the conventional quad-pol SAR, the AASR of each polarimetric channel after range-matched filtering is given by Villano *et al.* [15]:

$$\mathrm{AASR} \cong \frac{\sum_{k=-\infty,\neq 0}^{\infty} \int_{-B_a/2}^{B_a/2} G_a^2(f_\eta + kf_a) W^2(f_\eta) df_\eta}{\int_{-B_a/2}^{B_a/2} G_a^2(f_\eta) W^2(f_\eta) df_\eta}, \tag{10.32}$$

where $G_a^2(f_\eta)$ is the two-way antenna power pattern in azimuth. Such expression remains unchanged in the presence of diverse pulses.

10.3.2 Polarimetric probing strategy via combination of quasi-orthogonal NLFM waveforms and APC

Table 10.3 shows an intertwined combination scheme of quasi-orthogonal NLFM waveforms and APC method with the corresponding matched filtering outputs listed in Table 10.4. It can be seen that the periodic changes (at a given polarimetric channel) for the odd-numbered ambiguous signal corresponding to Tx-H and Tx-V are $\{S_1^*(f_t)S_2(f_t), S_1^*(f_t)S_2(f_t)e^{-j\pi}\}$ and $\{S_1(f_t)S_2^*(f_t), S_1(f_t)S_2^*(f_t)e^{j\pi}\}$, respectively. Therefore, the odd-numbered ambiguities experience an improved mitigation effect

Table 10.3 Combination of the quasi-orthogonal NLFM waveforms and APC method

Tx polarization	**H**	**V**	**H**	**V**
Transmitted signal	$S_1(f_t)$	$S_2(f_t)$	$S_1(f_t)e^{j\pi}$	$S_2(f_t)$
Odd-numbered ambiguous signal	$S_2(f_t)$	$S_1(f_t)$	$S_2(f_t)$	$S_1(f_t)e^{j\pi}$
Even-numbered ambiguous signal	$S_1(f_t)e^{j\pi}$	$S_2(f_t)$	$S_1(f_t)$	$S_2(f_t)$
Range-matched filter	$S_1^*(f_t)$	$S_2^*(f_t)$	$S_1^*(f_t)e^{-j\pi}$	$S_2^*(f_t)$

Table 10.4 Matched filter output (in the frequency domain) for the desired signal and range ambiguous signal

Tx polarization	H	V	H	V
Desired signal	$\vert S_1(f_t)\vert^2$	$\vert S_2(f_t)\vert^2$	$\vert S_1(f_t)\vert^2$	$\vert S_2(f_t)\vert^2$
Odd-numbered ambiguous signal	$S_1^*(f_t)S_2(f_t)$	$S_1(f_t)S_2^*(f_t)$	$S_1^*(f_t)S_2(f_t)e^{-j\pi}$	$S_1(f_t)S_2^*(f_t)e^{j\pi}$
Even-numbered ambiguous signal	$\vert S_1(f_t)\vert^2 e^{j\pi}$	$\vert S_2(f_t)\vert^2$	$\vert S_1(f_t)\vert^2 e^{-j\pi}$	$\vert S_2(f_t)\vert^2$

produced by the joint use of waveform diversity and APC. The resulting RASRs are given in (10.33), and the AASR is the same with (10.32):

$$\mathrm{RASR}_{\mathrm{HH}}^{\mathrm{APC}} \cong \frac{\sum\limits_{k=\mathrm{even},\neq 0}^{N_A} \dfrac{\sigma_{\mathrm{HH}}^0(\phi_k)\int\limits_{-Ba/2}^{Ba/2} G_r^2(\theta_k,f_n)\,W^2(f_n)df_n}{R_k^3\sin(\phi_k)}}{\dfrac{\sigma_{\mathrm{HH}}^0(\phi_0)\int_{-Ba/2}^{Ba/2} G_r^2(\theta_0,f_n)\,W^2(f_n)df_n}{R_0^3\sin(\phi_0)}} + \frac{\sum\limits_{k=\mathrm{odd}}^{N_A} \dfrac{\sigma_{\mathrm{HV}}^0(\phi_k)\int\limits_{-\tau m}^{\tau m} |g(\tau;\boldsymbol{b}_2;\boldsymbol{b}_1)|^2 d\tau \int\limits_{-Ba/2}^{Ba/2}\left[G_r^2\left(\theta_k,f_n+\dfrac{f_a}{2}\right)+G_r^2\left(\theta_k,f_n-\dfrac{f_a}{2}\right)\right]W^2(f_n)df_n}{\int_{-\tau m}^{\tau m}|h(\tau;\boldsymbol{b}_1)|^2 d\tau R_k^3\sin(\phi_k)}}{\dfrac{\sigma_{\mathrm{HH}}^0(\phi_0)\int_{-Ba/2}^{Ba/2} G_r^2(\theta_0,f_n)\,W^2(f_n)df_n}{R_0^3\sin(\phi_0)}}$$

$$\mathrm{RASR}_{\mathrm{HV}}^{\mathrm{APC}} \cong \frac{\sum\limits_{k=\mathrm{even},\neq 0}^{N_A} \dfrac{\sigma_{\mathrm{HV}}^0(\phi_k)\int\limits_{-Ba/2}^{Ba/2} G_r^2(\theta_k,f_n)\,W^2(f_n)df_n}{R_k^3\sin(\phi_k)}}{\dfrac{\sigma_{\mathrm{HV}}^0(\phi_0)\int_{-Ba/2}^{Ba/2} G_r^2(\theta_0,f_n)\,W^2(f_n)df_n}{R_0^3\sin(\phi_0)}} + \frac{\sum\limits_{k=\mathrm{odd}}^{N_A} \dfrac{\sigma_{\mathrm{HH}}^0(\phi_k)\int\limits_{-\tau m}^{\tau m} |g(\tau;\boldsymbol{b}_1;\boldsymbol{b}_2)|^2 d\tau \int\limits_{-Ba/2}^{Ba/2}\left[G_r^2\left(\theta_k,f_n+\dfrac{f_a}{2}\right)+G_r^2\left(\theta_k,f_n-\dfrac{f_a}{2}\right)\right]W^2(f_n)df_n}{\int_{-\tau m}^{\tau m}|h(\tau;\boldsymbol{b}_2)|^2 d\tau R_k^3\sin(\phi_k)}}{\dfrac{\sigma_{\mathrm{HV}}^0(\phi_0)\int_{-Ba/2}^{Ba/2} G_r^2(\theta_0,f_n)\,W^2(f_n)df_n}{R_0^3\sin(\phi_0)}} \tag{10.33}$$

$$\mathrm{RASR}_{\mathrm{VH}}^{\mathrm{APC}} \cong \frac{\sum\limits_{k=\mathrm{even},\neq 0}^{N_A} \dfrac{\sigma_{\mathrm{VH}}^{0}(\phi_k) \int_{-Ba/2}^{Ba/2} G_r^2(\theta_k, f_n)\, W^2(f_n) df_n}{R_k^3 \sin(\phi_k)}}{\dfrac{\sigma_{\mathrm{VH}}^{0}(\phi_0) \int_{-Ba/2}^{Ba/2} G_r^2(\theta_0, f_n)\, W^2(f_n) df_n}{R_0^3 \sin(\phi_0)}} + \frac{\sum\limits_{k=\mathrm{odd}}^{N_A} \dfrac{\sigma_{\mathrm{VV}}^{0}(\phi_k) \int_{-\tau m}^{\tau m} |g(\tau; \boldsymbol{b}_2; \boldsymbol{b}_1)|^2 d\tau \int_{-Ba/2}^{Ba/2} \left[G_r^2\left(\theta_k, f_n + \dfrac{f_a}{2}\right) + G_r^2\left(\theta_k, f_n - \dfrac{f_a}{2}\right) \right] W^2(f_n) df_n}{\int_{-\tau m}^{\tau m} |h(\tau; \boldsymbol{b}_1)|^2 d\tau R_k^3 \sin(\phi_k)}}{\dfrac{\sigma_{\mathrm{VH}}^{0}(\phi_0) \int_{-Ba/2}^{Ba/2} G_r^2(\theta_0, f_n)\, W^2(f_n) df_n}{R_0^3 \sin(\phi_0)}}$$

$$\mathrm{RASR}_{\mathrm{VV}}^{\mathrm{APC}} \cong \frac{\sum\limits_{k=\mathrm{even},\neq 0}^{N_A} \dfrac{\sigma_{\mathrm{VV}}^{0}(\phi_k) \int_{-Ba/2}^{Ba/2} G_r^2(\theta_k, f_n)\, W^2(f_n) df_n}{R_k^3 \sin(\phi_k)}}{\dfrac{\sigma_{\mathrm{VV}}^{0}(\phi_0) \int_{-Ba/2}^{Ba/2} G_r^2(\theta_0, f_n)\, W^2(f_n) df_n}{R_0^3 \sin(\phi_0)}} + \frac{\sum\limits_{k=\mathrm{odd}}^{N_A} \dfrac{\sigma_{\mathrm{VH}}^{0}(\phi_k) \int_{-\tau m}^{\tau m} |g(\tau; \boldsymbol{b}_1; \boldsymbol{b}_2)|^2 d\tau \int_{-Ba/2}^{Ba/2} \left[G_r^2\left(\theta_k, f_n + \dfrac{f_a}{2}\right) + G_r^2\left(\theta_k, f_n - \dfrac{f_a}{2}\right) \right] W^2(f_n) df_n}{\int_{-\tau m}^{\tau m} |h(\tau; \boldsymbol{b}_2)|^2 d\tau R_k^3 \sin(\phi_k)}}{\dfrac{\sigma_{\mathrm{VV}}^{0}(\phi_0) \int_{-Ba/2}^{Ba/2} G_r^2(\theta_0, f_n)\, W^2(f_n) df_n}{R_0^3 \sin(\phi_0)}}$$

10.4 Simulation and performance analysis

In this section, by exploiting the LT-1 system parameters, detailed simulation experiments are carried out to corroborate the theoretical developments.

10.4.1 Waveform performance evaluation

The following example is developed by exploiting the signal parameters of the LT-1 system whose bandwidth is 60 MHz and pulse length is 80 μs. The vector $\boldsymbol{b}_{\mathrm{Initial}}$ ($N = 130$) is obtained performing the instantaneous frequency sampling of an NLFM waveform whose spectrum approximates a Taylor window with a maximum sidelobe level of −35 dB (the generation method can be found in [25]). The PSLR thresholds and the stopping criterion are set to −35 dB and $\delta = 10^{-6}$, respectively.

Comparisons of matched filter outputs among quasi-orthogonal NLFM, up-down chirp and OFDM-chirp waveforms [28] are shown† in Figure 10.3. Besides, it should be noted that for up-down chirp and the proposed NLFM waveforms, the mentioned quantities can be obtained specializing the theoretical expressions in (10.5) and (10.20) to the appropriate vectors $\boldsymbol{b}_1$ and $\boldsymbol{b}_2$. For the up-down chirp waveforms, $\boldsymbol{b}_1$ and $\boldsymbol{b}_2$ are, respectively, given in

†All waveforms exhibit the same bandwidth (60 MHz) and pulse length (80 ms).

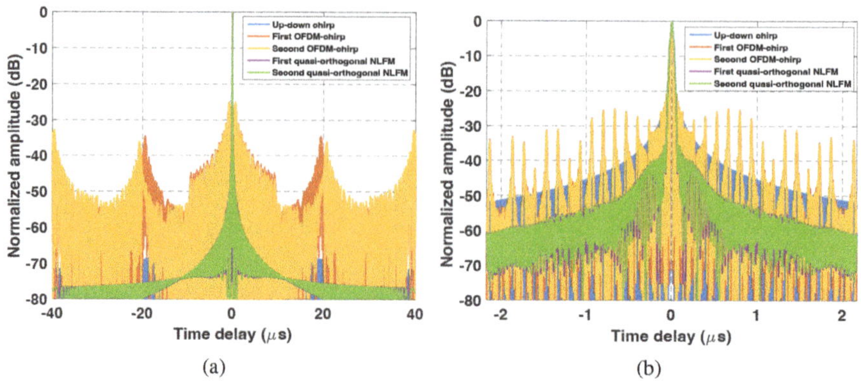

Figure 10.3 *Matched filter outputs in dB of different quasi-orthogonal waveforms. (a) Whole picture. (b) Enlarged picture around the main-lobe.*

$$b_{1,i} = -\frac{B}{2} + \frac{B}{N-1}(i-1), i = 1, 2, \cdots, N, \tag{10.34}$$

and

$$b_{2,i} = \frac{B}{2} - \frac{B}{N-1}(i-1), i = 1, 2, \cdots, N, \tag{10.35}$$

whereas for the proposed design they are provided by the devised algorithm [27]. Then, leveraging the theoretical (10.32), (10.23) and (10.24), all of PSLR, ISLR and CCE can be evaluated.

Moreover, the ambiguity functions of the designed quasi-orthogonal NLFM waveforms are illustrated in Figure 10.4. Finally, the main performance metric of the mentioned pair of waveforms, i.e., PSLR, ISLR, 3-dB main-lobe Width (MW) and CCE, are reported in Table 10.5. It can be seen from Figure 10.3 that the OFDM-chirp waveforms experience grating lobes and the generated quasi-orthogonal NLFM waveforms have both similar matched filter outputs (Figure 10.3) and similar auto-ambiguity functions (Figure 10.4). Inspection of Table 10.5 shows that the CCE of the quasi-orthogonal NLFM signals is reduced by 3.47 dB as compared to that of the up-down chirp. In addition, the PSLRs (already controlled at the design stage) and ISLRs of the synthesized quasi-orthogonal NLFM waveforms are low enough to meet the SAR imaging requirement.

10.4.2 Point target simulation

In this subsection, simulation experiments for a point-like target are conducted to provide insights on the ambiguity suppression capabilities of different Tx/Rx schemes. To this end, a typical quad-pol SAR system with parameters reported in Table 10.6 is considered and the range Doppler algorithm is employed for SAR imaging [31].

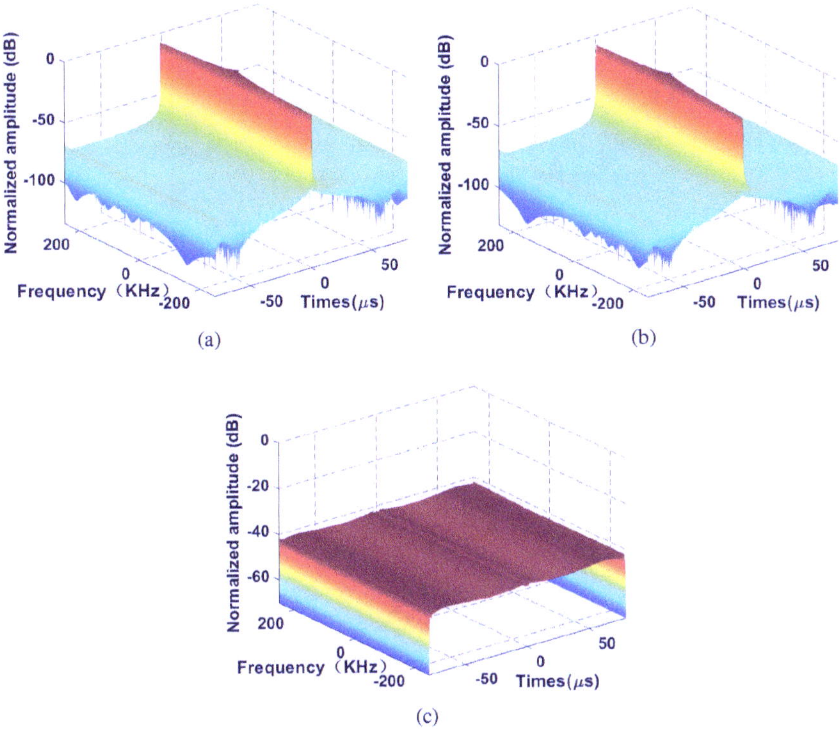

Figure 10.4 Ambiguity functions in dB of the proposed quasi-orthogonal NLFM waveforms. Auto-ambiguity functions for: the first-devised quasi-orthogonal NLFM waveform (a); the second-devised quasi-orthogonal NLFM waveform (b). (c) Cross-ambiguity function between the first and second quasi-orthogonal NLFM waveforms.

Table 10.5 Performance metrics for the different waveforms

Waveforms	PSLR (dB)	ISLR (dB)	3-dB MW (sample)	CCE* (dB)
Up-down chirp	−13.3	−9.7	1	0
Initial NLFM	−35	−26.6	1.33	0.91
First OFDM-chirp [28]	−13.26	−7.4	1	0.13
Second OFDM-chirp [28]	−13.29	−7.39	1	
First quasi-orthogonal NLFM	−35.7	−28	1.57	−3.47
Second quasi-orthogonal NLFM	−35.6	−28	1.56	

*The CCE relative to a chirp signal.

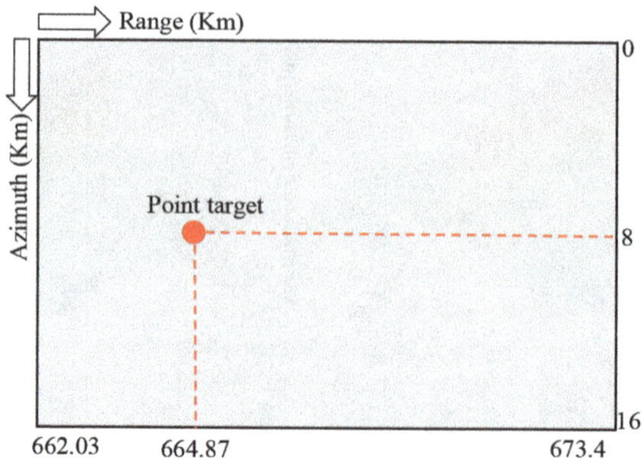

Figure 10.5 Location of a point target within the non-ambiguous imaging area

First, assuming a single polarization SAR system and using the synthesized signal $s_1(t)$, the 2-D impulse response function (IRF)‡ [31] of a point target whose location is shown in Figure 10.5, within the non-ambiguous image portion, is illustrated in Figure 10.6. It can be clearly seen that the point target has been focused correctly.

Then, aimed at assessing the interference at the first near ambiguous range induced by a point target whose location is shown in Figure 10.5, the 2-D IRFs (associated with the different transmission techniques) in correspondence of the first near range ambiguity portion of the image (HV polarization channel of a conventional quad-pol mode) are represented in Figure 10.7. In the simulation, the same energy is assumed for the raw signals and the azimuth signal is band limited. Besides, the effect of the antenna pattern gain, the backscattering reflectivity, the incidence angle and the slant range are not considered. The measured total ambiguity energies relative to a chirp signal are reported in Table 10.7. Because of the Doppler frequency

Table 10.6 Main simulation parameters of the LT-1 in quad-pol mode

Parameters	**Value**
Orbit height	600 Km
SAR pulse width	80μs
Bandwidth	60 MHz
PRF (f_a)	1660 Hz
Antenna type	Planar
Two-way antenna pattern in azimuth	$\text{sinc}^2(\cdot)$
Processed Doppler bandwidth B_a	1380 HZ
Azimuth processing window	Generalized Hamming (**α = 0.6**)
Backscatter model	Soil and Rock [35]

‡The number of pulses corresponding to the IRF is computed by the antenna beam and the PRF.

Table 10.7 Measured total ambiguity energy of a point target

Ambiguity	Tx scheme	Ambiguity energy* (dB)
Range	APC method	−0.81
	Quasi-orthogonal NLFM waveforms	−3.47
	Combination of quasi-orthogonal NLFM waveforms and APC	−4.29
Azimuth	Up chirp waveforms	0
	Up-down chirp waveforms	−0.003
	Quasi-orthogonal NLFM waveforms	−3.48

*Total ambiguity energy relative to a chirp signal.

shift induced by the APC and the azimuth filtering, the total range ambiguity energy, when APC is employed, is reduced by approximately 0.81 dB, with respect to the case of chirp waveform. If the quasi-orthogonal NLFM waveforms are used, the ambiguous signals are defocused and the energy is spread along the range dimension

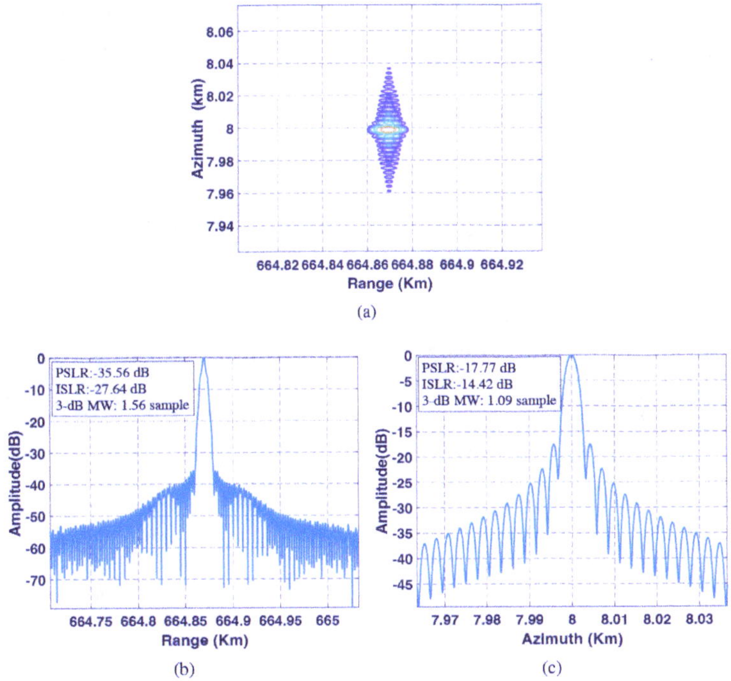

Figure 10.6 2-D IRF of a point target shown in Figure 10.5, in correspondence of the non-ambiguous image portion. The antenna pattern in elevation is flat and the antenna pattern in azimuth is $\mathrm{sinc}^2(\cdot)$. *(a) Contour plot in dB. (b) Range profile in dB. (c) Azimuth profile in dB.*

(see Figure 10.7(c)). Due to the decrease of CCE, the measured ambiguity energy is lessened by ~3.47 dB with respect to the chirp waveform. Moreover, the ambiguity energy for the combination scheme of quasi-orthogonal NLFM waveforms and APC is further diminished by approximately 4.29 dB with respect to the standard chirp. The residual ambiguities are also defocused, as shown in Figure 10.7(d).

Next, considering the implementation of quasi-orthogonal NLFM waveforms in the hybrid and $\pm\pi/4$ quad-pol SAR for azimuth ambiguity suppression, two modified versions referred to as generalized hybrid and generalized $\pm\pi/4$ quad-pol SARs can be defined [27]. The corresponding timing diagram for a generalized hybrid and a generalized $\pm\pi/4$ quad-pol SARs is defined in Figure 10.8(a) and (b), respectively.

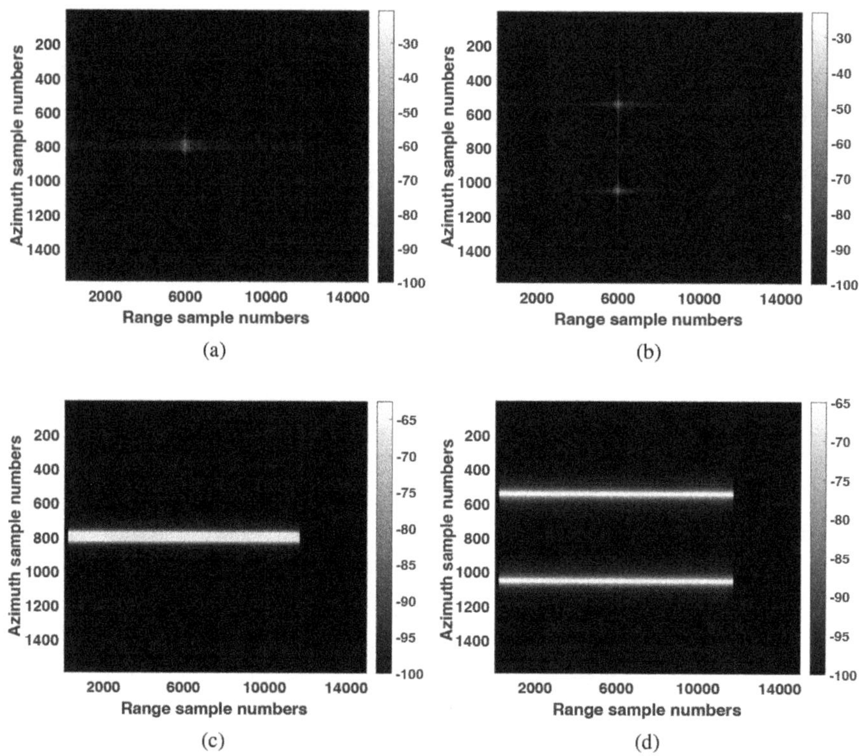

Figure 10.7 2-D IRFs (dB) in correspondence of the first near range ambiguity portion of the image (HV polarization channel of a conventional quad-pol mode) for: up chirp waveform (a), APC (b), quasi-orthogonal NLFM waveforms (c), combination of quasi-orthogonal NLFM waveforms and APC (d). The total ambiguity energies of (a), (b), (c) and (d) are −20.58, −21.39, −24.05 and −24.87 dB, respectively. Note that different colour scales have been used in different sub-plots.

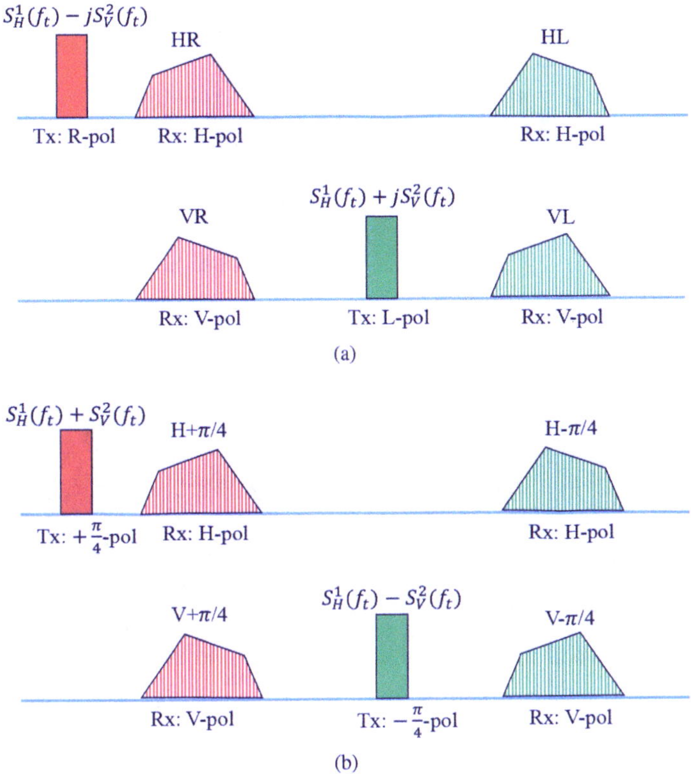

Figure 10.8 Timing diagrams. (a) Generalized hybrid quad-pol SAR. (b) Generalized $\pm\pi/4$ quad-pol SAR.

Table 10.8 Measured total ambiguity energy of distributed targets

Ambiguity	**Tx scheme**	**Ambiguity energy* (dB)**
Range	Up chirp waveforms	0
	APC method	−0.84
	Quasi-orthogonal NLFM waveforms	−3.51
	Combination of quasi-orthogonal NLFM waveforms and APC	−4.35
Azimuth	Up chirp waveforms	0
	Up-down chirp waveforms	−0.004
	Quasi-orthogonal NLFM waveforms	−3.48

*Total ambiguity energy relative to a chirp signal.

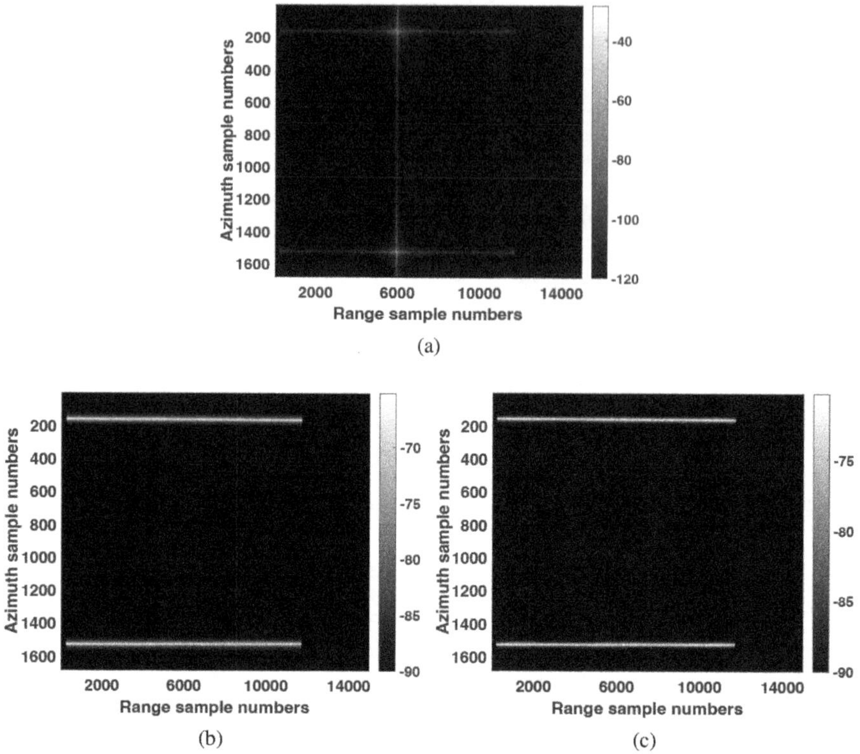

Figure 10.9 *2-D IRFs (dB) in correspondence of the first azimuth ambiguity portion of the image (HV polarization channel of a generalized hybrid or generalized π/4 quad-pol mode) for: up chirp waveform (a), up-down chirp waveforms (b) and quasi-orthogonal NLFM waveforms (c). The total ambiguity energies for (a), (b) and (c) are −21.98, −21.95 and −25.46 dB, respectively. Note that different black scales have been used in different sub-plots.*

The 2-D IRFs for the first azimuth ambiguity region of a generalized hybrid or generalized $\pm\pi/4$ quad-pol SAR (HV polarization channel) are given in Figure 10.9, where the same energy is assumed for the raw signals. Figure 10.9(a) shows that the azimuth ambiguities of the chirp waveform are focused to some extent. Besides, the azimuth ambiguities of the up-down chirp waveforms and quasi-orthogonal NLFM waveforms are defocused and the energies are spread along the range dimension. The measured azimuth ambiguities relative to a chirp signal are listed in Table 10.7. The measured total azimuth ambiguity energy of the up-down chirp pair and the standard chirp is approximately equal because up-down chirp waveforms are not able to reduce the CCE. Moreover, the ambiguity energy of the quasi-orthogonal NLFM waveforms is reduced by approximately 3.48 dB as compared with the standard chirp waveform.

10.4.3 Distributed target simulation

To evaluate the interference effects of ambiguous returns in the HV polarization channel (range ambiguous returns impairments for a conventional quad-pol mode and azimuth ambiguous returns impairments for either a generalized hybrid or generalized $\pm\pi/4$ quad-pol mode), distributed target experiments based on the system parameters of LT-1 are now conducted. The simulator for assessing the first near range ambiguity and the first azimuth ambiguity effects is shown in Figure 10.10. Considering the first sidelobe attenuation in elevation of the two-way antenna

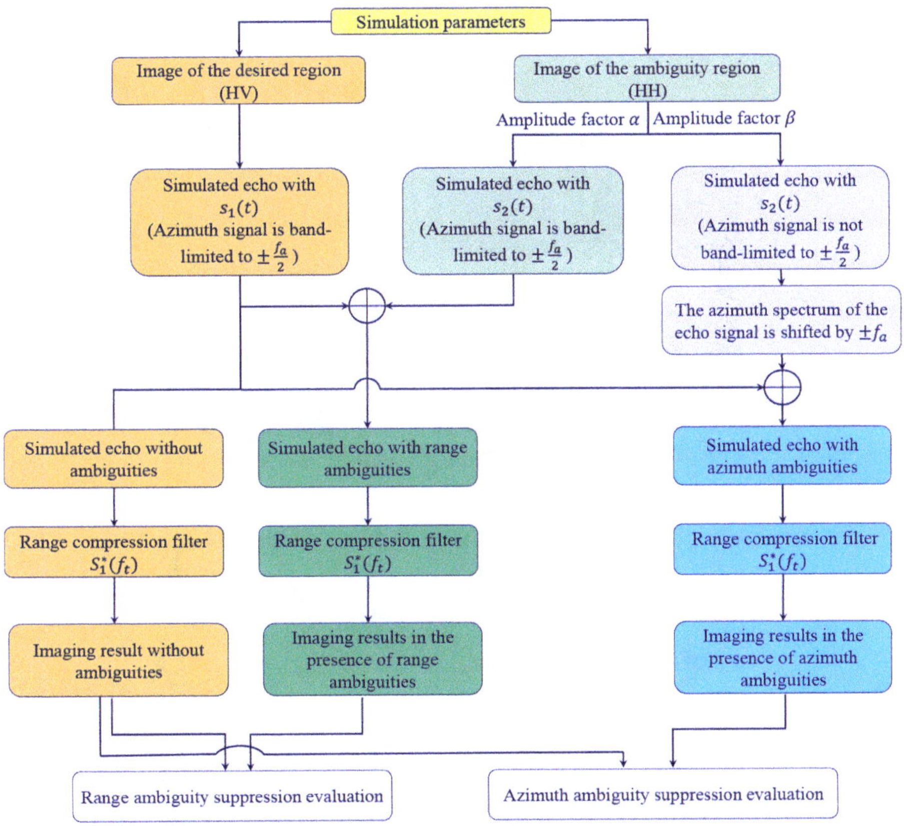

Figure 10.10 Simulator for assessing the first near range ambiguity suppression performance (HV polarization channel of a conventional quad-pol mode) and the first azimuth ambiguity suppression performance (HV polarization channel of a generalized hybrid or generalized $\pm\pi/4$ quad-pol mode). The simulation experiments are based on the LT-1 parameters, but the azimuth antenna pattern is not considered when assessing range ambiguity suppression performance. Besides, setting $s_2(t) = s_1(t)$ means there is no waveform diversity among the pulses.

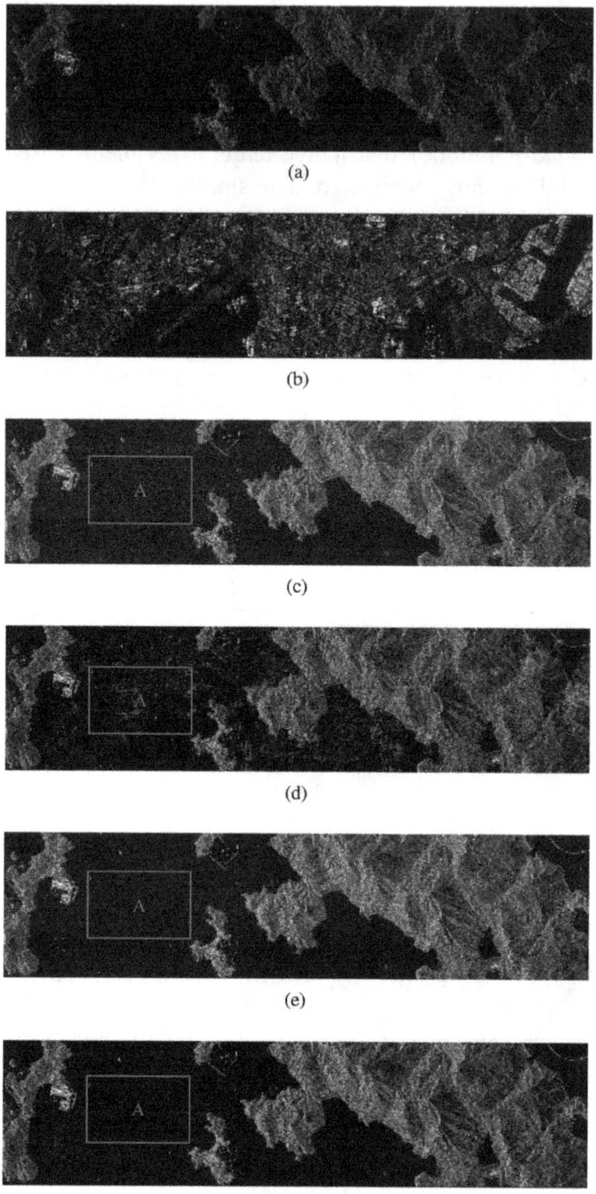

Figure 10.11 *Distributed scene simulation for the HV polarization (conventional quad-pol mode). (a) Image of the desired region (HV polarization). (b) Image of the range ambiguity region (HH polarization). Imaging result for: the up-down chirp waveforms (c), APC method (d), quasi-orthogonal NLFM waveforms (e), combination scheme of quasi-orthogonal NLFM waveforms and APC (f).*

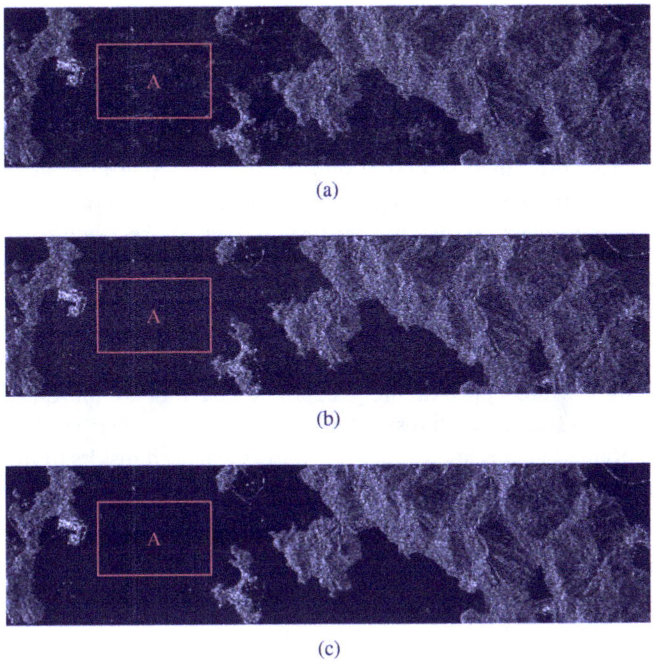

Figure 10.12 Distributed scene simulation for the HV polarization (generalized hybrid or generalized π/4 quad-pol mode). Imaging result for: the up chirp waveform (a), up-down chirp waveforms (b), quasi-orthogonal NLFM waveforms (c).

pattern equal to −26.6 dB and the backscattering reflectivity ratio between the HH and HV polarizations equal to 10 dB ($\sigma_{HH,dB} - \sigma_{HV,dB} = 10dB$), the amplitude ratio of the simulated range ambiguous echo is calculated as

$$\alpha = 10^{\frac{-26.6+10}{20}} = 0.1479 \tag{10.36}$$

The simulation results (concerning the range ambiguity interference assessment) for the different schemes are shown in Figure 10.11, where the horizontal axis is the range direction and the vertical axis is the azimuth direction. Figure 10.11(a) and (b) illustrate the images of the desired and the ambiguity regions, respectively. Moreover, Figure 10.11(c), (d), (e), and (f) show the imaging results of the different Tx/Rx schemes. As pointed out by the red rectangles labelled as 'A', it can be clearly seen that the range ambiguity suppression performance granted by the combination of quasi-orthogonal NLFM waveforms and APC method is the best. Moreover, the measured total ambiguity energies relative to those of a standard chirp signal are reported in Table 10.8. The obtained values are close to the measured values for the point target simulation experiments (Table 10.7).

With reference to the experiment aimed at assessing the first azimuth ambiguity effects, the amplitude factor of the simulated azimuth ambiguous echoes is computed as

$$\beta = 10^{\frac{10}{20}} = 3.16. \tag{10.37}$$

The results for the different waveforms are shown in Figure 10.12, where the horizontal axis labels the range direction and the vertical axis indicates the azimuth direction. The desired and the ambiguous regions are the same as the images used in the range ambiguity simulation experiment[§]. It can be observed from the red rectangles labelled as 'A' that the azimuth ambiguity suppression performance of the quasi-orthogonal NLFM pair is better than the counterparts. The ambiguity energies relative to the chirp signal of Figure 10.12(a), (b), and (c) are listed in Table 10.8. Moreover, the measured total interfering energies of the chirp waveforms and up-down chirp signals are approximately equal, while that of the quasi-orthogonal NLFM signals is −3.48 dB with respect to the conventional chirp pulse. As a consequence, the obtained values are quite close to the measured results of the point target simulation experiments (Table 10.7).

10.5 Conclusion

To further boost range ambiguity suppression in the conventional quad-pol mode, the APC method is jointly exploited with diverse quasi-orthogonal waveforms so as to develop a new and advanced polarimetric probing strategy. This chapter has considered the design and implementation of quasi-orthogonal NLFM waveforms to mitigate range ambiguity of the conventional quad-pol SAR and the azimuth ambiguity of the hybrid/$\pm\pi/4$ quad-pol SAR. Precisely, the proposed quasi-orthogonal NLFM waveforms grant a lower CCE than some counterparts available in the open literature, leading to a better ambiguity suppression performance. Furthermore, the designed waveforms exhibit low PSLRs necessary to ensure satisfactory imaging performance. To further boost range ambiguity suppression in the conventional quad-pol mode, the APC method is jointly exploited with diverse quasi-orthogonal waveforms so as to develop a new and advanced polarimetric probing strategy. Last but not least, this chapter has offered a systematic framework for the evaluation of the AASR and RASR for a conventional quad-pol SAR system using waveform diversity. At the analysis stage, detailed simulation experiments based on the LT-1 system parameters are carried out to illustrate the effectiveness of the devised signals and ambiguity suppression schemes.

Possible future researches might be focused on the design and implementation of phase-coding schemes for SAR systems. Besides, the development of appropriate calibration techniques for the hybrid and $\pm\pi/4$ quad-pol SARs with diverse waveforms, inheriting lectures learned by recent SAR polarimetric calibration strategies [36–43], is definitely an interesting and relevant research topic.

[§]Without loss of generality, both range and azimuth ambiguous returns are generated leveraging appropriately the backscattering characteristics illustrated in the image of Figure 10.11(b).

References

[1] Levanon N., Mozeson E. *Radar signals*. Hoboken, NJ, USA: Wiley; 2004 Jul 9. Available from http://doi.wiley.com/10.1002/0471663085

[2] Gini F., De Maio A., Patton L. *Waveform design and diversity for advanced radar systems*. Stevenage: Institution of Engineering and Technology; 2012 Jan 1. Available from https://digital-library.theiet.org/content/books/ra/pbra022e

[3] Alaee Kerahroodi M., Aubry A., De Maio A., Naghsh M.M., Modarres-Hashemi M. 'A coordinate-descent framework to design low PSL/ISL sequences'. *IEEE Transactions on Signal Processing*. 2017, vol. 65(22), pp. 5942–56.

[4] Wang W.-Q. 'Mitigating range ambiguities in high-PRF SAR with OFDM waveform diversity'. *IEEE Geosci Remote Sens Lett*. 2013, vol. 10(1), pp. 101–05.

[5] Jin G., Deng Y., Wang R., *et al*. 'Mitigating range ambiguities with advanced nonlinear frequency modulation waveform'. *IEEE Geoscience and Remote Sensing Letters*. 2019, vol. 16(8), pp. 1230–4.

[6] Krieger G., Gebert N., Moreira A. 'Multidimensional waveform encoding: a new digital beamforming technique for synthetic aperture radar remote sensing'. *IEEE Transactions on Geoscience and Remote Sensing*. 2008, vol. 46(1), pp. 31–46.

[7] Knapskog A.O. 'Range ambiguity suppression in space-borne SAR by up- and down-chirp modulation in combination with pseudo-random bi-phase coding'. *12th European Conference on Synthetic Aperture Radar*; Aachen, Germany, 2018. pp. 1–4.

[8] Krieger G. 'MIMO-SAR: opportunities and pitfalls'. *IEEE Trans Geosci Remote Sens*. 2014, vol. 52(5), pp. 2628–45.

[9] Moreira A., Prats-Iraola P., Younis M., Krieger G., Hajnsek I., Papathanassiou K.P. 'A tutorial on synthetic aperture radar'. *IEEE Geoscience and Remote Sensing Magazine*. 2013, vol. 1(1), pp. 6–43.

[10] Cloude S. *Polarisation: applications in remote sensing*. New York, NY: Oxford University Press; 2010. Available from https://academic.oup.com/book/11554

[11] Lee J.S., Pottier E. *Polarimetric radar imaging: from basics to applications*. Boca Raton, FL: CRC Press; 2009.

[12] Raney R.K., Freeman A., Jordan R.L. 'Improved range ambiguity performance in quad-pol SAR'. *IEEE Trans Geosci Remote Sens*. 2012, vol. 50(2), pp. 349–56.

[13] Raney R.K. 'Hybrid-polarity SAR architecture'. *IEEE Trans Geosci Remote Sens*. 2007, vol. 45(11), pp. 3397–404.

[14] Souyris J.-C., Imbo P., Fjortoft R., Sandra M., Jong-Sen L. 'Compact polarimetry based on symmetry properties of geophysical media: the /spl pi//4 mode'. *IEEE Trans Geosci Remote Sens*. 2005, vol. 43(3), pp. 634–46.

[15] Villano M., Krieger G., Moreira A. 'New insights into ambiguities in quad-pol SAR'. *IEEE Trans Geosci Remote Sens*. 2017, vol. 55(6), pp. 3287–308.

[16] Dall J., Kusk A. 'Azimuth phase coding for range ambiguity suppression in SAR'. *IEEE International Geoscience and Remote Sensing Symposium. vol. 3*; Anchorage, AK, 2004. pp. 1734–37.
[17] Bordoni F., Younis M., Krieger G. 'Ambiguity suppression by azimuth phase coding in multichannel SAR systems'. *IEEE Trans Geosci Remote Sens*. 2012, vol. 50(2), pp. 617–29.
[18] Jin G., Deng Y., Wang W., *et al*. 'On the SAR imaging performance analysis of alternate transmitting mode based on waveform diversity: theory and simulation'. *IEEE Geosci Remote Sens Lett*. 2020, vol. 17(9), pp. 1553–57.
[19] Fowle E. 'The design of FM pulse compression signals'. *IEEE Transactions on Information Theory*. 1964, vol. 10(1), pp. 61–67.
[20] Vizitiu I., Anton L., Popescu F., Iubu G. '2012 10th international symposium on electronics and telecommunications (ISETC)'. Timisoara, Timis, Romania, 2012.
[21] Kurdzo J.M., Cheong B.L., Palmer R.D., Zhang G. 'Optimized NLFM pulse compression waveforms for high-sensitivity radar observations'. *2014 International Radar Conference (Radar)*; Lille, France, 2014. pp. 1–6.
[22] Saeedi J., Faez K. 'Synthetic aperture radar imaging using nonlinear frequency modulation signal'. *IEEE Transactions on Aerospace and Electronic Systems*. 2016, vol. 52(1), pp. 99–110.
[23] Wang W., Wang R., Zhang Z., *et al*. 'First demonstration of airborne SAR with nonlinear FM chirp waveforms'. *IEEE Geosci Remote Sens Lett*. 2016, vol. 13(2), pp. 247–51.
[24] Wang W., Wang R., Deng Y., Zhang Z., Wu X., Xu Z. 'Demonstration of NLFM waveforms with experiments and doppler shift compensation for SAR application'. *IEEE Geosci Remote Sens Lett*. 2016, vol. 13(12), pp. 1999–2003.
[25] Jin G., Deng Y., Wang R., *et al*. 'An advanced nonlinear frequency modulation waveform for radar imaging with low sidelobe'. *IEEE Trans Geosci Remote Sens*. 2019, vol. 57(8), pp. 6155–68.
[26] Jin G., Liu K., Deng Y., *et al*. 'Nonlinear frequency modulation signal generator in LT-1'. *IEEE Geosci Remote Sens Lett*. 2019, vol. 16(10), pp. 1570–74.
[27] Jin G., Aubry A., De Maio A., Wang R., Wang W. 'Quasi-orthogonal waveforms for ambiguity suppression in spaceborne quad-pol SAR'. *IEEE Trans Geosci Remote Sens*. 2022, vol. 60, pp. 1–17.
[28] Wang W.-Q. 'MIMO SAR OFDM chirp waveform diversity design with random matrix modulation'. *IEEE Trans Geosci Remote Sens*. 2015, vol. 53(3), pp. 1615–25.
[29] Boersma J. 'Computation of fresnel integrals'. *Mathematics of Computation*. 1960, vol. 14(72), pp. 380–380. Available from https://www.ams.org/mcom/1960-14-072
[30] Ugray Z., Lasdon L., Plummer J., Glover F., Kelly J., Martí R. 'Scatter search and local nlp solvers: a multistart framework for global optimization'. *INFORMS Journal on Computing*. 2007, vol. 19(3), pp. 328–40.

[31] Hay W.W., Thilo E., Curlander J.B. 'Pulse oximetry in neonatal medicine' in *Clinics in perinatology*. Vol. 18; 1991. pp. 441–72.

[32] FK L., Johnson W.T.K. 'Ambiguities in spaceborne synthetic aperture radar systems'. *IEEE Transactions on Aerospace and Electronic Systems transactions on Aerospace and Electronic Systems*. 1983, vol. 19(3), pp. 389–97.

[33] Bayma R.W. 'Aperture size and ambiguity constraints for a synthetic aperture radar'. *IEEE International Radar Conference. vol. 21*; Arlington, Virginia, 1975. pp. 499–504.

[34] Villano M., Krieger G. 'Spectral-based estimation of the local azimuth ambiguity-to-signal based estimation of the local azimuth ambiguity-to-signal ratio in SAR images'. *IEEE Trans Geosci Remote Sens*. 2014, vol. 52(5), pp. 2304–13.

[35] Ulaby F., Dobson M.C., Álvarez-Pérez J.L. *Handbook of radar scattering statistics for terrain handbook of radar scattering statistics for terrain*. Dedham, MA: Artech House; 1989.

[36] Freeman A., van Zyl J.J., Klein J.D., Zebker H.A., Shen Y. 'Calibration of stokes and scattering matrix format polarimetric SAR data'. *IEEE Trans Geosci Remote Sens*. 1992, vol. 30(3), pp. 531–39.

[37] Villa A., Iannini L., Giudici D., Monti-Guarnieri A., Tebaldini S. 'Calibration of SAR polarimetric images by means of a covariance matching approach'. *IEEE Trans Geosci Remote Sens*. 2015, vol. 53(2), pp. 674–86.

[38] Ainsworth T.L., Ferro-Famil L. Jong-Sen L. Orientation angle preserving a posteriori polarimetric SAR calibration'. *IEEE Trans Geosci Remote Sens*. 2006, vol. 44(4), pp. 994–1003.

[39] Sarabandi K. 'Calibration of a polarimetric synthetic aperture radar using a known distributed target'. *IEEE Trans Geosci Remote Sens*. 1994, vol. 32(3), pp. 575–82.

[40] Quegan S. 'A unified algorithm for phase and cross-talk calibration of polarimetric data-theory and observations'. *IEEE Trans Geosci Remote Sens*. 1994, vol. 32(1), pp. 89–99.

[41] Kimura H., Mizuno T., Papathanassiou K.P. 'Improvement of polarimetric SAR calibration based on the quegan algorithm'. *IEEE International Geoscience and Remote Sensing Symposium. vol. 1*; Anchorage, AK, 2004. pp. 187.

[42] Shimada M., Isoguchi O., Tadono T., Isono K. 'PALSAR radiometric and geometric calibration'. *IEEE Trans Geosci Remote Sens*. 2009, vol. 47(12), pp. 3915–32.

[43] Shimada M. 'Model-based polarimetric SAR calibration method using forest and surface-scattering targets'. *IEEE Trans Geosci Remote Sens*. 2011, vol. 49(5), pp. 1712–33.

Chapter 11

Fully polarimetric monopulse spaceborne radar for space situational awareness

Marco Maffei[1], *Augusto Aubry*[1], *Antonio De Maio*[1], *and Alfonso Farina*[2]

11.1 Introduction and phenomenology for space situational awareness

The outer space around planet Earth comprises a variety of natural and man-made junk which pose a severe threat to strategic assets in space (e.g. orbital infrastructure, spacecraft and satellites) in case of collision [1]. In particular, several orbital regions are filled with debris populations, namely: the low Earth orbit (LEO) with the highest volume-density of items orbiting at significant angular velocities along different inclinations; the lower medium Earth orbit (lower MEO) housing a limited number of items with no specific inclination; the MEO with a limited number of items along specific inclinations; and the geostationary orbit (GEO) which accounts for a limited number of items moving at reduced angular velocities along specific inclinations. Considering such a volume of space around planet Earth, the cardinality of debris with an average diameter larger than 1 μm is estimated to exceed several hundred trillion items. Yet, when the average diameter is larger than 1 mm, the debris population size shrinks to roughly several hundred million items. Nevertheless, when the average diameter is larger than 1 cm, the debris population reaches several hundred thousand items. If the focus is on debris larger than 10 cm, the debris population count reduces to tens of thousands of objects (see [1–3]). From such a huge population cardinality, it appears useful to identify debris-related taxonomies. Thorough categorisations of debris populations appear in [3, 4], labelled by either the lack or abundance of knowledge to 'associate the root cause of the debris to a launch event'. Indeed, unidentified objects represent the former class whereas the latter class is further structured into several features and origin. Namely, payloads (PL), e.g. satellites instruments; payload mission (PM)-related objects, e.g. astronaut tools; payload fragmentation (PF) debris, e.g. items fragmented or released

[1]University of Napoli, Federico II, Italy
[2]Leonardo Company, Via Tiburtina Km 12.400 ZIP: 00131, Rome, Italy

during a specific event; payload debris (PD), e.g. items fragmented or released during an unknown event; rocket body (RB), e.g. released orbital stages belonging to a launcher; rocket mission (RM)-related objects, e.g. shrouds and engines; rocket fragmentation (RF) debris, e.g. objects created in case of a launch vehicle explosion; and rocket debris (RD), e.g. objects created from a rocket body during an unknown event. Moreover, a plethora of fragmentation events are also pinpointed as metadata related to the break-up cause such as 'accidental', 'aerodynamics', 'anomalous', 'collision', 'deliberate', 'electrical', 'propulsion' and 'unknown'. Also, items are grouped into two major populations: a 'large object population' whose debris size is roughly larger than 1 cm and a 'small object population' whose debris size is roughly larger than 1 μm, yet smaller than 1 cm. For example, micrometeoroids, solid rocket motor dust, paint flakes and ejecta belong to the small object population. Multi-layer insulation materials as well as launch- and mission-related objects including the items tracked in the two-line element (TLE) set of orbital items (publicly shared by the United States Air Force Space Command) belong to the large object population. Finally, explosion and collision fragments, sodium-potassium droplets and solid rocket motor slag belong to both classes.

According to this scenario, for near-Earth Space Situational Awareness (SSA), the translational and rotational motion of space debris* occur within an inertial (Euclidean) vector space (i.e. an inertial system is assumed to exist and can be referred to as 'distant' stars, e.g. the First Point of Aries, or the First Point of Libra). In particular, while the Earth oblate keeps revolving around the Sun (along with additional spinning, processing and nutating motions due to celestial mechanics), groups of meteoroids passing by the Earth or debris swarms periodically orbiting around the Earth move according to Newtonian and Eulerian mechanics, whereby the Earth exerts the main central force. Specifically, the motion of debris lies predominantly on elliptical orbits which result from fundamental integrals of the two-body problem [5, 6]. Perturbation drifts from such elliptical orbits are manifested by a disturbing acceleration $\boldsymbol{a}_d$ due to the superposition of nuisances. Namely, zonal and tesseral geopotential gradients; gravitational influences of the Sun and the Moon (with negligible gravitational effects from other solar planets); drifting effects due to atmospheric drags (predominantly in lower LEO regions); and Solar Radiation Pressure [5–8]. In other words, it is possible to model the nominal translational motion of an orbiting item as

$$\ddot{\boldsymbol{r}} = -\frac{\mu}{|\boldsymbol{r}|^3}\boldsymbol{r} + \boldsymbol{a}_d \tag{11.1}$$

*The "space debris" term, also known as "orbital debris" has been defined by the Inter Agency Space Debris Coordination Committee (IADC) as "…all man-made objects including fragments and elements thereof, in Earth orbit or re-entering the atmosphere, that are non functional" [1]. In fact, the IADC definition does not include natural (i.e. non man-made) meteoroids whose size can oftentimes be negligible. For the sake of clarification, both man-made and natural space junk could be properly, and more generally, referred to as "Meteoroid and Orbital Debris" (MOD). However, in order to adopt common acronyms and terminology within the aerospace community, both man-made and natural space junk can also be jointly referred to using the "MicroMeteoroid and Orbital Debris" (MMOD) term [1].

where $\boldsymbol{r}$ is the item's position, $\ddot{\boldsymbol{r}}$ is the second-order time derivative of $\boldsymbol{r}$, $|\cdot|$ indicates the modulus operator and μ is a gravitation-related constant [5, 6].

Within this framework, a plethora of interdisciplinary efforts are being devoted to increase SSA and tackle manifold types of debris-related challenges. Namely, positioning orbiting probes directly in situ for measuring possible impacts from debris particles with size ranging from 50 μm to 1 mm [9], installing tethered capturing systems via robots, harpoons and nets [10], monitoring debris shapes via optical and radar measurements [1] oftentimes sharing facilities tailored to surveillance [11–13] and radio-astronomy operations [14], as well as implementing debris recognition capabilities via machine learning [15]. To this end, physical insights on both channel and target phenomenology are deemed useful to frame radar detection and tracking capabilities to support SSA.

On one side, i.e. in order to address a suitable channel phenomenology, it must be stressed that the propagation of electromagnetic waves can be perturbed in plasma media [16–18]. Accordingly, assumptions on electromagnetic propagation through plasma slabs appear on works related to radar [19], one-way telemetry data links [20], radio science experiments [21] and global navigation satellite systems [22, 23]. In simple terms, plasma is a gas of charged particles whereby particles do not interact through collisions as per a neutral gas but rather via electromagnetic forces. The presence of external electromagnetic fields can thus influence the motion of these charged particles, whereas the motion of charged particles can, in turn, generate currents and therefore electromagnetic fields. From a radar perspective, a suitable description of plasma is that of a fluid slab of turbulent media made of two particle species, i.e. a so-called two-component plasma comprising electrons, e, and ions, i, and assuming the oversimplified approximation of rarefaction and thermodynamic equilibrium [24]. Consequently, depending on the spatial and temporal extent of the total electron content in the medium as well as in the presence of geomagnetic fields, undesired effects may disturb signals propagating through slabs of turbulent plasma [25]. Namely, absorption, refraction, dispersion, frequency offset, Faraday rotation, and amplitude and phase scintillation [16, 17, 25]. Within this context, the state of the channel can be estimated by experimental scintillation parameters [16, 26]. For example, the scintillation index s_4 [16, 26] as a function of the intensity $I(\lambda)$ of the electric field $\boldsymbol{E}(\lambda)$ at a specific wavelength λ in a given location and temporal window can be expressed as

$$s_4 = \sqrt{\frac{\mathbb{E}\left[I(\lambda)^2\right] - \mathbb{E}\left[I(\lambda)\right]^2}{\mathbb{E}\left[I(\lambda)\right]^2}} \tag{11.2}$$

where $\mathbb{E}[\cdot]$ is the statistical expectation operator, $I(\lambda) = |\boldsymbol{E}(\lambda)|^2$, and the wavelength dependence on s_4 has been omitted for simplicity, i.e. $s_4 = s_4(\lambda)$. Interestingly, a weak plasma turbulence allows simplifying the propagation model of an electromagnetic wave through a plasma slab predominantly as a forward scattering mechanism around a small cone with negligible attenuation and backscattering within the cone itself (thus excluding multiple scattering effects in the plasma slab) [25]. In this case, it is reasonable to assume a flat-flat channel during a short radar burst on a small bandwidth in which a fixed multiplicative complex term affects the complex envelope of the radar signal (see [19] and chapter 3 in [26]).

On the other side, i.e. in order to address a suitable target phenomenology, let us consider scattering mechanisms at a given λ to occur close to the high-frequency optical region [27]. Consequently, when a radar transmits an electromagnetic waveform impinging on the external surface of a debris target (i.e. either a debris cloud or, alternatively, a single debris possibly characterised by one or more scattering centres), the main scattering responses can be assumed to derive chiefly from randomly positioned items with a low spatial density [19]. In principle, the appearance of multiple scattering mechanisms among such individual entities might create electromagnetic coupling similar to chaff scattering behaviours [28, 29]. Yet, when spaced a few wavelengths apart in a random direction and orientation, negligible mutual coupling is likely to occur within the target. Moreover, if the target spatial extent is larger than λ, several phase offsets arise among the scattered field of each item within the target [27]. Elaborating on this aspect from [27], the radar cross section (RCS) σ_{RCS} of a debris target can be expressed as a function derived from, say N items (where the i-th item located at a distance R_i from the radar is characterized by an RCS σ_{RCS_i}, for $i = 1, \ldots, N$). More specifically, taking into account the i-th radar-item round trip distance as a phase offsets $\varphi_i = -4\pi R_i/\lambda$ along with an additional scattering phase offset φ_{s_i} (see, e.g. (5) in [30]), the RCS of a debris target can be formulated as

$$\sigma_{RCS} = \left(\sum_{i=1}^{N} \sqrt{\sigma_{RCS_i}}\, e^{-j\left(\frac{4\pi R_i}{\lambda} - \varphi_{s_i}\right)}\right)^2 \tag{11.3}$$

Thus, the single debris item (i.e. each i-th term in (11.3)) may possibly induce RCS fluctuations per se depending on the aspect angle since the coherent reflectivity of a debris is surely influenced by the electromagnetic roughness and relative permittivity of the constituents materials [19, 31, 32]. Furthermore, additional fluctuations may emerge from either a debris cloud or from a single debris (possibly characterised by one or more scattering centres) with peaks and nulls depending on the occurrence of a 0 or π phase offset, respectively, among the relevant scattering items in (11.3).

Interestingly, Reference 33 characterises the influence of both the aforementioned channel (i.e. plasma media in weak scintillation as a function of the scintillation index s_4) and target phenomenology (i.e. Rayleigh and Rice-based target fluctuations) on radar detection performance from a Neyman–Pearson perspective in terms of a degradation of the required signal-to-noise ratio (SNR).

Downstream this brief introduction to SSA from a radar surveillance perspective, the rest of the chapter is organised as follows. Section 11.2 introduces the concept of a novel fully polarimetric spaceborne radar (SBR) architecture for SSA. Section 11.3 addresses the sensor complex data hypercube structure and timing hierarchies in debris surveillance mode, along with the SBR finite state machine tasks for detection and tracking. Section 11.4 clarifies further details on the SBR sensor functional architecture. Section 11.5 provides a digression on the onboard Bayesian tracker relying on radar cognitivity whereas Section 11.6 outlines the conclusions.

11.2 Fully polarimetric SBR architecture for SSA

Current data fusion systems for SSA against hypervelocity debris orbiting at several km/s are fed with a variety of data measured predominantly by large ground-based radars as well as ground-based and in-situ optical instruments. Typical ground-based radar assets operate within the UHF band [1, 14, 34], L band [1, 12, 34, 35], S band [13], C band [35], X band [1, 11], Ku band [1, 11] along with novel measurements facilities from the UHF band [14] up to the W band [36]. In this case, surveillance operations are nominally aimed at monitoring large warped volumes of space relying on either beam park or scanning modes (along with additional experimental modes [12]) with an operative trade-off between available transmit peak-power and carrier frequency of the radar asset. On the other side, the idea of exploiting the Ka band on a real aperture SBR for debris detection and tracking was addressed in [37] as a novel archetype for supporting near-Earth SSA. For example, Figure 11.1 is a pictorial representation of an operative scenario from an SBR perspective made of

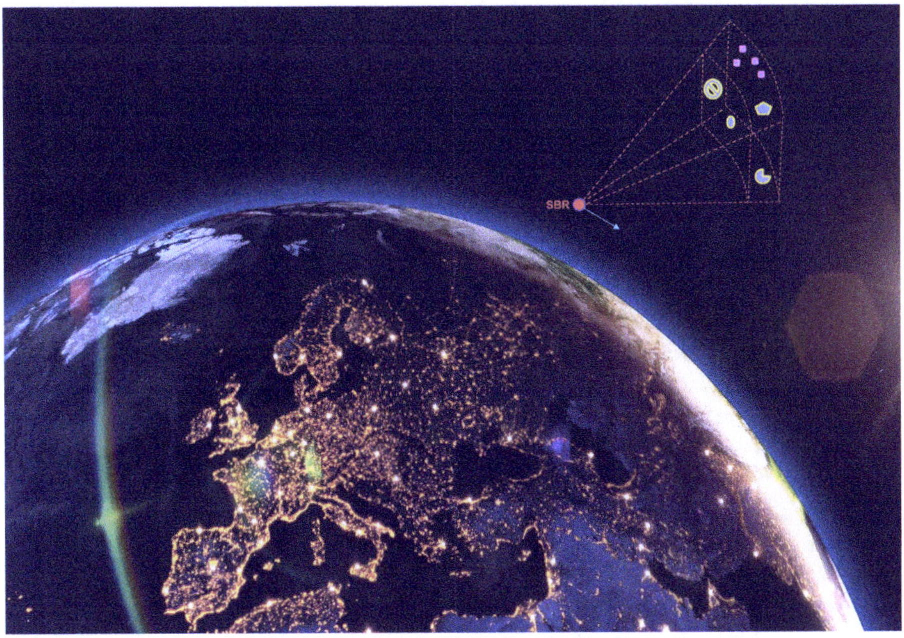

Figure 11.1 Pictorial view of an SBR (red point) whose FoV (red dashed lines) comprises elements of a large object population (yellow chunks) and a small object population (purple chunks)

individual debris targets belonging to a 'large object population' and a debris cloud from a 'small object population', possibly coexisting in the field-of-view (FoV).

In fact, the first digital-based SBR payload for synthetic aperture radar (SAR) reflectivity mapping was embarked as part of the Seasat satellite, managed by NASA Jet Propulsion Laboratory (JPL) for studying ocean physics [38] in 1978, along with different remote sensing instruments such as an altimeter, a scatterometer, as well as multi-spectral radiometers. Subsequently, early taxonomies for SBR classes [26] comprised Type I SBRs aimed at tracking for guidance control during rendezvous manoeuvres; Type II SBRs providing SAR reflectivity mapping; and Type III SBRs for early warning surveillance and moving target indicator functionalities. A couple of decades later, SBR classes [39] were also depicted as Earth observation SARs for geoscience services, sounders for planetary explorations and surveillance SBR for the homeland protection. Now, a real aperture SBR providing timely, accurate and reliable inference on debris populations may complement ground-based surveillance assets for SSA by tackling debris populations with an average size smaller than 10 cm, as well as by supporting a shorter, and more accurate, small debris conjunction prediction (e.g. one or two orbits in advance with respect to those needed by ground-based sensors, along with a smaller uncertainty). Interestingly, Reference 19 provided ontology for debris target and channel phenomenology with reasonable insights on space physics, debris fluxes environmental scenarios, orbital mechanics, along with ties to space-based radio science and general remote sensing experimental legacies. Also, Reference 19 pinpointed an SBR power-based sensitivity analysis aimed at indicating specific SNR regimes for possible onboard signal processing schemes tailored to hypervelocity debris detection and tracking. A scalable field of regard (FoR) was contrived with space-qualified enabling technologies predominantly depending on the complexity of the SBR payload architecture, with a focus on available transmit peak power and antenna type. Moreover, in [19], an SBR system for small debris detection and tracking was also envisaged as a suitable microwave instrument to endorse novel empirical mappings in the Ka band between possible radar backscattering measurements and key debris signatures related to the RCS. The results of the works in [19] and [37] led to ponder the possibility to carry out Laplacian orbit determinations [6] via SBRs estimating, at a given time epoch t_0, the position $\boldsymbol{p}$, and velocity $\dot{\boldsymbol{p}}$ of orbiting debris crossing the SBR FoR. Such SBR system concept was further addressed in [40] as built upon a digital core, and housed within a spacecraft interfacing the onboard computer, the power supply, and the data storage and downlink subsystems (Figure 11.2). Accordingly, the Ka band SBR system has been envisaged as a superheterodyne transceiver comprising an active electronically scanned array (AESA) subsystem, a radio frequency (RF) subsystem and a digital subsystem. For the sake of clarification, the rest of the SBR subsystems in Figure 11.2 will be further addressed throughout the rest of the chapter with ties to the functional aspects in Figure 11.3.

The SBR system architecture in Figure 11.2 allows implementing a bespoke monopulse-based radar as a suitable upgrade of legacy 'air-to-air missions' [41] to perform detection and parameter estimation of sprouting debris clouds in the SBR FoV. In this case, the AESA subsystem allows for digitally steerable radiation

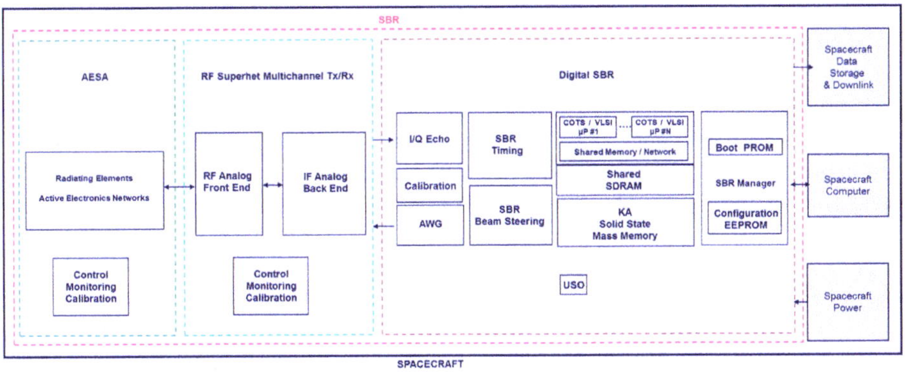

Figure 11.2 SBR system architecture

patterns in both azimuth and elevation planes with no need for true time delay lines (TTDL) due to the limited required operative bandwidth of a few MHz as indicated in [19]. Consequently, suitable onboard processing subsystems are identified in [42] as high-level block schemes where the complex envelope extraction process pertaining to a complex data hypercube occurs within an echo digitiser (i.e. see the I/Q echo subsystem in Figure 11.2 and the I/Q functional block scheme in Figure 11.3). Subsequently, the acquired samples in Figure 11.3 can be processed at every range gate through a Doppler filter bank implementing pulse compression (PC) via a cascade of phase rotators and matched filters, whereas the branch with maximum modulus value (built upon the non-coherent combining of a burst of pulse echoes) is fed to a constant false alarm rate-like (CFAR-like) block. Finally, the data stream can be fed to the data processor for tracking purposes.

Interestingly, considering gyrotropics effects in magnetized plasma media [17], the original polarisation of a general propagating transverse electromagnetic mode is affected by Faraday rotation angles (see, for instance, [43]). In addition, it is also

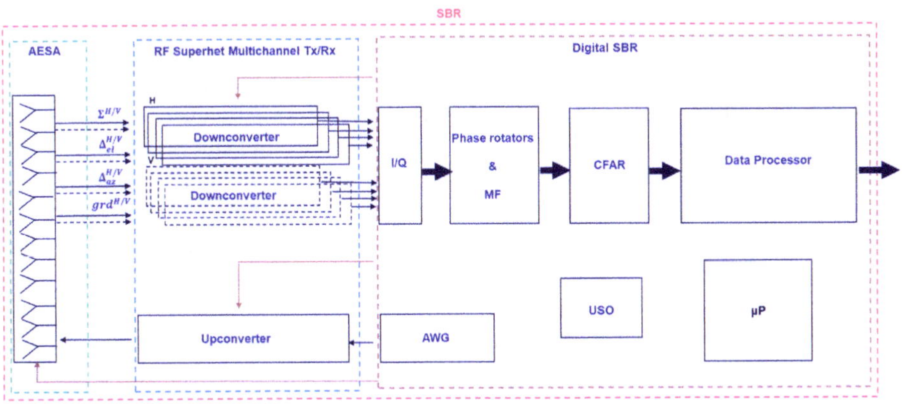

Figure 11.3 SBR monopulse-based functional architecture

worth mentioning that debris bodies spawn due to possible explosions, collisions, or fragmentation events whereas space weather may also induce an erosion of debris constitutive materials [19]. Consequently, a mild and slow tumbling or a staggered spinning behaviour may also emerge [5]. This is mostly true especially at lower orbits and on larger debris bodies, where the entries of the inertia matrix of an item are certainly dissimilar, thus inducing possible gravity torques [5]. The foregoing aspects drive the mandatory design of state-of-the-art radar systems for SSA encompassing polarimetric architectures [13, 42]. To this end, the SBR architecture in Figure 11.3 comprises both horizontal (H) and vertical (V) polarisation fluxes of data. Thus, for onboard tracking in limited time spans in the order of seconds [19], specific parameter estimates extracted in either dual or quad polarization from a set of complex data hypercubes pave the way for further Bayesian inference capabilities on small-size debris dynamic states as well as on scattering-related signatures. Namely,

a. the target range parameter computed via the monopulse sum-channels by scaling the delay parameter estimate.
b. The target elevation and azimuth angles parameters computed via the monopulse delta-channels (jointly with the sum-channels for gain normalisation) by adding the monopulse delta difference estimations and the a-priori known AESA off-broadside angles.
c. The target RCS parameter computed via the monopulse sum-channels (jointly with the monopulse delta-channels for ancillary target attributes) by time series analysis downstream the matched filter outputs.

In holistic terms, the SBR payload system in Figure 11.2 (along with its functional architecture in Figure 11.3) is conceived to act as a stand-alone active remote sensing instrument in the Ka band, embarked on a spacecraft orbiting around planet Earth with a standard platform agility for attitude control. In this case, we may assume that the SBR payload surveillance mode may rely solely on the steering capabilities of the AESA antenna, and whose operative orbit-duty-cycle is limited by power consumption and thermal conditioning thereof. Consequently, from a set of complex data hypercubes acquired in limited time spans in the order of seconds and deeply processed onboard the SBR for detection and tracking purposes, an efficient telemetry with a reduced amount of information bits can be formed. Subsequently, such a telemetry can be either transferred in downlink to the ground station for supporting SSA or, more dauntingly, for providing future spacecrafts with an autonomous early warning capability for direct collision avoidance manoeuvring [44]. Governments and space agencies might thus benefit from such a space-based instrument for augmenting SSA in a liaison with ground-based assets. From a kinematics perspective, the state-space liaison among possible debris tracks estimates (i.e. tracklets) obtained from an SBR and those tracklets estimated by a ground station can be tackled as a problem of particle kinematics with moving frames [5] for debris position and velocity state in relation to both a ground station and an SBR. Remarkably, this allows linking the contribution of SBR tracks estimates to ground-based data fusion operations needed for SSA. For the sake of completeness, it is also worth stressing that the assumed cardinality of a fleet of SBRs needed to properly support SSA would be, de facto, a mission-dependent aspect which is outside the scope of this work. Clearly, such a cardinality depends on the desired volume of interest (VoI) for the mission along with the operative SBR instantaneous FoV and overall FoR. Finally, the idea of a Ka band SBR based on an AESA with a 2-D surface area in the order of 1 m^2 can be conceived to be embarked on small-size spacecrafts launched by either dedicated airborne platforms or small rockets (e.g. Vega).

11.3 Polarimetric sensor data and operative strategies

This section delves with the sensor data and operative strategies. In particular, it outlines the time hierarchies and complex data hypercube as well as the low pulse repetition frequency (PRF) range and range rate search (RRRS) and pause while scan (PWS) collection strategy.

11.3.1 Time hierarchies and complex data hypercube

Considering an operative surveillance mode, the SBR raster scans a specific VoI to obtain an overall acquaintance on a set of sub-volumes (Figure 11.4), each identified as a cell under test (CUT) within the framework of binary hypothesis testing (in terms of whether or not a debris target or cloud might be present) along with possible parameters estimates thereof.

In this case, the SBR hops its AESA transmit beam according to a temporal hierarchy as per [41] made of a *frame* at its root (whose elapse comprises the spanning of the entire set of yellow spots in Figure 11.4), and parsed into a set of horizontal (or vertical) *bars* whereas each bar comprises a number of beam positions, each elapsing for a *dwell* period spanning one of the yellow spots in Figure 11.4. A dwell is temporally grouped into a number of *looks* where each look may span several

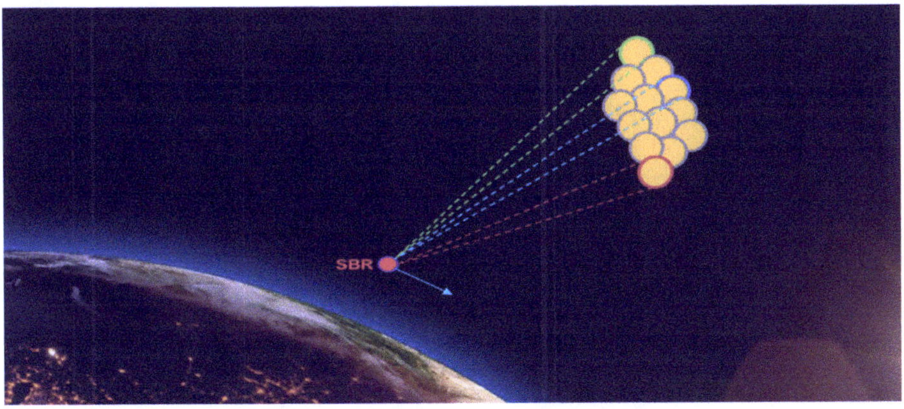

Figure 11.4 SBR raster scanning a specific VoI

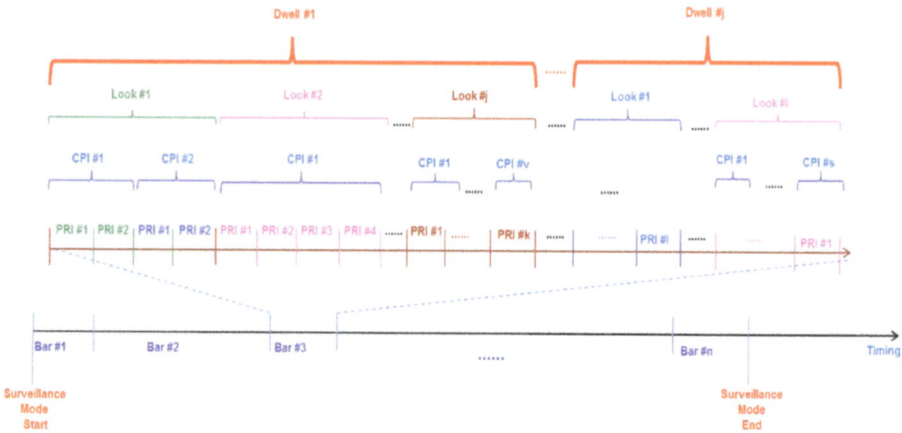

Figure 11.5 SBR temporal hierarchy in surveillance mode

coherent processing intervals (CPI) such that a CPI embeds a set of one (or more than one) *pulse repetition intervals* (PRI). For the sake of clarification, the foregoing temporal hierarchy is represented in Figure 11.5.

Now, at every dwell time, the set of in-phase (I) and quadrature (Q) components (extracted by the I/Q functional block scheme in Figure 11.3, i.e. the available samples downstream the echo complex envelope digitiser in Figure 11.2) can be organised into a complex data hypercube. More specifically, the complex data hypercube is depicted in Figure 11.6 and comprises a number of dimensions including the fast time ξ_{fast}, the slow time ξ_{slow}, as well as additional dimensions such as the receive polarization ξ_{pol}, the receive Doppler frequency ξ_{Dop} and the monopulse channel ξ_{mon}.

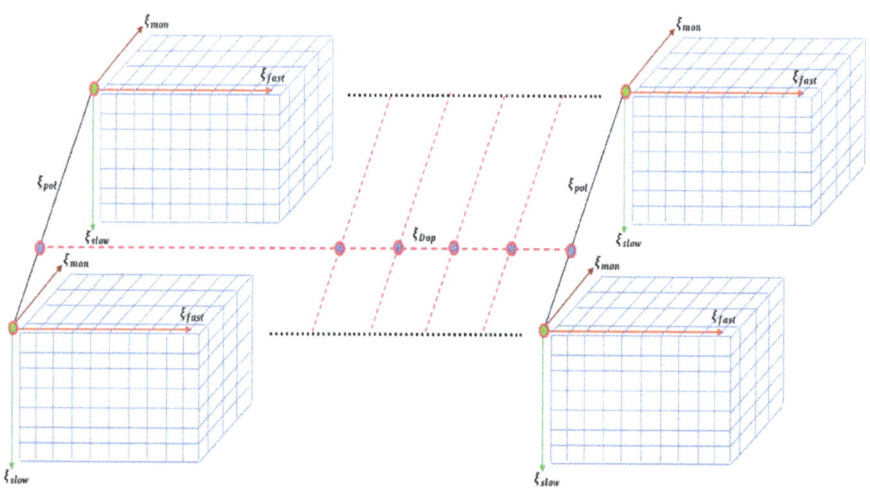

Figure 11.6 SBR complex data hypercube for SSA

That is, at every dwell time, the SBR radiation pattern points towards a specific angular direction via the AESA beam controller and acquires a 5-D complex data hypercube, i.e.

a. the fast time ξ_{fast} which pertains to the samples acquired during the sampling window length (SWL) within a PRI.
b. The slow time ξ_{slow} which indicates the number of transmitted pulses during a dwell.
c. The monopulse dimension ξ_{mon} which includes the sum Σ, elevation-difference Δ_{el}, azimuth-difference Δ_{az} and guard *grd* channels.
d. The receive polarisation ξ_{pol} which refers to the *H* and *V* channels.
e. The receive Doppler frequency ξ_{Dop} which is related to the number of programmable frequency offsets to be processed in parallel for selecting the most effective PC branch.

For the sake of clarification in Figure 11.6, the upper left data cube represents the (a), (b) and (c) dimensions at a given polarisation and minimum Doppler frequency offset; the lower left data cube represents the (a), (b) and (c) dimensions at the orthogonal polarisation and minimum Doppler frequency offset; the upper right data cube represents the (a), (b) and (c) dimensions at a given polarisation and maximum Doppler frequency offset; the lower right data cube represents the (a), (b) and (c) dimensions at the orthogonal polarisation and maximum Doppler frequency offset.

11.3.2 Low-PRF RRRS and PWS strategy

Taking into account programmable temporal hierarchies for acquiring a set of complex data hypercubes, a monopulse-based radar on an 'air to air mission' follows as per [41]. In addition, Reference 40 provides an insight on useful operative tasks for tracking via a Low-PRF RRRS jointly with a PWS collection strategy. That is, during a dwell in the order of a few milliseconds (e.g. 2 ms), a set of CUTs cascaded in slant range form an elongated volume of space, spanned by the solid angle of the SBR AESA beam after transmission of a burst of linear frequency modulation (LFM) waveforms at a low PRF (e.g. 2 kHz) and a medium-low transmit duty-cycle. Clearly, in this case, a target is range non-ambiguous in the fast-time and highly Doppler ambiguous in the slow-time. Despite such inherent pulse-to-pulse Doppler ambiguity, a bank of frequency offsets applied in parallel to the echo complex envelope in the fast-time allows enforcing Doppler tolerance on the LFM ambiguity function implemented by the matched filter. Remarkably, the frequency offset applied to the branch with minimum PC loss at the matched-filter output in the fast-time cues an estimate of the echo range-rate (embedding the relative radial velocity between the SBR and the debris). In addition, combining multiple samples in the slow-time (e.g. four pulses) allows for the build-up of the SNR before cascading any CFAR-like detection scheme. Such a strategy departs from the high PRF velocity search paradigm indicated in [41] and can be referred to as 'low-PRF RRRS'. Furthermore, by suitable design of the temporal hierarchies, a cascade of more than one RRRS allows introducing additional alert-confirm strategies combined with spoilt beams during the search task. It must be stressed that the FoV extent of such a surveillance mode based on a Low-PRF RRRS relies on the AESA transmit peak power as well as on beam shaping and steering capabilities.

Due to the imposing debris velocities, in a post-detection scenario, a track while scan (TWS) approach appears as a formidable collection strategy over a significant VoI such as the one represented in Figure 11.4. Interestingly, in a post-detection scenario pertaining to a sparsely populated debris environment [1, 19], the VoI can be immediately reduced in order to initiate an automatic tracking task. Remarkably, a fruitful approach can be characterised by adopting a Low-PRF RRRS in a PWS contacts-collection-strategy during an assumed near constant velocity (NCV) motion of a debris cloud for an elapse time up to several hundreds of milliseconds [19]. Indeed, despite the hyper-velocity of debris targets, changes in target motions are not expected events unless spawning among colliding debris modifies abruptly the trajectory of existing orbital items. That is, considering a dwell time no longer than, say 2 ms, in case a detection via RRRS on a CUT occurs, the SBR finite-state machine transitions into low-PRF RRRS PWS and acquires four complex data hypercubes during four additional dwells (i.e. in less than 8 ms) in a round robin fashion pertaining to the four beams surrounding the original beam in which a detection had occurred. Nevertheless, by extending the elapse time to no more than, say 16 ms (whereby the detected target is supposed to move for, say roughly 100 m), it is possible to envisage eight beams circumscribing in a round robin fashion the original beam in which a detection had occurred. Namely, during the first four dwells, the SBR acquires the complex data hypercubes of the North-West, North-East, South-East and South-West relative pointing directions, respectively, and then, during the next four dwells, the SBR acquires the complex data hypercubes pertaining to the West, North, East and South relative pointing directions, respectively. Indeed, from this point onwards, the onboard tracker can process measurements as either thresholded or unthresholded data upon one, or more than one, hyper-velocity targets.

11.4 Payload system architecture

The payload functional architecture addressed in Figure 11.3 is further described hereafter taking into account the subsystems shown in section 11.2.

11.4.1 Microprocessor and ultra-stable oscillator

The digital core of the SBR in Figure 11.3 implements a general purpose reduced instruction set computer fault tolerant microprocessor (μP) (e.g. LEON3FT [45]) whose micro-kernel could be housed on dedicated Programmable Read Only Memory (PROM) and Electrically Erasable PROM (EEPROM) along with specific algorithms and transceiver related constants. In this case, the overall system on chip (SOC) architecture can be based on a standard, modular and synchronous bus structure: the Advanced Microcontroller Bus Architecture (AMBA) which is the 'backbone' of the programmable SOC architecture with master/slaves communicating entities. Such a hardware-and-software partitioning would allow performing high-speed signal processing tasks of digital signal processing (DSP) slaves on Application Specific Integrated Circuit (ASIC) or Field Programmable Gate Array (FPGA) (e.g. complex envelope extraction, Doppler filter banks and PC in the custom logics), while carrying out advanced low-speed signal processing techniques (e.g. data processor) in software by employing the available μP or, eventually, a bespoke coprocessor (e.g. RC64 [46]). Moreover, the μP would control the custom DSP parameters while receiving, through the AMBA bus, proper DSP samples to be further processed in software. It is worth noting that the use of such a high-performance μP embedded in the architecture with suitable memory hierarchies (see the shared memories and knowledge-aided mass memory stack in Figure 11.2) could allow extending the transceiver processing capabilities as a representative example of a novel cognitive instrument devoted to debris detection and tracking applications. In addition, this μP-based approach allows for SBR management and data handling functions with the spacecraft and entails great flexibility in terms of functions, algorithms and design parameters.

Such a SBR architectural design (and related signal processing techniques) includes a reference ultra-stable oscillator (USO) for generating all the required local oscillator (LO) frequencies coherently locked to the USO. Furthermore, without changing the frequency reference of the USO, one may exploit the programmability of advanced phase-locked loop (PLL) synthesis to flexibly generate the upconversion and downconversion LOs on the SBR transmit and receive sections, respectively.

11.4.2 Arbitrary waveform generator

Alongside the μP and USO subsystems, another key subsystem is the arbitrary waveform generator (AWG) in Figure 11.3 aimed at synthesising a transmit LFM signal with a flexible programmability, high spectral purity in terms of phase noise and spurious free dynamic range, and high within-pulse and pulse-to-pulse stability. Such a design allows synthesising the single-pulse waveform directly on a carrier in the UHF band (e.g. $f_{UHF} \sim 500$ MHz) as

$$s_{UHF}(t) = A \cdot \cos\left[2\pi f_{UHF} t + 2\pi \frac{1}{2} k t^2 + \varphi_0\right] \cdot rect\left(\frac{t - T/2}{T}\right) \qquad (11.4)$$

where $rect\left(\frac{t}{T}\right)$ is 1 for $|t| \leq T/2$ and zero elsewhere. The synthesis of such a LFM waveform can be based on the direct digital synthesis (DDS) approach in [47] with high accuracy along with an efficient waveform descriptor by only four parameters: start phase φ_0, start frequency f_{UHF}, LFM slope k and pulse duration T (for the sake of completeness, the amplitude parameter A can be programmed as well).

Moreover, the DDS operational behaviour may benefit from a number of ancillary signal processing structures including waveform selection logics as well as compensation control logics in order to cope with undesired non-idealities in the transmit stages (Figure 11.7).

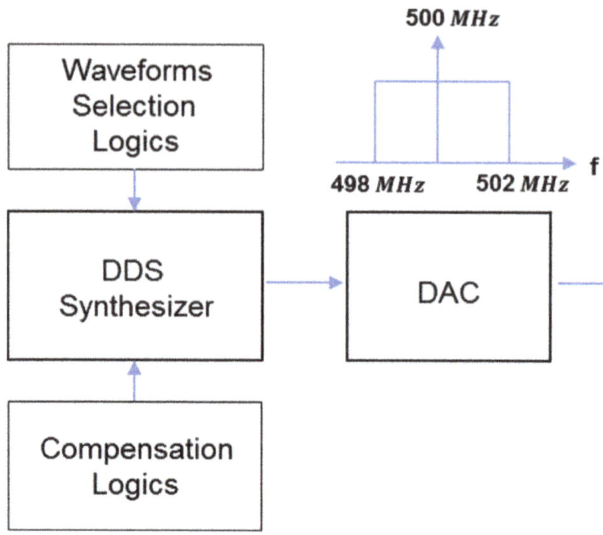

Figure 11.7 DDS-based waveform generator

11.4.3 RF upconversion

A number of strategies can be adopted for translating the UHF LFM radar signal in (11.4) to the Ka band before being distributed via a power divider to the transmit analog beam forming network (ABFN) of the AESA antenna in Figure 11.3 [47]. Hints, for example, include a single-side band modulator-based architecture, a PLL-based translation loop, or a superheterodyne-based architecture. Due to the limited required operative bandwidth indicated in [19] and the necessity of synthesising spectrally pure waveforms, a superhet architecture based on a double conversion stage with no need for a Q-fold band stretching represents a keystone for proper upconversion with mild band pass filter (BPF) transition bandwidths and phase noise containment. In comparison, this represents a $20\log(Q)$ improvement in phase noise compared to the use of stretching stages for wideband applications (e.g. see SAR applications in [30]). Interestingly, a superhet architecture based on a double-conversion stage can further deal with spectral purity demands by relying on a suitable selection of the USO and enforcing phase noise masks and Allan deviation performance [48, 49] directly on the LOs [50]. Figure 11.8 hints a possible frequency translation relying on a first upconversion stage in the C band (e.g. $f_C \sim 5.5$ GHz) and a second upconversion stage in the Ka band (e.g. $f_{Ka} \sim 35.5$ GHz). Consequently, the Ka band single-pulse signal input to the transmit power divider can be written as

$$s_{Ka}(t) = A_1 \cdot \cos\left[2\pi f_{Ka} t + 2\pi \tfrac{1}{2} k t^2 + \varphi_0\right] \cdot rect\left(\frac{t - T/2}{T}\right) \tag{11.5}$$

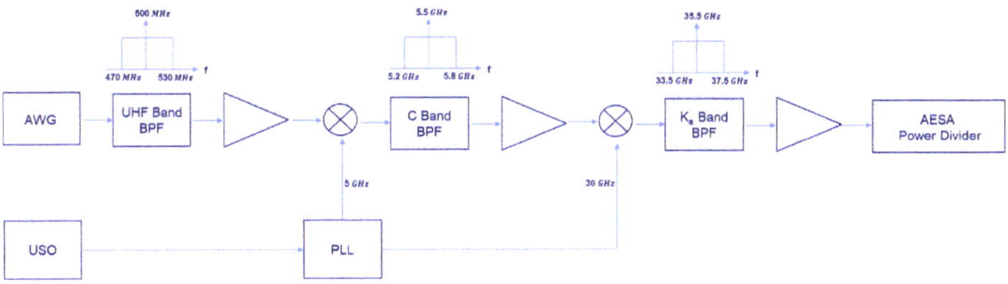

Figure 11.8 Upconverter frequency plan

where the A_1 parameter also accounts for the tunable output power to properly drive the AESA subsystem transmit line-up.

11.4.4 AESA subsystem

The upconverted signal in (11.5) is distributed from the transmit power divider towards the AESA antenna solid-state high-power amplifiers housed within transmit/receive (T/R) modules and finally transduced as a single-pulse low-duty-cycle pulsed electromagnetic wave in free space. The electrical-electromagnetic transduction occurs via the AESA radiators within a specific VoI as per the AESA array factor (AF), managed by the SBR beam steering controller in Figure 11.2. The limited required operative bandwidth of a few MHz in the Ka band [19] allows neglecting the adoption of TTDL against frequency dispersion, whereas a wide surveillance VoI needs significant steering angle capabilities in both azimuth and elevation dimensions. Consequently, steering angles larger than 50 deg with respect to broadside without arising grating lobes do require abandoning subarray grouping of radiating elements on a T/R module, thus mapping one T/R module to each radiating element. While transmit capabilities are required to illuminate a wider FoV with uniform tapering, an additional complexity burden arises on the receive side due to the multi-channel monopulse architecture which comprises, for each polarization, the quadruplet sum Σ, delta-elevation Δ_{el}, delta-azimuth Δ_{az} and guard *grd* channels, each tailored to optimally use a dedicated set of tapering coefficients for beam shaping with a considerable impact on AESA complexity. Accordingly, the amplitude and phase controls of a specific set of T/R modules (feeding radiators which are geometrically displaced on a planar lattice) determine the key characteristics of the AESA directivity in terms of beam-steering, beam-width and beam-shape. Similar considerations on T/R modules follow on receive, albeit with additional features, constraints and burdens. That is, echo signals spatially sampled by the AESA radiators are de facto fed (for each H and V polarization) to the T/R modules limiters and low-noise amplifiers (followed by low-level amplifiers to further increase power dynamics), modulated in amplitude and phase for beam control and passed through the receive power combiner towards the multi-channel downconverter in Figure 11.3. Indeed, while on transmit, a single polarisation is deemed sufficient for illuminating the environment for SSA; on receive, a dual polarisation allows paramount diversity to be further exploited during the debris detection and post-detection processes (see also the

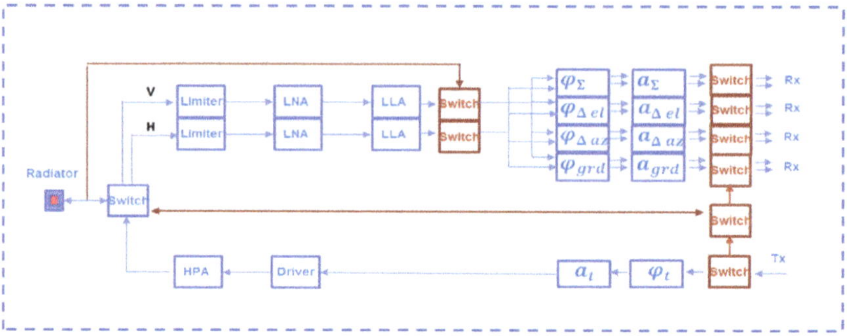

Figure 11.9 AESA radiator and T/R element

architectural design in [13]), especially against gyrotropic effects on signal propagation through plasma slabs as well as electromagnetic scattering from tumbling debris [19]. Clearly, for target classification, swapping a given transmit polarisation with its orthogonal counterpart on a pulse-by-pulse basis allows collecting on receive the complete polarimetric content of the target (i.e. the complex scattering matrix) every other PRI. Figure 11.9 depicts an AESA radiator (the red-blue square on the left) connected to a T/R module (the transmit branch at the bottom and the receive branches at the top) via a switch (instead of a circulator). The transmit branch (with signal flow from right to left) is characterised by the $\{a_t, \varphi_t\}$, amplitude-phase offsets whereas the receive branches (with signal flow from left to right for both the H and V polarisations) are characterized by the $\{a_\Sigma, \varphi_\Sigma\}$, $\{a_{\Delta el}, \varphi_{\Delta el}\}$, $\{a_{\Delta az}, \varphi_{\Delta az}\}$, $\{a_{grd}, \varphi_{grd}\}$, amplitude-phase offsets for the sum, delta-elevation, delta-azimuth and guard channels, respectively. For the sake of completeness, Figure 11.9 illustrates the AESA T/R element functional architecture not only for the transmit and receive chains (i.e. radiator elements, active electronics, amplitude and phase modulations) but also in terms of ancillary routings (the brown switches) for mandatory AESA calibration and blanking. Figure 11.10 completes the functional concept of the monopulse AESA subsystem whereby the whole set of radiators and T/R modules (the red-blue-white rectangles on the left) is linked to the power combiners of the sum, delta-elevation, delta-azimuth and guard channels (the green rectangles on the right for the V polarization and the beige rectangles on the right for the H polarization), as well as to the power divider of the transmit channel (the orange rectangle on the right).

The AESA antenna manufacturing is thus a formidable task per se, as it entails manifold subtleties in terms of electrical, electromagnetic, mechanical and thermal nuisances which may severely degrade, or rather hamper, the expected beam behaviours and AESA instrument performance [51, 52]. More specifically, the pitch accommodation, the possible mechanical distortion, the amount of mass, the electromagnetic coupling effects, and the heat dispersion combine as a design and development challenge for the radiating elements, the T/R modules, and the ABFN power divider/combiner integration, especially considering the small dimensions linked to the Ka band operating frequency and the aforementioned demands on the scanning

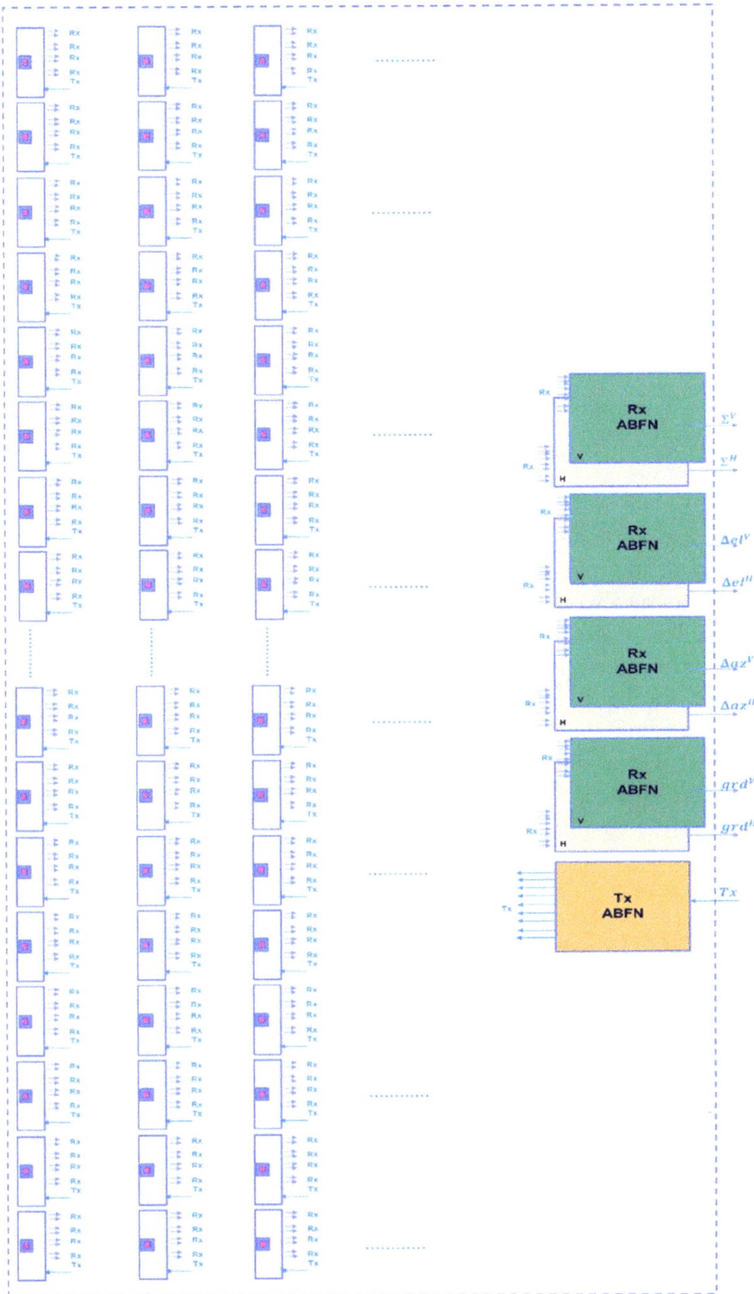

Figure 11.10 *AESA monopulse-based subsystem*

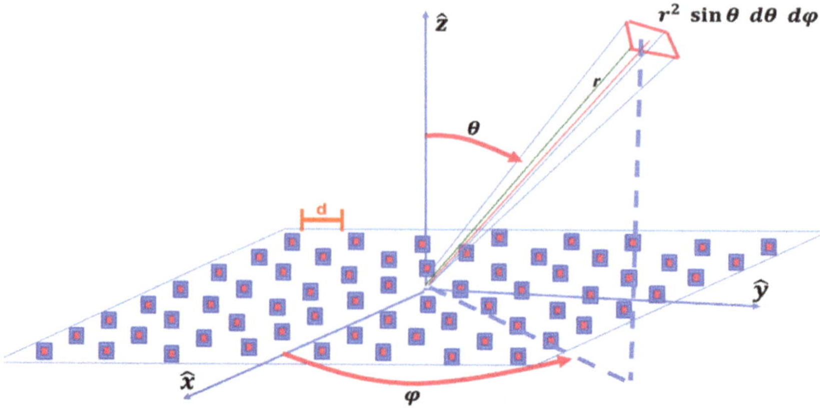

Figure 11.11 *AESA off-broadside angle*

angles. It is also worth mentioning that, albeit the transmit amplifiers operate in slight compression to improve efficiency and avoid arising amplitude modulations, the receive line-up must work linearly to properly exploit automatic gain control (AGC) along the downconversion section in Figure 11.3 as well as to preserve the superposition of echoes complex envelopes. Furthermore, stacking the AESA radiating elements and feeding two orthogonal polarisations (i.e. H and V in Figure 11.9) can be based, for example, on a single square horn element radiator with a septum polarizer cascaded with a differential phase shifter implementing an orthomode transducer aimed at splitting a signal into two balanced waveguides routes with a reduced cross-polarisation coupling [53]. Also, the T/R branches can be based, for instance, on gallium nitride (GaN)-based monolithic microwave integrated circuit implementing pseudo-morphic high electron mobility transistor (HEMT) processes with gains roughly more than 30 dB and noise figure of a few dB on the operative bandwidths. Instead of using burdensome circulators, single-pole double-throw switches can be adopted for signal routing and fast timing control [53]. In addition, the minimum element displacement d on the array lattice has a key influence on the maximum off-broadside scanning angle θ_0 without sprouting grating lobes. For example, let us consider in Figure 11.11 the geometry of an AESA whose radiating elements appear rectangularly displaced on the x, y axes with inter-element distance d.

A differential solid angle $d\Omega = \sin\theta\, d\theta\, d\varphi$ subtends the differential area $dA = r^2 \sin\theta\, d\theta\, d\varphi$ identified by the spherical triplet (r, θ, φ) whereas, in this case, at a given wavelength λ the displacement d limits the ideal off-broadside angle θ_0 from $\hat{z}$ (e.g. excluding degradations due to coupling [51]) as

$$d \leq \frac{\lambda}{1 + \sin\theta_0} \tag{11.6}$$

The bound in (11.6) also cues the scalability of the '*operative angular FoR*' depending on the design (and manufacturing capability) of the inter-element distance d. In parallel, the scalability of the '*operative range FoR*' depends on the sum of all T/R

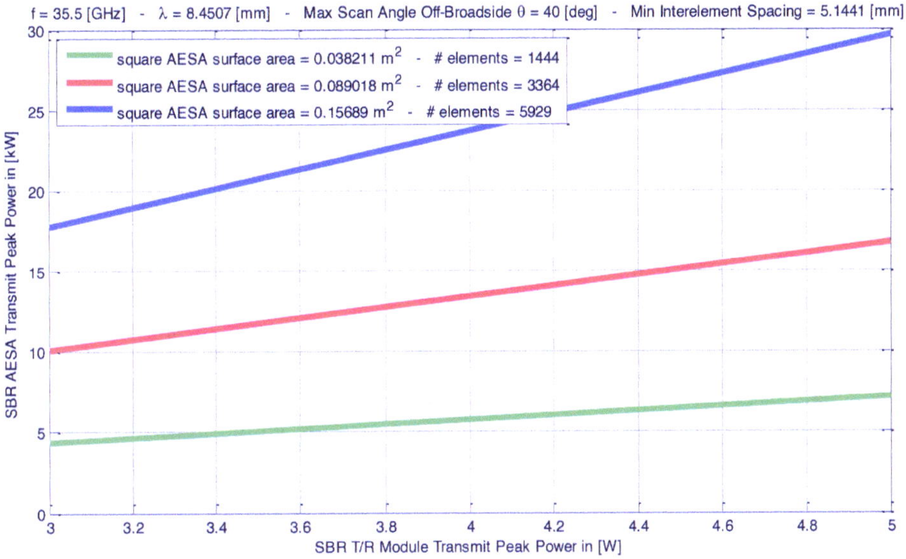

Figure 11.12 *AESA peak power with max off-broadside angle $\theta_0 = 40$ deg*

modules transmit peak power allocated on the AESA. Thus, for a given antenna area, one can obtain the SBR total transmit peak power by adding all T/R modules transmit peak powers and, consequently, the directivity and beam solid angle thereof [51]. The plots reported in Figures 11.12 and 11.13 show the SBR transmit peak power at 35.5 GHz for a set of AESA square-surface areas pertaining to a number of T/R modules

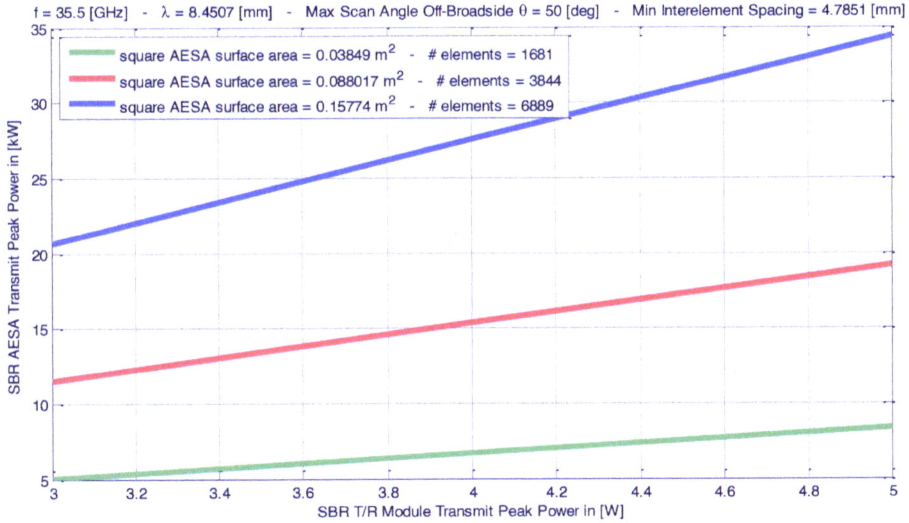

Figure 11.13 *AESA peak power with max off-broadside angle $\theta_0 = 50$ deg*

(whose lattice inter-element spacing attains the upper bound in (11.6) with equality) with respect to several T/R module transmit peak power and maximum off-broadside angle θ_0 without occurrence of grating lobes.

Interestingly, Figures 11.12 and 11.13 highlight several insights on the pros and cons of a Ka band AESA antenna for debris detection and tracking purposes. Indeed, large operative transmit peak powers could be obtained not by focusing efforts on T/R modules designs with larger and larger transmit peak powers but rather on tighter inter-element pitch spacing and related T/R modules integration for accommodating an increased number of elements on the lattice (thus relaxing thermal requirements on heat pipes). In this case, a moderate AESA area enlargement would allow for a significantly larger transmit peak power of the SBR whereas smaller element pitch spacing can provide important benefits for the SBR surveillance over larger off-broadside scanning angles without occurrence of grating lobes (e.g. in the previous example, a 5.1 mm pitch spacing would allow for a 40 deg maximum off-broadside angle, whereas a 4.8 mm pitch spacing would allow for a 50 deg maximum off-broadside angle). Finally, a dedicated set of complex excitation coefficients for each channel (i.e. each amplitude attenuators and phase shifters addressed as the φ, a cascaded doublet in Figure 11.9) represent a separable (i.e. individual) configurability of the AESA transmit and receive beams ensued from different optimal monopulse radiation patterns (e.g. a uniform tapering on the transmit channel, a Taylor tapering on the receive sum channel and a Bayliss tapering on the receive difference channels [52]), yet to the detriment of additional AESA complexity. With respect to this latter aspect, modern literature is abundant with unconventional array architectures, topologies and design methodologies for implementing desired beams with reduced complexities. In general terms, Reference 54 provides a review of clustered, thinned, sparse and time-varying arrays. While legacy monopulse techniques were based on implementing sum and difference channels mostly at RF with a limited number of simultaneous lobing [55], advances in AESA lattice topologies as well as in array processing in hostile environments have pushed modern monopulse approaches towards digital multi-channel adaptive architectures [56]. For ease of design and manufacturing feasibility, the monopulse paradigm proposed for the SBR is framed by the multi-channel cardinality in Figure 11.3 (i.e. four RF channels for each polarisation) whereby simultaneous lobing occurs directly in the Ka band. From such a standard monopulse scheme at a given CUT, the joint processing of sum and difference channels in the digital domain in Figure 11.3 is entrenched in maximum likelihood estimation (see the covariance-matrix-free estimator in [57] (15)] for an individual target which is pulse-to-pulse fluctuating, or see the covariance-matrix-based estimator in [58 (13)] for an individual target).

Within this framework, for monopulse radar applications Reference 59 indicates convex-optimization-based approaches aimed at sharing common weights on the periphery of the array while maintaining independent excitations of sum and difference channels in the rest of the array. As a remarkable step forward, Reference 60 proposes a compromise solution which maximizes the directivity of the sum pattern subject to several constraints (including one on the slope of the difference patterns) and shares the excitation coefficients of the sum and difference patterns. Accordingly, a number of possible AESA layout quadrant lattices layouts boundaries arise as pertaining to monopulse constraints on the AESA amplitude-phase offsets represented by the complex excitation coefficient I_{mn} positioned on the discrete lattice of Figure 11.11 in the $\hat{x}$, $\hat{y}$ plane as per the subscript mn. That is, the AF for a planar AESA centred in the $\hat{x}$, $\hat{y}$ plane (see Figure 11.11) with an even number of elements $2M$ on the $\hat{x}$-axis and $2N$ on the $\hat{y}$-axis can be written as

$$AF(u,v) = \sum_{m=1}^{2M} \sum_{n=1}^{2N} I_{mn} e^{j\frac{2\pi}{\lambda}\left[\left(m-M-\frac{1}{2}\right)d_x u + \left(n-N-\frac{1}{2}\right)d_y v\right]} \tag{11.7}$$

where $u = \sin\theta\cos\varphi$, $v = \sin\theta\sin\varphi$ are the directional cosines; d_x, d_y are the lattice inter-element spacing along the $\hat{x}$, $\hat{y}$-axis, respectively; $I_{mn} = a_{mn}e^{j\varphi_{mn}}$ with a_{mn} and φ_{mn} representing the amplitude and phase of the complex excitation coefficient I_{mn}, respectively, such that, in a non-hostile environment, the φ_{mn} terms are used to set the beam scanning towards the direction (u_0, v_0) as per the phase shifts $\varphi_{mn} = -\frac{2\pi}{\lambda}\left[(m-1)\,d_x u_0 + (n-1)\,d_y v_0\right]$. Following [59] for instance and considering a square lattice (i.e. $d_x = d_y$ and $M = N$), one may then impose quadrant symmetry constraints on the whole set of monopulse-based a_{mn}, i.e. $a_{t_{mn}}$, $a_{\Sigma_{mn}}$, $a_{\Delta\mathrm{el}_{mn}}$, $a_{\Delta\mathrm{az}_{mn}}$ and $a_{grd_{mn}}$.

11.4.5 RF downconversion

Another variety of strategies can be adopted for translating the Ka band LFM radar echo signal from the receive power combiner output (from each H, V polarisation and from each Σ, Δ_{el}, Δ_{az}, grd monopulse sum, delta-elevation, delta-azimuth and guard channel, respectively) into the downconverter subsystem in Figure 11.3 all the way down to a suitable intermediate frequency (IF) before feeding the digital section. For example, an interesting multi-channel architecture relies on a modular approach where the whole set of monopulse channels can be assigned to work under the shared control of an AGC loop providing the instrument dynamic range with over 50 dB of gain control while the receiver chain applies a double-conversion superheterodyne scheme [61]. The use of a double-conversion configuration is caused by the difficulties of applying a analog-to-digital converter (ADC) directly at such high frequencies, whereas a single conversion would inevitably hamper image frequency rejection. In addition, a double-conversion would allow spreading the gain control over larger sections of the receiver. Figure 11.14 indicates a possible frequency translation relying on a first downconversion stage in C band (e.g. $f_C \sim 5$ GHz) and a second downconversion in the high-frequency (HF) band (e.g. $f_{HF} \sim 25$ MHz). Without lack of generalisation, the proposed frequency plan in Figure 11.14 assumes a Ka band BPF centred at 35.515 GHz in order to exploit integer-N PLLs for downconversion to a fixed IF frequency centred at 25 MHz. For the sake of clarification, if compared to the transmit frequency plan centred at 35.5 GHz in Figure 11.8, the additional 15 MHz is an agile programmable task on the AWG. Moreover, it is worth noting that the IF filter in the HF band in Figure 11.14 has a slightly larger percentage of bandpass

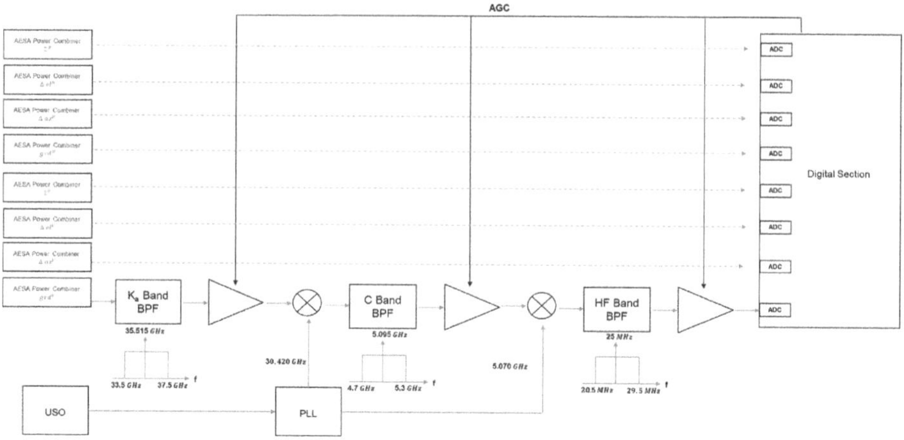

Figure 11.14 Downconverter frequency plan

compared to the other BPFs in order to accommodate variable operative bandwidths and could be based on surface acoustic waves (SAW) devices implementing anti-aliasing functions before the signal is digitised.

11.4.6 Multichannel complex envelope acquisition

The receive digital section in Figure 11.3 is based on harmonic-sampling the IF signal using a single ADC and producing I and Q samples without phase and amplitude imbalances of a quadrature analog demodulator. Considering relative SBR-debris velocities up to 15 km/s in LEO, the Doppler effect on the received echoes amounts to a few MHz (e.g. as a rough upper bound one may consider that a 16 km/s relative velocity at 36 GHz would result in a Doppler frequency around 3.8 MHz). Consequently, the HF single pulse echo signal at ADC input (see Figure 11.14) for a point-like target with a τ_R delay as well as with a f_D Doppler offset is fed as input to the digital section as per the following real signal

$$s_{HF}(t) \sim C\cdot\cos\left[2\pi\left(f_{HF}+f_D\right)\left(t-\tau_R\right)+2\pi\tfrac{1}{2}k\left(t-\tau_R\right)^2+\varphi_0\right]\cdot rect\left(\frac{(t-\tau_R)-T/2}{T}\right) \tag{11.8}$$

The single-pulse point-like target echo approximation in (11.8) is obtained by neglecting the scaling of the amplitude and the stretching of the time axis by the $\gamma = 1+\frac{f_D}{f_{Ka}}$ factor [62, 63] and abides by space-time factorability [64]. For the sake of clarification, a brief digression follows for both aspects. Considering a CUT resolution, References 62 and 63 provide key bounds for employing either narrowband or wideband representations of the signal and, consequently, of the signal ambiguity function for estimating range and range rate parameters. By referring to c as the speed of light in the propagation media, for a radial velocity v and a signal with a time-bandwidth product $T\cdot B$, a narrowband representation of the signal relies on the following assumption

$$v \ll \frac{c}{BT} \tag{11.9}$$

whereas for a radial acceleration $\dot{v}$ a narrowband representation of the signal relies on the following inequality to hold

$$\dot{v} \ll \frac{c}{BT^2} \tag{11.10}$$

During limited time spans in the order of a second, debris detection and tracking can be tackled assuming an NCV model for both the SBR and debris motion models [19]. In this case, with a PRF on the order of a few kHz, a low transmit duty cycle and a few MHz of bandwidth B, the inequalities in (11.9) and (11.10) are both attained even for extremely fast micrometeoroids. Indeed, $\dot{v} \cong 0$ by NCV assumption whereas, considering a micrometeoroid flying at an imposing 70 km/s relative radial velocity v with respect to the SBR [1], for a 5 MHz bandwidth B and a 50 μs pulse duration T (11.9) results in $0.06 \ll 1$. Interestingly, taking into account a 16 km/s relative velocity v with a 2.5 MHz bandwidth B (11.9) still holds as $0.27 \ll 1$ when replacing the 50 μs pulse duration T with a 2 ms CPI. Therefore, the adoption of a narrowband signal model for range and range rate parameters estimation entails neglecting the γ factor for all Doppler terms pertaining to the received echo analytical representation in (11.8) with the exception of the carrier phase term. In parallel, the additional modelling aspect on the point-like target echo in (11.8) pertains to pondering its space-time factorability [64] (i.e. considering the x, y variables, the *x-y factorability* of a function $f(x, y)$ results in $f(x, y) = f_1(x) f_2(y)$). For an SBR with an AESA with 1-D dimension L (i.e. either along the $\hat{x}$ or, equivalently, the $\hat{y}$ axis in Figure 11.11), the condition on the uncoupled representation of the temporal and spatial terms in the point-like target echo for analogue beamforming within the AESA subsystem would ensue as sufficient condition (see, for instance, Reference [64] (18))

$$\frac{L}{c} \ll \frac{1}{B} \tag{11.11}$$

which in our narrowband case holds for an AESA with a 1 dimension up to several tens of metres (e.g. for a square AESA with 1 dimension L of 3 m and a 5 MHz bandwidth B the condition in (11.11) results in $0.05 \ll 1$). Indeed, the condition in (11.11) allows neglecting the use of TTDL (see [64 (19–21))] in favour of phase-shifts. Besides the AESA subsystem, the digression on space-time factorability is in a subtle way still intertwined within (11.8) with profound implications for the possible non-ambiguous estimates of parameters embedded within the observables. Indeed, considering the point-like target echo acquired from the AESA main beam, the target range-rate parameter v_{target} would be ambiguously embedded in $f_D = -\frac{2v}{\lambda}$ in (11.8) due to the inherent coupling between the mutually unknown radial target velocity v_{target} and SBR velocity v_{SBR}, i.e. $v = v_{SBR} - v_{target}$. As a by-product, this would imply the lack of factorability of the SBR ambiguity function into separate spatial and temporal ambiguity functions (see [64 (10 and 48)]). For example, Figure 11.15 depicts three scenarios with ambiguities arising from several mutual

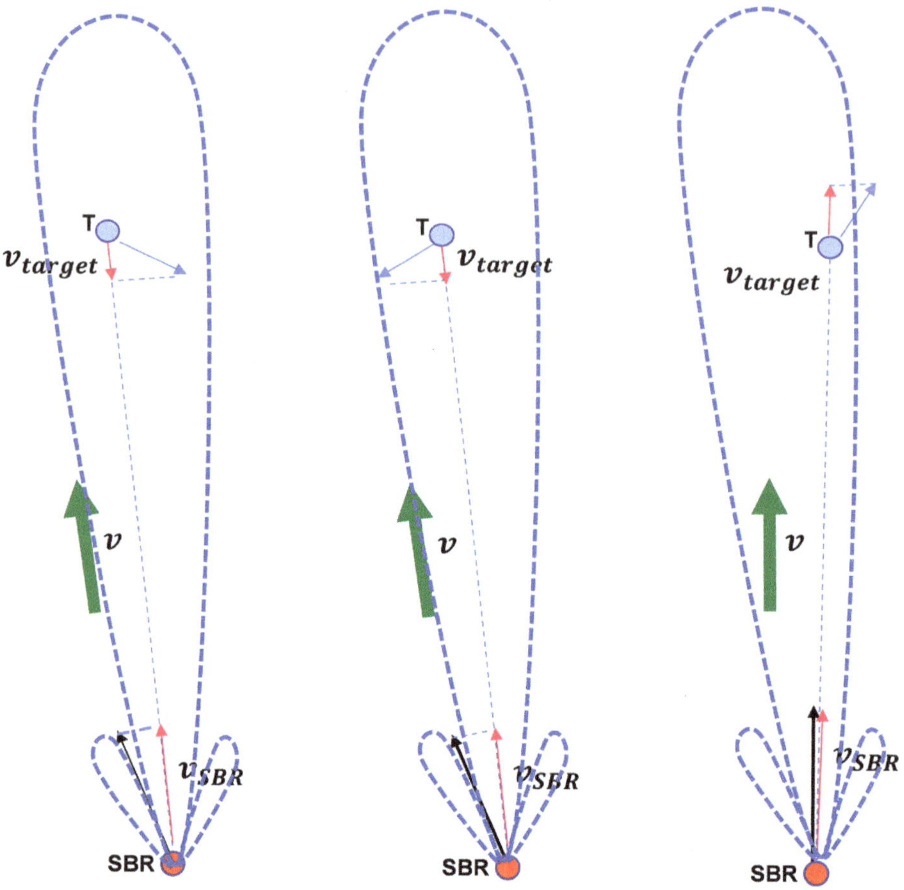

Figure 11.15 *Target range rate ambiguities embedded in* f_D

AESA and target radial velocities configurations (i.e. purple arrows) leading to the same relative AESA-target radial velocity $v = v_{SBR} - v_{target}$ (i.e. green arrow) whereby an outward target motion is defined as positive (i.e. $v_{target} > 0$).

By abandoning the intent to design an onboard tracker which bases its inference on target range-rate estimates, it is possible to define a resolution cell for a parameter vector made of range, azimuth angle, elevation angle and amplitude. In this specific case, depending on the proper design of a low sidelobe radiation pattern as well as on the guard channel radiation pattern lying strictly above the sidelobe pattern, the signal in (11.8) can be modelled as factorable in space-time. Then, the CUT can be factored into its angular components (i.e. azimuth angle and elevation angle) and temporal components (i.e. delay and amplitude). Downstream such a clarification, the parameter C in (11.8) can be factored into an angular component (i.e. the AF in (11.7)) and its temporal component (i.e. the point-like target echo amplitude occurring at a τ_R delay). Now, the separation of I and Q from (11.8) can be efficiently obtained

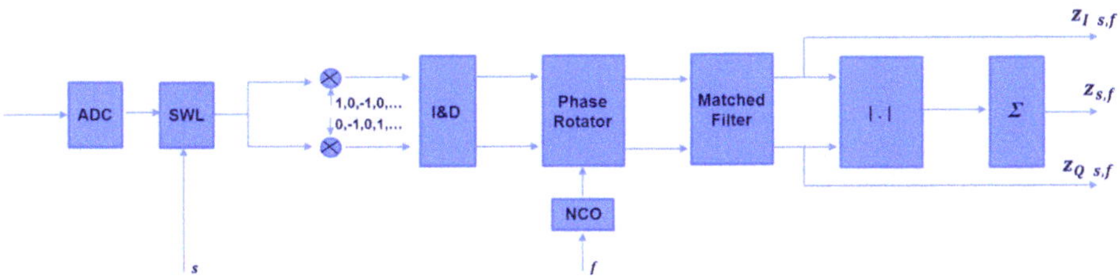

Figure 11.16 Bespoke matched filter with Doppler offset f *and range gate* s

by sampling the HF signal centred at 25 MHz with a 20 MHz ADC (thus handling the HF passband in Figure 11.14) and multiplying the ADC output by the 1, 0, −1, 0,… sequence and 0, −1, 0, 1, 0, … sequence for the I and Q branches, respectively, followed by an integrate and dump (ID) low-pass filter [65]. Remarkably, due to the imposing f_D Doppler offset with respect to the required operative bandwidth of a few MHz, at every range gate s it is possible to deal with the inherent limits of Doppler tolerance of the LFM-based ambiguity function. In particular, a set of filter banks, each bearing a dedicated frequency offset f, allows estimating within-pulse the echo Doppler frequency f_D from the branch with minimum matched filter loss (as cued by the branch with maximum modulus value obtained after the non-coherent combining of a train of echoes at the matched filter output). Such a bespoke design structure relies on an ADC with at least 14 bits (or, eventually, 16 bits [66] for managing an even larger instantaneous power dynamics) as well as on efficient COordinate Rotational DIgital Computer (CORDIC) [67] for phase rotators driven by numerically controlled oscillators (NCO) with a frequency offset granularity in the order of 100 kHz. For a given Doppler offset f (driving the NCO) and range gate s (driving the SWL), Figure 11.16 represents such a bespoke matched filter which is a key structure to be employed to form the SBR complex data hypercube.

More specifically, the modulus value of the matched filter output after a suitable non-coherent combining is indicated as $z_{s,f}$ whereas the matched filter complex output (i.e. $z_{I\ s,f}$ and $z_{Q\ s,f}$) can be further processed within the receive digital chain. A proof of concept for the basic onboard processing structure indicated in Figure 11.16 is shown in Figures 11.17–11.22. That is, Figure 11.17 sketches a simulated point-like target echo at ADC output at a given range gate s with a 4 dB SNR and a 1 MHz bandwidth along with a Doppler frequency of roughly 2.5 MHz. In particular, Figure 11.17(a) and (b) represents the temporal and spectral echo behaviour, respectively, whereby (without lack of generality) the temporal reference system is centred at the true target delay. After enforcing a suitable Doppler offset f, Figure 11.18 reports the non-coherent combining output $z_{s,f}$ built upon a burst of 4 echoes with interleaved upchirp-downchirp LFM waveforms, i.e. in Figure 11.18(a), the linear magnitude on odd pulses; in Figure 11.18(b), the linear magnitude on even pulses; in Figure 11.18(c), the dB magnitude on odd pulses; in Figure 11.18(d), the dB magnitude on even pulses. Clearly, an interleaved upchirp-downchirp LFM waveforms burst-transmit-strategy is enticing per se, as it allows soothing the LFM-inherent range estimation bias and squelching even-time-around echoes, if any.

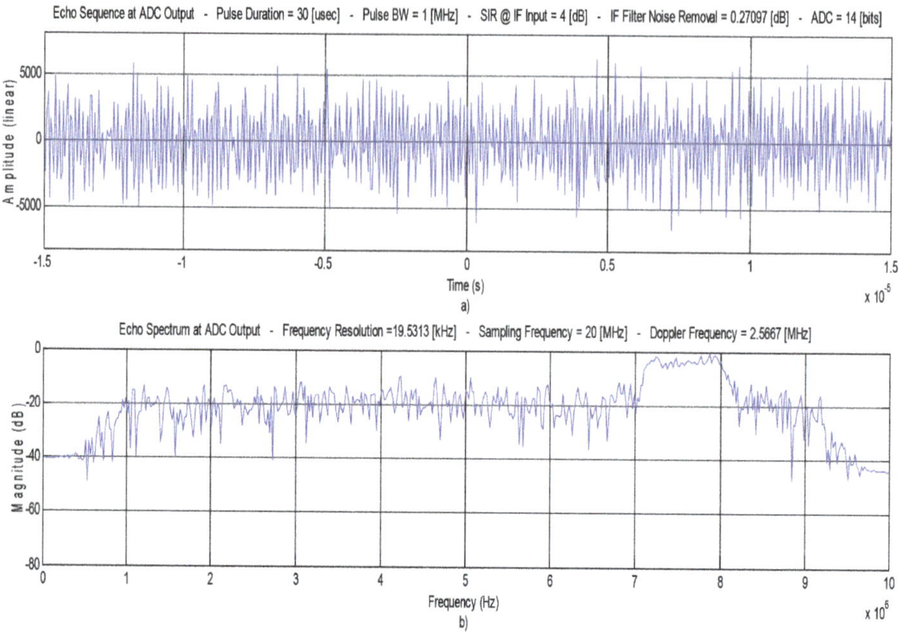

Figure 11.17 *Point-like target echo at ADC output at a given range gate* s: *(a) temporal echo; (b) spectral echo*

For the sake of clarification of the digital receiver line-up, Figure 11.19 shows the simulated point-like target echo behaviours, i.e. temporal echo in Figure 11.19(a) and spectral echo in Figure 11.19(b) at the phase rotator input of Figure 11.16. Interestingly, the mild echo image at roughly −2.5 MHz also appears (distorted and attenuated) due to digital mixing and I&D transfer function with respect to the useful echo at roughly 2.5 MHz. Figures 11.20 and 11.21 depict the echo appearance (i.e. temporal echo in Figures 11.20(a) and 11.21(a) and spectral echo in Figures 11.20(b) and 11.21(b) downstream the phase rotator output when either a suitable or unsuitable Doppler offset f is applied, respectively. That is, while in Figure 11.20, the residual echo Doppler offset appears successfully enforced to fit within a ± 50 kHz frequency span and, therefore, is suitably tolerated by the matched filter centre frequency at 0 Hz; in Figure 11.21, the residual echo Doppler offset is much larger (i.e. roughly 1.9 MHz) and, therefore, unbearable by the matched filter.

Indeed, in case of a significant residual echo Doppler offset, the matched filter output degradation is evident in Figure 11.22 (with a residual echo Doppler offset of roughly 1.9 MHz) with respect to the example depicted in Figure 11.18 (with a residual echo Doppler offset of roughly −30 kHz). Again, Figure 11.22 reports the non-coherent combining output $z_{s,f}$ built upon a burst of four echoes with interleaved upchirp-downchirp LFM waveforms, i.e. in Figure 11.22(a) the linear magnitude on odd pulses; in Figure 11.22(b) the linear magnitude on even pulses; in

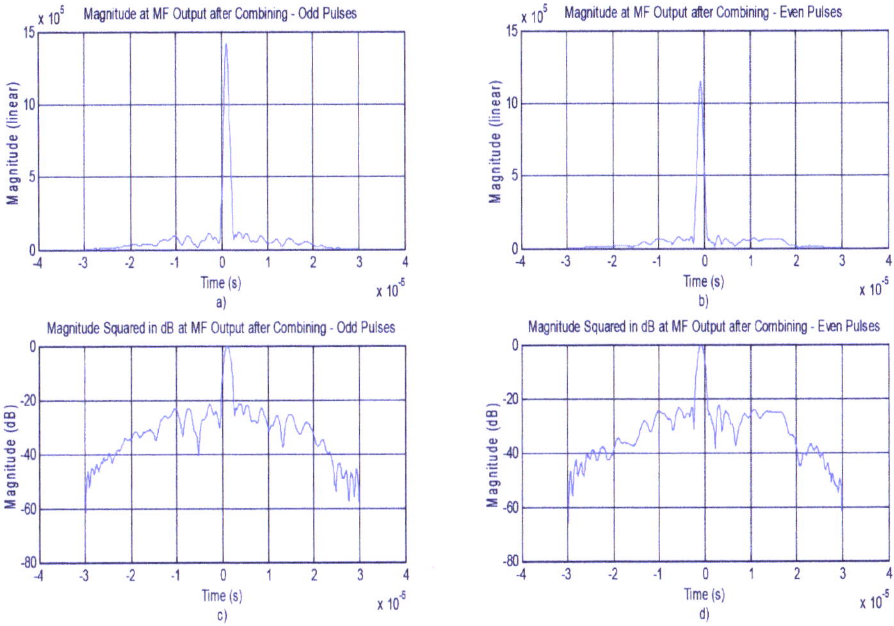

Figure 11.18 $z_{s,f}$ *in case of a negligible residual echo Doppler offset. (a) Linear magnitude on odd pulses; (b) linear magnitude on even pulses; (c) normalised dB magnitude on odd pulses; (d) normalised dB magnitude on even pulses*

Figure 11.22(c) the dB magnitude on odd pulses; in Figure 11.22(d) the dB magnitude on even pulses.

The simulation results presented in Figure 11.17 through Figure 11.22 are solely aimed at stressing the controllable Doppler tolerance of the transceiver complex envelope acquisition scheme. Thus, further echo acquisition variabilities and inherent degradations due to endogenous (e.g. phase noise [68, 69] and numerical quantization [70]) and exogenous (e.g. channel scintillation and target scattering assumptions [19]) impairments are outside the scope of this work. Clearly, the effectiveness of a target echo extraction during a dwell via the signal processing chain in Figure 11.16 relies on the availability of a suitable Doppler offset f and range gate s. However, these quantities are unknown, and to overcome this shortcoming, the bespoke matched filter structure in Figure 11.16 is replicated in Figure 11.23 for a set of programmable Doppler offsets f (i.e. applied in the horizontal dimension between the f_{min} and f_{max} values) and a set of range gates s (i.e. applied in the vertical dimension between the s_{min} and s_{max} values), thus paving the way for target echo extraction within the environmental scenario during a dwell.

The eight monopulse channels implementing the bespoke matched filter for all Doppler offsets and range gates upstream the CFAR block are finally depicted in

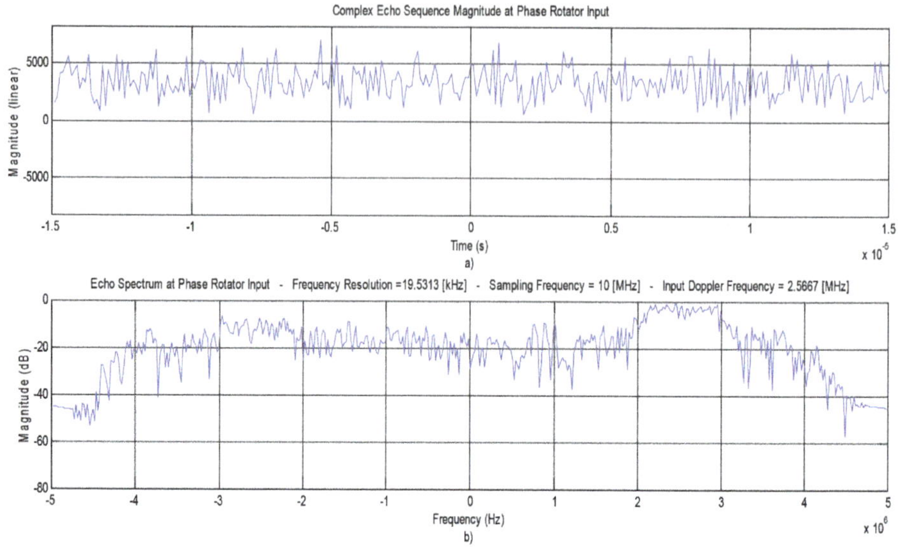

Figure 11.19 *Point-like target echo at phase rotator input: (a) temporal echo; (b) spectral echo*

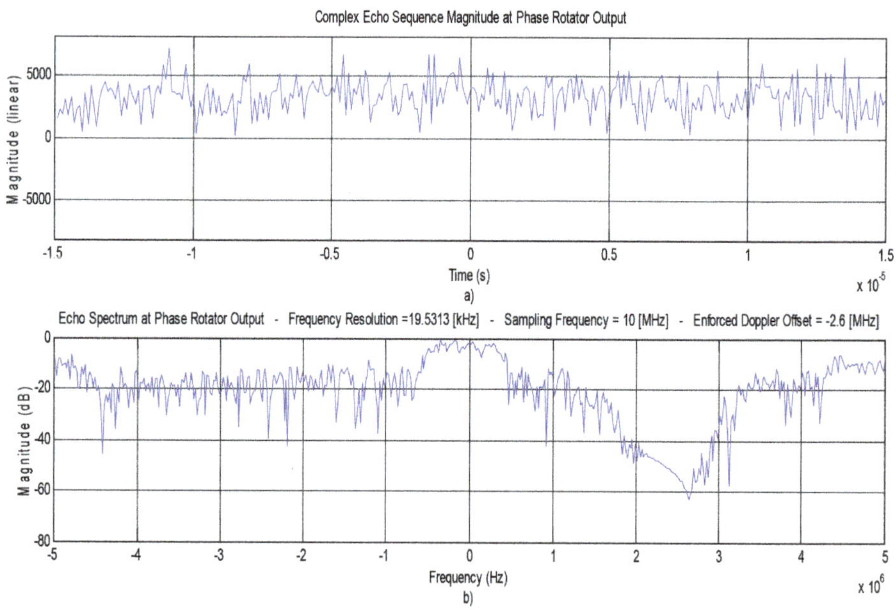

Figure 11.20 *Point-like target echo at phase rotator output (suitable Doppler offset): (a) temporal echo; (b) spectral echo*

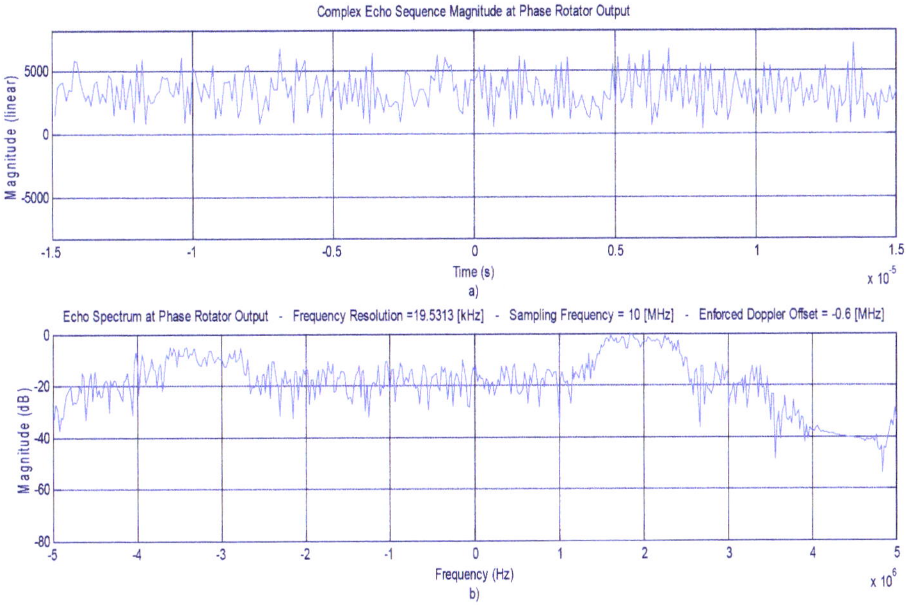

Figure 11.21 *Point-like target echo at phase rotator output (unsuitable Doppler offset): (a) temporal echo; (b) spectral echo*

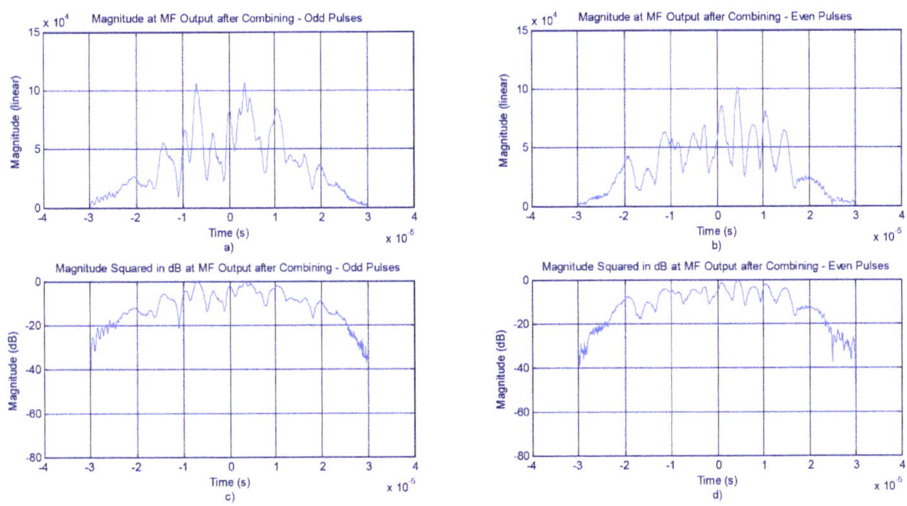

Figure 11.22 $z_{s,f}$ *in case of a significant residual echo Doppler offset: (a) linear magnitude on odd pulses; (b) linear magnitude on even pulses; (c) normalised dB magnitude on odd pulses; (d) normalised dB magnitude on even pulses*

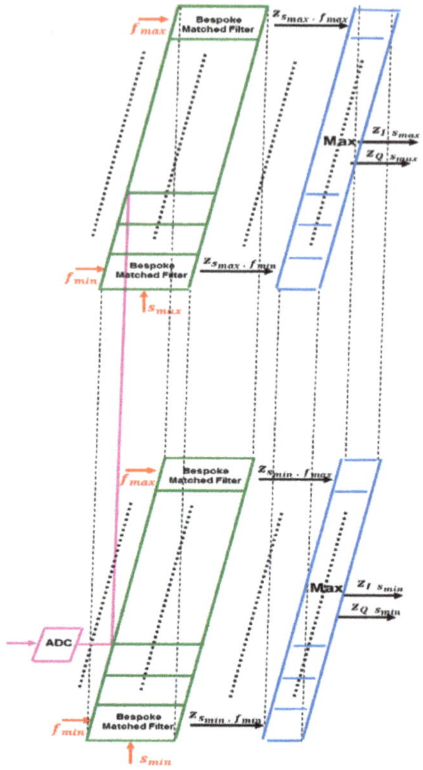

Figure 11.23 *Bespoke matched filter for all Doppler offsets f and range gates s*

Figure 11.24. For both the *H* and *V* polarisations, the *grd* and the Σ channels are routed to the CFAR-like block.

The CFAR-like block in Figure 11.3 includes a decision rule whose detection threshold for a given false alarm probability is (possibly) independent of the nuisance parameters variability. This is aimed at enforcing robustness in the binary-hypothesis test whereby the probability density function (pdf) under the noise-only hypothesis is independent of nuisance parameters. In our case, besides the target echo Doppler frequency f_D, additional nuisance parameters variability accounts for the system temperature change (in case the antenna radiation pattern is pointed towards either the Sun, the Earth, or the outer space) or the event of an imposing echo return spike from a large clutter discrete (e.g. from an orbital infrastructure [19]).

Considering that the building blocks in Figure 11.16 represent those basic fixed functions that enable the formation of the complex data hypercube, the use of a space qualified 28 nm ASIC [71] appears as an interesting solution for implementing the multichannel structure represented in Figure 11.24. On the contrary, a programmable FPGA according to possible evolving ameliorations can represent a viable

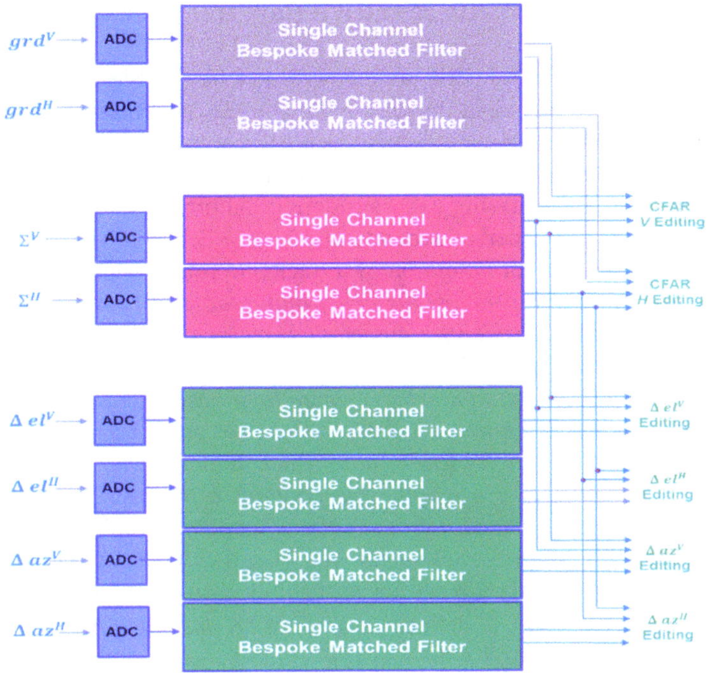

Figure 11.24 *Multi-channel bespoke matched filter structure*

design paradigm for the CFAR-like block as well as for the implementation of dedicated editing functionalities such as the *grd* channel-based excision-control-logics.

Finally, the data processor in Figure 11.3 can be a software-based functionality for tracking purposes, along with support editing transformations such as the target delay-range conversion for the Σ channel, monopulse angular conversion look-up tables for the Δ_{el} and Δ_{az} channels jointly with the Σ channel for gain normalisation, and attributes formation for unresolved detection of closely spaced targets [72, 73].

11.4.7 Notes on the guard channel

The structure in Figure 11.3 includes the implementation of outliers-excision control-logics editing functionalities exploiting the guard-channel. A comparison occurs between the signal from the pencil beam of the sum channel and the signal from the broader beam of the guard channel to discern if the echo derives from a signal in the main beam or from the sidelobes of the sum channel. Ideally, the broad beam pattern of the guard channel should lie above the pencil beam sidelobe pattern of the sum channel. The control-logics for the outliers-excision is outside the scope of this work. In simple terms, when the sum channel decides for a detection in a CUT, the detection is blanked if the guard channel statistic is larger than the sum channel statistic ([74] and [75]). In this case, chances are that the tentative detection in the sum channel is being triggered by the presence of a clutter discrete in the sidelobes.

11.5 Digression towards cognitive Bayesian tracking

A digression follows on the SBR architecture sketched in Figures 11.2 and 11.3 for a cognitive-based Bayesian tracking tailored to SSA. Remarkably, in [19], the modelling keystone for tracking debris via a SBR sprouted as an approximation of debris motion models moving along straight lines in 3-D at constant hyper-velocities with no manoeuvring and considering 'automatic tracking' [76] in limited time spans in the order of seconds in which a debris might (or might not) cross the SBR FoR. A neat definition of a track and related automatic tracking is given by [76]: '*A track represents the belief that a target is present and has actually been detected by the radar. An automatic tracking radar system forms a track when enough radar detections are made in a believable enough pattern to indicate a target is actually present...and when enough time has passed to allow accurate calculation of the target's kinematic state... Thus, the goal of tracking is to transform a (time-lapse) detection picture into a track picture...The result of the automatic tracking process is a track file that contains a track state for each target detected by the radar*'. Armed with such 'track' and 'tracking' definitions, let us assume that reliable '*debris-tracks-estimates*' (i.e. the $\hat{\boldsymbol{X}}_K(\boldsymbol{Z}_K)$ tracklets) are suitably inferred by a SBR via a posterior probability distribution on target state. In particular, $\boldsymbol{Z}_K \triangleq \{z_1, .., z_K\}$ is defined as a set of '*K-instants-SBR-measurements*' and the state $\boldsymbol{X}_K \triangleq \{\boldsymbol{x}_1, .., \boldsymbol{x}_K\}$ is defined as a set of '*K-instants-debris-states*' of one (or more than one) debris from the time instant t_1 to the time instant t_K. Within such an elapse time, $\boldsymbol{x}_i = \{\boldsymbol{p}_{i,w}\}_{w=1}^{W_i}$ is the '*i*th-*instant-debris-state-vector*' (for $i = 1, .., k$), which includes a number W_i of '*debris-entity-state-vectors*' (i.e. the cardinality of the debris-entity-state-vector set is possibly variable during the $[t_1,\ t_K]$ elapse time), each identified by a 3-D Cartesian position $\boldsymbol{p}_{i,w}$ and representing (in the most general sense) either an individual debris or a point of a debris cloud. Moreover, $z_i = \{z_{i,w}\}_{w=1}^{W_i'}$ is the '*i*th-*instant-debris-measurement-vector*' (for $i = 1, .., k$), which includes a number W_i' of '*debris-singleton-measurement-vectors*' $z_{i,w}$ (i.e. the cardinality of the debris-singleton-measurement-vector set is possibly variable during the $[t_1,\ t_K]$ elapse time). Accordingly, z_i can be processed by the Bayesian tracker (either as thresholded or unthresholded data [77]), as per the following structure of the debris-singleton-measurement-vector

$$z_{i,w} = \left[\hat{\boldsymbol{r}}_{i,w}^T, \hat{\boldsymbol{\alpha}}_{el_{i,w}}^T, \hat{\boldsymbol{\alpha}}_{az_{i,w}}^T, \left[vec\left\{\boldsymbol{Z}_{MF_{i,w}}\right\}\right]^T\right]^T \tag{11.12}$$

where ^{T} is the matrix transpose operator and $vec\{\cdot\}$ is the matrix column stacking operator. Thus, at the *i*th time instant for the *w*th debris-entity (i.e. at $i = 1, .., k$, for w=1,.., W_i'); $\hat{\boldsymbol{r}}_{i,w} \in \mathbb{R}^{2\times 1}$ is the range estimation for the H and V polarizations, respectively; $\hat{\boldsymbol{\alpha}}_{el_{i,w}} \in \mathbb{R}^{2\times 1}$ is the centroid elevation angle estimation for the H and V polarizations, respectively; $\hat{\boldsymbol{\alpha}}_{az_{i,w}} \in \mathbb{R}^{2\times 1}$ is the centroid azimuth angle estimation for the H and V polarizations, respectively; and $\boldsymbol{Z}_{MF_{i,w}} \in C^{6\times N_i}$ is a complex matrix (considering six rows pertaining to the Σ, Δ_{el}, and Δ_{az} channels, each bearing the H and V polarizations) whose column-wise entries represent a set of N_i complex matched filter outputs $z_{I\ i,w,s,f} + jz_{Q\ i,w,s,f}$ (pertaining to N_i pulses which began during a dwell at the ith measurement time) with estimated Doppler offset f and range gate s. In fact, the complex matched filter outputs of the Σ, Δ_{el}, and Δ_{az} channels embody additional attribute measurements [72, 73] useful to the onboard tracker, related to the possible unresolved detection of closely spaced targets within the CUT in terms of joint centroid-extent estimation (see the footnote of page 131 in [72] and Figure 11.11 in [58]). As hinted in [19], considering limited time spans in the order of a second, it is possible to enforce an NCV model on $\hat{\boldsymbol{X}}_K(\boldsymbol{Z}_K)$. Thus, from each tracklet cued by the position set $\{\boldsymbol{p}_{i,w}\}$ within $\hat{\boldsymbol{X}}_K(\boldsymbol{Z}_K)$ whose w index occurs during a sufficient elapse time $[t_{w\ initial},\ t_{w\ end}]$, i.e. pertaining to the time instants $i \in [t_{w\ initial},\ t_{w\ end}] \subseteq [t_1,\ t_K]$, it is possible to estimate $\{\boldsymbol{p}_w(t_0),\ \dot{\boldsymbol{p}}_w(t_0)\}$ at a given time instant $t_0 \in [t_{w\ initial},\ t_{w\ end}]$ and, consequently, estimate a complete set of ephemerides at t_0. Interestingly, space objects manoeuvring [78] can be neglected, thus abandoning the need for banks of parallel tracking filters or interacting multiple models [79, 80]. Nevertheless, the debris motion relying on the linear continuous white noise acceleration (CWNA) model [79, 80] allows representing the NCV scenario for debris crossing the SBR FoR. In addition, by properly constraining the CWNA model parameters, it is also possible to further characterise mild changes in debris velocity components during an observation interval. It is also worth noting that a Bayesian multi-target tracking (MTT) functionality [77] can thus be contrived in case of a multitude of debris appearing within the instantaneous SBR FoV. In this case, it is possible to consider approaches relying either on measurement to track associations as tailored to multiple *individual debris* or on point process models as tailored to *debris clouds*. In the specific case of an SBR aimed at SSA via Bayesian inference, the objective of the SBR may coincide with estimating the tracklets $\hat{\boldsymbol{X}}_K$ from a MTT perspective and according to a Bayesian formulation of the inference problem. Namely, allowing for the capability to exploit a-priori distributions for environmental debris density, target motion models and interference; the capability to form a suitable likelihood-function based on the SBR sensor data acquisition; and finally, the capability to combine the a-priori distributions and likelihood-functions into a suitable posterior distribution on target state [77]. Traditionally, the MTT-based state estimate could be accomplished by a so-called feed-forward chain. According to the feed-forward paradigm, the SBR, after transmitting electromagnetic waveforms in the environment, acquires in its sensor possible scattered signals from debris present in the environmental scene and then feeds the processor input with the sensor output information $\boldsymbol{Z}_K$ in cascade. An advanced, and more complex, approach would allow the processor decisions to be supported by an additional controller, which perceives feedback data from the processor itself and, by virtue of a constrained optimization framework, performs in closed loop a retroaction on both the sensor and processor future behaviour in terms of transmit waveform and radiation pattern (Figure 11.25).

The foregoing tracking concept based on message passing and feedback between the sensor and processor has been hinted in [72] with clear ties to modern radar lines of research [81, 82] relying on the so-called perception-action-cycle for target tracking inspired by the cognitivity of the human brain as well as by sonar echolocating capabilities of several mammalians. Both feedforward- and feedback-based

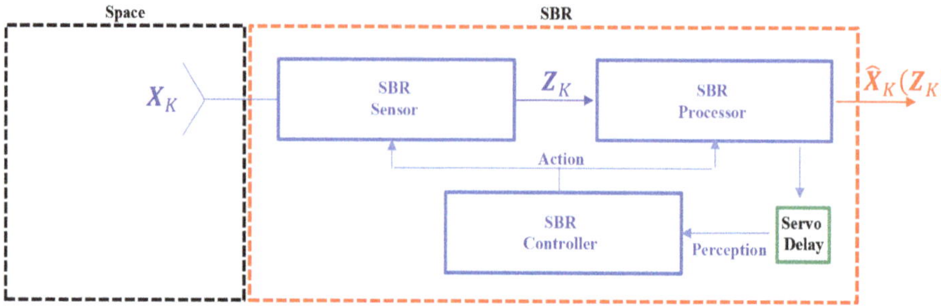

Figure 11.25 SBR perception action cycle

paradigms can be implemented via the SBR architecture hinted in Figures 11.2 and 11.3, as pertaining to a novel radar payload transceiver with Bayesian inference capabilities on small-size debris dynamic states and time series analysis on scattering signatures pertaining to the RCS.

11.6 Conclusions

This chapter has outlined a novel SBR payload functional architecture for SSA as a fully polarimetric monopulse-based single input multiple output (SIMO) radar in the Ka band taking into account space-qualified technologies in both digital and RF domains. The envisaged architecture in Figure 11.3 acquires a complex data hypercube and comprises a filter bank with a group of Doppler frequency offsets not for estimating a target radial velocity (due to the inherent ambiguity of the echo range-rate in cueing the debris range-rate) but rather as a means to enforce Doppler tolerance on the PC scheme and avoid straddle losses. The adaptivity of such a radar architecture allows including robust and selective debris detection schemes tailored to CFAR-like paradigms. Finally, specific parameter estimates from a burst of pulse echoes make provision for further Bayesian inference capabilities on small-size debris dynamic states as well as RCS-related signatures via time series analysis. For this latter purpose, the acquisition of echoes related to the motion of a debris for an elapse time up to several hundreds of milliseconds could be operatively extended to a few seconds, thus augmenting the time on target with additional measurement and gauging perspectives. By selecting an optimal transceiver configuration such that the SBR AESA transmit beam points the debris target minimizing a cost function (e.g. as per a joint waveform and beam control optimisation with ties to [83]), it would be possible to refine radiometric signatures insights. In particular, this would entail performing a time series analysis on RCS signatures embedding polarimetric scattering descriptors [84] and exploiting an adaptive polarization design for tracking (see also section 16.4 in [85]). Last but not least, it is worth stressing that the SIMO SBR concept discussed in this chapter has been tailored to a PWS contact collection strategy. Interestingly, Reference 86 extends the complexity of the SBR payload with a Code Division Multiplexing Multiple Input Multiple Output (MIMO) configuration entailing a Track While Simultaneous Search contacts collection strategy.

References

[1] Committee for the Assessment of NASA's Orbital Debris Programs *Limiting future collision risk to spacecraft – an assessment of NASA's meteoroid and orbital debris programs*. Washington, D.C: THE NATIONAL ACADEMIES PRESS; 2011.

[2] Klinkrad H. *Space debris models and risk analysis*. Berlin, Germany: 2006: Springer-Verlag; 2006.

[3] Horstmann A., Kebschull C., Müller S, *et al.* 'Survey of the current activities in the field of modeling the space debris environment at Tu Braunschweig'. *Aerospace*. 2018, vol. 5(2), p. 37.

[4] *ESA website on SSA* [online]. Available from http://www.esa.int/ssa-sst

[5] Schaub H., Junkins J.L. *Analytical Mechanics of Space Systems [online]*. 3rd edition. Washington, DC: AIAA; 2014. Available from http://arc.aiaa.org/doi/book/10.2514/4.102400

[6] Montenbruck O., Gill E. *Satellite Orbits [online]*. Berlin, Heidelberg: Springer; 2000. Available from http://link.springer.com/10.1007/978-3-642-58351-3

[7] Mallick M., Rubin S., Vo B.-N. 'An introduction to force and measurement modeling for space object tracking'. 16th International Conference on Information Fusion; Istanbul, Turkey, 2013.

[8] Jones B.A., Bryant D.S., Vo B.-T., Vo B.-N. 'Challenges of multi-target tracking for space situational awareness'. 18th International Conference on Information Fusion; Washington, DC, 2015.

[9] Tsao M.A., Ngo H.T., Corsaro R.D., Anderson C.R. 'An in situ measurement system for characterizing orbital debris'. *IEEE Transactions on Instrumentation and Measurement*. 2016, vol. 65(12), pp. 2758–72.

[10] Zhao Y., Huang P., Zhang F., Meng Z. 'Contact dynamics and control for tethered space net robot'. *IEEE Transactions on Aerospace and Electronic Systems*. 2019, vol. 55(2), pp. 918–29.

[11] Weiss H.G. 'The millstone and haystack radars'. *IEEE Transactions on Aerospace and Electronic Systems*. 2001, vol. 37(1), pp. 365–79.

[12] Wilden H., Ben Bekhti N., Hoffmann R., *et al.* 'GESTRA - recent progress, mode design and signal processing'. *2019 IEEE International Symposium on Phased Array System & Technology (PAST)*; Waltham, MA, USA, USA, 2019. pp. 1–8. Available from https://ieeexplore.ieee.org/xpl/mostRecentIssue.jsp?punumber=8982099

[13] Haimerl J.A., Hudson B., Fonder G.P., Lee D.K. 'Overview of the large digital arrays of the space fence radar'. *2016 IEEE International Symposium on Phased Array Systems and Technology (PAST)*; Waltham, MA; 2016. pp. 1–8.

[14] Muntoni G., Schirru L., Montisci G, *et al.* 'A space debris-dedicated channel for the P-band receiver of the Sardinia radio telescope: a detailed description

and characterization'. *IEEE Antennas and Propagation Magazine*. 2020, vol. 62(3), pp. 45–57.

[15] Yang X., Nan X., Song B. 'D2N4: a discriminative deep nearest neighbor neural network for few-shot space target recognition'. *IEEE Trans Geosci Remote Sens*. 2020, vol. 58(5), pp. 3667–76.

[16] Yakovlev O.I. *Space radio science*. New York, NY, USA: Taylor & Francis; 2002. Available from https://www.taylorfrancis.com/books/9781482265002

[17] Staelin D.H., Morgenthaler A.W., Kong J.A. *Electromagnetic waves*. London, U.K: Pearson; 1993.

[18] ECSS Space Engineering - Space Environment 'ECSS-E-ST-10-04C'. *ESA Requirements and Standards Division*; 2008.

[19] Maffei M., Aubry A., De Maio A., Farina A. 'An ontology for spaceborne radar debris detection and tracking: channel-target phenomenology and motion models'. *IEEE Aerospace and Electronic Systems Magazine*. 2021, vol. 36(6), pp. 18–42.

[20] Feria Y., Belongie M., McPheeters T., Tan H. *Solar scintillation effects on Telecommunication links at Ka-Band and X-band* [NASA JPL TDA Progress Report]. 1997.

[21] Morabito D.D. 'Solar corona-induced fluctuations on spacecraft signal amplitude observed during solar superior conjunctions of the cassini spacecraft'. *Radio Science*. 2007, vol. 42(3).

[22] Humphreys T.E., Psiaki M.L., Hinks J.C., O'Hanlon B., Kintner P.M. 'Simulating ionosphere-induced scintillation for testing GPS receiver phase tracking loops'. *IEEE Journal of Selected Topics in Signal Processing*. 2009, vol. 3(4), pp. 707–15.

[23] Vila-Valls J., Closas P., Fernandez-Prades C., Curran J.T. 'On the mitigation of ionospheric scintillation in advanced GNSS receivers'. *IEEE Transactions on Aerospace and Electronic Systems*. 2018, vol. 54(4), pp. 1692–708.

[24] Eriksson A.I. *Waves in space plasmas*. Uppsala Univ., Uppsala, Sweden: Lecture Notes, Swedish Institute of Space Physics – Department of Astronomy and Space Physics; 2004.

[25] Yeh K.C., Liu C.-H. 'Radio wave scintillations in the ionosphere'. *Proceedings of the IEEE*; 1982. pp. 324–60.

[26] Cantafio L.J. *Space based radar handbook*. Norwood, MA, USA: Artech House; 1989.

[27] Knott E.F., Shaeffer J.F., Tuley M.T. *Radar cross section*. Rijeka, Croatia: SciTech; 2004 Jan 1. Available from https://digital-library.theiet.org/content/books/ra/sbra026e

[28] Peebles P. 'Bistatic radar cross sections of chaff'. *IEEE Transactions on Aerospace and Electronic Systems*. 1984, vol. AES-20(2), pp. 128–40.

[29] Peebles P. 'Bistatic radar cross sections of horizontally oriented chaff'. *IEEE Transactions on Aerospace and Electronic Systems*. 1984, vol. AES-20(6), pp. 798–809.

[30] Moreira A., Prats-Iraola P., Younis M., Krieger G., Hajnsek I., Papathanassiou K.P. 'A tutorial on synthetic aperture radar'. *IEEE Geoscience and Remote Sensing Magazine*. 2013, vol. 1(1), pp. 6–43.
[31] Ulaby F.T., Long D.G. *Microwave Radar and Radiometric Remote Sensing*. Ann Arbor, MI: University of Michigan Press; 2013.
[32] De Roo R.D., Ulaby F.T. 'Bistatic specular scattering from rough dielectric surfaces'. *IEEE Trans Antennas Propag*. 1994, vol. 42(2), pp. 220–31.
[33] De Maio A., Maffei M., Aubry A., Farina A. 'Effects of plasma media with weak scintillation on the detection performance of spaceborne radars'. *IEEE Trans Geosci Remote Sens*. 2022, vol. 60, pp. 1–13.
[34] Klinkrad H., Alby F., Alwes D., Portelli C., Tremayne-Smith R. 'Space debris activities in Europe'. *Fourth European Conference on Space Debris*; Darmstadt, Germany, 2005.
[35] Baskakov A.I., Komarov A.A., Ruban A.V. 'Estimation of the energy characteristics of a multi-position radar system for the control of small-sized space debris for various orbital zones'. *2018 Progress in Electromagnetics Research Symposium (PIERS-Toyama)*; Toyama, Japan, 2018. pp. 470–75.
[36] Usoff J.M. 'Haystack ultra-wideband satellite imaging radar antenna'. *2014 IEEE Benjamin Franklin Symposium on Microwave and Antenna Sub-systems for Radar, Telecommunications, and Biomedical Applications (BenMAS)*; Philadelphia, PA, USA, 2014. pp. 1–3.
[37] Maffei M., Aubry A., Maio A.D., Farina A. 'On the Exploitability of the Ka Band for Spaceborne Radar Debris Detection and Tracking Measurements'. *2019 IEEE 5th International Workshop on Metrology for AeroSpace (MetroAeroSpace)*; Torino, Italy; 2019. pp. 355–60.
[38] Raney R.K., M.i S. 'Chapter 18 of the radar handbook' in Skolnik M.I. (ed.). *Space-based remote sensing radars*. 3rd Edition. Skolnik, Ed. New York, NY, USA: McGraw Hill; 2008.
[39] Pillai S.U., Li K.Y., Himed B. *Space based radar*. New York, NY, USA: McGraw Hill; 2008.
[40] Maffei M., Aubry A., De Maio A., Farina A. 'Spaceborne radar functional architecture for debris Bayesian inference'. *2020 IEEE 7th International Workshop on Metrology for AeroSpace (MetroAeroSpace)*; Pisa, Italy, 2020. pp. 1–6.
[41] Stralka J.P., Fedarko W.G. 'Chapter 4 of the radar handbook' in Skolnik M.I. (ed.). *Pulse doppler radar*. 3rd Edition. Skolnik, New York, NY, USA: Mc Graw Hill; 2008.
[42] Maffei M., Aubry A., De Maio A., Farina A. 'Spaceborne radar sensor architecture for debris detection and tracking'. *IEEE Trans Geosci Remote Sens*. 2021, vol. 59(8), pp. 6621–36.
[43] Brookner E., Hall W., Westlake R. 'Faraday loss for L-band radar and communications systems'. *IEEE Transactions on Aerospace and Electronic Systems*. 1985, vol. AES-21(4), pp. 459–69.
[44] Briskman R. 'Autonomous satellite orbital debris avoidance system and method'. *US 8,333,702 B2*; 2014.
[45] Available from www.gaisler.com

[46] Ginosar R., Aviely P., Israeli T., Meirov H. 'RC64: high performance rad-hard manycore'. *2016 IEEE Aerospace Conference*; Big Sky, MT, USA, 2016. pp. 1–9.
[47] Salerno N., Simone L., Cocchi S., Piloni V., Maffei M., Cocciolillo O. Wideband arbitrary waveform generator for enhanced spaceborne SAR. *2008 European Radar Conference*; Amsterdam; 2008. pp. 416–9.
[48] Allan D.W. 'Time and frequency (time-domain) characterization, estimation, and prediction of precision clocks and oscillators'. *IEEE Transactions on Ultrasonics, Ferroelectrics, and Frequency Control*. 1987, vol. 34(6), pp. 647–54.
[49] Allan D.W., Levine J. 'A historical perspective on the development of the Allan variances and their strengths and weaknesses'. *IEEE Transactions on Ultrasonics, Ferroelectrics, and Frequency Control*. 2016, vol. 63(4), pp. 513–19.
[50] Simone L., Maffei M., Gelfusa D., Boscagli G. '*10th International Workshop on Signal Processing for Space Communications (SPSC)*'. Rhodes Island, Greece, 2008. pp. 1–10.
[51] Mailloux R.J. *Phased array antenna handbook*. Norwood, MA, USA: Artech House; 2018.
[52] Haupt R.L. *Antenna arrays [online]*. Hoboken, NJ: Wiley; 2010. Available from http://doi.wiley.com/10.1002/9780470937464
[53] Pozar D.M. *Microwave engineering*. 4th Ed. Hoboken, NJ, USA: Wiley; 2011.
[54] Rocca P., Oliveri G., Mailloux R.J., Massa A. 'Unconventional phased array architectures and design methodologies—A review'. *Proceedings of the IEEE*; 2016. pp. 544–60.
[55] Sherman S.M., Barton D.K. *Monopulse principles and techniques*. Norwood, MA, USA: Artech House; 2011.
[56] Nickel U.R.O., Chaumette E., Larzabal P. 'Statistical performance prediction of generalized monopulse estimation'. *IEEE Transactions on Aerospace and Electronic Systems*. 2011, vol. 47(1), pp. 381–404.
[57] Mosca E. 'Angle estimation in amplitude comparison monopulse systems'. *IEEE Transactions on Aerospace and Electronic Systems*. 1969, vol. AES-5(2), pp. 205–12.
[58] Greco M., Gini F., Farina A. 'Joint use of sum and delta channels for multiple radar target DOA estimation'. *IEEE Transactions on Aerospace and Electronic Systems*. 2007, vol. 43(3), pp. 1146–54.
[59] Rocca P., Morabito A.F. 'Optimal synthesis of reconfigurable planar arrays with simplified architectures for monopulse radar applications'. *IEEE Trans Antennas Propag*. 2015, vol. 63(3), pp. 1048–58.
[60] Gottardi G., Poli L., Rocca P., Montanari A., Aprile A., Massa A. 'Optimal monopulse beamforming for side-looking airborne radars'. *IEEE Antennas and Wireless Propagation Letters*. 2017, vol. 16, pp. 1221–24.
[61] Croci R., Delfino A., Marchetti F. 'Space based radar technology evolution'. *European Radar Conference (EuRAD)*; Rome, 2009. pp. 601–04.

[62] Rihaczek A. 'Radar resolution of moving targets'. *IEEE Transactions on Information Theory*. 1967, vol. 13(1), pp. 51–56.

[63] Rihaczek A.W. 'Delay-doppler ambiguity function for wideband signals'. *IEEE Transactions on Aerospace and Electronic Systems*. 1967, vol. AES-3(4), pp. 705–11.

[64] Pasupathy S., Venetsanopoulos A.N. 'Optimum active array processing structure and space-time factorability'. *IEEE Transactions on Aerospace and Electronic Systems*. 1974, vol. AES-10(6), pp. 770–78.

[65] Maffei M., Simone L. '*10th International Workshop on Signal Processing for Space Communications (SPSC)*'. Rhodes Island, Greece, 2008. pp. 1–12.

[66] *STMicroelectronics [online]*. Available from www.st.com

[67] Volder J.E. 'The CORDIC trigonometric computing technique'. *IRE Transactions on Electronic Computers*. 1959, vol. EC-8(3), pp. 330–34.

[68] Aubry A., Maio A.D., Carotenuto V., Farina A. 'Radar phase noise modeling and effects-part I: MTI filters'. *IEEE Transactions on Aerospace and Electronic Systems*. 2016, vol. 52(2), pp. 698–711.

[69] Aubry A., Carotenuto V., Farina A., De Maio A. 'Radar phase noise modeling and effects-part II: pulse doppler processors and sidelobe blankers'. *IEEE Transactions on Aerospace and Electronic Systems*. 2016, vol. 52(2), pp. 712–25.

[70] Widrow B., Kollár I. *Quantization Noise [online]*. Cambridge University Press; 2008 Jul 3. Available from https://www.cambridge.org/core/product/identifier/9780511754661/type/book

[71] Technologies for European Non-Dependence and Competitiveness (Critical Space Technologies) - SPACE-10-TEC-2018 - Ref. Ares(2017)5361666 -

[72] Blackman S.S. *Multiple-target tracking with radar applications*. Norwood, MA, USA: Artech House; 1986.

[73] Blackman S., Popoli R. *Design and analysis of modern tracking systems design and analysis of modern tracking systems*. Norwood, MA, USA: Artech House; 1999.

[74] Farina A. *Antenna based signal processing techniques for radar systems*. Norwood, MA, USA: Artech House; 1992.

[75] Cui G., De Maio A., Piezzo M., Farina A. 'Sidelobe blanking with generalized Swerling-Chi fluctuation models'. *IEEE Transactions on Aerospace and Electronic Systems*. 2013, vol. 49(2), pp. 982–1005.

[76] Bath W.G., Trunk G.V. 'Chapter 7 of the radar handbook' in Skolnik M.I. (ed.). *Automatic detection, tracking, and sensor integration automatic detection, tracking, and sensor integration*. 3rd Edition. Norwood, MA, USA: Mc Graw Hill; 2008.

[77] Stone L.D., Streit R.L., Corwin T.L., Bell K.L. *Bayesian multiple target tracking*. Norwood, MA, USA: Artech House; 2014.

[78] Li X.R., Jilkov V.P. 'Survey of maneuvering target tracking. part II: motion models of ballistic and space targets'. *IEEE Transactions on Aerospace and Electronic Systems*. 2010, vol. 46(1), pp. 96–119.

[79] Bar-Shalom Y., Li X. -Rong, Kirubarajan T. *Estimation with Applications to Tracking and Navigation: Theory, Algorithms and Software*; 2004.

[80] Studer F.A., Farina A. 'Research studies press'. [Letchworth, U.K] *Radar Data Processing. 1—Introduction and Tracking*. 1985, p. 325.

[81] Bell K.L., Baker C.J., Smith G.E., Johnson J.T., Rangaswamy M. 'Cognitive radar framework for target detection and tracking'. *IEEE Journal of Selected Topics in Signal Processing*. 2015, vol. 9(8), pp. 1427–39.

[82] Farina A., De Maio A., Haykin S. *The Impact of Cognition on Radar Technology [online]*. Scitech; 2017 Nov 21. Available from https://digital-library.theiet.org/content/books/ra/sbra520e

[83] Benavoli A., Balleri A., Farina A. 'Joint waveform and guidance control optimization for target rendezvous'. *IEEE Transactions on Signal Processing*. 2019, vol. 67(16), pp. 4357–69.

[84] Ferro-Famil L., Pottier E. 'Chapter 21 of academic press library in signal processing' in Chellappa R., Theodoridis S. (eds.). *Radar polarimetry basics and selected earth remote sensing applications*. Vol. 2. ISBN 9780123965004. Academic Press; 2014. p. 1119.

[85] Gini F., De Maio A., Patton L. *Waveform Design and Diversity for Advanced Radar Systems [online]*. IET; 2012. Available from https://digital-library.theiet.org/content/books/ra/pbra022e

[86] Maffei M., Aubry A., De Maio A., Farina A. 'MIMO SBR via code division multiplexing for track while simultaneous search'. *IEEE Transactions on Geoscience and Remote Sensing*. 2022, vol. 60, pp. 1–14.

Chapter 12

Polarization information processing in insect radar

Cheng Hu[1,2], *Weidong Li*[1,2], *and Rui Wang*[1,2]

Migratory insects are usually small in size, and many of them fly at night at high altitude. Traditional methods for monitoring migratory insects include optical observation, ovarian dissection, net trapping, and light trapping. Although these methods provide a preliminary understanding of the migration mechanism of migratory insects, they are time-consuming and laborious and can only monitor limited range. More importantly, such methods may interfere with the flight of insects and make the interpretation of observations more difficult [1].

An insect radar is a bespoke remote sensing instrument used for monitoring migratory insects. The radar backscattering characteristics of insects are similar to raindrops as their body contains large amounts of water, thus, resembling suitable radar targets. Compared with traditional methods, radar has the advantages of all-day, all-weather, scalable detection range (depending on the transmit capabilities of the instrument) and does not interfere with swarm flights. Therefore, an insect radar is one of the most effective tools for monitoring insect migration.

Since the first attempt to exploit radar to observe insect migration in 1968, radars have been used to observe insect migration for more than half a century. The early insect radars were scanning systems. With the help of scanning insect radars, many insects migration phenomena which cannot be obtained by traditional methods have been found, such as layers in stable nighttime atmospheres or nocturnal common orientation [1, 2]. X-band vertical-looking radars (VLRs) were subsequently developed in the 1990s relying on noncoherent zenith-pointing linear polarized conical scan configurations [3]. Based on VLRs, routine and continuous observation of insect migration was realized, from which the orientation strategies and the annual airborne insect biomass have been revealed [4, 5]. Due to the rapid development of radar techniques, advanced radar techniques and signal processing methods are emerging constantly, such as phase coherence and polarization. Fully polarimetric

[1]Radar Research Lab, School of Information and Electronics, Beijing Institute of Technology, China
[2]Advanced Technology Research Institute, Beijing Institute of Technology, China

insect radars have now been developed, which extend and improve the capability and performance of current insect radars [6–10].

This chapter focuses on radar with applications to entomology by first introducing the history of insect radar and the role of polarization in insect radar. Then, insect parameters (including the orientation, mass, and body length) estimation based on the polarization characteristics is presented. Finally, a fully polarimetric insect radar system is outlined.

12.1 History of insect radar

12.1.1 Scanning insect radar

The first insect radar was a scanning system adopted for insect swarm observation. It appeared in the 1960s, eventually remodeled by an X-band commercial marine radar. Using this radar, a trial study was conducted in Niger whereby a large number of desert locusts migrating at night were successfully observed [11]. This trial proved the success of the insect radar. Since then, researchers from the United Kingdom, Australia, the United States, and China have carried out a series of studies on insect migration behavior based on various types of insect radar.

Most of the scanning insect radars are based on a horizontal linear polarization configuration. Experiments indicate that for a horizontal flight insect, when the insect is illuminated with a horizontal linear polarized radar from broadside (in which case, the polarization direction has a large component parallel to the insect body axis), the echo radar cross-section (RCS) of the insect reaches the maximum, while from head or tail, the echo RCS is smaller [2, 12]. Based on this principle, the scanning insect radar can be used to measure the common orientation (a common phenomenon of insect migration) of migrating insect swarm. Insect swarm flying with a common orientation would be expected to produce a nonuniform radar plan position indicator (PPI) display (with more dots visible in case of a side-on radar insect gauging perspective) [13]. Thus, the common orientation appears as a dumb bell–shaped pattern perpendicular to the direction of orientation on a PPI display.

However, the time-consuming nature of data procedures and the limited capability of insect biological parameters (which is critical for species identification with the insect radar) measurement precluded the use of scanning insect radar for routine, long-term monitoring.

12.1.2 Vertical looking insect radar

Ever since the 1990s, scanning insect radars have been gradually replaced with VLRs, which achieve routine and long-term observations and can provide more detailed information for insect individuals [14]. The VLR is vertical looking and used for observing the insect individual flight across the radar beam. VLR has experienced two generations of development. Built in the mid-1970s, the first-generation VLRs employed a vertical beam with rotating linear polarization and could measure the amplitude of the target echo signal at different polarization directions (called "polarization pattern"), from which the information about the individual insect's speed, height, orientation, and mass was estimated [15]. However, since the radar did not have the ability of angle measurement, the insect position in the beam could not be determined. Moreover, as far as the antenna gain depends on positions of beam, the insect RCS could not be retrieved. In the 1990s, the second-generation VLRs were built. They employed the same polarization type, but the beam center deviates from the vertical axis by a small angle and rotates around the vertical axis with high speed [16–18]. Echo modulations of insects arising from this change allowed determining the target position within the beamwidth. This allows the insect RCS as well as the displacement direction to be estimated [16, 19, 20]. The insect RCS is usually calibrated by measuring the echo signal of a target with known RCS at a known range [19]. The second generation of VLRs is based on a digital design relying on specific signal processing schemes and algorithms to process and save the data on the computer. This design provides the ability of automatic long-term observation.

The ability of measuring insect orientations makes the VLR a powerful tool for studying the flight behavior of migrating species [5]. Based on experimental campaigns, it was found that the RCS reaches its maximal value while the polarization direction of the radar antenna is parallel to the insect body axis. This result was used as an assumption to estimate the insect orientation from polarization pattern with VLRs [21]. However, for some insects (usually large body size insects) whose RCS reaches its maximal value when the polarization direction of the radar antenna is perpendicular to the insect body axis, this assumption is not applicable [22]. These insects are referred to as "perpendicular insects (PE)," and the first type for which the assumption is true is referred to as "parallel insects (PA)." A 90° error will arise if this assumption is used to estimate the orientation of a PE insect.

Mass and body length of insects are also critical information for species identification of migrating insect observed with an insect radar. At X-band, the scattering of insects is in the beginning of the resonance region (large insects) or the Rayleigh region (small insects). It was found that the mean RCS over different polarization directions (i.e., the mean value of polarization pattern, denoted as a_0) has a monotonic mapping relationship with the mass of insect. Based on the measured insect RCS and mass (40–4 000 mg) data and through parabola fitting, the empirical equation of the insect mass is obtained [21, 23]. This empirical relationship is accurate within a factor of 2 [24]. Based on more insect specimens, the mass–RCS relationship was improved by undertaking a multiple-regression analysis with insect mass as the dependent variable, and a_0 and α_2 – a ratio represents the effective shape of the polarization pattern (it will be clarified in (12.47)) as the independent variables. This method yielded a ±40% uncertainty for mass estimation. To realize accurate species identification, the precision of mass estimation should be improved.

Compared with the single linear polarization scanning insect radar, the parameter measurement ability of VLR with rotating polarization has been greatly improved, and both the orientation and mass of individual insect can be obtained. However, VLRs are noncoherent, the phase information of insect echo reflection is not available; the ability of polarization measurement for a rotating polarization radar is limited, thus only partial polarization information of insects can be obtained.

12.1.3 Fully polarimetric insect radar

Amplitudes, phases, polarization, and frequency constitute the complete set of information of a single target that can be obtained from radar observation [25, 26]. These information are completely included in the target's scattering matrix (SM), which is a 2 × 2 complex matrix and represents all four linear polarization receive–transmit backscattering combinations (HH, HV, VH, VV) [27].

The fully polarimetric radar can directly measure the target SM by successively or simultaneously transmitting orthogonally polarized signals, while simultaneously receiving the target return through two orthogonally polarized receive channels [28, 29]. However, due to amplitude and phase differences, and the cross-coupling between two orthogonally polarized channels, system errors are included in the measured raw SM. These errors are usually eliminated by the polarization and RCS calibration procedure [30].

The polarization pattern measured by rotating polarization VLR can be derived from SM, which means the target information contained in the polarization pattern can be derived from the SM [8, 9]. However, the reverse is not true, e.g., the phases and polarization information of a target are not completely included in the polarization pattern. Therefore, the fully polarimetric insect radar can obtain more information on the target echo (such as phase and polarization), thereby has a stronger potential ability to further improve the estimation accuracy of insect parameters.

The fully polarimetric insect radar has been investigated and developed by the Radar Research Laboratory of Beijing Institute of Technology since 2014. The first fully polarimetric insect radar prototype was developed in 2017, and the orientation of free flying insect individual at night was observed successfully in the first field observation in Langfang, Hebei [31]. As of January 2022, four fully polarimetric insect radars were deployed in Lancang, Jiangcheng and Xundian in Yunnan, and Dongying in Shandong, for long-term automatic observation of migratory insects.

Studies on the application of full polarization to measurement of biological and behavioral parameters of insects showed that the use of polarization information can improve the accuracy of orientation and mass estimates, enable body length estimation, and discriminate the 90° orientation error of PE insects [7–10]. As far as the basic target data that a fully polarimetric radar provides are the SM, all these full polarization methods exploit SM. It was found that the invariant target parameters of the SM can be used to estimate the insect body length and mass with good performance [8]. The relative phase of the SM eigenvalues can be used to discriminate between PA and PE insects with high confidence [9]. The dominant eigenvector of the SM can be used to estimate the insect orientation with better performance at low signal-to-noise ratio (SNR) than the traditional method that extracts the orientation based on the model of polarization pattern [7].

12.2 Insect parameters estimation based on the polarization characteristics

12.2.1 Polarization echo model of insects

The insect SM can be modeled as follows:

$$S = \begin{bmatrix} s_{11} & s_{12}e^{j\beta} \\ s_{21}e^{j\beta'} & s_{22}e^{j\gamma} \end{bmatrix} \tag{12.1}$$

where s_{11}, s_{12}, s_{21}, and s_{22} represent the amplitude of each element, respectively; β, β', and γ represent the phases that are related to target properties. The phase of s_{11} term is neglected since it is not needed for polarimetric purposes. For monostatic radar, $s_{12}=s_{21}$ and $\beta=\beta'$. For insect radar, s_{11}, s_{12}, s_{21}, and s_{22} are defined as the square roots of the RCSs (unit: m^2) in the HH, HV, VH, and VV polarization [21, 24].

The polarization pattern measured by the rotating polarization VLR can be derived from the SM. Rotating polarization insect radars are monostatic and employ linear polarization. The direction of the linear polarization rotates continuously. The H-polarization is chosen as the reference direction, whose normalized effective length can be written as follows:

$$\boldsymbol{h}_0 = \begin{bmatrix} 1 \\ 0 \end{bmatrix} \tag{12.2}$$

The normalized effective antenna length at angle α (with respect to the horizontal direction) can be obtained by rotating the direction of polarization:

$$\boldsymbol{h}(\alpha) = \boldsymbol{U}(\alpha)\,\boldsymbol{h}_0 \tag{12.3}$$

where $U(\alpha)$ represents the rotating matrix

$$\boldsymbol{U}(\alpha) = \begin{bmatrix} \cos\alpha & -\sin\alpha \\ \sin\alpha & \cos\alpha \end{bmatrix} \tag{12.4}$$

By substituting (12.2) and (12.4) into (12.3), the normalized effective antenna length at polarization direction α is given as follows:

$$\begin{aligned} \boldsymbol{h}(\alpha) &= \begin{bmatrix} \cos\alpha & -\sin\alpha \\ \sin\alpha & \cos\alpha \end{bmatrix} \begin{bmatrix} 1 \\ 0 \end{bmatrix} \\ &= \begin{bmatrix} \cos\alpha \\ \sin\alpha \end{bmatrix} \end{aligned} \tag{12.5}$$

Supposing that the normalized effective antenna length of transmitting antenna is $\boldsymbol{h}(\alpha)$, the scattering echo from a target with SM, $\boldsymbol{S}$ can be represented as follows:

$$e_s(\alpha) = \boldsymbol{S}\boldsymbol{h}(\alpha) \tag{12.6}$$

For a monostatic radar, the normalized effective antenna length of receiving and transmitting antennas is identical. Thus, the insect backscattering observed by the radar can be written as

$$\begin{aligned} e(\alpha) &= \boldsymbol{h}^T(\alpha)\, e_s(\alpha) \\ &= \boldsymbol{h}^T(\alpha)\, \boldsymbol{S}\boldsymbol{h}(\alpha) \\ &= \begin{bmatrix} \cos\alpha & \sin\alpha \end{bmatrix} \begin{bmatrix} s_{11} & s_{12}e^{j\beta} \\ s_{21}e^{j\beta'} & s_{22}e^{j\gamma} \end{bmatrix} \begin{bmatrix} \cos\alpha \\ \sin\alpha \end{bmatrix} \end{aligned} \tag{12.7}$$

As current VLRs are noncoherent instruments, it is the power of the echo signal that is measured. The reflected power related to the RCS variation with α, i.e., the polarization pattern, is written as follows [21]:

$$\begin{aligned} \sigma(\alpha) &= |e(\alpha)|^2 \\ &= \left| \begin{bmatrix} \cos\alpha & \sin\alpha \end{bmatrix} \begin{bmatrix} S_{11} & S_{12}{}^{e^{j\beta}} \\ S_{21}{}^{e^{j\beta'}} & S_{22}{}^{e^{j\gamma}} \end{bmatrix} \begin{bmatrix} \cos\alpha \\ \sin\alpha \end{bmatrix} \right|^2 \\ &= a_0 + a_1\cos 2(\alpha - \theta_1) + a_2\cos 4(\alpha - \theta_2) \\ &= a_0 + a_{11}\cos 2\alpha + a_{12}\sin 2\alpha + a_{21}\cos 4\alpha + a_{22}\sin 4\alpha \end{aligned} \tag{12.8}$$

where

$$a_0 = \frac{1}{8}\left(3s_{11}^2 + 3s_{22}^2 + 4s_{12}^2 + 2s_{11}s_{22}\cos\gamma\right) \tag{12.9}$$

$$a_1 = \sqrt{a_{11}^2 + a_{12}^2} \tag{12.10}$$

$$a_2 = \sqrt{a_{21}^2 + a_{22}^2} \tag{12.11}$$

$$\theta_1 = \frac{1}{2}\arctan\frac{a_{12}}{a_{11}}, \quad -90^\circ < \theta_1 < 90^\circ \tag{12.12}$$

$$\theta_2 = \frac{1}{4}\arctan\frac{a_{22}}{a_{21}}, \quad -45^\circ < \theta_2 < 45^\circ \tag{12.13}$$

$$\begin{cases} a_{11} = \frac{1}{2}\left(s_{11}^2 - s_{22}^2\right) \\ a_{12} = s_{12}\left[s_{11}\cos\beta + s_{22}\cos(\beta - \gamma)\right] \\ a_{21} = \frac{1}{8}\left(s_{11}^2 + s_{22}^2 - 4s_{12}^2 - 2s_{11}s_{22}\cos\gamma\right) \\ a_{22} = \frac{1}{2}s_{12}\left[s_{11}\cos\beta - s_{22}\cos(\beta - \gamma)\right] \end{cases} \tag{12.14}$$

The polarization pattern represents an important RCS scattering characteristic of insects. Its shape and absolute values are related to the size (represented usually by mass or length), form (represented by, e.g., its length-to-width ratio), and attitude of insect.

12.2.2 Orientation estimation

12.2.2.1 Polarization characteristics of insect orientation

The polarization characteristics of insects are correlated with radar wavelength (or, equivalently, with frequency). To study this relationship, insect multifrequency SMs are required. From June to September 2018 and June to July 2019, using a multifrequency fully polarimetric rig constructed with a vector network analyzer (R&S ZVA 40, developed by Rohde & Schwarz, Germany) and dual-polarized horn antennas, insect multifrequency SM measurements were made in a microwave anechoic chamber (shielding frequency ranges from 400 MHz to 110 GHz). The step-frequency continuous wave signal was applied, and the measurement bandwidth was 500 MHz. A steel ball with diameter of 3.2 cm was measured for polarization and RCS calibrations [30]. More details about the digital signal processing are described in [32]. A total of 159 specimens representing 23 different species were measured. Their masses ranged from 20.2 to 964 mg, body lengths from 10.3 to 47 mm, and body widths from 2.2 to 14 mm. The specimens were fresh dead in the measurements, and the measurements were completed within half an hour (thus, the scattering of the dead insect was assumed to closely resemble that of one still alive). In the measurements, the direction of the H-polarization of the antenna was chosen as the reference direction, and the specimens were parallel to it so that the insect's orientation was ~0°. For each specimen, a series of SMs at X-band (8.25–11.75 GHz, 50 MHz interval) and Ku-band (12.25–17.75 GHz) were measured [9, 10]. These data have been used for analyzing the polarization characteristics of insect orientation.

Figure 12.1 shows the polarization patterns of four insects with different sizes at 9.4 GHz. The angles 0° and 90° represent the directions parallel and perpendicular to the insect's body axis, respectively. It can be seen that the polarization pattern varies regularly with the increasing insect size. The polarization pattern of a small insect is similar to the shape of number "8," and the maximal RCS direction (MRD)

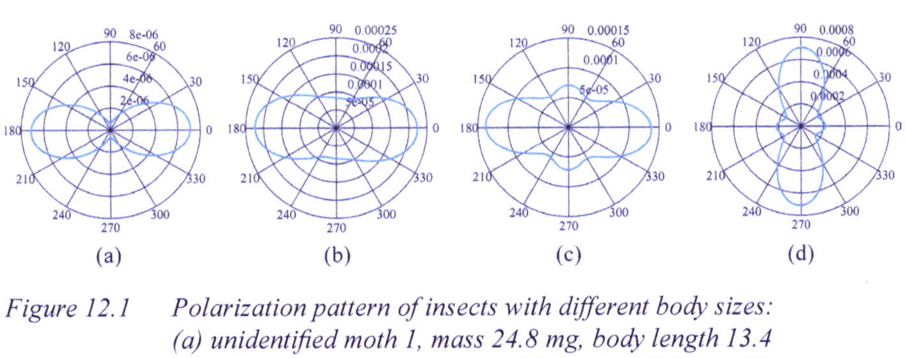

Figure 12.1 Polarization pattern of insects with different body sizes: (a) unidentified moth 1, mass 24.8 mg, body length 13.4 mm; (b) Helicoverpa armigera (Hubner), 109.7 mg, 17 mm; (c) unidentified moth 2, 391.6 mg, 28.4 mm; and (d) Agrius convolvuli (Linnaeus), 908.6 mg, 41.5 mm

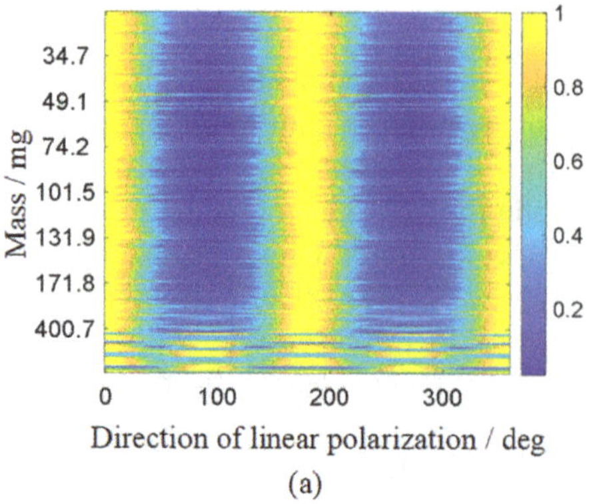

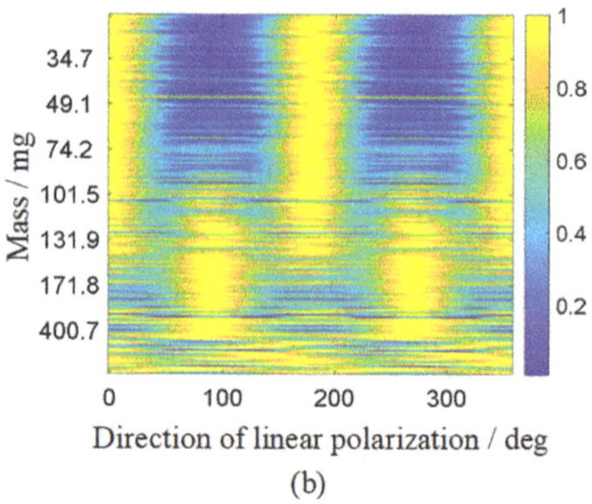

Figure 12.2 *The normalized polarization pattern of all insect specimens at: (a) X-band (9.4 GHz) and (b) Ku-band (16.2 GHz)*

is parallel to the body axis of insect (Figure 12.1a). With the increase of insect size, the RCS at direction perpendicular to the axis of the insect gradually increases, and the polarization pattern approximates the shape of a racetrack (Figure 12.1b) and then a cross (Figure 12.1c). As size increases further, this RCS will exceed that for the head–tail direction (Figure 12.1d), and MRD is perpendicular to the insect body axis.

The result indicates that MRD is parallel or perpendicular to the insect body axis. To verify this conclusion, the normalized polarization patterns of all insect specimens are shown in Figure 12.2. In Figure 12.2, each row represents the expanded and normalized (relative to the maximal RCS of the insect) polarization pattern of an insect. The horizontal axis indicates the polarization direction. The color of each pixel indicates the normalized amplitude of the echo signal. The longitudinal axis represents the masses of 159 insects in ascending order (not uniform). For X-band (Figure 12.2a), considering insects with masses less than about 400 mg, MRD appears at 0° and 180°, which means that MRD is parallel to the body axis, and these insects are PA insects. For insects with mass larger than 400 mg, the MRDs could be parallel or perpendicular (PE insects) to the body axis. For Ku-band (Figure 12.2b), the relationship between MRD and body size is similar to that of X-band; however, the mass of insects decreases, about 100 mg. In addition, at Ku-band, for insects with mass larger than about 660 mg, the MRD of some insects could not be parallel or perpendicular to body axis.

The above analysis showed that except for insects with mass larger than 660 mg at Ku-band, MRDs of insects can be only approximately parallel or perpendicular to the body axis of insects. This conclusion makes it possible to estimate the orientation of an insect body axis based on radar measurements. Therefore, the orientation of insects can be estimated based on the assumption that the orientation of PA insects is parallel to MRD, while that of PE insects is perpendicular to MRD. This assumption is different from that of VLRs, for which an orientation error of 90° will obviously arise when the target is a PE insect. Based on this assumption, the orientation can be extracted in two steps: first, discriminate between PA insect and PE insects; and second, estimate the MRD and obtain the orientation according to the insect class.

12.2.2.2 Discrimination between PA and PE insects

The phase information of polarimetric radars has played important roles for target interpretation in synthetic aperture radar and weather radar [28, 33, 34]. Here, it will be shown that phase information of the fully polarimetric insect radar can be used to distinguish PA and PE insects.

12.2.2.2.1 Identification method

The insect SM has two complex eigenvalues, which can be denoted as μ_1 and μ_2. It can be assumed that $|\mu_1| \geq |\mu_2|$, so that they can be calculated from SM as follows:

$$\begin{aligned}\mu_1 &= \tfrac{1}{2}\left(s_{11} + s_{22}e^{j\gamma}\right) + \tfrac{1}{2}\sqrt{\left(s_{11} - s_{22}e^{j\gamma}\right)^2 + 4s_{12}s_{21}e^{j(\beta+\beta')}} \\ &= |\mu_1|\, e^{j\phi_1}\end{aligned} \tag{12.15}$$

$$\begin{aligned}\mu_2 &= \tfrac{1}{2}\left(s_{11} + s_{22}e^{j\gamma}\right) - \tfrac{1}{2}\sqrt{\left(s_{11} - s_{22}e^{j\gamma}\right)^2 + 4s_{12}s_{21}e^{j(\beta+\beta')}} \\ &= |\mu_2|\, e^{j\phi_2}\end{aligned} \tag{12.16}$$

where $|\mu_1|$ and $|\mu_2|$ represent the amplitudes of μ_1 and μ_2, respectively, and ϕ_1 and ϕ_2 represent the phase, respectively.

For PA insects, the eigenvalues μ_1 and μ_2 correspond to the echo signals when the polarization direction is parallel and perpendicular to the insect's body axis, respectively. For PE insects, the physical meanings of the eigenvalues μ_1 and μ_2 are exchanged.

The relative phase between μ_1 and μ_2 can be represented as follows:

$$\Delta\phi = \phi_1 - \phi_2 + 2k\pi,\ \Delta\phi \in (-\pi, \pi],\ k = 0, \pm 1 \tag{12.17}$$

where ϕ_1 and $\phi_2 \in (-\pi, \pi]$ are introduced in order to make $\Delta\phi \in (-\pi, \pi]$.

Figure 12.3 shows the relationships of $\Delta\phi$ and MRD with radar frequency for different size insects. $\Delta\phi$ changes sign (positive-to-negative or vice versa) at all frequencies where MRD jumps 90°. That is, they jump at exactly the same frequency. Another important phenomenon in Figure 12.3 is that the sign of $\Delta\phi$ is positive for the PE insect and negative for the PA insect. This suggests that the sign of $\Delta\phi$ can be used for discrimination among PA and PE insects, i.e.,

$$\text{Insect class} = \begin{cases} \text{PA} & \Delta\phi < 0 \\ \text{PE} & \Delta\phi > 0 \end{cases} \tag{12.18}$$

The proposed PA and PE insects discrimination method is based on the sign of $\Delta\phi$. When $\Delta\phi$ of an insect is 0°, this method is inapplicable. In addition, this method should be used carefully when $\Delta\phi$ is close to 0°, because $\Delta\phi$ may be easily

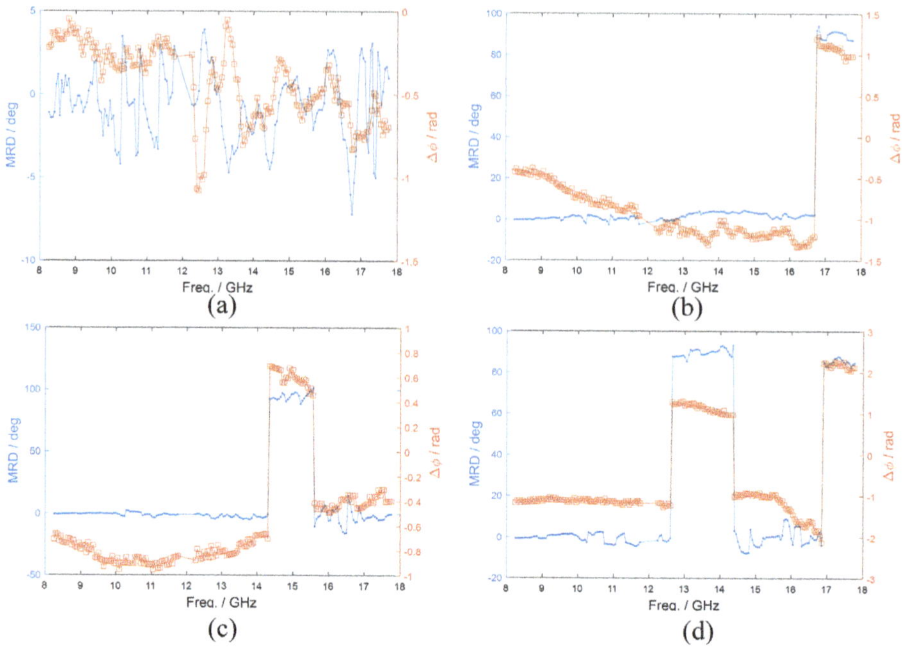

Figure 12.3 *Variations of Δϕ and MRD with respect to radar frequency of (a) Diaphania perspectalis (Walker), body length 14.2 mm, body width 3.5 mm, mass 30.0 mg; (b) Helicoverpa armigera (Hubner), 16.5 mm, 4.1 mm, 114.9 mg; (c) Psilogramma menephron, 34.2 mm, 9.4 mm, 387.7 mg; and (d) Agrius convolvuli (Linnaeus), 46 mm, 12.8 mm, 702.8 mg*

affected by noise and change sign. Fortunately, this can be distinguished by the value of $\Delta\phi$.

It should be noticed that the physical meaning of $\Delta\phi$ for PE and PA insects is different, because $\Delta\phi$ is defined as the phase of μ_1 minus that of μ_2, and the physical meanings of μ_1 and μ_2 for PE and PA insects are exchanged. Specifically, $-\Delta\phi$ of PE insects and $\Delta\phi$ of PA insects have the identical physical meaning, and they are both negative. Thus, for both PE and PA insects, the phase of the echo signal when the polarization direction is parallel to the body axis minus the phase when the polarization direction is perpendicular to the body axis (i.e., $-\Delta\phi$ for a PE insect and $\Delta\phi$ for a PA insect) is always negative. This indicates that the relative relationship between the phases of the echo signals when the polarization direction is parallel and perpendicular to the body axis is fixed, which does not change with the 90° jump of the MRD.

12.2.2.2.2 Performance at different radar frequencies

Based on 159 specimen samples, the correct rate (ratio of the number of correctly identified specimens to the total number) of the proposed method at different radar frequencies is shown in Figure 12.4. For X-band frequencies (8.25–11.75 GHz), the correct rates range from 99.4% to 100%. The correct rate for Ku-band frequencies (12.25–17.75 GHz) slightly decreases as the frequency increases. Over the frequency range of 12–14 GHz, the correct rate varies between 97.5% and 98.8%, while for 14–17 GHz and 16–17.75 GHz, they vary between 95% and 97.5% and 97.5% and 98%, respectively. The results indicate that the proposed method has high correct rates at both X- and Ku-band frequencies.

The misidentified specimens were analyzed, and three possible reasons that may cause the misidentification were found:

1. Random disturbance of $\Delta\phi$. Figure 12.5a shows $\Delta\phi$ of a typical insect with this type of error. It can be seen that $\Delta\phi$ fluctuates continuously with frequency and may cross the 0 rad threshold at some frequencies, which leads to a misidentification. Further analyses highlighted that insects with this type error are the lightest of the insect specimens, with masses ranging from 25.6 to 28.8 mg. This indicates that the disturbance may be caused by low SNR.
2. Phase wrapping. The domain of $\Delta\phi$ is defined as $(-\pi, \pi]$. However, for some insects, the real value of $\Delta\phi$ may exceed this range. The residual part will be folded into the $(-\pi, \pi]$ range (due to phase wrapping), thus including a sign change of $\Delta\phi$. For example, the second jump of $\Delta\phi$ in Figure 12.5b is clearly

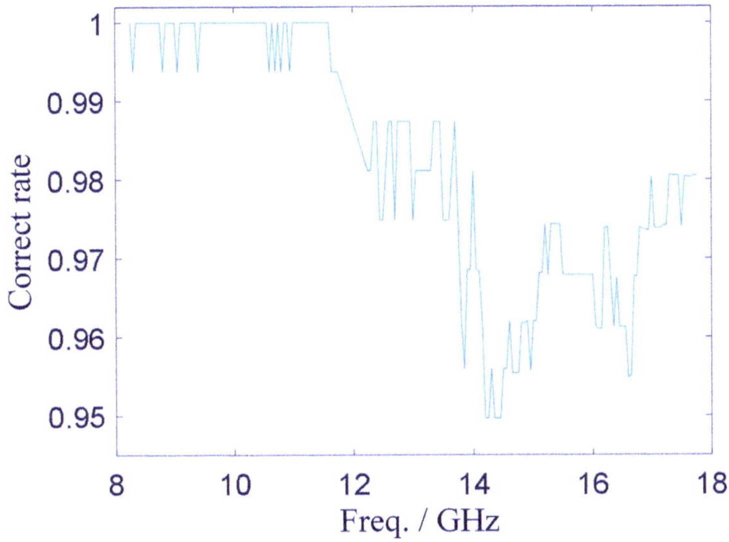

Figure 12.4 The correct rate for different radar frequencies calculated based on 159 specimen samples

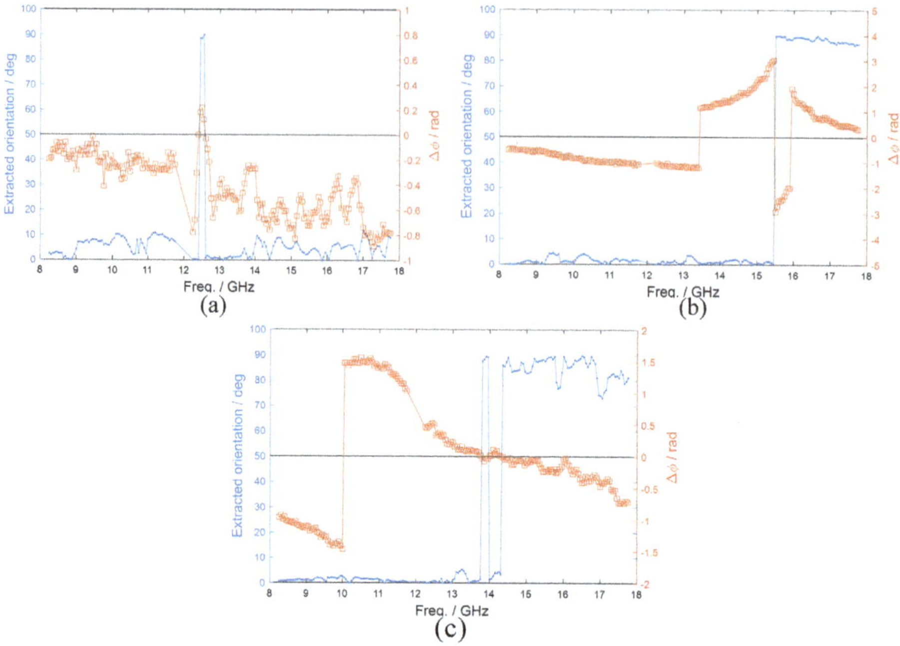

Figure 12.5 *The estimated orientation (blue curves: the estimated orientations about 90° indicate misidentification because the true value is about 0°) and the corresponding Δϕ (brown curves) of three typical misidentified insects: (a) an unidentified moth, mass 20.2 mg, body length 13.2 mm, and body width 3.0 mm; (b) Psilogramma menephron, 310.9 mg, 35.8 mm, 9.2 mm; and (c) an unidentified moth, 391.6 mg, 28.4 mm, 10 mm*

due to domain limitations. Unfortunately, phase wrapping will cause the misidentification of all the frequencies after the shift frequency.

3. $\Delta\phi$ passes through 0 rad while varying continuously with frequency. Figure 12.5c shows a typical insect with this error type. It can be seen that misidentification will occur at all the frequencies after $\Delta\phi$ passes through 0 rad.

12.2.2.2.3 Performance at different SNRs

The data of insect specimens are measured in the microwave anechoic chamber, where the SNRs of most target radar echoes are larger than 30 dB. The fully polarimetric radar directly measures the four elements of the SM. The measured SM $\boldsymbol{S}_m$ is modeled as follows:

$$\boldsymbol{S}_m = \boldsymbol{S} + \boldsymbol{N}_1 \tag{12.19}$$

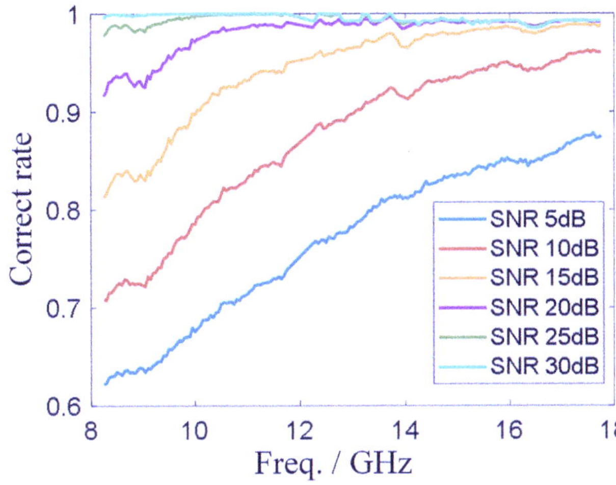

Figure 12.6 Relationship between correct rate and SNR for fully polarimetric measurement

where $\boldsymbol{S}$ is the complex SM; $\boldsymbol{N}_1$ is the complex circular Gaussian matrix. Here, SNR is defined as the ratio of the modulus square of first element of the SM (HH) to the mean noise power, and it was assumed that the mean noise levels of the four polarization channels of the radar are equal. As the amplitudes of the four elements of the SM are usually not equal, the SNRs of the four elements are usually different.

Recalculate $\Delta\phi$ for different SNRs, the correct rates vary with the radar frequencies at different SNRs were obtained (Figure 12.6). At a given SNR, the correct rate increases with increase of radar frequency. At a given radar frequency, the correct rate increases with increasing SNR. To achieve a correct rate greater than 0.9, the SNR is required to be higher than 19 dB at 8 GHz, or 15 dB at 10 GHz. For the Ku-band radar frequencies, nearly all correct rates are higher than 0.9 when the SNR is greater than 10 dB.

To determine why the correct rate increases with increasing radar frequency, the relationship between the mean $|\Delta\phi|$ (over 159 specimens) and radar frequency is analyzed (Figure 12.7). The result shows that $|\Delta\phi|$ (the distance from $\Delta\phi$ to 0°) increases with increasing radar frequency. At X-band frequencies, $\Delta\phi$ is more prone to crossing over 0° due to noise than at Ku-band frequencies, and such a sign change will result in a misidentification. Therefore, the performance of this method at Ku-band is better than that at X-band.

12.2.2.3 Estimation of orientation

Based on the assumption that the orientation of PA insects is parallel to MRD, while that of PE insect is perpendicular to MRD, the orientation of an insect can be represented as follows:

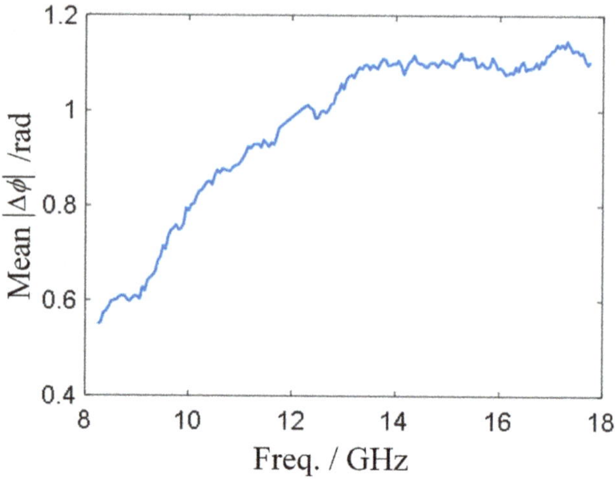

Figure 12.7 *Relationship between mean $|\Delta\phi|$ and frequency. $|\Delta\phi|$ means the absolute value of $\Delta\phi$.*

$$\varphi = \begin{cases} \alpha_m & \text{PA insect} \\ \alpha_m + 90^\circ & \text{PE insect} \end{cases} \tag{12.20}$$

where α_m represents MRD. By substituting (12.18) into (12.20), we can get

$$\varphi = \begin{cases} \alpha_m & \Delta\phi < 0 \\ \alpha_m + 90^\circ & \Delta\phi > 0 \end{cases} \tag{12.21}$$

Note that the extracted orientation is ambiguous at 180°, because the head direction and tail direction are indistinguishable.

There are two methods that can be used to estimate the MRD of an insect. The first one is based on modeling polarization pattern (used in VLRs). The second one is based on an eigenanalysis of the SM, which is mainly used in the fully polarization insect radar.

12.2.2.3.1 MRD estimation based on polarization pattern model

Rewriting (12.8) as

$$\sigma(\alpha) = a_0 + a_1 \cos 2(\alpha - \theta_1) + a_2 \cos 4(\alpha - \theta_2) \tag{12.22}$$

one obtains the simplified model of polarization pattern. In this model, a_0 is the mean RCS over all polarization directions (360°), where a_1 and a_2 are two RCS parameters representing an elongated and a cruciform component of polarization pattern [35]. According to [27], θ_1 represents the orientation of "maximum polarization," i.e., when the direction of polarization is θ_1, the received power density is maximal (therefore, θ_1 is the MRD of an insect).

The basic data that the traditional VLRs provide are the RCSs at different directions of linear polarization. θ_1 and the other parameters (a_0, a_1, a_2, and θ_2) can be estimated based on parameter estimation methods (e.g., via least squares) [16, 20, 21].

For a fully polarimetric radar, this method can also be used because θ_1 can be represented with the parameters of SM (refer to (12.12) and (12.14)). Substituting (12.14) into (12.12), we can obtain

$$\alpha_m = \theta_1 = \frac{1}{2}\arctan\frac{2s_{12}\left[s_{11}\cos\beta + s_{22}\cos\left(\beta-\gamma\right)\right]}{s_{11}^2 - s_{22}^2}, \quad -90^\circ < \theta_1 < 90^\circ \tag{12.23}$$

where α_m is the notation of MRD.

12.2.2.3.2 MRD estimation based on the main eigenvector of SM

According to radar polarization theory [36], the returned echo power could be maximal when the polarization of radar is consistent with the dominant eigenvector of SM. The polarization of radar can be modeled as a polarization ellipse where the direction of polarization is included. The direction of polarization when the returned echo power is maximal is identical to MRD. Therefore, MRD can be extracted by making the polarization of radar equal to the dominant eigenvector of SM.

The main eigenvalue of SM has been calculated in (12.15). Its corresponding unit eigenvector can be derived as follows:

$$\boldsymbol{v}_1 = \frac{1}{\sqrt{\left(\left|\frac{s_{12}e^{j\beta}}{\lambda_1 - s_{11}}\right|\right)^2 + 1}}\begin{bmatrix} \frac{s_{12}e^{j\beta}}{\lambda_1 - s_{11}} & 1 \end{bmatrix}^T \tag{12.24}$$

The model of the polarization ellipse can be represented as follows:

$$\boldsymbol{a}(\alpha, \tau_m, A) = A\begin{bmatrix} \cos\alpha & -\sin\alpha \\ \sin\alpha & \cos\alpha \end{bmatrix}\begin{bmatrix} \cos\tau_m \\ j\sin\tau_m \end{bmatrix} \tag{12.25}$$

where α represents the polarization orientation; τ_m is the ellipticity angle, and A is the magnitude.

If the polarization of radar is consistent with $\boldsymbol{v}_1$, $\boldsymbol{v}_1$ can be rewritten in the form of unit polarization ellipse as follows:

$$\begin{aligned} \boldsymbol{v}_1 &= \frac{\boldsymbol{a}(\alpha,\tau_m,A)}{|\boldsymbol{a}(\alpha,\tau_m,A)|} \\ &= \frac{1}{\cos 2\tau_m}\begin{bmatrix} \cos\alpha\cos\tau_m - j\sin\alpha\sin\tau_m \\ \sin\alpha\cos\tau_m + j\cos\alpha\sin\tau_m \end{bmatrix} \end{aligned} \tag{12.26}$$

Equation (12.26) is the normalization form of (12.25). By making (12.24) and (12.26) equal, α can be obtained (note that τ_m in (12.26) is an unknown variable that will be eliminated as shown hereafter).

Rewrite (12.24) as follows:

$$\boldsymbol{v}_1 = \begin{bmatrix} v_x \\ v_y \end{bmatrix} \tag{12.27}$$

where

$$\begin{cases} v_x = \dfrac{\frac{s_{12}e^{j\beta}}{\lambda_1 - s_{11}}}{\sqrt{\left(\left|\frac{s_{12}e^{j\beta}}{\lambda_1 - s_{11}}\right|\right)^2 + 1}} \\ v_y = \dfrac{1}{\sqrt{\left(\left|\frac{s_{12}e^{j\beta}}{\lambda_1 - s_{11}}\right|\right)^2 + 1}} \end{cases} \tag{12.28}$$

By making (12.26) and (12.27) equal, we can obtain:

$$\begin{cases} v_x = \dfrac{\cos\alpha\cos\tau_m - j\sin\alpha\sin\tau_m}{\cos 2\tau_m} \\ v_y = \dfrac{\sin\alpha\cos\tau_m + j\cos\alpha\sin\tau_m}{\cos 2\tau_m} \end{cases} \tag{12.29}$$

From (12.29), it can be found that

$$\frac{\mathrm{Re}\left(v_y\right)}{\mathrm{Re}\left(v_x\right)} = \frac{\dfrac{\sin\alpha\cos\tau_m}{\cos 2\tau_m}}{\dfrac{\cos\alpha\cos\tau_m}{\cos 2\tau_m}} = \tan\alpha \tag{12.30}$$

Substituting (12.28) into (12.30), we can get

$$\tan\alpha = \frac{\mathrm{Re}\left(v_y\right)}{\mathrm{Re}\left(v_x\right)} = \mathrm{Re}\left(\frac{1}{\sqrt{\left(\left|\frac{s_{12}e^{j\beta}}{\lambda_1 - s_{11}}\right|\right)^2 + 1}}\right) \Bigg/ \mathrm{Re}\left(\frac{\frac{s_{12}e^{j\beta}}{\lambda_1 - s_{11}}}{\sqrt{\left(\left|\frac{s_{12}e^{j\beta}}{\lambda_1 - s_{11}}\right|\right)^2 + 1}}\right) \tag{12.31}$$

Then α can be easily obtained as follows:

$$\alpha = \arctan\left[\mathrm{Re}\left(\frac{1}{\sqrt{\left(\left|\frac{s_{12}e^{j\beta}}{\lambda_1 - s_{11}}\right|\right)^2 + 1}}\right) \Bigg/ \mathrm{Re}\left(\frac{\frac{s_{12}e^{j\beta}}{\lambda_1 - s_{11}}}{\sqrt{\left(\left|\frac{s_{12}e^{j\beta}}{\lambda_1 - s_{11}}\right|\right)^2 + 1}}\right)\right] \tag{12.32}$$

It should be noted that if $s_{12} = 0$, (12.32) will not hold. In this case, MRD is equal to 0° when $\lambda_1 - s_{11} = 0$, and MRD is equal to 90° when $\lambda_1 - s_{11} \neq 0$. Therefore, MRD can be represented as follows:

$$\alpha_m = \begin{cases} 0^\circ & \lambda_1 - s_{11} = 0 \;\&\; s_{12} = 0 \\ 90^\circ & \lambda_1 - s_{11} \neq 0 \;\&\; s_{12} = 0 \\ \arctan\left[\dfrac{\mathrm{Re}\left(1 \Big/ \sqrt{\left(\left|\frac{s_{12}e^{j\beta}}{\lambda_1 - s_{11}}\right|\right)^2 + 1}\right)}{\mathrm{Re}\left(\frac{s_{12}e^{j\beta}}{\lambda_1 - s_{11}} \Big/ \sqrt{\left(\left|\frac{s_{12}e^{j\beta}}{\lambda_1 - s_{11}}\right|\right)^2 + 1}\right)}\right] & \text{other} \end{cases} \tag{12.33}$$

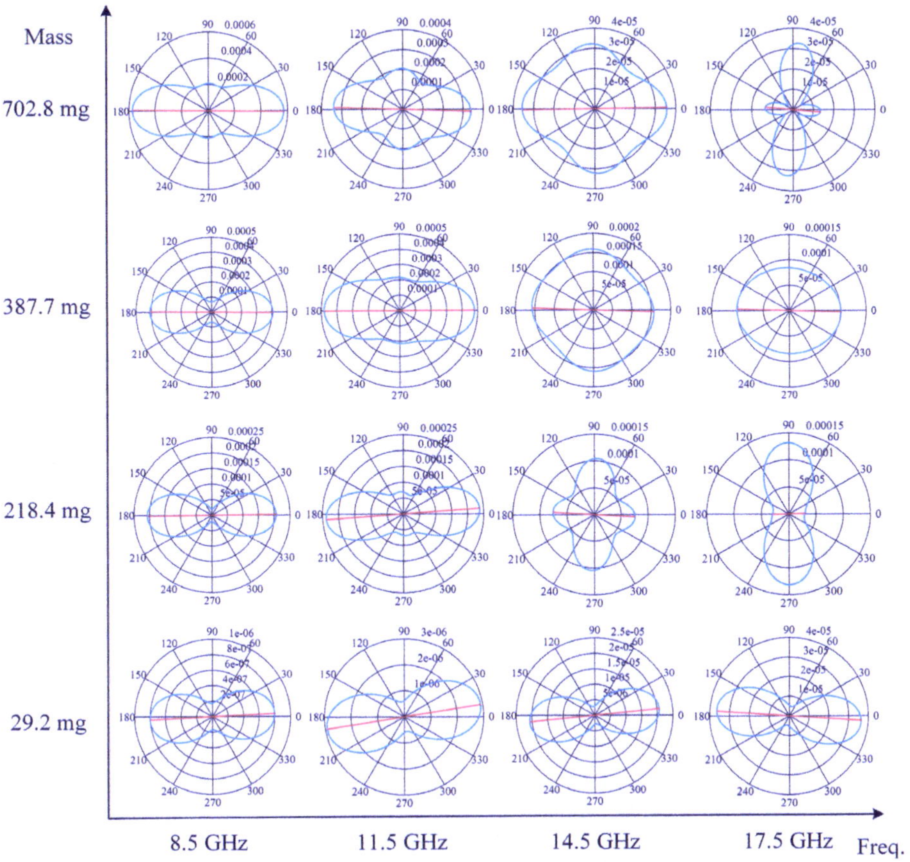

Figure 12.8 *The estimated orientations of insects with different masses at different frequencies. Blue curves represent the polarization pattern. The red lines represent the estimated orientation.*

12.2.2.3.3 Verification of orientation estimation based on measured insect data

Based on the measured insect specimens, the orientation estimation method was verified. Figure 12.8 shows the estimated orientations of four insects with different masses at four radar frequencies, where the horizontal axis indicates the radar frequencies, and the longitudinal axis represents the insect masses. The true values of orientation of these insects are 0°, and the angle between the red line and the angle 0° is the estimation error of orientation. The species of the insects are the unidentified moth (20.2 mg, marked as insect 1), *Agrotis ipsilon* (Rottemberg) (218.4 mg, insect 2), *Psilogramma menephron* (Cramer) (387.7 mg, insect 3), and *Agrius convolvuli* (Linnaeus) (702.8 mg, insect 4). The blue curves represent the polarization pattern. The red lines represent the estimated orientation based on (12.21) where the MRD estimation is based on the main eigenvector of SM. It can be seen that the shapes of the polarization pattern of insects with different masses at different frequencies are different, and one insect at different frequencies could belong to different classes (PA or PE). Insect 2 at frequencies of 14.5 and 17.5 GHz, and insect 4 at frequency of 17.5 GHz are PE insects as their MRDs (around 90°) are perpendicular to the true value (0°) of orientation. All estimated orientations for these cases are around 0°, which means their classes are correctly discriminated and the 90° errors are compensated. Insects in other cases are all PA insects, and all estimated orientations are around the true value. Therefore, orientations of all PE and PA insects are correctly estimated. We note that some of the MRDs and the estimated orientations have small errors. Four possible reasons may cause the errors:

1. Small error of the basic assumption. Due to the complexity of insect body structure, the MRD of polarization pattern may not be strictly parallel or perpendicular to the body axis (i.e., small error may exist).
2. Noise and system error. The noise and the residual error after polarization calibration of the fully polarimetric rig may cause the measurement error in the measured SMs, and finally result in the MRD and orientation errors.
3. Experimental error. In the experiment, the orientation of the insect axis may not be strictly 0°.
4. Error of MRD estimation. Small error exists in the MRD estimated based on the methods introduced in this chapter.

12.2.2.3.4 Performance at different radar frequencies

The performances of the orientation estimation at different frequencies are analyzed, and the two orientation estimation methods are compared. Figure 12.9 gives the variations of orientation estimation error with radar frequency and insect mass based on 159 specimens, where Figure 12.9a represents the result of the method based on the main eigenvector of SM, and Figure 12.9b represents that based on the model of polarization pattern. The orientation estimation error is defined as the difference between the estimated value and the true value (0°). The color of each pixel indicates the error value in degrees (see color scale at right). The values of each row represent error of an insect at different frequencies, and the longitudinal axis represents the masses of the 159 insects in ascending order (not uniform). The mass ranges from 20.2 mg to 964 mg. For the method based on the main eigenvector of SM (Figure 12.9a), orientation errors of most insects are less than 5° at X-band frequencies. At Ku-band frequencies, for some of the insects with mass larger than 400 mg, large errors (up to 45°) occur; for others with mass less than 400 mg, the errors are similar to insects at X-band frequencies.

For the method based on the model of polarization pattern (Figure 12.9b), large errors also occur for large insect at high frequency. However, large errors occur at smaller masses and lower frequencies than that of the method based on the main eigenvector of SM. Therefore, the estimation accuracy of the method based on the

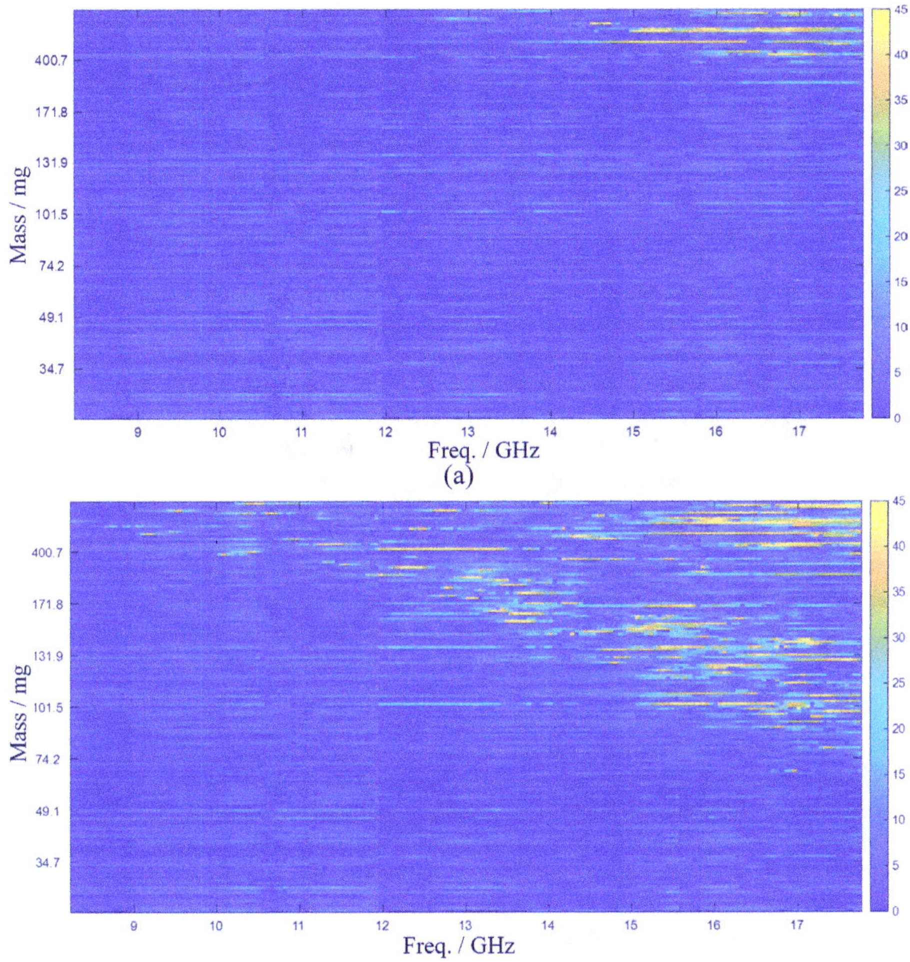

Figure 12.9 *The variations of orientation estimation error with radar frequency and insect mass based on 159 specimens: (a) MRD estimation based on the main eigenvector of SM and (b) MRD estimation based on the model of polarization pattern*

main eigenvector of SM is higher than those pertaining to the method based on polarization pattern.

12.2.2.3.5 Performance at different SNRs

Based on the measured 159 insect specimens and Monte-Carlo simulation, the performances of orientation estimation methods at different SNRs are analyzed. The definition of SNR and the model of measured SM are similar to that introduced in

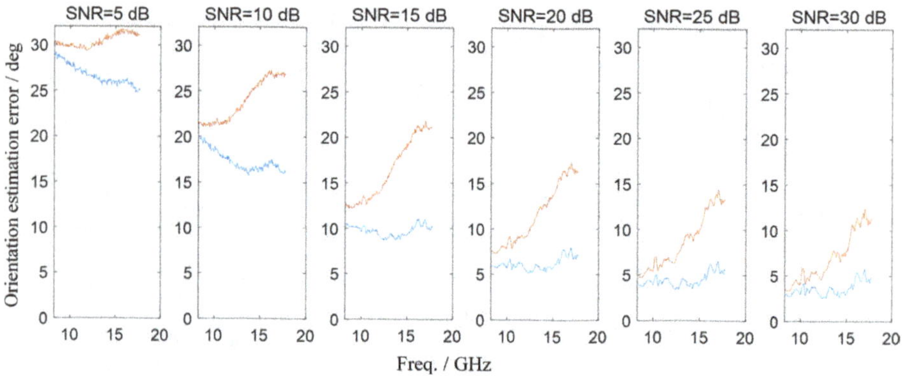

Figure 12.10 *Orientation estimation errors at different SNR levels. The brown and blue curves represent orientation errors of methods based on the model of polarization pattern and the main eigenvector of SM, respectively.*

section 12.2.2.2. SNRs range from 5 to 30 dB with a 5-dB interval. Figure 12.10 gives the orientation errors of two methods at different SNR levels and different frequencies. It can be seen that orientation errors decrease with increasing SNR. For a certain SNR, the orientation error of the method based on the model of polarization pattern increases with increasing frequency. For the method based on the main eigenvector of SM, the orientation error decreases with increasing frequency when the SNR is less than 10 dB; when the SNR is larger than 10 dB, the frequency has little influence on the orientation error at X-band, but the error is slightly larger at Ku-band than at X-band.

12.2.3 Mass/body length estimation

12.2.3.1 Invariant target parameters

When a rigid target is illuminated by a radar, it is clear that its size and shape are constant, regardless of the type of radar polarization. However, the SM of that rigid target with a constant physical exposure is not constant but depends on the type of radar polarization. To estimate the size and shape of a rigid target with radar measurement, it is necessary to extract parameters independent of the type of radar polarization from SM. These parameters, from which the undesirable variables have been removed, are called invariant target parameters that are related directly to the target's geometrical properties [37, 38].

Suppose that the SM defined on the polarization basis $(\hat{x}, \hat{y})$ is $S(XY)$. The Graves matrix [39] can be defined as follows:

$$\boldsymbol{G}(XY) = \boldsymbol{S}^{\dagger}(XY)\,\boldsymbol{S}(XY) \tag{12.34}$$

Here the superscript $(.)^{\dagger}$ indicates the conjugate and transpose operator.

Let $\left(\hat{a}, \hat{b}\right)$ be another polarization base, and $\boldsymbol{V}$ is the transition matrix between $(\hat{x}, \hat{y})$ and $\left(\hat{a}, \hat{b}\right)$. Then, the Graves matrix under the new polarization basis can be written as follows:

$$\boldsymbol{G}(AB) = \boldsymbol{V}^{\dagger}\boldsymbol{G}(XY)\,\boldsymbol{V} \tag{12.35}$$

Through eigenvalue decomposition, $\boldsymbol{G}(XY)$ can be written as

$$\boldsymbol{G}(XY) = \boldsymbol{V}_0^{\dagger}\boldsymbol{\Lambda}\boldsymbol{V}_0 \tag{12.36}$$

where

$$\boldsymbol{\Lambda} = \begin{bmatrix} \lambda_1 & 0 \\ 0 & \lambda_2 \end{bmatrix} \tag{12.37}$$

$$\boldsymbol{V}_0^{\dagger} = \begin{bmatrix} \boldsymbol{V}_{01} & \boldsymbol{V}_{02} \end{bmatrix} \tag{12.38}$$

λ_1 and λ_2 are two eigenvalues of $\boldsymbol{G}(XY)$, and $\lambda_1 \geq \lambda_2 \geq 0$. $\boldsymbol{V}_{01}$ and $\boldsymbol{V}_{02}$ are the corresponding unit eigenvectors of λ_1 and λ_2, respectively.

Substituting (12.36) into (12.35), we can obtain

$$\begin{aligned} \boldsymbol{G}(AB) &= \boldsymbol{V}^{\dagger}\boldsymbol{G}(XY)\,\boldsymbol{V} \\ &= \boldsymbol{V}^{\dagger}\boldsymbol{V}_0^{\dagger}\boldsymbol{\Lambda}\boldsymbol{V}_0\boldsymbol{V} \\ &= (\boldsymbol{V}_0\boldsymbol{V})^{\dagger}\boldsymbol{\Lambda}(\boldsymbol{V}_0\boldsymbol{V}) \end{aligned} \tag{12.39}$$

Equation (12.39) indicates that the eigenvalues of $\boldsymbol{G}(XY)$ and $\boldsymbol{G}(AB)$ are identical. That is, the eigenvalues of the Graves matrix are independent of the polarization basis. Therefore, the eigenvalues of Graves matrix are invariant target parameters [40, 41].

Substituting (12.1) into (12.34), we can obtain

$$\boldsymbol{G} = \boldsymbol{S}^{\dagger}\boldsymbol{S} = \begin{bmatrix} g_{11} & g_{12} \\ g_{21} & g_{22} \end{bmatrix} \tag{12.40}$$

where

$$\begin{cases} g_{11} = s_{11}^2 + s_{12}^2 \\ g_{12} = s_{11}s_{12}e^{j\beta} + s_{12}s_{22}e^{j(\gamma-\beta)} \\ g_{21} = s_{11}s_{12}e^{-j\beta} + s_{12}s_{22}e^{-j(\gamma-\beta)} \\ g_{22} = s_{12}^2 + s_{22}^2 \end{cases} \tag{12.41}$$

Its two eigenvalues can be written as follows:

$$\lambda_1 = \frac{(g_{11} + g_{22}) + \sqrt{(g_{11} - g_{22})^2 + 4g_{12}g_{21}}}{2} \tag{12.42}$$

$$\lambda_2 = \frac{(g_{11} + g_{22}) - \sqrt{(g_{11} - g_{22})^2 + 4g_{12}g_{21}}}{2} \tag{12.43}$$

The eigenvalues λ_1 and λ_2 correspond to the maximum of the insect polarization pattern and the value (the minimum or local maximum) perpendicular to this maximum direction, respectively [8].

According to the study in section 12.2.2.1, the body axis of a PA insect is parallel to MRD, while that of PE insect is perpendicular to MRD. Thus, the physical meanings of λ_1 and λ_2 for PA and PE insects are different. For the PA insect, λ_1 and λ_2 represent the RCSs when the polarization direction is parallel and perpendicular to the insect body axis, respectively. However, for PE insects, the physical meanings of λ_1 and λ_2 are exchanged. To represent the consistent physical meaning, a new variable is defined:

$$\nu = \begin{cases} \lambda_2 & \Delta\phi < 0 \\ \lambda_1 & \Delta\phi > 0 \end{cases} \tag{12.44}$$

where $\Delta\phi$ is used for insect class discrimination (PA or PE). This equation considers λ_2 for PA insect and λ_1 for PE insect. Thus, ν represents the RCS when the direction of polarization is perpendicular to the insect body axis.

The determinant of the Graves matrix (denoted as Det$\boldsymbol{G}$) can be expressed as the product of the two eigenvalues:

$$\text{Det}\boldsymbol{G} = \lambda_1\lambda_2 \tag{12.45}$$

To make the dimension consistent with λ_1 and λ_2, the square root determinant is considered as a new variable

$$d = \sqrt{\text{Det}\boldsymbol{G}} = \sqrt{\lambda_1\lambda_2} \tag{12.46}$$

Obviously, both ν and d are invariant target parameters as both λ_1 and λ_2 are invariant target parameters.

With insect radar observations, constant ventral aspect is usually assumed. The fully polarimetric radar transmits and receives orthogonally polarized signals (i.e., H- and V-polarization signals). The type of polarization and frequency of the radar are constant. In that case, the SM varies with the insect orientation in the plane perpendicular to the radar line of sight. Therefore, the physical parameters of a target cannot be extracted immediately from the SM. If the orientation of an insect is taken as reference, the change of insect orientation is equivalent to the rotation of radar polarization direction, which is a form of polarization variation. The meaning of the invariant target parameters for an insect radar is that they do not vary with the insect orientation and can be related to the insect's size and form.

In this sense, the polarization parameters a_0, a_1, a_2, and θ_2 are orientation-invariant quantities as the orientation has been extracted as θ_1 (12.22). The average RCS a_0 provides a measure of the target's effective size (at X-band), and it is convenient to introduce the ratio

$$\alpha_2 = \begin{cases} \frac{a_1}{a_0} & \Delta\phi < 0 \\ -\frac{a_1}{a_0} & \Delta\phi > 0 \end{cases} \quad (12.47)$$

that represents the target's effective shape of polarization pattern.

Next, the relationships between the insect size and the invariant target parameters ν and d, and polarization pattern parameters a_0 and α_2 are studied. The applications of these parameters in insect size estimation are introduced.

12.2.3.2 Mass estimation

In this study, 340 insect specimens measured at X-band were utilized, including 159 insects representing 23 different species measured by the fully polarimetric rig introduced in section 12.2.2.1 and 181 insects representing 53 different species published in [42] of the measured specimens. These specimens were either freshly dead, anesthetized, or freshly thawed after having been frozen [8]. Since water is thought to be the main component of insect for reflecting radar wave [22], scattering of these specimens should differ little from that of live insect in migration. Their masses ranged from 1.83 to 4120 mg, body lengths from 4.9 to 64 mm, and body widths from 1 to 14 mm. Most of the measured specimens were PA insects (297 specimens); only 43 specimens were PE insects.

As introduced in section 12.2.2.1, the shape of the polarization pattern is related to the size of an insect. Figure 12.11 gives polarization patterns of four insects with different masses measured at X-band. The insects were oriented towards 0°. The radial scale is RCS (units m^2). For small insects, the RCS reaches the maximum λ_1 and the extremum λ_2 when the polarization direction is parallel and perpendicular to the body axis, respectively (Figure 12.11a). As the insect size increases, the extremum λ_2 changes from a minimum to a local maximum (Figure 12.11b and c). With further increase of insect size, this RCS will exceed that for the head–tail direction (Figure 12.11d; since it is defined that $\lambda_1 \geq \lambda_2$, the RCS when the polarization direction is perpendicular to the body axis is represented as λ_1 and that for the head–tail direction as λ_2). This indicates that when the polarization direction

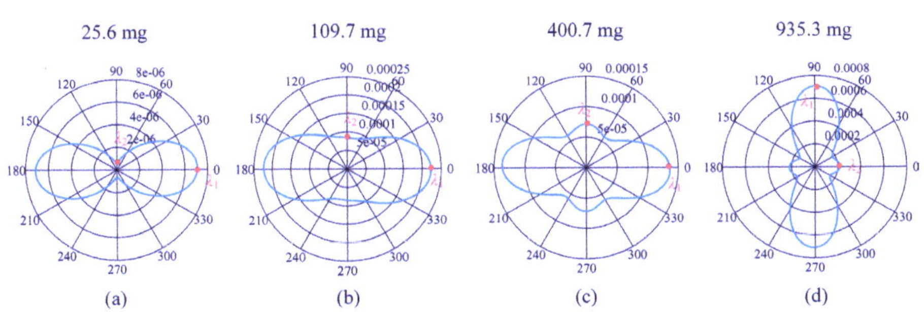

Figure 12.11 Measured insect polarization patterns at X-band: (a) the unidentified moth 1; (b) Helicoverpa armigera (Hubner); (c) the unidentified moth 2; and (d) Agrius convolvuli (Lepidoptera)

is perpendicular to the body axis, the RCS (defined as ν, which represents λ_2 for small insect or λ_1 for large insect) increases monotonically with the size of insect. Therefore, ν may be a good parameter for estimating body length or mass.

Figure 12.12 shows the scatterplots of $\lg \lambda_1$ (lg operator is defined as a log base 10) and $\lg \lambda_2$ against insect $\lg[\text{mass (mg)}]$, respectively. R represents the Pearson correlation coefficient, and the P value is used to evaluate the correlation between the estimated and the true values. The MRE (representing mean relative error that is used to evaluate the accuracy of the estimates) is defined as follows:

$$\text{MRE} = \frac{1}{N}\sum_{i=1}^{N}\frac{|M_i - E_i|}{M_i} \times 100\% \tag{12.48}$$

where M_i represents the true value, E_i represents the estimate, and N is the number of values.

It can be seen that the eigenvalues and masses of PE insects are usually greater than those of the PA ones. The scatterplots can be divided into two parts: the lower left comprising the PA insects and the upper right PE insects. λ_1 and λ_2 are strongly correlated with the masses for both classes. However, the correlations for PA and PE insects are different.

Figure 12.12 shows that both λ_1 and λ_2 have good relationships with the PA insects, while λ_1 has a better relationship with mass for PE insects. In order to quantify this tendency, the linear fits of masses to λ_1 and λ_2 for the two insect classes were separately made, respectively (Figure 12.12). The fitting results verify the stronger correlations between mass and λ_1 for the PE insects, and the strong correlations between mass and either λ_1 or λ_2 for the PA insects. Apparently, λ_1 is a good parameter for the estimation of body lengths and masses. However, the

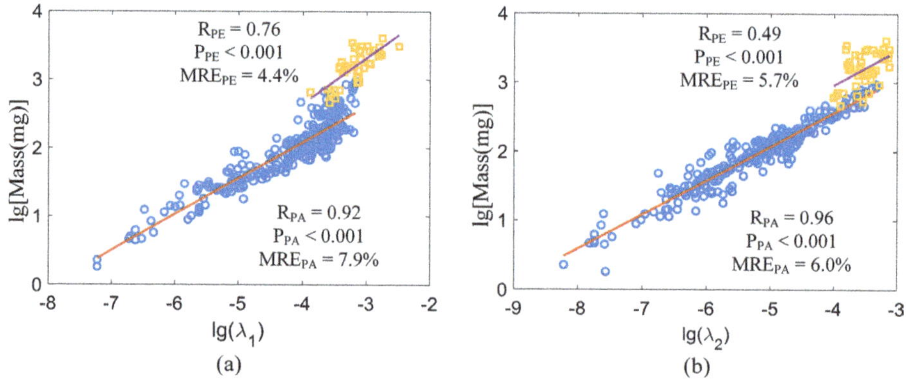

Figure 12.12 Relations of the two eigenvalues to body length and mass for the study data set: (a) λ_1 to mass and (b) λ_2 to mass. All variables are log-transformed. Points for PA insects shown blue, for PE insects yellow.

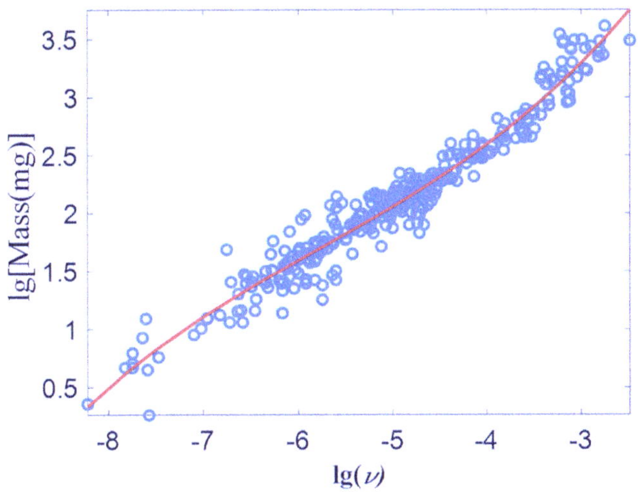

Figure 12.13 Relationship between mass and lg ν

physical meanings of λ_1 for PA and PE insects are different. Additionally, it is shown in Figure 12.12 that the relationships between mass and λ_1 for PA and PE insects are quite different. Consequently, λ_1 is not be the best parameter for mass estimation.

It was shown earlier that when the polarization direction is perpendicular to the body axis, the received RCS ν appears to increase monotonically with insect size; therefore, ν has the potential to be a single estimator quantity for insect mass. The relationship of ν and insect mass is shown in Figure 12.13. lg ν is strongly correlated with the insect logarithm-transformed masses. The mapping relationships between mass and ν can be described using mathematical equations, which can be used as empirical equations to estimate an insect mass.

Polynomial fitting can be made to obtain the empirical formulas for mass estimation. A proper order for the fitting is required. An improper fitting order will cause underfitting (too low) or overfitting (too high), and the correlation coefficient (R) between the fitting curve and the true value will be small. To determine a proper fitting order, the insect masses and lg ν were fitted via polynomials of different orders (first and fifth), as shown in Figure 12.14. It can be seen in the upper right of Figure 12.14a and b, the first- and second-order fitting curves are underfitting for the large insects. The third- and fourth-order fittings (Figure 12.14c and Figure 12.14d) are similar, and both perform well. As for the fifth-order fitting curve (Figure 12.14e), some unnatural fluctuations appear in the fitting curve, showing slight overfitting. Thus, both the third- and fourth-order fitting curves are proper. In this case, the lower order is preferred. Therefore, the third-order fitting is selected.

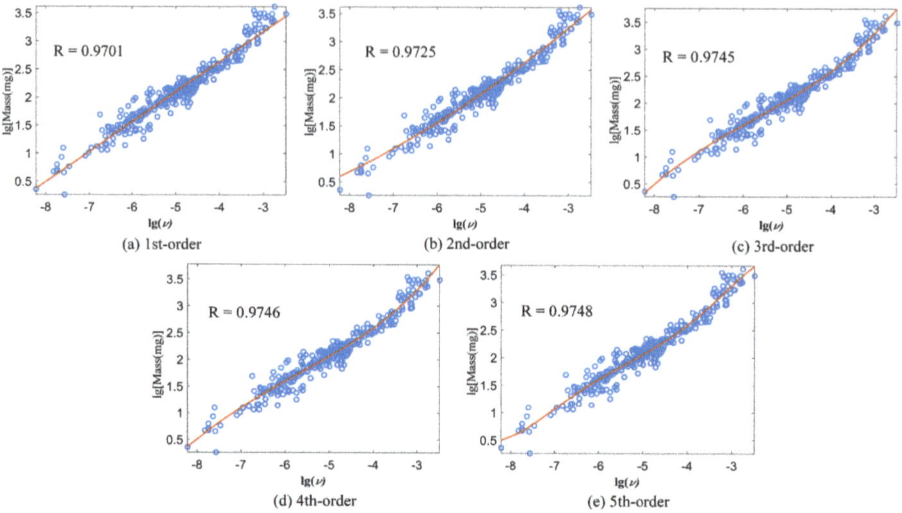

Figure 12.14 The fitted insect masses and lg ν with first- to fifth-order polynomials. The dots represent the 340 insect specimens.

The logarithm-transformed masses are fitted with third-order polynomials in lg ν. The correlation coefficient R for the masses is 0.9745 ($P < 0.001$). The fitted curve can be represented as follows:

$$\begin{aligned} M_\nu\left(mg\right) &= 10^{0.0161\cdot(\lg\nu)^3+0.2771\cdot(\lg\nu)^2+2.0486\cdot\lg\nu+7.3832}, \\ &\lg\nu \in \left[-8.23, -2.48\right] \end{aligned} \tag{12.49}$$

This could be used as an empirical formula for insect mass estimation. The MRE for insect masses estimation is 24.35%. Therefore, the parameter ν can be used to estimate insect masses with good accuracy.

For comparison, the insect masses are also fitted with third-order polynomials in λ_1, as shown in Figure 12.15. The correlation coefficient R for the masses is 0.9308 ($P < 0.001$), and the MREs are 40.94%. The calculated indices verify the improved relationship with mass for the parameter ν.

Besides ν, the parameters d and a_0, and parameter-pair a_0 & α_2 are also good parameters for estimation of insect mass. The relationships between the masses and the parameters d, a_0, and a_0 & α_2 are presented as scatter diagrams in Figure 12.16. It can be seen that all these parameters are strongly correlated with masses. The relationship between the masses and lg d is fitting via polynomials of third order, and so is for lg a_0. For the parameter-pair a_0 & α_2, the fitting orders of lg a_0 and α_2 are both first. The fitting results are shown as curves and the wireframe mesh in Figure 12.16. These fitting results can be represented as follows:

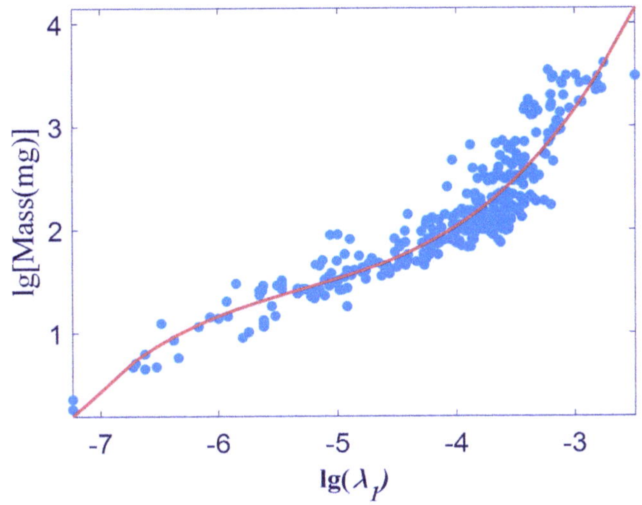

Figure 12.15 *Relationship between mass and lg λ_1*

$$M_d\,(mg) = 10^{0.0508\cdot(\lg d)^3+0.862\cdot(\lg d)^2+5.281\cdot\lg d+12.974},\quad \lg d \in [-7.73, -2.80] \tag{12.50}$$

$$M_{a_0}\,(mg) = 10^{0.069\cdot(\lg a_0)^3+1.155\cdot(\lg a_0)^2+6.798\cdot\lg a_0+15.41},\quad \lg a_0 \in [-7.55, -2.77] \tag{12.51}$$

$$M_{a_0\&\alpha_2}\,(mg) = 10^{0.555\cdot\lg a_0-0.459\alpha_2+4.91},\quad \lg a_0 \in [-7.55, -2.77] \tag{12.52}$$

The fitting methods, correlation coefficients R, and MREs of these parameters are listed in Table 12.1. It can be seen that the ν method has the best performance. The d and a_0 & α_2 are methods closely followed and they have similar good performance. The a_0 method has a correlation coefficient R of 0.954 and MRE of 33.63%.

12.2.3.3 Body length estimation

The parameters ν, d, a_0, and the parameter-pair a_0 & α_2 are used to estimate the insect mass. As the mass and body length of an insect are strongly correlated, these parameters may also have good relationships with insect body length. To verify this, the relationships between the body lengths and these parameters are plotted as scatter diagrams in Figure 12.17. It can be seen that all these parameters are strongly correlated with body length, and therefore, can be used for the insect body length estimation. The relationship between body length and lg a_0 is fitting via polynomials of third order and so is for lg ν and lg d. For the parameter-pair method, the regression analysis is used, and the proper orders of lg a_0 and α_2 are third and first,

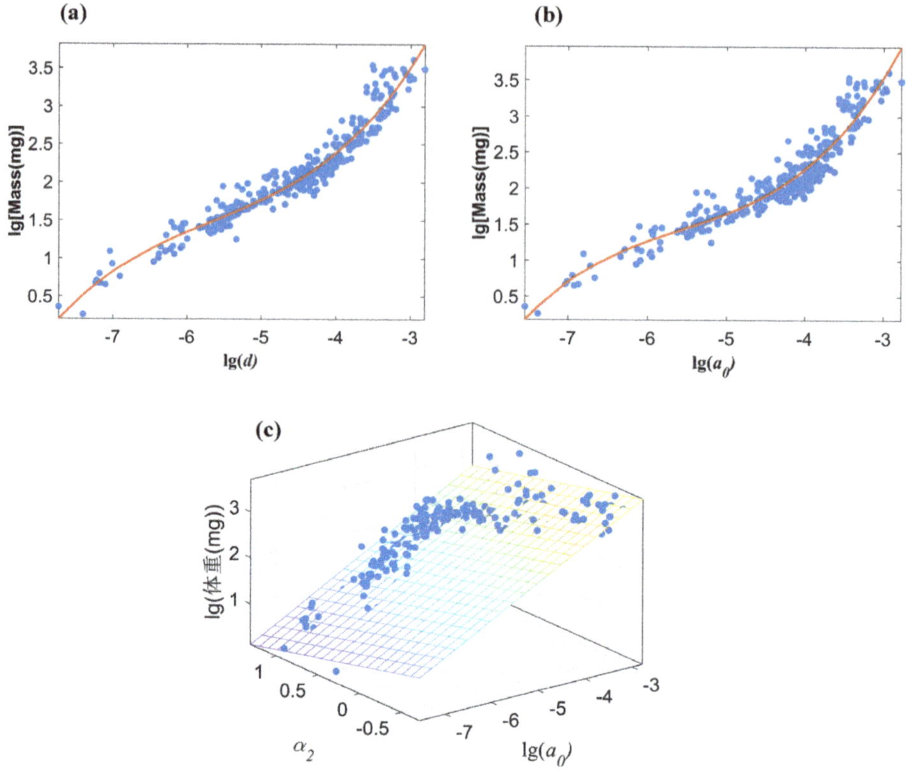

Figure 12.16 *Relations of insect masses to parameters: (a) d; (b) a_0; and (c) a_0 & α_2. The dots represent the 340 insect specimens. The curves and the wireframe mesh represent the fits to masses.*

respectively. The fitted results (shown as curves or curved surfaces in Figure 12.17) can be represented as follows:

$$\begin{aligned} L_\nu &= 10^{0.0115\cdot(\lg \nu)^3+0.198\cdot(\lg \nu)^2+1.256\cdot\lg \nu+4.01} \\ &\lg \nu \in [-8.23, -2.48] \end{aligned} \tag{12.53}$$

Table 12.1 *Comparison of polarimetric parameters for insect mass estimation*

Parameter	**Fitting method**	***R* (*P* value)**	**MRE (%)**
ν	Third-order polynomial	0.975 (*P* <0.001)	24.35
d	Third-order polynomial	0.968 (*P* <0.001)	26.95
a_0	Third-order polynomial	0.954 (*P* <0.001)	33.63
a_0 & α_2	Regression analysis	0.969 (*P* <0.001)	27.15

$$L_d = 10^{0.028\cdot(\lg d)^3+0.458\cdot(\lg d)^2+2.64\cdot\lg d+6.343}$$
$$\lg d \in [-7.73, -2.80] \tag{12.54}$$

$$L_{a_0} = 10^{0.034\cdot[\lg a_0]^3+0.549\cdot[\lg a_0]^2+3.095\cdot[\lg a_0]+7.050}$$
$$\lg a_0 \in [-7.55, -2.77] \tag{12.55}$$

$$L_{a_0\&\alpha_2} = 10^{0.024\cdot[\lg a_0]^3+0.392\cdot[\lg a_0]^2+2.226\cdot[\lg a_0]-0.095\cdot[\lg \alpha_2]+5.578}$$
$$\lg a_0 \in [-7.55, -2.77] \tag{12.56}$$

These equations can be used as empirical formulas to estimate insect body lengths (unit: mm).

The fitting methods, correlation coefficient R, and MRE for the body length estimation are listed in Table 12.2. The ν method has the best performance and is closely followed by the d and a_0 & α_2 methods. These three methods have comparably good performance. The a_0 method has a correlation coefficient R of 0.897 and MRE of 17.13%.

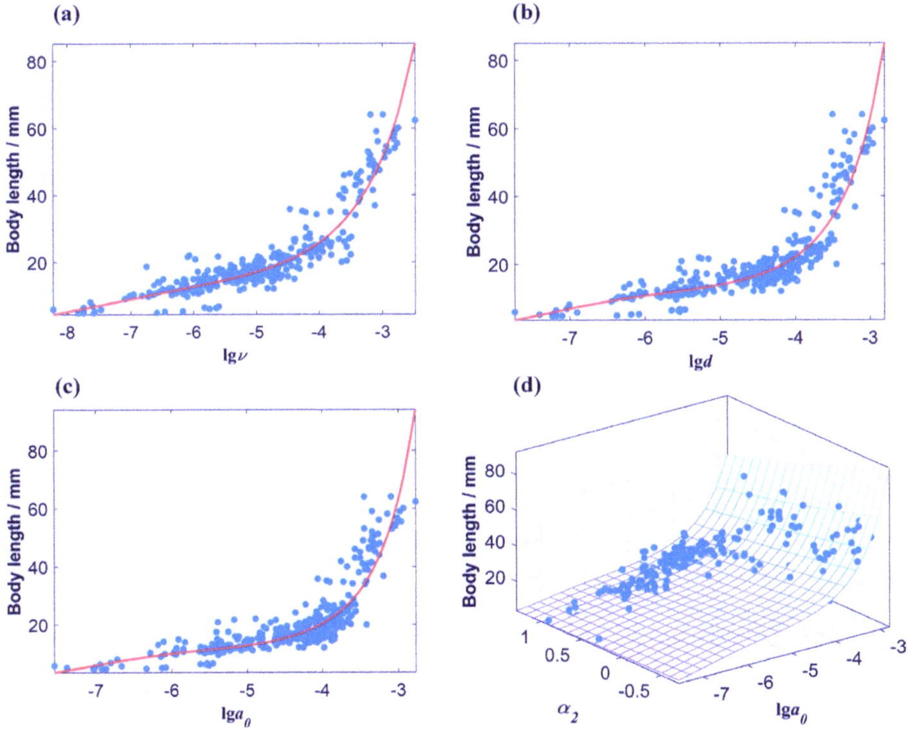

Figure 12.17 *Relations of insect body lengths to parameters: (a) ν; (b) d; (c) a_0; and (d) a_0 & α_2*

Table 12.2 Comparison of polarimetric parameters for insect body length estimation

Parameter	Fitting method	**R** (**P** value)	MRE (%)
ν	Third-order polynomial	0.936 (P <0.001)	15.20
d	Third-order polynomial	0.915 (P <0.001)	15.84
a_0	Third-order polynomial	0.897 (P <0.001)	17.13
a_0 & α_2	Regression analysis	0.930 (P <0.001)	15.92

12.3 The fully polarimetric insect radar system

12.3.1 The fully polarimetric insect radar prototype

The fully polarimetric insect radar prototype had been developed in 2017 by the Radar Research Laboratory, Beijing Institute of Technology. The radar works in a vertical looking mode and observes insects crossing the radar beam. The main operating parameters are indicated in Table 12.3. As shown in Figure 12.18, the radar consists of a dual-polarized parabolic antenna with a 1-m diameter, a radio frequency (RF) unit for transmitting and receiving RF signal, a digital unit for generating the radar waveform (synthesizing the high resolution range profile (HRRP)) and preprocessing digital signal (for target detection), a PC for control, data postprocessing and display, a servo turntable for azimuth and elevation rotation, and a radome.

The radar is a phase coherent system, employs dual-orthogonal polarization waveform, and allows measuring all elements of the insect SM simultaneously in one coherent processing interval. The polarization of the radar is calibrated based on the single target calibration technique by measuring a steel ball [30]. After calibration, the cross-polarization isolation of the radar is larger than 30 dB; the magnitude and phase differences between two polarization channels are within 0.5 dB and within 5°, respectively.

For a 90% detection probability at a 1×10^{-6} false alarm probability (considering a 13 dB SNR), the maximum detection range of the radar for insects with mass of 100 mg (RCS: −45 dBsm) is ~1 100 m. The range resolution of the radar is ~0.2 m. To realize this resolution, the stepped frequency chirp waveform is employed. The radar transmits a series of narrowband subchirp pulses sequentially and synthesizes the return echo signals to a broadband signal, and then the HRRP is obtained. The stepped number and frequency of the subchirp are 40 and 20 MHz, respectively (thus, the synthesized broadband is 800 MHz, whereas the instantaneous bandwidth processed by the receiver is 20 MHz). This waveform does not increase the burden of radar bandwidth and sampling rate, but decreases the interframe interval in the slow time domain as a frame is synthesized with 40 subchirp pulses.

As the insect echo signal is weak, long-time integration is required for increasing the SNR before echo detection. However, the range walk induced by the insect

Table 12.3 Main operating parameters of the fully polarimetric insect radar prototype

Parameters	Values
Frequency	Ku @ 16.2 GHz
Transmitted peak power	30 W
Pulse width	1 μs
Pulse repetition frequency	40 kHz
Harmonic suppression	−30 dBc
Spurious suppression	−50 dBc
Phase noise	−83 dBc/Hz @ 10kHz −88 dBc/Hz @ 100kHz −92 dBc/Hz @ 1MHz
Waveform	Stepped frequency chirp
Step number	40
Step frequency	20 MHz
Synthesized bandwidth	800 MHz
Range resolution	0.2 m
Antenna type	Parabolic reflector with diameter of 1 m
Antenna gain	41 dBi
Beamwidth	1.5°
Noise figure	5 dB
Dynamic range	≥60 dB
Sampling rate of ADC (Analog-to-Digital Converter) in the fast time domain	160 MHz
Inter frame time interval of HRRP	2 ms/500 Hz
Detection range	1.1 km @ −45 dBsm
Polarization	HH/HV/VH/VV

radial (i.e., vertical, in this gauging radar configuration) might possibly limit the maximum allowed integration. Therefore, it appears necessary to analyze the vertical motion characteristics of the insects. Two million insect targets observed with the radar deployed in Lancang of Yunnan Province from April to September 2019 were analyzed. The first- (i.e., linear), second-, and third-order polynomials were used to fit the cell range of the insect tracks. As shown in Figure 12.19, the vertical motion of most insects (96.8%) is linear; the proportion of insects with second-order motion is 1.8%; less than 1.4% of the motion is third or higher order. Consequently, it appears that long-time integration is a suitable strategy.

12.3.2 Experimental validation of the measurement of the fully polarimetric insect radar

To verify the effectiveness of the fully polarimetric insect radar, a field experiment was conducted based on two cooperative flying unmanned aerial vehicles (UAVs), whereby the orientation, mass, and body length of insect were measured with the fully polarimetric insect radar prototype. The experimental scene is shown in

Figure 12.18 The fully polarimetric insect radar prototype

Figure 12.20. The distance between the two UAVs (DJI M600 Pro) is 120 m. A PE thread with 0.1 mm diameter is suspended between the UAVs. The back of insect is glued to the middle of the PE thread, so that the insect is upright and horizontal.

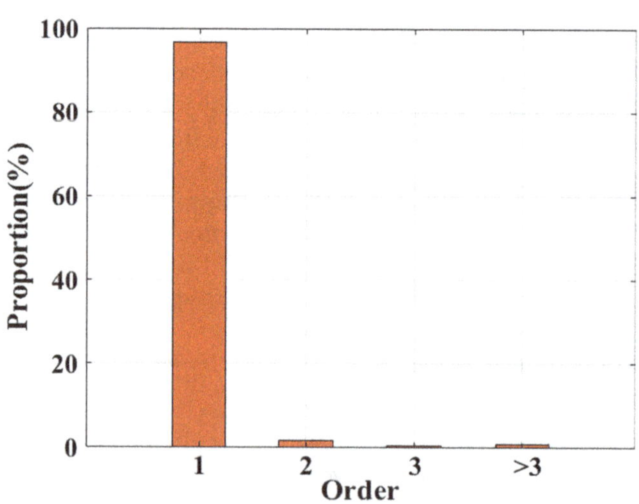

Figure 12.19 The vertical motion characteristics of the insects

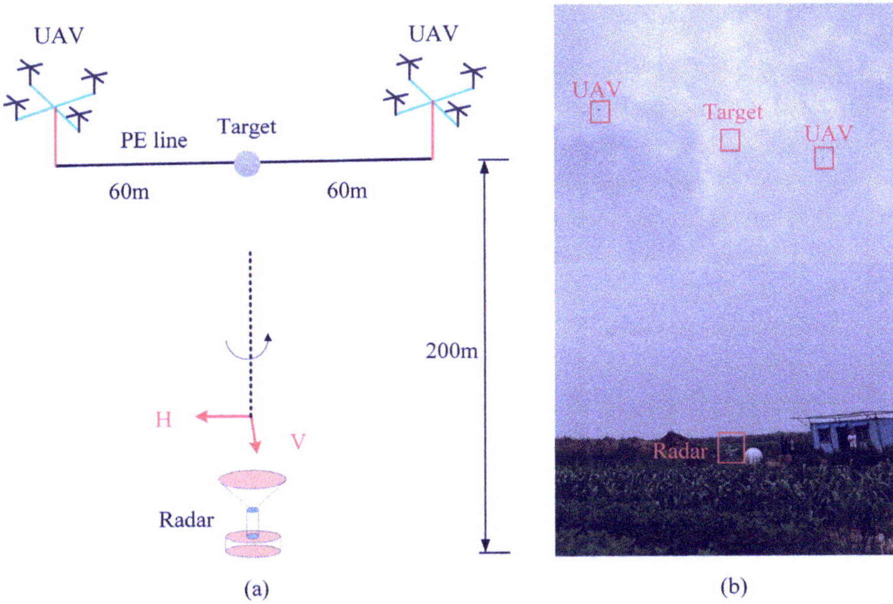

Figure 12.20 *Experimental scene of measurement: (a) diagram and (b) scene. H and V represent the directions of radiation, H-polarization and V-polarization, respectively.*

The radar is vertical looking. By controlling the UAVs to hover, the insect is directly above the radar. The distance between the insect and the radar is about 200 m. The insect specimens were live in the measurement. Since possible vibrations of the UAVs along with wind effects may affect the stability of the insect during the measurement (thus introducing undesired modulations on its echo), only the strongest echo above the radar was selected. The orientation of an insect is defined as the angle between the body axis and the H-polarization direction. Thus, by rotating the direction of the H-polarization, insects at different orientations can be measured. For calibration purposes, a steel ball was measured along with the insect.

In the experiment, 44 insect specimens with orientations of 0°, 30°, 45°, 60°, and 90° were measured, with masses ranging from 19.8 to 274.1 mg, and body lengths from 8.5 to 22.1 mm. Based on the measured insect specimens, the correct rate of discrimination of PA and PE insects, the accuracies of measurement of orientation, the mass and body length were analyzed taking into account.

12.3.2.1 Discrimination of PA and PE insects

As introduced in section 12.2.2.2, the insect class can be discriminated based on the sign of $\Delta\phi$. The sign of $\Delta\phi$ is expected to be negative for the PA insect and positive for the PE insect. Figure 12.21 shows $\Delta\phi$ of insects at different orientations. For 0° orientation, the signs of $\Delta\phi$ of two PA insects are positive (thus causing a

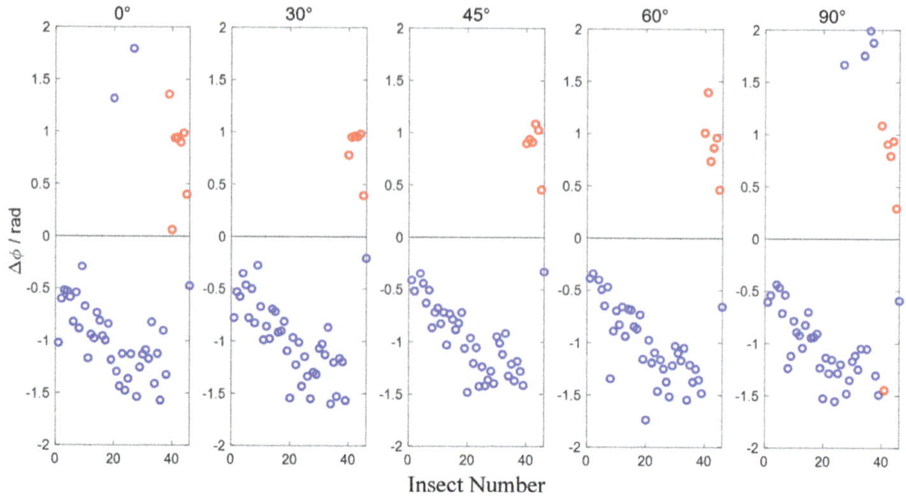

Figure 12.21 *Δϕ of insects at different orientations. The red and blue circles represent PE and PA insects, respectively. The true orientations of the insects are shown on the top of each figure.*

discrimination error). For 30°, 45°, and 60°, the signs of all insects are correct. For 90° orientation, the signs of $\Delta\phi$ of four PA and one PE insects are not correct. The total correct rate for insect class discrimination at different orientations is 96.82%. Therefore, the insect class can be discriminated with good precision.

12.3.2.2 Orientation estimation

Based on the insect class discrimination, the insect orientation can be estimated via (12.20) (where the MRD estimation is based on the main eigenvector of SM). As the accuracy of the insect class discrimination has been analyzed in the previous paragraph, only the accuracy of orientation estimation is analyzed here with the assumption that the classes of all insects are correctly identified. The boxplot of the orientation errors at different orientations is shown in Figure 12.22. The medians of the orientation errors at all orientations range from −1° to 1°. The orientation errors of more than half of the insects are within ±2°. For most insects, the errors are within ±3°. The standard deviation of the orientation errors of all insects is 1.7°. In general, the accuracy of the orientation measurement satisfies the requirements for radar observation of migratory insects.

12.3.2.3 Mass and body length estimation

The masses and body lengths of all insect specimens are estimated based on the ν method. The relative errors are shown in Figure 12.23. It can be seen that most of the relative errors of mass estimation are less than 40%, with the mean value of 22.5%.

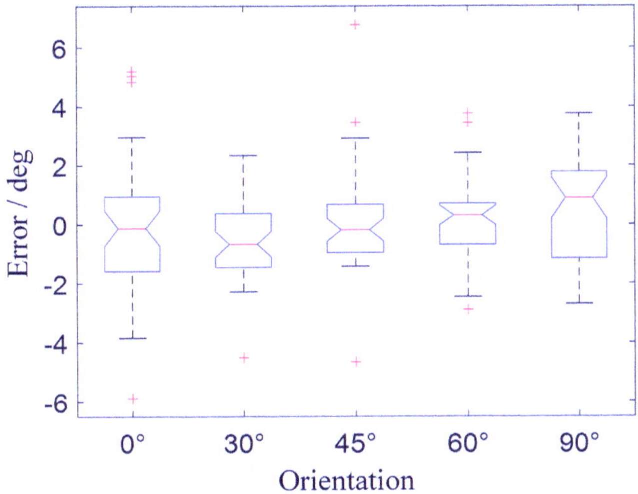

Figure 12.22 *Boxplot of orientation errors at different orientations*

For the body length estimation, most of the relative errors are less than 20%, and the mean value is 12.99%. The errors of insects less than 12 mm are larger than that of the larger insects. Therefore, both masses and body lengths can be estimated with useful precision.

This experiment indicates that the insect parameters of orientation, mass, and body length can be measured with useful accuracy by a fully polarimetric radar. Therefore, the fully polarimetric insect radar is effective for the parameter measurement of insects.

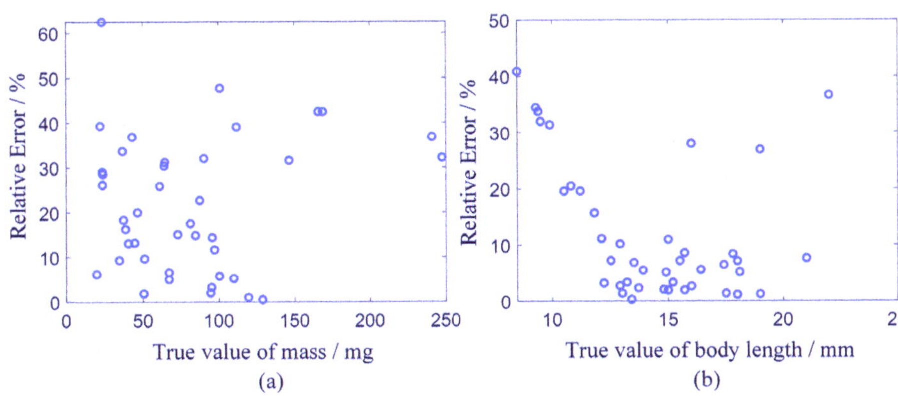

Figure 12.23 *Relative errors of: (a) mass and (b) body length measurement*

12.4 Conclusion and future prospects

In this chapter, polarization-based signal processing techniques via insect radars have been introduced. Three types of insect radars have been outlined: scanning insect radar, vertical insect radar, and fully polarimetric insect radar. The scanning insect radar was used for insect swarm measurements; the vertical insect radar and fully polarimetric insect radar were designed for insect individual measurements. When a swarm of insects fly simultaneously cross the beam, the insect individuals in different range units can be measured separately. However, when more than one insect appears in the same range unit (0.2 m resolution), they cannot be separated, and the estimated insect parameters may be wrong. Fortunately, according to measured data in long-time monitoring, this will not happen for most cases.

To this end, polarization-based information extraction has been playing an important role in insect radar measurements. For the scanning radar, the polarization is used for common orientation measurements, whereby the RCS of an insect illuminated with a horizontal linear polarized radar and the echo RCS of the insect from the broadside illumination are maximum. By rotating the polarization during the measurement, the vertical looking insect radar is able to measure the orientation and mass of insects. With a fully polarimetric insect radar, it is also possible to measure orientation, mass, and body length with higher accuracy. To this end, a prototype of a fully polarimetric insect radar has been developed for measuring parameters of insects. These measurement capabilities of insect radar have allowed understanding insect migration phenomena that were not explained by traditional methods. However, there are many unsolved mysteries about insect migration, whereas more advanced insect radar instruments are required to investigate further mechanisms related to insect migration.

For example, with the development of radar and microwave technology, multiple-frequency measurements are an important research area in radar system [43], thus extending the capability of information acquisition for insect radars.

Supported by the National Natural Science Foundation of China with a fund of about RMB 90 million, the Radar Research Laboratory of Beijing Institute of Technology has been researching and developing a novel insect radar system termed "high-resolution and multidimensional measurement radar system", which consists of a high-resolution phased array radar (HRPAR) and three multifrequency fully polarimetric radars (MFFPRs) (Figure 12.24) [14].

The HRPAR works at Ku-band with a 0.2-m range resolution. It is a scanning radar with mechanical scanning in azimuth and electronic scanning in elevation, whereby the precise position of an insect is measured via monopulse techniques. The detection coverage is designed as a cylinder with a 1.5-km height and a 1-km radius for insects with a 10-mg mass. The HRPAR is designed to simultaneously track and measure all individual insects appearing in the coverage, such that the precise position of a target of interest is sent to the three MFFPRs for collaborative fine measurements.

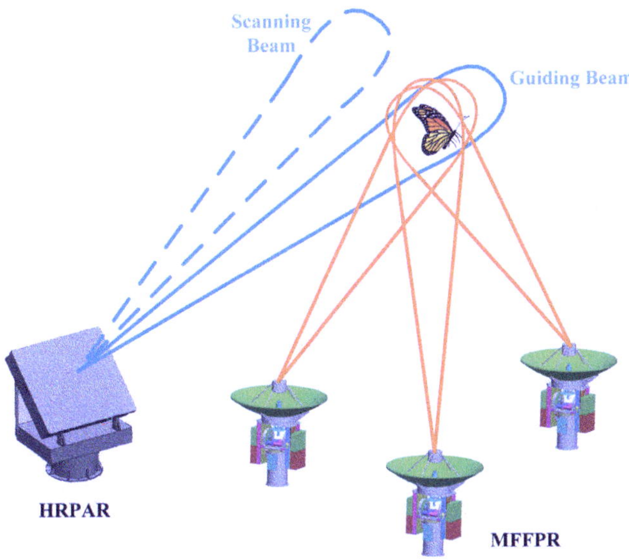

Figure 12.24 *High-resolution and multidimensional measurement radar system [14]*

A MFFPR works simultaneously at X-, Ku-, and Ka-bands again with a 0.2-m range resolution, performing tracking tasks, and supporting fully polarimetric measurement (HH/HV/VH/VV) at all three bands. The monopulse technique is applied at Ka-band for precise target position measurement. Thus, the MFFPR can provide multidimensional insect backscattering information consisting of amplitude, phase, polarization, and frequency, which support insect biological parameters (such as orientation, mass, body length, wingbeat frequency, etc.) estimation and species identification. In the cooperative measurement mode, based on the target position information sent by the HRPAR, all three MFFPRs observe the selected insect from different aspects. The aforementioned multiaspect observation setup allows obtaining the 3D orientation of insects [14].

References

[1] Drake V.A., Reynolds D.R. *Radar entomology: observing insect flight and migration*. UK: CABI; 2012 Jan. Available from http://www.cabidigitallibrary.org/doi/book/10.1079/9781845935566.0000

[2] Schaefer G.W. 'Radar observations of insect flight'. *Symposia of the Royal Entomological Society of London*. 1976, vol. 7, pp. 157–97.

[3] Smith A.D., Reynolds D.R., Riley J.R. 'The use of vertical-looking radar to continuously monitor the insect fauna flying at altitude over southern England'. *Bulletin of Entomological Research*. 2000, vol. 90(3), pp. 265–77.

[4] Hu G., Lim K.S., Horvitz N., *et al.* 'Mass seasonal bioflows of high-flying insect migrants'. *Science.* 2016, vol. 354(6319), pp. 1584–87.
[5] Chapman J.W., Nesbit R.L., Burgin L.E., *et al.* 'Flight orientation behaviors promote optimal migration trajectories in high-flying insects'. *Science.* 2010, vol. 327(5966), pp. 682–85.
[6] Hu C., Li W., Wang R., Li Y., Li W., Zhang T. 'Insect flight speed estimation analysis based on a full-polarization radar'. *Science China Information Sciences.* 2018, vol. 61(10), p. 109306.
[7] Hu C., Li W., Wang R., Liu C., Zhang T., Li W. 'Accurate insect orientation extraction based on polarization scattering matrix estimation'. *IEEE Geosci Remote Sens Lett.* 2017, vol. 14(10), pp. 1755–59.
[8] Hu C., Li W., Wang R., Long T., Liu C., Drake V.A. 'Insect biological parameter estimation based on the invariant target parameters of the scattering matrix'. *IEEE Trans Geosci Remote Sens.* 2019, vol. 57(8), pp. 6212–25.
[9] Hu C., Li W., Wang R., Long T., Drake V.A. 'Discrimination of parallel and perpendicular insects based on relative phase of scattering matrix eigenvalues'. *IEEE Trans Geosci Remote Sens.* 2020, vol. 58(6), pp. 3927–40.
[10] Li W., Hu C., Wang R., Kong S., Zhang F. 'Comprehensive analysis of polarimetric radar cross-section parameters for insect body width and length estimation'. *Science China Information Sciences.* 2021, vol. 64(2), p. 122302.
[11] Schaefer G.W. 'Radar studies of locust, moth and butterfly migration in the Sahara'. *Proceedings of the Royal Entomological Society of London.* 1969, vol. 34(33), pp. 39–40.
[12] Riley J.R. 'Angular and temporal variations in the radar cross-sections of insects'. *Proceedings of the Institution of Electrical Engineers.* 1973, vol. 120(10), p. 1229.
[13] Riley J.R. 'Collective orientation in night-flying insects'. *Nature.* 1975, vol. 253(5487), pp. 113–14.
[14] Long T., Hu C., Wang R., *et al.* 'Entomological radar overview: system and signal processing'. *IEEE Aerospace and Electronic Systems Magazine.* 2020, vol. 35(1), pp. 20–32.
[15] Riley J.R., Reynolds D.R. 'Radar-based studies of the migratory flight of grasshoppers in the middle Niger area of Mali'. *Proceedings of the Royal Society of London. Series B, Biological Sciences.* 1979, vol. 204(1154), pp. 67–82.
[16] Smith A.D., Riley J.R., Gregory R.D. 'A method for routine monitoring of the aerial migration of insects by using A vertical-looking radar'. *Philosophical Transactions of the Royal Society of London. Series B, Biological Sciences.* 1993, vol. 340(1294), pp. 393–404.
[17] Drake V.A. 'Automatically operating radars for monitoring insect pest migrations'. *Insect Science.* 2002, vol. 9(4), pp. 27–39.
[18] Chapman J.W., Reynolds D.R., Smith A.D. 'Vertical-looking radar: a new tool for monitoring high-altitude insect migration'. *BioScience.* 2003, vol. 53(5), 503.

[19] Hobbs S., Allsopp K., Wolf W. *Signal analysis for an entomological radar with a vertical nutating beamsignal analysis for an entomological radar with a vertical nutating beam*. Cranfield: Cranfield University; 2003.

[20] Harman I.T., Drake V.A. 'Insect monitoring radar: analytical time-domain algorithm for retrieving trajectory and target parameters'. *Computers and Electronics in Agriculture*. 2004, vol. 43(1), pp. 23–41.

[21] Aldhous A.C. *An investigation of the polarisation dependence of insect radar cross sections at constant aspect an investigation of the polarisation dependence of insect radar cross sections at constant aspect*. Dissertation Cranfield: Cranfield Institute of Technology; 1989.

[22] Riley J.R. 'Radar cross section of insects'. *Proceedings of the IEEE*. 1985, vol. 73(2), pp. 228–32.

[23] Chapman J.W., Smith A.D., Woiwod I.P., Reynolds D.R., Riley J.R. 'Development of vertical-looking radar technology for monitoring insect migration'. *Computers and Electronics in Agriculture*. 2002, vol. 35(2–3), pp. 95–110.

[24] Hobbs S.E., Aldhous A.C. 'Insect ventral radar cross-section polarisation dependence measurements for radar entomology'. *IEE Proceedings - Radar, Sonar and Navigation*. 2006, vol. 153(6), p. 502.

[25] Zhuang Z., Xiao S., Wang X. *Radar polarization information processing and application*. Beijing: National Defense Industry Press; 1999.

[26] Giuli D. 'Polarization diversity in radars'. *Proceedings of the IEEE*. 1986, vol. 74(2), pp. 245–69.

[27] Huynen J.R. 'Measurement of the target scattering matrix'. *Proceedings of the IEEE*. 1965, vol. 53(8), pp. 936–46.

[28] Bringi V.N., Chandrasekar V. *Polarimetric doppler weather radar*. Cambridge: Cambridge University Press; 2001 Aug 30. Available from https://www.cambridge.org/core/product/identifier/9780511541094/type/book

[29] Krasnov O.A., Ligthart L.P., Li Z., Babur G., Wang Z., van der Zwan F. 'Parsax: hhigh-resolution doppler-polarimetric FMCW radar with dual-orthogonal signals'. *The 18th International Conference on Microwave Radar and Wireless Communications*; Vilnius, Lithuania, 2010.

[30] Sarabandi K., Ulaby F.T. 'A convenient technique for polarimetric calibration of single-antenna radar systems'. *IEEE Trans Geosci Remote Sens*. 1990, vol. 28(6), pp. 1022–33.

[31] Li W., Hu C., Wang R., Liu C., Li W. 'Experimental validations of insect orientation extraction based on fully polarimetric measurement'. *The Journal of Engineering*. 2019, vol. 2019(21), pp. 7954–57. Available from https://onlinelibrary.wiley.com/toc/20513305/2019/21

[32] Kong S., Hu C., Wang R., Zhang F., Wang L. 'Insect multifrequency polarimetric radar cross section: experimental results and analysis'. *IEEE Trans Geosci Remote Sens*. 2020, vol. 99, pp. 1–13.

[33] Lee J.S., Pottier E. *Polarimetric radar imaging: from basics to applications polarimetric radar imaging: from basics to applications*. Boca Raton, FL: CRC Press; 2009.

[34] Melnikov V.M., Istok M.J., Westbrook J.K. 'Asymmetric radar echo patterns from insects'. *Journal of Atmospheric and Oceanic Technology*. 2015, vol. 32(4), pp. 659–74.

[35] Wolf W.W., Vaughn C.R., Harris R., Loper G.M. 'Insect radar cross-sections for aerial density measurements and target classification'. *Transactions of the ASAE*. 1993, vol. 36(3), pp. 949–54.

[36] Huynen J.R. *Phenomenological theory of radar targets*. Dissertation Delft: Delft University of Technology; 1970.

[37] Bickel S.H. 'Some invariant properties of the polarization scattering matrix'. *Proceedings of the IEEE*. 1965, vol. 53(8), pp. 1070–72.

[38] Huynen J.R. 'Invariant radar target parameters'. *1966 Antennas and Propagation Society*. 1966, pp. 486–87.

[39] Graves C. 'Radar polarization power scattering matrix'. *Proceedings of the IRE*. 1956, vol. 44(2), pp. 248–52.

[40] Boerner W.-M., Yan W.-L., Xi A.-Q., Yamaguchi Y. 'On the basic principles of radar polarimetry: the target characteristic polarization state theory of Kennaugh, Huynen's polarization fork concept, and its extension to the partially polarized case'. *Proceedings of the IEEE*. 1991, vol. 79(10), pp. 1538–50.

[41] Kostinski A., Boerner W. 'On foundations of radar polarimetry'. *IEEE Trans Antennas Propag*. 1986, vol. 34(12), pp. 1395–404.

[42] Drake V.A., Chapman J.W., Lim K.S., Reynolds D.R., Riley J.R., Smith A.D. 'Ventral-aspect radar cross sections and polarization patterns of insects at X band and their relation to size and form'. *International Journal of Remote Sensing*. 2017, vol. 38(18), pp. 5022–44.

[43] Leinonen J., Lebsock M.D., Tanelli S., Suzuki K., Yashiro H., Miyamoto Y. 'Performance assessment of a triple-frequency spaceborne cloud–precipitation radar concept using a global cloud-resolving model'. *Atmospheric Measurement Techniques*. 2015, vol. 8(8), pp. 3493–517.

Chapter 13

Polarimetric weather radar signal processing

F. Gekat[1], R. Hannesen[1], and P. Gölz[1]

13.1 Introduction

The weather radar database of the World Meteorological Organization (WMO) [1] provides more than 1 000 entries by the end of 2020. Almost all radars are operating in S-, C-, or X-band, and more than 340 are polarimetric. Examples are shown in Figure 13.1 and Figure 13.2. Considering that the real numbers are larger because the database is not complete and not up-to-date, the number of ground-based weather radar installations probably exceeds the number of all other types of large-scale, civil-used ground-based radars, including primary radars.

It should be noted here that the number of airborne weather radars exceeds that of ground-based systems by far since almost each commercial aircraft carries a weather radar under its nose radome. However, these radars are not addressed in this chapter because they use single linear polarization.

The weather radar has become an indispensable remote sensor for weather services and airports. Networks of weather radars provide countrywide coverage, as shown in Figure 13.3.

Their data are displayed by dedicated smartphone apps and are used by farmers, construction companies, sport event managers, and individuals for planning open-air activities. The weather services are using the data for weather nowcasting up to 6 h by ingesting the data into numerical weather prediction models [2, 3]. Airports rely on weather radars for the detection of dangerous turbulences, microbursts, and gust fronts [4, 5].

The operational availability of radar polarimetry has made the data even more valuable. The acquisition and processing of polarimetric data allow the correction of major measurement errors and even the determination of the type of hydrometeors (rain, snow, graupel, and hail). The basic principle of operation as well as the signal and data processing of modern operational weather radars will be explained in the following sections.

[1]LEONARDO Germany GmbH, Neuss, Germany

Figure 13.1 *Mobile polarimetric X-band Doppler weather radar with a 1.8-m antenna*

Figure 13.2 *S-band Doppler weather radar with a 8.5-m antenna in bird-bath position, radome removed*

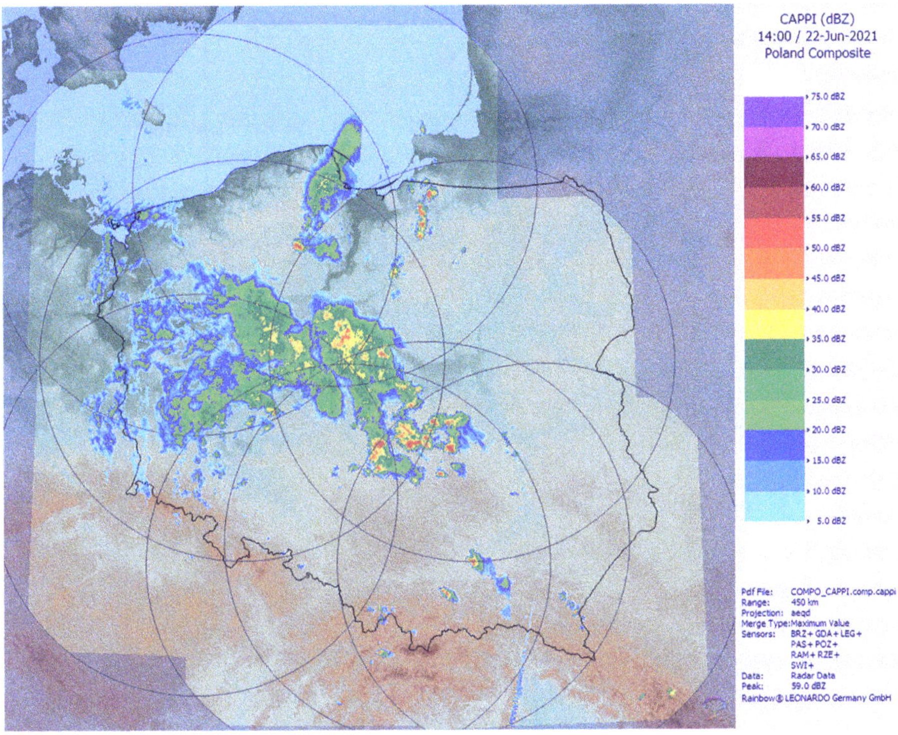

Figure 13.3 Reflectivity constant altitude plan position indicator. Data courtesy of Institute of Meteorology and Water Management—National Research Institute, Poland.

13.2 Principle of operation

A weather radar is a measuring rather than a detection device. It measures the amplitude and phase of the signal backscattered by hydrometeors and clutter scatterers. "Clutter" in the radar meteorological sense means all kinds of non-meteorological objects like ground scatterers, planes, birds, insects, etc. A simplified block diagram of a polarimetric Doppler weather radar system is shown in Figure 13.4.

The best compromise between the level of the backscattered signal and its attenuation is found in the S- (2.7–3.0 GHz), C- (5.2–5.9 GHz), and X- (9.3–9.5 GHz) radar bands. Lower frequencies do not provide sufficient backscatter amplitude, and shorter wavelengths are subject to excessive attenuation. X-band radars are mainly used for short-range surveillance up to 50 km. C-band radars are chosen by weather services in moderate climatic regions such as Europe, whereas all countries facing strong precipitation events are using S-band radars.

Almost all operational polarimetric Doppler weather radars are using transmitters featuring either magnetron or klystron tubes. The rationale is cost and the

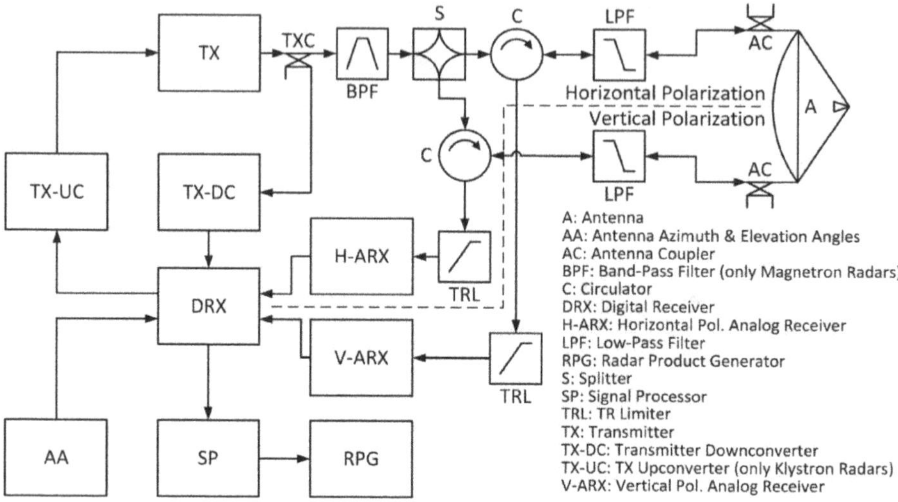

Figure 13.4 Polarimetric Doppler weather radar with a single transmitter

so-called "blind range" associated with pulse compression techniques. The receiver of a pulse compression radar is blinded during the transmission of the long pulse. This blind range, which can easily extend up to 15 km, is sampled by a short, unmodulated pulse. This reduces the signal-to-noise ratio (SNR) in the blind range by up to 20 dB, which is fatal in particular for wind profile scanning modes performed at high elevations. These modes sample the troposphere over its full height of 15 km even during fair weather conditions using, e.g., aeroplankton (small insects and spiders) as scatterers, which have been detected flying at heights up to 5790 m, and in mountain snowfields at 6700 m [6].

The necessary coherence and SNR of the radar systems are mainly established by sampling each transmitter pulse using a dedicated coupler (TXC), calculating the complex matched filter, and applying it to the received signal.

The band-pass filter (BPF) attenuates the spurious emissions of magnetron transmitters. It is required by many national radio bandwidth regulations, e.g., by the Radio Equipment Directive (RED) 2014/53/EU of the European Union.

The transmitter signal is divided equally between the horizontal and vertical polarization channel by the splitter (S) that can be a Magic T or a hybrid. Since both polarizations are transmitted simultaneously, the mode of operation is called "hybrid" [7] or STAR (simultaneous transmit and receive) mode.

After passing the circulators (C), the signals are filtered by low-pass filters (LPF) that attenuate the harmonic emissions generated by the transmitter tube and the circulators, again because of radio bandwidth regulations.

The antenna (A) is crucial for the performance of the radar and the quality of the data. Most radars feature prime focus feed reflector antennas with orthomode transducers. Besides the basic parameters like gain, beamwidth, and sidelobe suppression, the cross polarization and the parameter balance between both polarizations are critical. A high mechanical symmetry is therefore required. The radome covering the antenna features sandwich panels with a pseudo-random cut in order to spread the unavoidable errors caused by scattering of the panel joints equally on both polarizations.

The beam pointing direction of the antenna in azimuth and elevation is controlled with a two-axis positioner mounted on a pedestal. High precision in both pointing and angle measurement is necessary in order to assign the received data to the correct volume element. Thus, a weather radar is basically a 3D radar although with a mechanical scanner.

The received signals pass the TR limiters (TRL) and are amplified and downconverted by the analog receivers (H-ARX and V-ARX). The digital receiver (DRX) digitizes the intermediate frequency (IF) signals from the receivers as well as the transmitter signal sample provided by the transmitter downconverter (TX-DC). The data are tagged with the azimuth and elevation antenna angles (AA). If a klystron transmitter is used, the DRX provides the IF transmitter signal which is upconverted by the transmitter upconverter (TX-UC).

The data are routed to the signal processor (SP). The complex matched filtering is either performed by the DRX or by the SP. Clutter filtering, spectral moment covariance estimation, and polarimetric observable estimation are performed by the SP. The estimates are sent to the radar product generator (RPG) that performs the 2D and 3D data processing such as rain rate estimation, wind field generation, and hydrometeor classification.

Weather radars with solid-state transmitters allow a dual-channel design. As can be seen in Figure 13.5, the radar features two completely independent channels, one for each polarization. This is possible because solid-state transmitters are combining

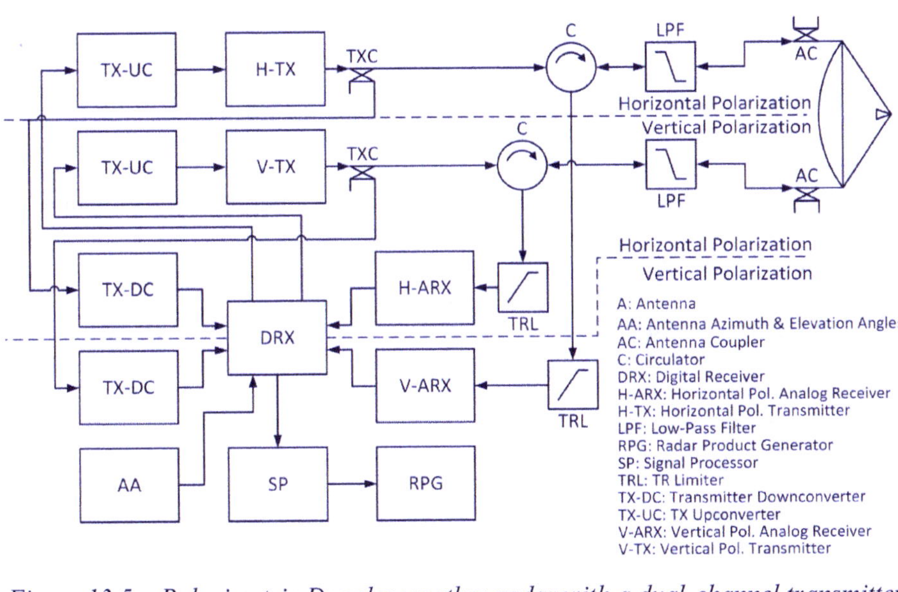

Figure 13.5 Polarimetric Doppler weather radar with a dual-channel transmitter

the power of several amplifier modules operating in parallel. Instead of splitting the power of a single high-power transmitter into two channels, two transmitters are used.

Each channel features its own upconverter (TX-UC), transmitter signal sampling coupler (TXC), and transmitter signal downconverter (TX-DC). The remaining signal flow is the same as described above.

Besides the hybrid mode of operation, the dual-channel design allows the alternating transmission of horizontal and vertical polarized pulses. This mode is called alternating or switched mode. Because also in the switched mode both polarizations are received simultaneously, the depolarization can be measured.

13.3 Signal processing

13.3.1 Overview

Though different transmission and reception schemes for polarimetric radars are possible, this chapter puts the focus on the hybrid mode (also called STAR mode, i.e., simultaneous transmission and reception). This is the most common mode of operational dual-polarization weather radars. Reception of echo signals in the hybrid mode is done simultaneously for both polarization channels.

After analog-to-digital (AD) conversion and baseband downconversion in the digital receiver, the signal processor receives time series of samples for each transmitter pulse.

The output of signal processing are estimates of single polarimetric observables for both polarization channels separately and estimates of dual-polarimetric observables (sometimes also called dual polarization variables). Each observable estimate describes properties of an atmospheric resolution volume, also called range bin or range gate.

Estimates are reported as radials or rays for each observable. A radial describes a sequence of estimates in polar coordinates, i.e., evenly spaced values along a range with angular bounds that correspond to the antenna movement. A plane position indicator (PPI) scan refers to an antenna movement along an azimuth angle with fixed elevation. The corresponding scan with fixed azimuth and movement along elevation is called range height indicator (RHI). A pointing mode scan with fixed azimuth and elevation position is also possible.

Radials may be visualized in real time and can be collected into volume data used for offline post-processing and product generation.

The signal processing flow diagram is depicted in Figure 13.6. The signal processing performed for both polarization channels is shown only for the horizontal polarization channel.

Please note that the complete single-channel and polarimetric covariance processing including the subsequent steps is performed also for unfiltered data.

13.3.2 Digital downconversion and matched filtering

High-resolution (16 Bit), high-bandwidth (approx. 200 MS/s) analog/digital converters sample the signals from the analog receivers (H-ARX and V-ARX) and the

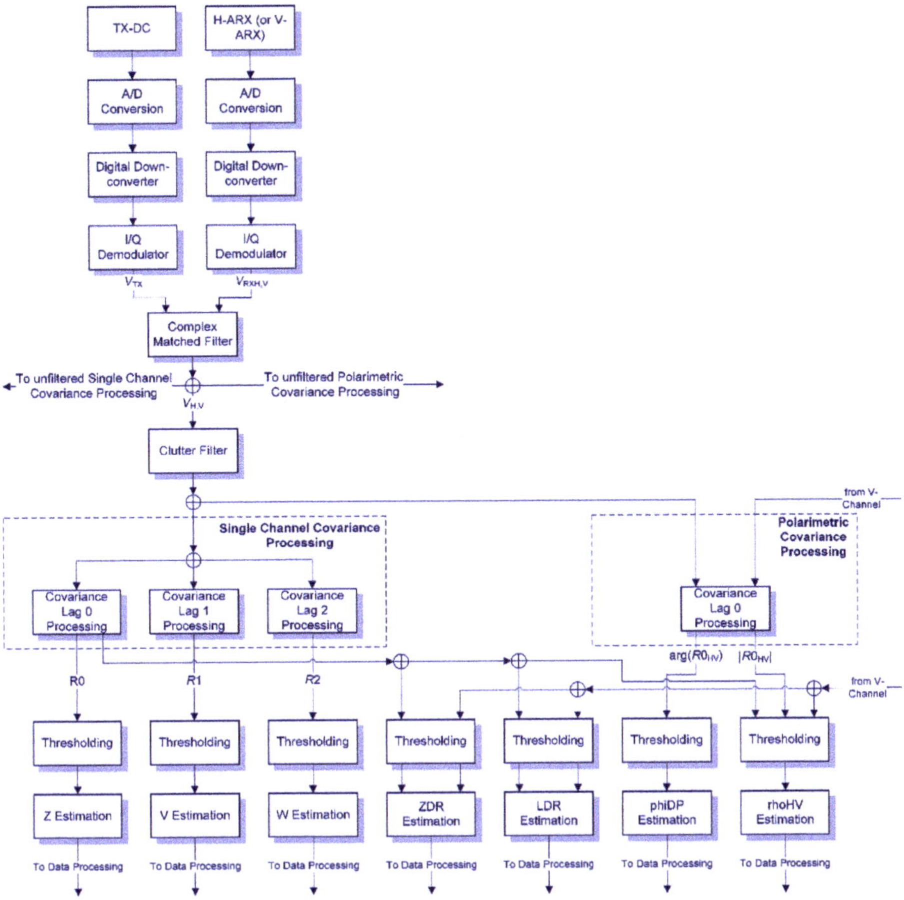

Figure 13.6 Dual polarization channel signal processing flow diagram (vertical channel is not indicated)

transmitter signal sample provided by the transmitter downconverter (TX-DC) at IF in a higher Nyquist interval. Downconversion and I/Q demodulation are performed by a complex multiplication or a Hilbert transformation. The transmitter signal sample is used to calculate a complex matched filter for each pulse width.

13.3.3 I/Q sample collection

In the first step, the signal processor collects all samples that refer to all range bins of the current radial. Equation (13.1) describes the sample sequences for a single range bin:

$$V_{Hm} = Re\left(V_{Hm}\right) + jIm\left(V_{Hm}\right),\ V_{Vm} = Re\left(V_{Vm}\right) + jIm\left(V_{Vm}\right) \tag{13.1}$$

where V_{Hm} and V_{Vm} are the complex signal voltages for horizontal and vertical polarization. The pulse index m refers to the transmitter pulses and runs from 1 to M. The number of samples M collected per radial is determined by the transmitter PRF, the angular extension of the range bin, and the antenna rotation speed. The time period related to the transmission and reception of M signal samples at constant transmitter PRF is also called the coherent processing interval (CPI).

The I/Q sample collection includes a correction for pulse-to-pulse variations of the transmitter phase and amplitude. It is important to note that each sample contains noise.

13.3.4 Ground clutter processing

Scattering from non-meteorological ground-based scatterers results in signal components that interfere with echo signals from hydrometeors. Sources of ground clutter are non-moving ground structures like towers, mountains, etc., but also echoes from moving scatterers [sea clutter, wind turbine (WT) blades] give rise to clutter interference. This section restricts to clutter processing that arises from non-moving ground clutter.

Removal of clutter components and a possible subsequent reconstruction of precipitation signal components may introduce a bias during signal estimation. Therefore, the ultimate aim of I/Q-based clutter processing is the removal of clutter signals while keeping the precipitation signal undisturbed. This goal can be achieved if a clutter detection step restricts clutter filtering to range bins where significant clutter was identified.

13.3.4.1 Clutter detection

A clutter detection algorithm is a tool to limit the application of a filter to range bins containing clutter.

Static clutter maps do not include effects like a seasonal change in the clutter environment and clutter echoes due to anomalous propagation. Therefore, dynamic maps that automatically identify ground clutter from the current signal properties are preferred. Clutter mitigation decision (CMD) [8] is a commonly used algorithm for ground clutter identification. It combines several signal parameters in a fuzzy logic resulting in a binary clutter decision. Application in dual polarization systems allows CMD to use information from both polarization channels. This improves the clutter identification probability compared to single-polarization systems.

Beneath CMD, there are other I/Q clutter processing algorithms that contain a detection step (clutter environment analysis using adaptive processing (CLEAN-AP) [9] and spectrum-time estimation and processing (STEP) [10]). In addition to I/Q--based ground clutter detection, a clutter identification is also possible during post-processing (see section 13.6.3).

13.3.4.2 Clutter filtering

As the Doppler shift of non-moving ground clutter signals is located near 0 Hz, spectral filters are used to suppress such clutter signal components. Simple time-domain filters just remove all signal components inside a predefined spectral range, whereas more advanced filters try to dynamically identify the spectral extension of the clutter signal and additionally restore the part of the weather signal lost due to the filter activity. GMAP [11] is such an algorithm; other procedures are described, e.g., in [9] and [10]. GMAP iteratively tries to reconstruct a Gaussian-shaped weather signal inside a dynamically defined clutter notch. The combination of CMD and GMAP is an established clutter processing algorithm.

For dual-polarization weather radars, the stream of I/Q data for both polarizations runs through separate clutter filter instances. It is therefore possible to define filter parameters like, e.g., the spectral extension of the clutter notch for each polarization channel independently or use combined information from both polarization channels to determine common filter parameters.

13.3.5 Weather radar observables

Weather radar observables or estimates are the result of the radar signal processing and provide the basis for the downstream radar data processing. Beneath estimates of reflectivity, radial velocity, and spectrum width for each polarization channel, polarimetric radar observables contain information from both polarizations. The following paragraphs list the most commonly used single- and dual-polarization observables in hybrid mode and their estimation from time-domain signal covariances. The hat notation characterizing estimates is used in equations but omitted in the text.

13.3.6 Covariance processing

The clutter filtering is followed by the covariance processing of the signal series from the respective range gate. Single and dual polarimetric radar observables can be estimated from discrete-time series using signal covariances. Covariances may be derived from time- or frequency-domain signal representations.

The covariance processing consists of the estimation of the first lags of the autocorrelation function of the complex signal series [12, 13], followed by time integration and, if activated, range integration.

Each estimate R of a specific lag of the autocorrelation can be calculated from the complex signal series by applying:

$$\hat{R}(L) = \frac{1}{M-L}\sum_{m=1}^{M-L} V_m V_{m+L}^* \tag{13.2}$$

where M is the number of signal samples from a specific range gate, L is the number of lags between the two signal series, and the subscript m indicates the mth sample from the range gate under consideration. T_L is the time lag between two pulses and is equal to the inverse of the pulse repetition frequency f_{PRF}:

$$T_L = 1/f_{PRF} \tag{13.3}$$

Equation (13.2) does already consider the time integration, which is simply the summation of the autocorrelation products. This process is sometimes also referred to as time or azimuth averaging because the sum is divided by the number of integrated products.

13.3.6.1 Single-channel covariance Processing

The autocorrelation coefficient at lag 0 is required for the signal power estimation:

$$\hat{R}(0) = \frac{1}{M}\sum_{m=1}^{M} V_m V_m^* \tag{13.4}$$

For the estimation of the mean radial velocity, the autocorrelation coefficient at lag 1 must be estimated:

$$\hat{R}(1) = \frac{1}{M-1}\sum_{m=1}^{M-1} V_m V_{m+1}^* \tag{13.5}$$

The spectral width estimate, which is a measure for turbulence, can be calculated from the autocorrelation at lag 2:

$$\hat{R}(2) = \frac{1}{M-2}\sum_{m=1}^{M-2} V_m V_{m+2}^* \tag{13.6}$$

Details of the single-polarization channel observable estimation may be found in [7] and [14].

13.3.6.2 Polarimetric covariance processing and observable estimation

The differential phase Φ_{DP} is the argument of the lag 0 covariance between V_H and V_V. It represents the average phase difference between the received H and V signal voltages:

$$\hat{\Phi}_{DP} = \arg(\hat{R}_{HV}(0)) = \arg\left(\sum_{m=1}^{M} V_{H_m} V_{V_m}^*\right) \tag{13.7}$$

Measured Φ_{DP} values include the phase difference between H and V voltage signals regarding transmit and receive hardware, wave propagation, and backscattering. If the propagation path contains oblate hydrometeors, then Φ_{DP} typically increases with range.

The variable Φ_{DP} is used for K_{DP} determination (13.8) and attenuation correction (section 13.6.1). It is not necessary to apply a correction for the hardware-related phase difference because only the change of Φ_{DP} along the range is of interest during product generation.

The specific differential phase K_{DP} is the range derivative of Φ_{DP}:

$$\hat{K}_{DP} = \frac{1}{2}\frac{\mathrm{d}\hat{\Phi}_{DP}(r)}{\mathrm{dr}} \approx \frac{1}{2}\frac{\hat{\Phi}_{DP}(r+\Delta r) - \hat{\Phi}_{DP}(r)}{\Delta r} \tag{13.8}$$

where Δr is a configurable range interval and the factor 1/2 corrects for the two-fold propagation path between radar and scattering volume. K_{DP} values can be used to estimate rain rates (section 13.6.2) with the advantage that phase differences do not depend on the system calibration. A meaningful determination of K_{DP} should be restricted to the availability of sufficiently qualified Φ_{DP} values and should include spatial Φ_{DP} filtering to avoid noise contamination [15]. An offline K_{DP} calculation is also possible during product generation.

The dimensionless copolar correlation coefficient ρ_{HV} is the normalized modulus of the cross correlation between the signals of H and V polarization at lag 0:

$$\hat{\rho}_{HV} = \frac{\left|\left(\hat{R}_{HV}(0)\right)\right|}{\sqrt{\left(\hat{R}_H(0) - \hat{P}_{N,H}\right)\left(\hat{R}_V(0) - \hat{P}_{N,V}\right)}} \tag{13.9}$$

$R_H(0)$and $R_V(0)$ are the power estimates in the horizontal and vertical channels. The corresponding noise estimates are $P_{N,H}$ and $P_{N,V}$. Noise correction in the denominator is necessary to avoid a bias at low SNR. The ρ_{HV} observable reflects the shape variability of scatterers, its expected value range is [0;1]. Noisy signals result in low ρ_{HV} values, whereas strong correlated signals from light rain with a high SNR have a ρ_{HV} near 1. The copolar correlation is also used for hydrometeor classification (section 13.6.3).

The differential reflectivity Z_{DR} can be derived from the logarithmic ratio between the noise-corrected power of horizontal and vertical polarization channels:

$$\hat{Z}_{DR} = 10 \log\left(\frac{\hat{R}_H(0) - \hat{P}_{N,H}}{\hat{R}_V(0) - \hat{P}_{N,V}}\right) + bias_{ZDR} \tag{13.10}$$

The constant $bias_{ZDR}$ accounts for the transmit and receive path imbalance between H and V channels.

The noise correction in (13.10) is necessary to avoid a bias at low SNRs. Noise power estimates are usually obtained from range bins without precipitation echoes, i.e., volumes above the typical precipitation layer. This can be done in dedicated calibration procedures or during operational data collection. Ivic [16] proposed a method to extract the noise power during operational data collection for each radial even in the presence of precipitation signals. This method tries to compensate the effect of spatial and temporal noise power variations.

$Z_{DR}$is related to the shape of hydrometeors and can be used for hydrometeor classification and rain rate estimation. In case of equal transmitted and backscattered power in both polarization channels, Z_{DR} should be 0 dB. This result is used to calibrate Z_{DR} in adequate precipitation (section 13.5.2.2).

The linear depolarization ratio LDR is the logarithmic power ratio between V and H signals for horizontally polarized transmission only. LDR is not accessible in hybrid mode but needs a waveguide switch that restricts transmission to horizontally polarized radiation. Its determination must include a system constant $bias_{LDR}$ that contains the imbalance of the complete receive path between V and H polarization:

$$\widehat{LDR} = 10 \log \left(\frac{\hat{R}_H(0) - \hat{P}_{N,H}}{\hat{R}_V(0) - \hat{P}_{N,V}} \right) + bias_{LDR} \tag{13.11}$$

LDR relates to the shape of hydrometeors. By definition, scattering from isotropic particles should give *LDR* values of negative infinity because the nominator in (13.11) becomes zero, but the system isolation between the two polarization channels defines a lower limit for *LDR* measurements. A more extensive background regarding the estimation of dual-polarization observables may be found in [7] and [17].

Several authors [18, 19] propose the use of multilag estimators for dual polarimetric observables. Noise samples are not correlated; therefore, estimators that only use correlations at lags>1 have the advantage that no noise correction is necessary.

13.3.7 Threshold checks

For all radar observables, signal processing performs threshold checks to avoid low-quality results. For example, if the SNR in a range bin is below a user-defined threshold, then no estimates for this range bin are created. Other threshold checks use the signal coherency of the amount of clutter rejection.

13.3.8 Beam filling effects on weather radar observables

The weather radar estimates as derived in the previous paragraphs refer to all scatterers in the measurement volume. They represent a kind of average for this volume. When a measurement volume is inhomogeneously filled with scatterers, the true values at a particular point in the volume can differ significantly from these averages. This becomes important, e.g., when a part of the measurement volume is blocked, e.g., by a building or a distant mountain. If the blocked part contains significantly different scatterers than the unblocked part (e.g., a light rain compared to a hailstorm), the above estimates represent only the unblocked part.

In most cases of partial beam blocking, however, the meteorological scatterers in the blocked part are similar to those in the unblocked part. In such conditions, the above estimates are only marginally biased (unless the blocking reduces SNR too much). Differential phase Φ_{DP} and specific differential phase K_{DP} are derived from the phase difference between H and V signals. If a measurement volume is homogeneously filled with precipitation, phase differences do not depend on the amount of blocking. Differential reflectivity Z_{DR} and linear depolarization ratio *LDR* are relative quantities and thus do not primarily depend on the amount of scatterers. The copolar correlation coefficient ρ_{HV} mainly depends on shape, orientation, and temporal change of scatterers and not on their amount.

Reflectivity, on the other hand, depends on the amount of scatterers and thus is significantly biased in case of partial beam blocking.

13.4 The meteorological radar equation

For all power estimates, which have passed the threshold criteria, the reflectivity Z can be estimated by applying the meteorological radar equation, which is the basis for radar meteorological power measurements. It uses the power of the received signal for the estimation of the reflectivity Z of the hydrometeors. Equation (13.12) represents the weather radar equation as derived in section 4.4.5 of [14]:

$$P_{RX}(r) = \frac{\pi^3 P_{TX} G^2 G_{RX} \theta_A \phi_A c\tau \, |K|^2 Z}{2^{10} \ln(2) \lambda^2 r^2 L^2 L_{MF}} \tag{13.12}$$

The parameters, which were not explained earlier, are provided in Table 13.1.

The radar equation as provided in (13.12) is based on some important assumptions. The antenna beam must have a circular symmetric shape and the observed volume must be completely filled with precipitation.

For further discussions, some terms of the radar equation are rewritten in order to get a more practical expression. By applying $c = \lambda \cdot f$, the wavelength λ is replaced by the frequency f. The equation is rearranged for the calculation of the reflectivity from the received signal power and grouped in terms, which refer directly to the different radar subsystems:

$$Z = \underbrace{\frac{2^{10}\ln(2)c}{\pi^3 |K|^2}}_{\text{constant}} \cdot \underbrace{\frac{1}{f^2 P_{TX}\tau}}_{\text{transmitter}} \cdot \underbrace{\frac{1}{G^2 \theta_A \, \phi_A}}_{\text{antenna}} \cdot \underbrace{\frac{1}{G_{RX}}}_{\text{receiver}} \cdot L_{MF} \cdot r^2 \cdot L^2 \cdot P_{RX} \tag{13.13}$$

All parameters of the radar equation are referred to the antenna coupler AC, which is indicated in Figure 13.4. Therefore, the attenuation of the waveguide that connects the antenna with the transmitter and the receiver must be considered with the transmitter power and the receiver gain.

The factor $|K|^2$ is related to the propagation medium. Usually, it is assumed that the beam propagates through liquid precipitation, for which $|K|^2$ is 0.93. During normal operation, the parameters of the radar subsystems are not changed. Therefore,

Table 13.1 Parameters of the meteorological radar equation

P_{TX}	Transmitter peak power at the antenna
G	Antenna gain
G_{RX}	Total receiver gain, considering also path losses between antenna and electronics
θ_A	−3 dB antenna elevation beam width
ϕ_A	−3 dB antenna azimuth beam width
$\|K\|^2$	Constant based on refractive index, 0.93 for precipitation and 0.2 for ice
τ	Transmitter pulsewidth
r	Range
L	One-way transmission loss
L_{MF}	Matched filter losses

(13.13) can be simplified by multiplying all constant factors into the radar constant C_{R_R}:

$$Z = C_R \cdot r^2 \cdot L_{MF} \cdot L^2 \cdot P_{RX} \tag{13.14}$$

The physical dimension of Z in SI units is m^3. However, the dimension used in radar meteorology is mm^6/m^3. Because of the huge dynamic range covered by the radar signal, logarithmic units are commonly used for displays. The log of Z is dBZ:

$$\text{dBZ} = 10\log\left(Z/\left(\text{mm}^6/\text{m}^3\right)\right) \tag{13.15}$$

Although dBZ is a unit, the expression "dBZ" is sometimes also incorrectly used as a variable.

13.5 Radar calibration and monitoring

A weather radar is a measuring instrument, but without a laboratory standard maintained by a national institute of standards. Therefore, the application of terms like "calibration" and "accuracy" has to be used with great care. Since any radar meteorological observable is estimated from a certain number of measurement samples, the estimation accuracy is based on the applied algorithm and the number of samples. On the other hand, the measurement accuracy of the samples is related to the instrument, i.e., the radar, which leads to the term "calibration accuracy." Calibration in the context of radar meteorology means either the determination (or subsequent removal) of an instrument offset or the calibration of parts of the instrument by associating a dimension to an arbitrary unit, e.g., a voltage to the binary output of an A/D converter. Thus, the calibration accuracy is associated to the calibration process, i.e., the way the offset is determined or the accuracy of the instrument that measures the voltage applied to the A/D converter.

13.5.1 Engineering calibration

The accurate measurement of Z depends on the accurate determination of the range r, the one-way transmission loss L, and of the radar constant C_R, as can be seen from (13.14). The one-way transmission loss includes the attenuation of the radome, of atmospheric gases, of clouds, and of rain or other hydrometeors. The gaseous attenuation is usually taken from international references like [20]. The radome attenuation can be measured; however, it increases significantly when the radome is wet.

"Engineering Calibration" refers to all methods that are used to measure the radar constant C_R. At first glance, it seems to be the simplest and most accurate method to measure a target with a known backscatter cross-section like a sphere. However, there are several constraints for an accurate calibration [21]. The measurement range must be free of ground clutter, which requires that a certain height is required. This can be achieved by hanging the sphere under a tethered balloon. The sphere must be in the far-field of the antenna, and the distance must allow the TR limiter to open completely. The sphere must not oscillate and the gate needs to sample the peak of the return which is problematic. Finally, the antenna must point directly to it. All these limitations make it difficult to use this method operationally.

The situation can be improved by using a simple active transponder, which receives the radar signal, amplifies, and returns it. The power should exceed the clutter power in order to reduce the measurement error. The returned signal can also be phase modulated to simulate a velocity and avoid the competition with clutter power [22]. However, such a transponder must also be calibrated and has to provide an exceptional stability.

The engineering calibration method applied to almost all operational weather radars is the individual calibration of the radar subsystems contributing to the radar constant C_R. The calibration consists of the measurement of the transmitter power, of the receiver gain, of the losses of the dry radome, and of the gain and beam width of the antenna. All these measurements are well-established procedures using microwave measurement instruments. For each procedure, an uncertainty budget can be calculated. All uncertainties can then be used to compute the total uncertainty of a power measurement. For instance, the standard deviation of the received power measurement of a C-band weather radar is 0.6 dB according to [23].

Since a stable cw signal is used for the power calibration of the radar receive chain the matched filter loss L_{MF} is not considered. The loss depends on the actual TX pulse waveform and the spectral shape of the matched filter. The amount of loss may be calculated or measured, e.g., using the transmission loss of the TX signal sample running through the matched filter.

13.5.2 Polarimetric calibration and monitoring

This section presents selected calibration and monitoring methods related to dual-polarization weather radars with a focus on Z_{DR} calibration. An overview about Z_{DR} calibration and monitoring methods is given in [24]. A Z_{DR} accuracy between 0.1 and 0.2 dB [25] is necessary for usage in dual-polarization systems.

13.5.2.1 Solar calibration and monitoring

The use of solar microwave radiation for calibration purposes was originally proposed by Whiton *et al.* [26] and Pratte and Ferraro [27].

In single-polarization weather radars, solar signals are used to calibrate the antenna positioning, to characterize the reflector beam, and to monitor the receiver chain stability. The non-polarized nature of solar radiation additionally allows the evaluation of the receive path imbalance of dual-polarization weather radars. If a weather radar antenna with equal beam characteristics for both polarizations is positioned to the center of the solar disc, then the received signal power in both polarization channels, after correction for receive path imbalances, should be equal. This does not necessarily mean that the measured Z_{DR} is 0 dB because differential reflectivity determination additionally includes the transmit path imbalance [see (13.10)]. Using solar signals only the receive path imbalance of dual-polarization weather radar systems can be calibrated.

Table 13.2 Accuracy of Z_{DR} monitoring using solar signals for C-band weather radars

Country	Accuracy (dB)	Remarks on reported accuracy	References
Finland	≤0.05	Standard deviation derived from fit on daily power measurements	[30]
Switzerland	≤0.06	Standard deviation of daily median values	[31]
Italy	<0.2	Standard deviation of daily measurements	[32]

Offline sun scans for Z_{DR} calibration and monitoring were presented by Melnikov [28], the reported variation of solar Z_{DR} measurements during a period of several months was 0.3 dB, and the associated standard deviation was 0.08 dB.

The monitoring of differential reflectivity in operational mode (detection of "solar hits") has been developed by Holleman *et al.* [29]. Table 13.2 lists the selected Z_{DR} monitoring results using this method.

13.5.2.2 Bird-bath calibration and monitoring

When viewed at 90° elevation, raindrops from light precipitation are expected to have a circular shape and the corresponding average Z_{DR} should be 0 dB. Any residual Z_{DR} average away from 0 dB can be used as a calibration offset for future Z_{DR} measurements. This method, also called "bird-bath" calibration, was proposed by Gorgucci *et al.* [33].

The selection of measurements used to create a bird-bath Z_{DR} measurement result must be restricted by censoring criteria. These criteria typically relate to data evaluation inside a radial range above the radar. Thresholding of reflectivity and ρ_{HV} is used to restrict data evaluation to echo signals from light precipitation. A sufficient data coverage along range and azimuth is an additional precondition to create a final bird-bath Z_{DR} result. Averaging Z_{DR} over 360° azimuth is recommended to minimize the effects of, e.g., residual canting, sidelobe clutter, or any other hardware-related structures that induce non-rotational symmetry in signal transmission/reception along the azimuth. Table 13.3 lists the selected Z_{DR} monitoring results using this method.

Table 13.3 Accuracy of Z_{DR} monitoring from bird-bath measurements

Country	Accuracy (dB)	Remarks on reported accuracy	References
Finland	≤0.06	Standard deviation of daily average	[30]
Germany	<0.2	Interquartile range of median measurements	[34]

13.6 Polarimetric data processing

For many decades, data processing algorithms for single-polarization radar have been used to obtain, e.g., precipitation intensity, cloud tops, and wind and turbulence data, to derive severe weather information, or to track thunderstorms. Some algorithms are used to correct radar observables for undesired artifacts, e.g., to correct for attenuation or remove non-meteorological echoes. All these algorithms for single-polarization radar can be used for polarimetric radar as well. But polarimetric radar provides more observables than single-polarization radar, and this offers new algorithms and better final data and final results. This section describes three key areas of radar data processing, namely, the correction for attenuation from precipitation, the derivation of quantitative precipitation estimates, and the classification of radar scatterers.

Like in most fields of meteorology, radar meteorology does not always use SI units. For example, rainfall rates are typically given in mm/h (which corresponds to water mass deposit rates in kg/m^2h), rather than in m/s, as the latter does not provide values suitable for the general public. This section uses the observables in the following units, in particular in equations, unless otherwise specified:

- reflectivity Z in mm^6/m^3
- differential reflectivity Z_{DR} as logarithmic quantity in dB
- differential phase Φ_{DP} in deg
- specific differential phase K_{DP} in deg/km
- linear depolarization ratio LDR as logarithmic quantity in dB

13.6.1 Attenuation correction

Radar returns from meteorological scatterers are subject to attenuation from four main sources:

- attenuation during propagation through precipitation
- attenuation by the air
- attenuation by precipitation on the antenna dish or, if covered, on the radome
- attenuation in waveguides and electronic components

This section is about the first source of attenuation. The attenuation caused by rain introduces the largest uncertainty to the reflectivity measurement. The rain attenuation is relatively small for S-band, which is the reason why S-band is preferred for regions with strong precipitation events. Only through the availability of polarimetric measurement methods, it was possible to reduce the error caused by rain attenuation in particular for C- and X-band radars to an acceptable level.

Attenuation by air is fairly constant for a given radar wavelength (dependency on air density, temperature, and humidity can usually be neglected). Many research has focused on attenuation by precipitation on the radar dish or radome (the latter being commonly used with weather radar). A review of such methods is beyond the scope of this book. Attenuation within the radar system itself is accounted for in the radar calibration and monitoring.

13.6.1.1 Attenuation correction of reflectivity

Path-integrated attenuation by precipitation (*PIA*, here expressed in dB) is a quantity accumulating with range; close echoes are less attenuated than distant ones. To properly correct reflectivity for attenuation at any range, it is necessary to calculate the specific attenuation A for each range, which is, when expressed in dB/km, defined as ([7] eqn 1.133d)

$$A = 4.34 \times 10^3 \int \sigma_E N\left(D\right) dD \tag{13.16}$$

where σ_E is the extinction (attenuation) cross section for particles of size D and $N(D)$ is the particle number density.

For polarimetric radars, attenuation can be different for horizontal and vertical polarization, respectively, due to different σ_E; i.e., $\sigma_{E,H} \neq \sigma_{E,H}$.

The amount of path-integrated attenuation for a particular range r_2, i.e., the amount of reflectivity correction, is then the integral

$$PIA\left(0, r_2\right) = \int_0^{r2} A\left(r\right) dr \tag{13.17}$$

Attenuation correction usually has to be performed independently for each radial, because the distribution of attenuating precipitation in neighbor radials may be significantly different. Most attenuation algorithms are designed for rain, where σ_E is well known, being a function of drop size D and radar wavelength. For single-polarization radar, a common method is the Hitschfeld–Bordan algorithm [35], which uses a power–law relationship between A (in dB/km) and the true radar reflectivity Z

$$A = aZ^b \tag{13.18}$$

Here, values for a and b have been suggested based on empirical observations, such as drop size distributions. The main disadvantage of the formula is its computational instability: For increasing range, reflectivity has to be iteratively corrected to apply a corrected Z in the formula for calculating A for the next range step. Over- or underestimations in initial steps of the correction (e.g., from mis-calibrated radar or other reflectivity bias) will magnify during processing with range. It is thus typically necessary to limit the total path-integrated attenuation or to set a constraint for that from independent measurement.

Polarimetric radar offers several possibilities to overcome the disadvantages of the reflectivity-based attenuation correction methods. Theoretical calculations based on relations between drop size and drop shape, as well as observations using radar, have shown that an almost linear relation exists between the attenuation (for both horizontal and vertical polarization) and the differential phase shift from propagation through precipitation. This means $PIA(r_1, r_2) = \alpha\,(\Phi_{DP}(r_2) - \Phi_{DP}(r_1)) = \alpha\,\Delta\Phi_{DP}$, when using the measured phase Φ_{DP} as defined in (13.7). The proportionality factor α depends on the radar wavelength. The better the linearity, the larger the considered range intervals $r_2 - r_1$. Correction algorithms, resulting from these findings, either take a configurable, but fixed α (i.e., depending only on the radar wavelength) or use empirical α (depending also on measured reflectivity and Φ_{DP}).

Another approach is to use the total *PIA*, derived from Φ_{DP}, as a constraint for reflectivity-based corrections. The ZPHI algorithm, for example, combines Φ_{DP} -derived *PIA* with the Hitschfeld–Bordan algorithm in a way that the factor a in (13.18) is automatically derived from the Φ_{DP} constraint [36]. The *PIA* constraint of ZPHI can be used either for an entire radial in one step or can be applied separately for each precipitation system covering a radial (the latter in particular in convective cases).

While the factor a in (13.18) is derived automatically in ZPHI, the exponent b has to be selected properly. Also the factor α in the Φ_{DP}–*PIA* relation is a configurable parameter. The latter can be derived automatically in an iterative ZPHI algorithm, where the corrected reflectivity profile along the range is used to recalculate the Φ_{DP} profile. α is subsequently modified until the reconstructed and the measured Φ_{DP} profiles match best.

Table 13.4 lists four attenuation correction algorithms that follow the aforementioned descriptions. Figure 13.7 gives an example of X-band data from a storm in Frankfurt, Germany, at 2013-08-06T13:50:00Z. The reflectivity-based Hitschfeld–Bordan correction tends to overcorrect. The polarimetric correction, using ZPHI, performs much better. Toward southwest, the signal was completely lost at distances larger than about 30 km due to heavy attenuation. The polarimetric correction was around 15 dB in that area.

Table 13.4 Methods to correct reflectivity for attenuation

Method	Main advantages	Main disadvantages	References
Linear Φ_{DP} with fixed α	Simplicity; easy to implement in real time	Unconstrained; can over- or underestimate attenuation	[37]
Empirical linear Φ_{DP} with average α Estimated from measured Z_H and Φ_{DP}	Moderately simple Tends to correct for mean attenuation	Not straightforward for real-time implementation Inaccuracies due to non-homogeneous precipitation	[38, 39]
Standard ZPHI	Constrained Does not depend on Z calibration	α is fixed *a priori*	[36]
Iterative ZPHI	Constrained α is estimated optimally Does not depend on Z calibration	More computation intensive	[40]

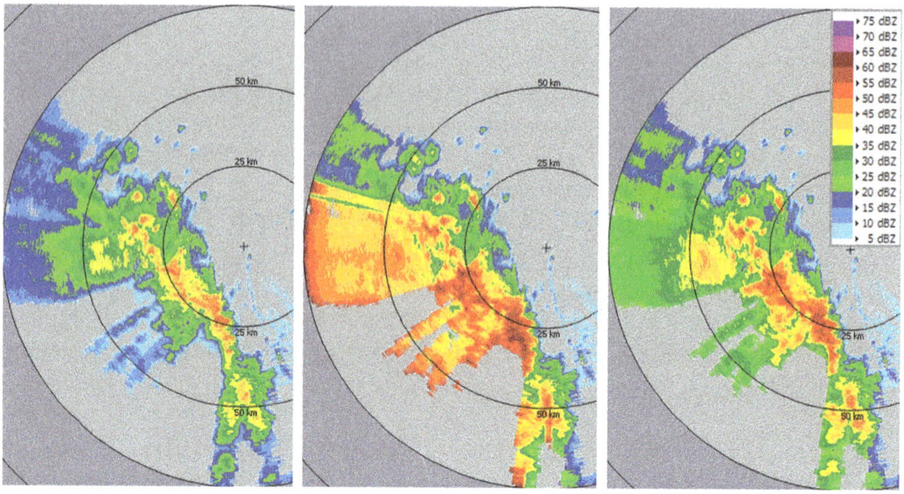

Figure 13.7 Reflectivity PPI of (left) original data, (center) Hitschfeld–Bordan corrected data, and (right) corrected data using polarimetric data

13.6.1.2 Attenuation correction of differential reflectivity

As discussed above, specific attenuation and PIA can be different for horizontal and vertical polarization, respectively, due to different σ_E. For rain, the horizontal attenuation is usually larger than the vertical one, due to the oblate shape of drops. As a result, differential reflectivity Z_{DR} is usually negatively biased. Theoretical considerations and observations have shown that often a linear relationship exists between horizontal and vertical attenuation, e.g., in [41]. Once specific attenuation A_H and PIA_H for horizontal reflectivity have been derived, differential specific attenuation A_D and PIA_D are calculated by multiplication of A_H and PIA_H with a factor β, which is usually depending on radar wavelength only. As an alternative, β can be derived automatically once the total differential PIA_D can be obtained independently. This can, e.g., be done in case of weak to moderate rain at the most distant ranges, when true Z_{DR} can be approximated from reflectivity measurements, as suggested in [42]. The difference between this true Z_{DR} estimation and the measured Z_{DR} is then PIA_D.

Figure 13.8 gives an example of X-band data (same case as in Figure 13.7). The polarimetric correction of Z_{DR} performs very well. Toward southwest, the amount of correction was around 3 dB for ranges beyond 25 km.

13.6.2 Rain rate estimation

Rainfall estimates based on reflectivity data are commonly calculated using a power-law Z–R relationship $Z = aR^b$, R representing the rainfall intensity in mm/h. In the past decades, hundreds of parameter sets a and b have been suggested, taking into account theoretical considerations, e.g., precipitation microphysics, and observations such as drop size distribution measurements. Whatever conversion formula

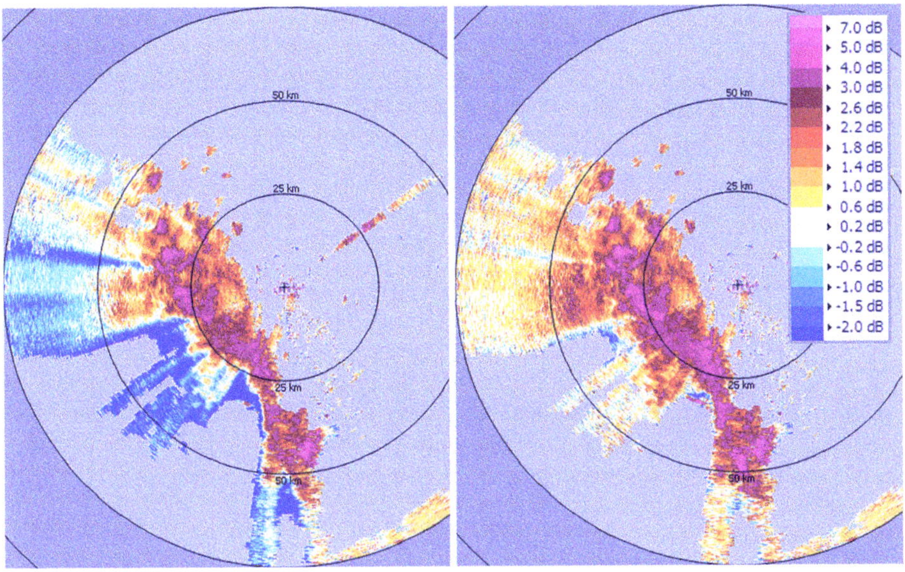

Figure 13.8 Differential reflectivity PPI of (left) original data, and (right) corrected data using polarimetric data

was used to derive rain rate from reflectivity, errors remain which are often of the same magnitude as the rain rate.

With the advent of polarimetric radar, the relation between rain rate and polarimetric observables was investigated. Physical considerations taking into account various drop size–shape relations (in particular, how oblate is a falling rain drop depending on its diameter) and again observations showed that relationships between rain rate and specific differential phase K_{DP} lead to smaller rainfall errors than Z–R relationship. The use of K_{DP} for rainfall estimation has some additional advantages: It is

- independent of radar calibration
- not affected by attenuation
- almost independent of partial beam blocking
- almost unbiased by ground clutter contamination and hail

On the other hand, K_{DP}, being not a direct observable but the radial derivative of Φ_{DP}, requires Φ_{DP} filtering. The higher K_{DP}, the smaller is the relative rain rate error from $K_{DP} - R$ relationships. However, even for small rainfall rates with low K_{DP}, differential phase measurements can be used for reliable rainfall estimates in the areal Φ_{DP} methods. Integrating rain rates with range to obtain rainfall estimates for areas such as river catchments avoids the necessity of deriving K_{DP}; the measured Φ_{DP} is used instead (e.g., [43, 44]). Table 13.5 lists some $K_{DP} - R$ relationships of field campaigns, where rain gauge data were used for validation.

Table 13.5 Parameters of $K_{DP} - R$ relationships

Wavelength	a K_{DP}^{b}	Location	References
X-band	$19.6\ K_{DP}^{0.82}$	Tokyo	[45]
X-band	$12.3\ K_{DP}^{0.81}$	Wallops Island, Virginia, USA	[46]
X-band	$18.9\ K_{DP}^{0.85}$	Darwin, Australia	[47]
C-band	$32.4\ K_{DP}^{0.83}$	Darwin, Australia	[43]
C-band	$34.6\ K_{DP}^{0.83}$	100 km North of Darwin, Australia	[48]

As an alternative to $K_{DP} - RR$ relations, K_{DP} and Z_{DR} measurements can be used to tune reflectivity–rainfall relationships. Again, several methods, including power-law formulas with three or four variables, have been proposed. The following set of equations, for example, is the NEXRAD algorithm of the U.S. NWS radars [49, 50]:

$$\begin{aligned}
&R = \frac{R(Z)}{0.4 + 0.5|Z_{dr} - 1|^{1.3}} && \text{for } R(Z) < 6 \\
&R = \frac{R(K_{DP})}{0.4 + 3.5|Z_{dr} - 1|^{1.7}} && \text{for } 6 < R(Z) < 50 \\
&R = R(K_{DP}) && \text{for } R(Z) > 50 \\
&R(Z) = 0.017Z^{0.717} \\
&R\left(K_{DP}\right) = 44\left|K_{DP}\right|^{0.822}
\end{aligned} \tag{13.19}$$

In this equation, differential reflectivity Z_{dr} is not used as logarithmic quantity in dB but as linear quantity without dimension.

Among such methods are those that derive drop size distributions, e.g., fitting a gamma DSD model to observed polarimetric observables. Some of these algorithms are described, e.g., in [51–54].

Table 13.6 gives the pros and cons of various polarimetric rain rate algorithms. Figure 13.9 shows an example of radar and rain gauge comparison (where the rain gauge locations are at the gauge symbol bottoms). The reflectivity-based rain rate data underestimate the gauge rainfall. Polarimetric rainfall data match better to gauge data. Figure 13.10 gives an overview of algorithm performances, where radar and rain gauge data are compared for all gauge locations and an areal average.

13.6.3 Echo and hydrometeor classification

In order to use polarimetric weather radar data, it is usually necessary to remove non-meteorological echoes as much as possible. In most cases, ground clutter filter or cancellation algorithms are applied during signal processing. Most signal processors also have the capability to filter interference from other transmitting devices. However, several returns of non-meteorological origin pass the signal processor. Most of such echoes are residual ground clutter, e.g., from anomalous propagation, and returns from moving scatterers such as birds, insects, ships, or sea surface

Table 13.6 Pros and cons of various rain rate algorithms at S-band

Algorithm class	Main advantages	Main disadvantages
$R(K_{DP})$	• Unaffected by absolute calibration • Unaffected by hail or bright-band • Not sensitive to partial beam blockage	• Considerable smoothing is required at expense of range resolution • Sensitive to mean drop shape relation
$R(Z,Z_{DR})$	• Robust and accurate with respect to DSD variations • No smoothing technique required	• Need for high accuracy for Z (±1 dB) and Z_{DR} (±0.2 dB) • Need to identify hail or other ice particles • Affected by partial beam blockage
$R(K_{DP},Z_{DR})$	• Unaffected by reflectivity calibration • Less sensitive to drop shape relation and DSD variations	• Need for high accuracy for Z_{DR} (±0.2 dB) • Poorer resolution due to the need for comparable smoothing intervals for K_{DP} and Z_{DR} • Need to identify hail or other ice particles
$R(Z, K_{DP}, Z_{DR})$	• High accuracy • Robust with respect to absolute calibration, DSD variations, drop shapes, and hail contamination	• Need for high accuracy for Z_{DR} (±0.2 dB) • Poorer resolution due to the need for comparable smoothing intervals for K_{DP} and Z, Z_{DR}

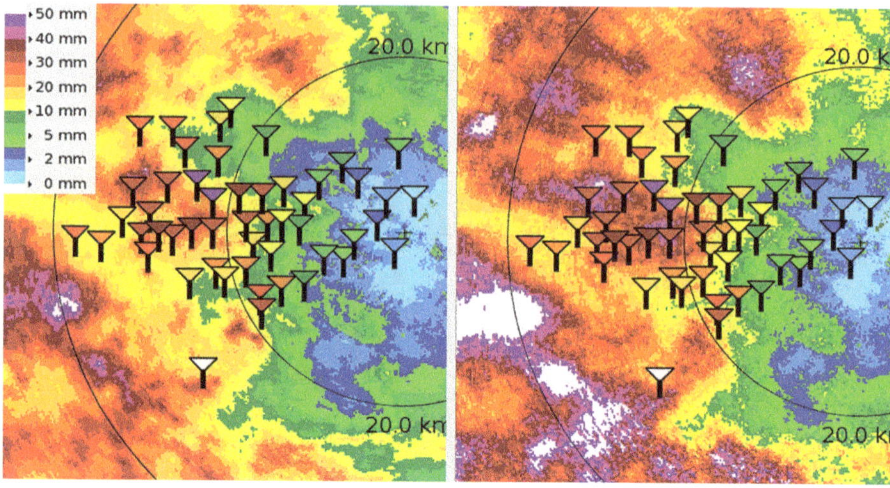

Figure 13.9 7-h rainfall accumulation from rain gauge data and S-band radar, using (left) reflectivity data only and (right) polarimetric data and the combined algorithm of (13.19)

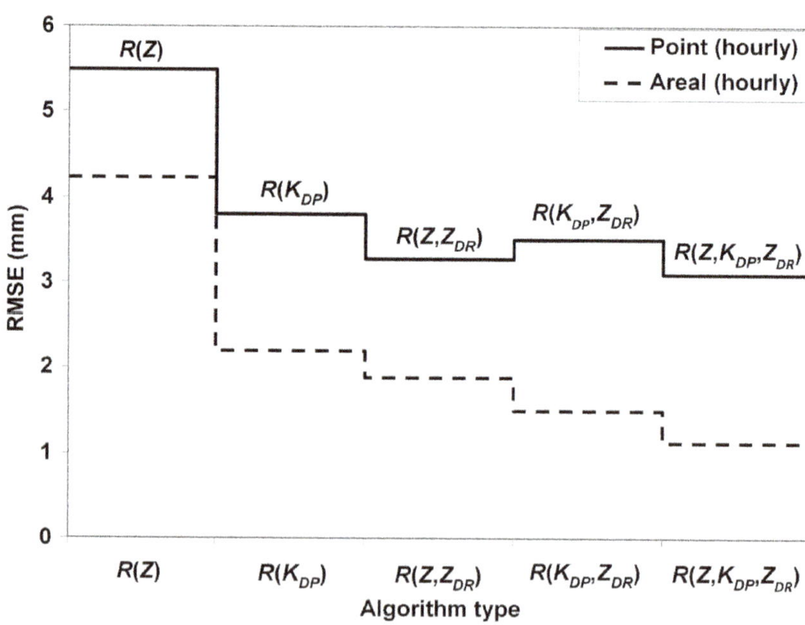

Figure 13.10 RMS errors of point and areal hourly rainfall accumulations for various algorithms (adapted from [50]) © American Meteorological Society. Used with permission.

waves. In recent years, wind farm clutter (resulting from the moving WT blades) has become a severe issue for many weather radar operators.

Several algorithms have been developed to automatically identify and, typically, remove such echoes by pattern analysis (where pattern refers to the radar observables and their variation with space and, occasionally, time). Various such algorithms were designed not only to separate between meteorological and non-meteorological echoes but also to further distinguish them. An important additional information for meteorologists, for example, is the identification of the precipitation type, such as rain, snow, and hail. But also non-meteorological returns are often of interest: weather radar has been used, e.g., to detect volcanic eruptions (e.g., [55–57]) or migrating birds (e.g., [58] and [59]).

The methods of echo classifications range from simple threshold decision to sophisticated algorithms including fuzzy logic, Bayesian approaches, neural networks, and other machine learning and artificial intelligence methods. Classification schemes may be based on a single observable. For example, in the arctic cold season, a reflectivity value of more than, say, 50 dBZ, is likely non-meteorological and can be removed or filtered. But the more radar observables are available, the more classification possibilities exist and the higher their quality.

13.6.3.1 Threshold and fuzzy-logic approaches

The simplest classification method is by thresholds, i.e., the possible range of observations is sub-divided into intervals belonging to particular echo classes:

$$C = C_i \ \forall \, x \in \{x_{i,S}; x_{i,P}\} \tag{13.20}$$

using the following notation:

C, C_i	Radar echo class
x	Radar observation (consisting of one or more observables)
$x_{i,S}$	Smallest (or start) value of observation x for class C_i
$x_{i,P}$	Largest (or stop) value of observation x for class C_i

In case of more than one radar observable, or if derived data such as the texture of a measured observable are used, the observation x has n dimensions, $x \in \mathbb{R}^n$. For threshold approaches, the intervals $\{x_{i,S}; x_{i,P}\}$ do not overlap, i.e., a unique class C_i can always be identified for an observed x. An example is the Höller scheme [60] developed for the DLR C-band radar, using Z_{DR}, LDR, and the height difference to the melting layer height as observations (i.e., $x \in \mathbb{R}^3$) to classify between nine hydrometeor types. The threshold values are given in Table 13.7.

A similar approach is fuzzy-logic methods. As for the threshold approach, classes C_i belong to observation intervals $\{x_{i,S}; x_{i,P}\}$ of the observation x. But other than for the threshold approach, these intervals do not have sharp but fuzzy limits. This is realized by assigning a weight (or probability) P to any class C_i for any x,

Table 13.7 Limits of Z_{DR} and LDR used for hydrometeor identification using the Höller scheme. H_{melt} refers to the melting level height.

Hydrometeor type	Z_{DR} threshold	LDR threshold	Height threshold
Small raindrops	$Z_{DR} < +1$ dB	$LDR < -35$ dB	$H \leq H_{melt}$
Large raindrops	$Z_{DR} \geq +1$ dB	$LDR < -25$ dB	
Graupel (dry, small), snow	$Z_{DR} \leq +1$ dB	$LDR < -35$ dB	$H > H_{melt}$
Graupel (wet, melting, small), graupel (dry, large), hail (dry, small)	$-1 \leq Z_{DR} < +1$ dB	$-35 < LDR \leq -25$ dB	
Hail (dry)	$-1 \leq Z_{DR} < +1$ dB	$-25 < LDR \leq -20$ dB	
Hail (wet)	$-1 \leq Z_{DR} \leq +1$ dB	$-20 < LDR \leq -15$ dB	
Large wet hail (spongy or water shell)	$Z_{DR} \leq +1$ dB	$LDR > -15$	
Rain + small hail	$Z_{DR} \geq +1$ dB	$-25 < LDR \leq -20$ dB	
Rain + large hail	$Z_{DR} \geq +1$ dB	$LDR > -20$ dB	

which has high values inside the interval $\{x_{i,S}; x_{i,P}\}$ and low values outside. The relations defining P for any class C_i for any x are often called membership functions. As another difference to the threshold approach, the intervals of fuzzy-logic algorithms are allowed to overlap: For example, rain and snow can both occur for reflectivity dBZ around 30, so both C_i are high there.

Fuzzy-logic methods require more than one observable, i.e., $x \in \mathbb{R}^n$ with $n > 1$. Another difference to the threshold method is that, in spite of $x \in \mathbb{R}^n$ with $n > 1$, the relations between subsets j of the observation and the assigned weights P_{ij} for each class C_i may be one-dimensional functions. Figure 13.11 gives an example: The two classes C_i are "rain" and "hail," and the observation x is three-dimensional, namely, consists of reflectivity, differential reflectivity, and LDR. Instead of using one relation each in the $\mathbb{R}^3$ space for both classes, it uses three relations each in the $\mathbb{R}$ space for both classes.

Some fuzzy-logic algorithms combine one-dimensional and multi-dimensional subsets of the observation x. For the example of Figure 13.11, one could use two sets of relations instead: one for a combination of reflectivity and Z_{DR} (i.e., combining the upper two diagrams in the figure) and a second one for LDR (i.e., the lowest diagram in the figure). Multi-dimensional sub-sets, i.e., combination of observables into one relation, are an appropriate choice when these observables have systematical dependencies for the majority of the classes C_i. This is, for example, the case for reflectivity and Z_{DR} for the class C_i = "rain": In case of weak rain, and in particular in case of drizzle, reflectivity is low and Z_{DR} is around or just slightly above 0 dB. In case of strong rain, larger drops dominate both observables; reflectivity is high and Z_{DR} is significantly above zero. So for rain, reflectivity can range from low to high values, and differential reflectivity from around zero to several dB, but a case with high reflectivity and zero Z_{DR} at the same point is likely not from rain (but, e.g. from hail).

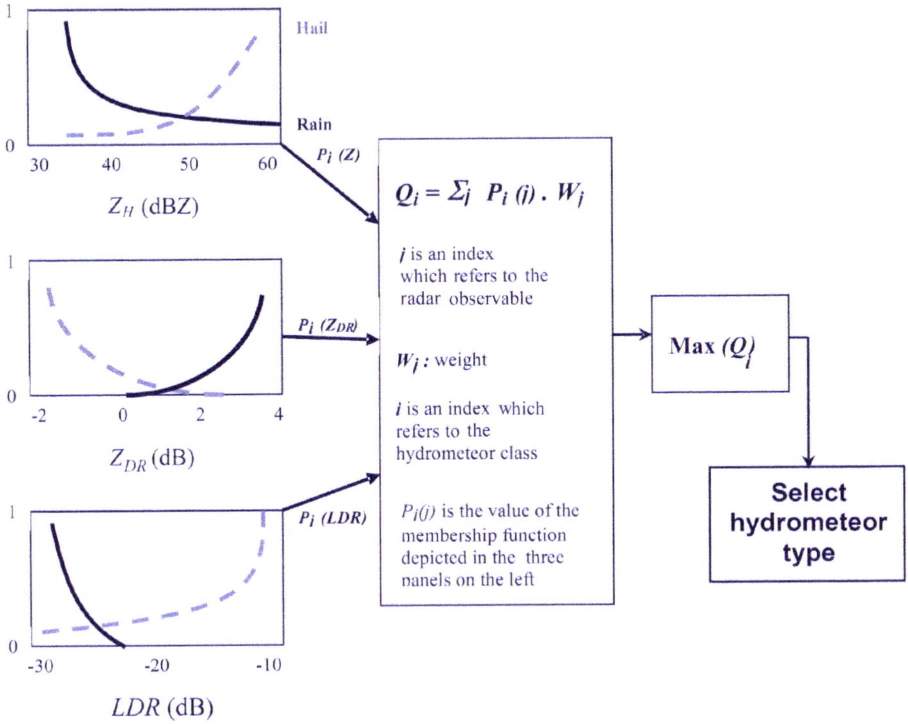

Figure 13.11 Conceptual block diagram of a fuzzy-logic classifier for rain and hail (adapted from [61]) © American Meteorological Society. Used with permission.

In an aggregation step, the observation x is finally classified to belong to a particular class C_P. Therefore, the fuzzy-logic weights of all observations (or subsets) are combined for each class C_i, and the class with the highest score Q_i is chosen. Combination may be sum or average with or without additional weights. In several fuzzy-logic algorithms, this combination is a sophisticated logic that may apply additional thresholds. For example, any frozen precipitation type can be excluded in areas with high temperatures.

The above algorithms are typically applied independently to reach radar range gate. Noisy measurements can then result in noisy classification data; in particular, when two classes C_i reach a similar score in certain areas. This can be avoided by spatial smoothing (e.g., using a median filter or a running-window average) of either the radar observations x or the weights P_{ij}. Finally, a post-processing can be applied after the classification of the radar observations at all range gates. For example, "rain above snow" is not a reasonable result, so vertical consistency checks can improve the results [62].

Figure 13.12 shows an example of fuzzy-logic echo classification. The upper four images give PPI of radar observables at 7° elevation angle from

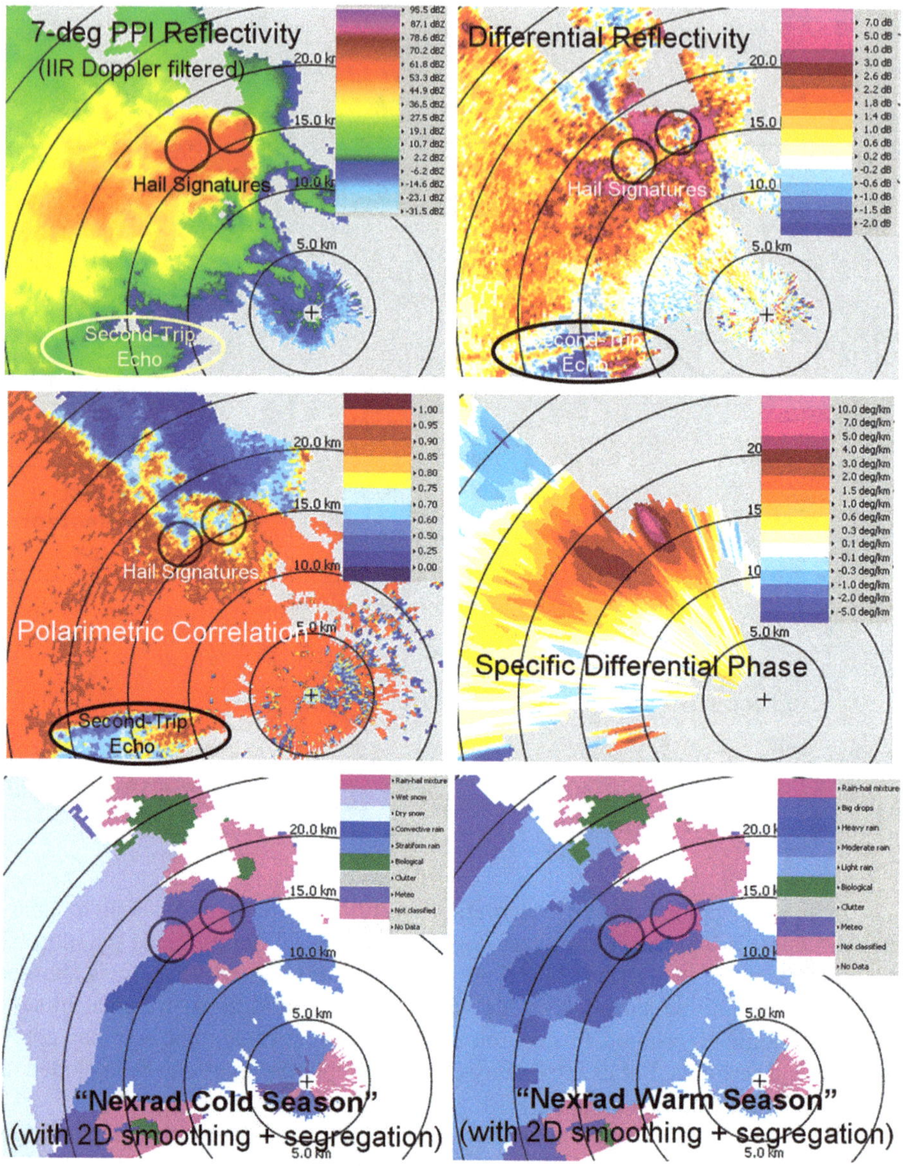

Figure 13.12 Example of polarimetric radar observations and classification results from two different fuzzy-logic algorithms (adapted from [62])

a hail storm in Neuss, Germany, at 2008-05-30T03:45:00Z. The lower two images contain the results from two different versions of the NEXRAD fuzzy-logic hydrometeor classification [63] with application of spatial smoothing and consistency checks in a post-processing [62].

13.6.3.2 Other approaches

An alternative approach to fuzzy-logic algorithm is the Bayesian approach. Similar to fuzzy logic, it allows a radar observation x to belong to several echo classes C_i (with $x \in \mathbb{R}^n$ for a radar observation consisting of n different observables). In the Bayesian approach, that class C_i with the highest probability for a given radar observation x is chosen, using Bayes' theorem, which reads for the echo classification as follows:

$$P\left(C_i|x\right) = \frac{P\left(x|C_i\right) P\left(C_i\right)}{P\left(x\right)} \tag{13.21}$$

In this equation, the unknown conditional probability $P\left(C_i|x\right)$ of class C_i, given the radar observation x, is expressed by a term using the following probabilities:

- The conditional probability $P\left(x|C_i\right)$ of observing x given the class C_i. This can be calculated from theoretical considerations (e.g., starting from the scattering matrix for scatterers belonging to C_i) or derived from observations. In principle, such conditional probability relations have shapes similar to the membership functions of fuzzy-logic algorithms, as, e.g., shown in the probability diagrams on the left part of the conceptual block diagram in Figure 13.11.
- The *a priori* probability $P\left(C_i\right)$ of occurrence of a scatterer belonging to class C_i. This can be obtained, e.g., from climate statistics. However, this must be chosen carefully: For example, the overall *a priori* probability of snow may be quite low in low altitudes in, say, central Europe. But in case of a cold outbreak in the winter season, it may approach 100 percent of all radar scatterers in the same geographical area.
- The *a priori* probability $P\left(x\right)$ of occurrence of an observation x. This can be obtained from radar data sets covering a sufficient amount of time.

Echo classifications using Bayesian approaches are widely applied, for example, using single-polarization radar data to remove echoes from non-meteorological scatterers as in [64], identifying hydrometeor types as in [65] and [66], or quantifying volcanic ash detections as in [67].

Finally, neural networks have also been used for radar echo classification, for example, using convolutional neural networks as in [68], or using a neural network to obtain the best membership functions in fuzzy-logic algorithms, as in [69].

13.7 Wind turbine clutter mitigation

13.7.1 Introduction

The United Nations Framework Convention on Climate Change set as a target that global warming shall not exceed 2°C compared to the preindustrial age. This can only be achieved by a reduction of greenhouse gas emissions caused also by power plants using fossil fuels. One possible alternative is the generation of electrical power by WTs, which must be installed in large numbers in order to achieve the required power level. The increasing density and size of WT parks on- and offshore are causing interference for radar systems which are difficult to mitigate.

13.7.2 Wind turbine interference for weather radar

A WT is not a simple clutter scatterer like a building. The blades are rotating, and the axis of rotation depends on the prevailing direction of the wind. The rotational speed and the pitch of the blades are depending on the wind speed. For these reasons, the WT radar signature with respect to reflectivity, velocity, spectral width, and polarimetry changes permanently.

WTs are often installed in farms, which increases the variability of the signature even further. Moreover, the radar signal is reflected between the WTs of a park several times causing multipath reflections with a similar variability. An example is shown in Figure 13.13. Please note the multipath echoes in the radar shadow of the wind farm.

The mitigation of these interferences is a two-step process consisting of the identification and the actual mitigation. This affects only range cells containing WTs. Range cells contaminated by multipath echoes have to be processed by methods as described in [72, 73].

Not much work has been published yet regarding the interaction of WTs with polarimetric radars. In [74], it is shown that a WT has a significant impact on Z_{DR}, which might be used in the future as an additional parameter for identification and mitigation procedures. Most of the weather radar WT clutter mitigation methods are based on non-polarimetric signal processing techniques. Significant research and development efforts are required before these methods can be used operationally.

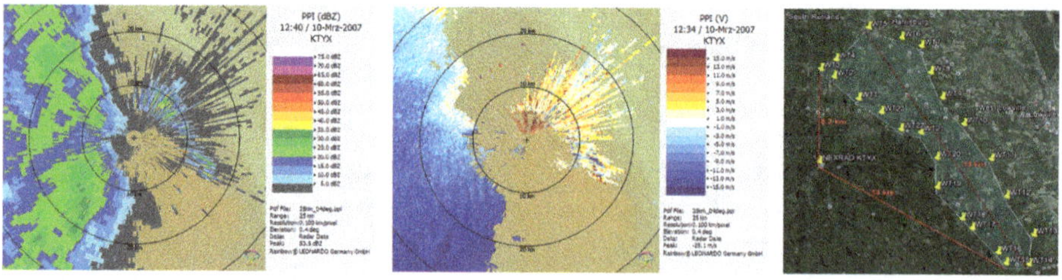

Figure 13.13 Reflectivity and velocity PPIs of NEXRAD KTYX, showing interference from a WT park located NW of the radar. Data sources: [70, 71].

13.7.3 Identification of WTs

An important source for the identification of WTs in the vicinity of a weather radar is the careful optical inspection of the radar horizon and publicly available databases providing the coordinates of WTs. Identification using radar signal processing is only required to detect new installations and the operational status of a WT.

Most identification methods are based on the time/Doppler frequency and range/Doppler frequency signatures of a WT. Examples are shown in Figure 13.14 and Figure 13.15.

The signatures depend on the rotation rate and the rotational axis of the WT, as well as on its size and distance since the radar horizon may block the lower part

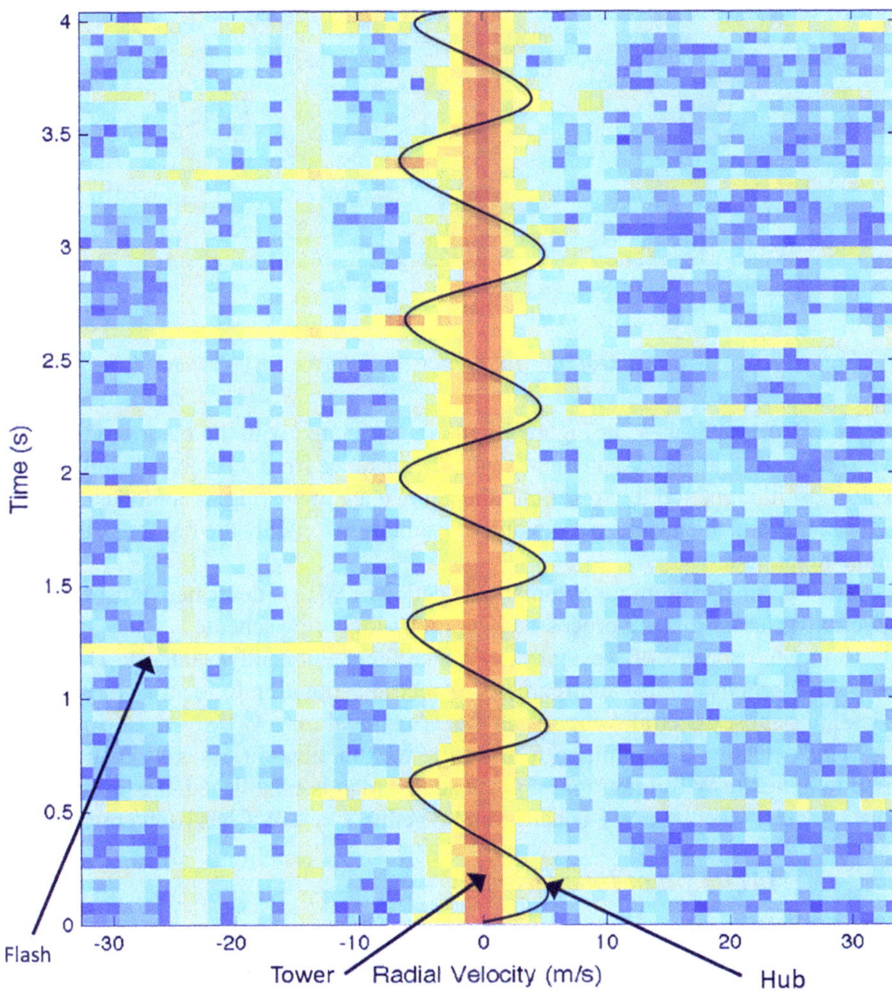

Figure 13.14 Time/Doppler frequency signature of a WT, taken from [75]

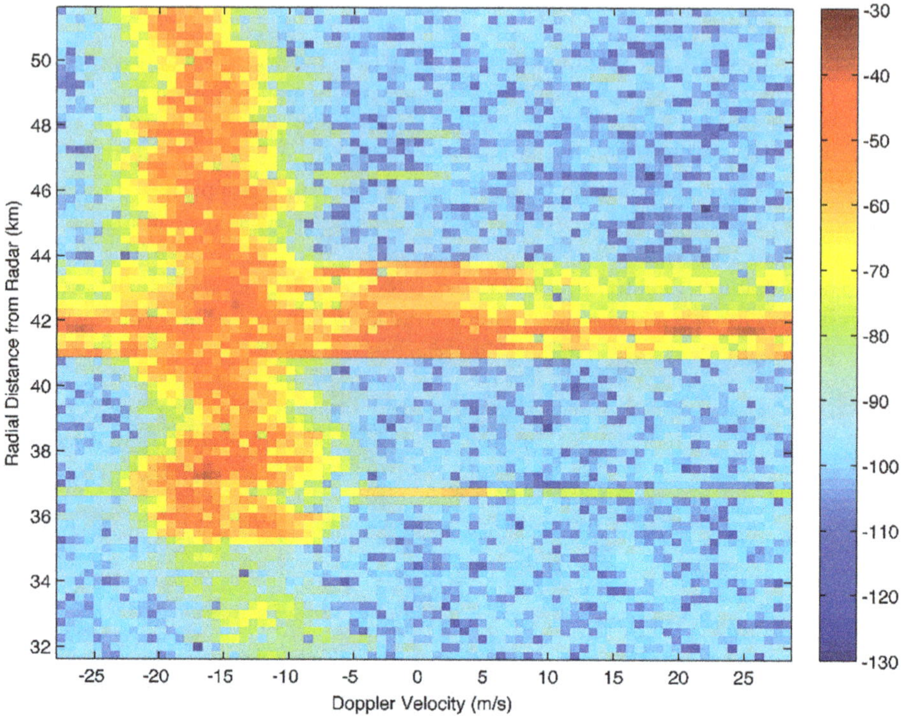

Figure 13.15 Range/Doppler frequency signature of a WT, taken from [75]

of the WT. The "flashes" in Figure 13.14 are associated with the blades of the WT, whose radar radial velocity increases with the distance from the hub. The hub causes a slow oscillation, and the tower and the nacelle are appearing as stationary clutter. If the tower and the nacelle are beyond the radar horizon only the flashes will appear in the radar data.

In [76], an algorithm for automatic detection of WTs is described. It uses several spectral variables that are processed by a fuzzy-logic rule scheme.

Since the spectra of weather and WT differ significantly, Gallardo-Hernando *et al.* [77] propose to use autoregressive spectral estimation to identify range gates containing those spectral components that are typical for a WT.

This detection algorithm performs quite well in clear-air cases and can thus be used to generate WT clutter maps. Once such a map is established, it is only necessary to detect whether a WT is active and whether a new WT has been erected.

13.7.4 Mitigation of interferences

13.7.4.1 Distance

Based on an investigation reported in [78], organizations representing the interests of weather services like EUMETNET and the WMO issued guidelines for the co-siting of weather radars and WTs [79]. These guidelines define protection areas where the

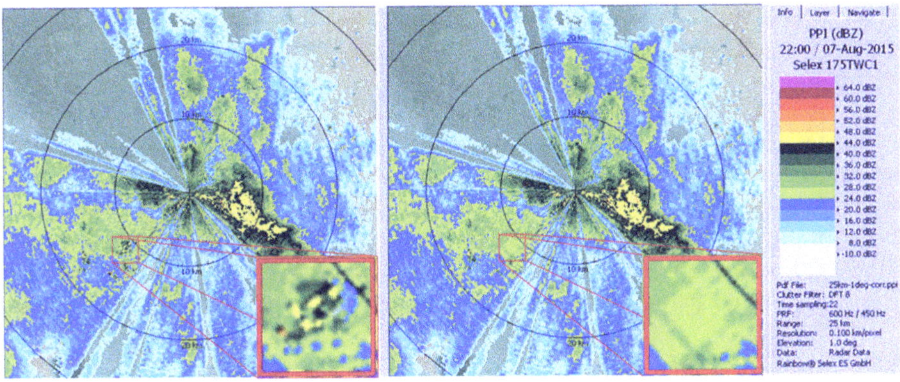

Figure 13.16 Mitigation of WT interferences using interpolation

erection of WTs should be prohibited and consultation areas requiring a detailed investigation of the potential impact of the planned WT on the weather radar.

13.7.4.2 Interpolation

If the contaminated range cell is detected, the interference can be mitigated by discarding all signals from that range cell and interpolating the data from uncontaminated cells. An example of a simple 2D interpolation is shown in Figure 13.16 [79].

A more advanced 3D interpolation technique is described in [80]. The methods described in [81] and [82] are interpolating disturbed range gates in the range-Doppler matrix shown in Figure 13.15.

Common to all interpolation techniques is the assumption of spatial continuity of the weather signals and the necessity of high-quality meteorological data close to the discarded cells that can be used as a support point for the interpolation. The larger the wind farm the more artifacts appear since the assumption of continuity is no longer completely true. Therefore, interpolation methods are only suited for small wind farms. It should be noted, however, that compressed sensing is promising for the recovery of data even for large discarded cells [83].

References

[1] *World meteorological organization (WMO) radar database [online].* Available from https://wrd.mgm.gov.tr/Home/Wrd

[2] *Deutscher wetterdienst data assimilation [online].* Available from https://www.dwd.de/EN/research/weatherforecasting/num_modelling/02_data_assimilation/data_assimilation_node.html

[3] MacPherson B., Lindskog M., Ducrocq V. 'Assimilation of radar data in numerical weather prediction (NWP) models' in Meischner P. (ed.). *Weather radar principles and advanced applications*. Heidelberg: Springer; 2004.

[4] Chan P.W., Zhang P. 'Aviation applications of doppler radars in the alerting of windshear and turbulence' in Bech J. (ed.). *Doppler radar observations - weather radar, wind profiler, ionospheric radar, and other advanced applications*. London: IntechOpen; 2012.

[5] Mahapatra P.R. *Aviation weather surveillance systems*. London: The IET; 1999.

[6] Berenbaum M. 'Frequent flyer miles'. *American Entomologist*. 2010, vol. 56(1), pp. 4–5.

[7] Bringi V.N., Chandrasekar V. *Polarimetric doppler weather radar*. Cambridge: Cambridge University Press; 2001 Aug 30. Available from https://www.cambridge.org/core/product/identifier/9780511541094/type/book

[8] Hubbert J.C., Dixon M., Ellis S.M., Meymaris G. 'Weather radar ground clutter. part II: real-time identification and filtering'. *Journal of Atmospheric and Oceanic Technology*. 2009, vol. 26(7), pp. 1181–97.

[9] Torres S.M., Warde D.A. 'Ground clutter mitigation for weather radars using the autocorrelation spectral density'. *Journal of Atmospheric and Oceanic Technology*. 2014, vol. 31(10), pp. 2049–66.

[10] Cao Q., Zhang G., Palmer R.D., Knight M., May R., Stafford R.J. 'Spectrum-time estimation and processing (step) for improving weather radar data quality'. *IEEE Trans Geosci Remote Sens*. 2012, vol. 50(11), pp. 4670–83.

[11] Siggia A., Passarelli R. 'Gaussian model adaptive processing (GMAP) for improved ground clutter cancellation and moment calculation'. *Proc. Third European Conf. on Radar in Meteorology (ERAD)*; Visby, Sweden, 2004. pp. 67–73.

[12] Gekat F., Meischner P., Friedrich K. 'The state of weather radar operations, networks and products' in Meischner P. (ed.). *Weather radar principles and advanced applications*. Heidelberg: Springer; 2004.

[13] Keeler R.J., Passarelli R.E. 'Signal processing for atmospheric radars processing for atmospheric radars' in Atlas D. (ed.). *Radar in meteorology*. Boston, MA: AMS; 1990.

[14] Doviak R.J., Zrnic D.S. *Doppler radar and weather observations*. 2nd Ed. San Diego, CA: Academic Press; 1993.

[15] Hubbert J., Bringi V.N. 'An iterative filtering technique for the analysis of copolar differential phase and dual-frequency radar measurements'. *Journal of Atmospheric and Oceanic Technology*. 1995, vol. 12(3), pp. 643–48.

[16] Ivić I.R., Curtis C., Torres S.M. 'Radial-based noise power estimation for weather radars'. *Journal of Atmospheric and Oceanic Technology*. 2013, vol. 30(12), pp. 2737–53.

[17] Zhang G. *Weather radar polarimetry*. Boca Raton, FL: CRC Press; 2016. Available from https://www.taylorfrancis.com/books/9781439870600

[18] Melnikov V.M., Zrnić D.S. 'Autocorrelation and cross-correlation estimators of polarimetric variables'. *Journal of Atmospheric and Oceanic Technology*. 2007, vol. 24(8), pp. 1337–50.

[19] Lei L., Zhang G., Doviak R.J., *et al.* 'Multilag correlation estimators for polarimetric radar measurements in the presence of noise'. *Journal of Atmospheric and Oceanic Technology*. 2012, vol. 29(6), pp. 772–95.

[20] Rec. ITU-R P.676-12 (08/2019) Attenuation by atmospheric gases and related effects

[21] Koistinen J.R., King R., Saltikoff E., Harju A. 'Monitoring and assessment of systematic measurement errors in the NORDRAD network'. *29th International Conference on Radar Meteorology*; Montreal, Quebec, 1999. pp. 765–68.

[22] Drury W.H., Frankovich J.M. *The terminal doppler weather radar (TDWR) moving target simulator (MTS) at Orlando, Florida*. Lexington, MA: MIT Lincoln Laboratory Project Report ATC-188; 1992.

[23] Gekat F., Hille M., Niese H., Pool M. 'Accuracy of the engineering calibration of weather radars'. *IEEE International Geoscience and Remote Sensing Symposium (IGARSS 2010)*; Honolulu, Hawaii, 2010.

[24] Ryzhkov A.V., Zrnic D.S. 'Radar polarimetry for weather observations' in *Radar polarimetry for weather observations*. Cham: Springer; 2019.

[25] Ryzhkov A.V., Giangrande S.E., Melnikov V.M., Schuur T.J. 'Calibration issues of dual-polarization radar measurements'. *Journal of Atmospheric and Oceanic Technology*. 2005, vol. 22(8), pp. 1138–55.

[26] Whiton R.C., Smith P.L., Harbuck A.C. 'Calibration of weather radar systems using the sun as a radio source'. *17th Conference on Radar Meteorology*; Boston, MA, 1976. pp. 60–65.

[27] Pratte J.F., Ferraro D.G. 'Automated solar gain calibration'. Presented at 24th Conference on Radar Meteorology; Tallahassee, FL, 1989. Boston, MA: American Meteorological Society,

[28] Melnikov V.M., Zrnic D.S., Doviak R.J., Carter J.K. *Calibration and performance analysis of NSSL's polarimetric WSR-88D*. NOAA/NSSL Report; 2003. p. 77.

[29] Holleman I., Huuskonen A., Gill R., Tabary P. 'Operational monitoring of radar differential reflectivity using the sun'. *Journal of Atmospheric and Oceanic Technology*. 2010, vol. 27(5), pp. 881–87.

[30] Huuskonen A., Kurri M., Holleman I. 'Improved analysis of solar signals for differential reflectivity monitoring'. *Atmospheric Measurement Techniques*. 2016, vol. 9(7), pp. 3183–92.

[31] Gabella M., Sartori M., Boscacci M., Germann U. 'Vertical and horizontal polarization observations of slowly varying solar emissions from operational swiss weather radars'. *Atmosphere*. 2014, vol. 6(1), pp. 50–59.

[32] Vaccarono M., Bechini R., Chandrasekar C.V., Cremonini R., Cassardo C. 'An integrated approach to monitoring the calibration stability of operational dual-polarization radars'. *Atmospheric Measurement Techniques*. 2016, vol. 9(11), pp. 5367–83.

[33] Gorgucci E., Scarchilli G., Chandrasekar V. 'A procedure to calibrate multiparameter weather radar using properties of the rain medium'. *IEEE Trans Geosci Remote Sens*. 1999, vol. 37(1), pp. 269–76.

[34] Frech M., Hubbert J. 'Monitoring the differential reflectivity and receiver calibration of the German polarimetric weather radar network'. *Atmospheric Measurement Techniques*. 2020, vol. 13(3), pp. 1051–69.
[35] Hitschfeld W., Bordan J. 'ERRORS inherent in the radar measurement of rainfall at attenuating wavelengths'. *Journal of Meteorology*. 1954, vol. 11(1), pp. 58–67.
[36] Testud J., Le Bouar E., Obligis E., Ali-Mehenni M. 'The rain profiling algorithm applied to polarimetric weather radar'. *Journal of Atmospheric and Oceanic Technology*. 2000, vol. 17(3), pp. 332–56.
[37] Bringi V.N., Chandrasekar V., Balakrishnan N., Zrnić D.S. 'An examination of propagation effects in rainfall on radar measurements at microwave frequencies'. *Journal of Atmospheric and Oceanic Technology*. 1990, vol. 7(6), pp. 829–40.
[38] Ryzhkov A., Zrnic D.S. 'Precipitation and attenuation measurements at a 10-cm wavelength'. *Journal of Applied Meteorology*. 1995, vol. 34(10), pp. 2121–34.
[39] Carey L.D., Rutledge S.A., Ahijevych D.A., Keenan T.D. 'Correcting propagation effects in C-band polarimetric radar observations of tropical convection using differential propagation phase'. *Journal of Applied Meteorology*. 2000, vol. 39(9), pp. 1405–33.
[40] Bringi V.N., Keenan T.D., Chandrasekar V. 'Correcting C-band radar reflectivity and differential reflectivity data for rain attenuation: a self-consistent method with constraints'. *IEEE Trans Geosci Remote Sens*. 2001, vol. 39(9), pp. 1906–15.
[41] Park S.-G., Maki M., Iwanami K., Bringi V.N., Chandrasekar V. 'Correction of radar reflectivity and differential reflectivity for rain attenuation at X band. part II: evaluation and application'. *Journal of Atmospheric and Oceanic Technology*. 2005, vol. 22(11), pp. 1633–55.
[42] Smyth T.J., Illingworth A.J. 'Correction for attenuation of radar reflectivity using polarization data'. *Quarterly Journal of the Royal Meteorological Society*. 1998, vol. 124(551), pp. 2393–415. Available from http://doi.wiley.com/10.1002/qj.v124:551
[43] Bringi V.N., Huang G.-J., Chandrasekar V., Keenan T.D. 'An areal rainfall estimator using differential propagation phase: evaluation using a C-band radar and a dense gauge network in the tropics'. *Journal of Atmospheric and Oceanic Technology*. 2001, vol. 18(11), pp. 1810–18.
[44] Ryzhkov A., Zrnić D., Fulton R. 'Areal rainfall estimates using differential phase'. *Journal of Applied Meteorology*. 2000, vol. 39(2), pp. 263–68.
[45] Park S.G., Maki M., Iwanami K., Bringi V.N. 'Correction of radar reflectivity and differential reflectivity for rain attenuation and estimation of rainfall at X-band wavelength'. *6th International Symposium on Hydrogen Applications of Weather Radar*; Melbourne, Australia, 2004.
[46] Matrosov S.Y., Clark K.A., Martner B.E., Tokay A. 'X-band polarimetric radar measurements of rainfall'. *Journal of Applied Meteorology*. 2002, vol. 41(9), pp. 941–52.

[47] Maki M., Park S.G., Bringi V.N. 'Statistical error of rain rate estimation due to natural variations of rain drop size distributions for 3 cm wavelength polarimetric radar'. *Journal of the Meteorological Society of Japan*. 2005, vol. 83(5), pp. 871–93.

[48] May P.T., Keenan T.D., Zrnić D.S., Carey L.D., Rutledge S.A. 'Polarimetric radar measurements of tropical rain at a 5-cm wavelength'. *Journal of Applied Meteorology*. 1999, vol. 38(6), pp. 750–65.

[49] Ryzhkov A. *Rainfall measurements with the polarimetric WSR-88D radar, CIMMS report*. Norman, OK: National Severe Storms Laboratory; 2003.

[50] Ryzhkov A.V., Giangrande S.E., Schuur T.J. 'Rainfall estimation with a polarimetric prototype of WSR-88D'. *Journal of Applied Meteorology*. 2005, vol. 44(4), pp. 502–15.

[51] Gorgucci E., Chandrasekar V., Bringi V.N. 'Drop size distribution retrieval from polarimetric radar measurements'. *2nd European conference on Radar in Meteorology and Hydrology (ERAD)*; Delft, The Netherlands, 2002.

[52] Vulpiani G., Marzano F.S., Chandrasekar V., Berne A., Uijlenhoet R. 'Polarimetric weather radar retrieval of raindrop size distribution by means of a regularized artificial neural network'. *IEEE Transactions on Geoscience and Remote Sensing*. 2006, vol. 44, pp. 3262–75.

[53] Anagnostou M.N., Anagnostou E.N., Vulpiani G., Montopoli M., Marzano F.S., Vivekanandan J. 'Evaluation of X-band polarimetric-radar estimates of drop-size distributions from coincident S-band polarimetric estimates and measured raindrop spectra'. *IEEE Trans Geosci Remote Sens*. 2008, vol. 46(10), pp. 3067–75.

[54] Cao Q., Zhang G., Brandes E.A., Schuur T.J. 'Polarimetric radar rain estimation through retrieval of drop size distribution using a Bayesian approach'. *Journal of Applied Meteorology and Climatology*. 2010, vol. 49(5), pp. 973–90.

[55] Marzano F.S., Barbieri S., Vulpiani G., Rose W.I. 'Volcanic ash cloud retrieval by ground-based microwave weather radar'. *IEEE Trans Geosci Remote Sens*. 2006, vol. 44(11), pp. 3235–46.

[56] Marzano F.S., Picciotti E., Montopoli M., Vulpiani G. 'Inside volcanic clouds: remote sensing of ash plumes using microwave weather radars'. *Bulletin of the American Meteorological Society*. 2013, vol. 94(10), pp. 1567–86.

[57] Vulpiani G., Ripepe M., Valade S. 'Mass discharge rate retrieval combining weather radar and thermal camera observations'. *Journal of Geophysical Research*. 2016, vol. 121(8), pp. 5679–95.

[58] Buler J.J., Diehl R.H. 'Quantifying bird density during migratory stopover using weather surveillance radar'. *IEEE Trans Geosci Remote Sens*. 2009, vol. 47(8), pp. 2741–51.

[59] Dokter A.M., Liechti F., Stark H., Delobbe L., Tabary P., Holleman I. 'Bird migration flight altitudes studied by a network of operational weather radars'. *Journal of the Royal Society, Interface*. 2011, vol. 8(54), pp. 30–43.

[60] Höller H., Hagen M., Meischner P.F., Bringi V.N., Hubbert J. 'Life cycle and precipitation formation in a hybrid-type hailstorm revealed by polarimetric

and doppler radar measurements'. *Journal of the Atmospheric Sciences*. 1994, vol. 51(17), pp. 2500–22.

[61] Vivekanandan J., Ellis S.M., Oye R., Zrnic D.S., Ryzhkov A.V., Straka J. 'Cloud microphysics retrieval using S-band dual-polarization radar measurements'. *Bulletin of the American Meteorological Society*. 1999, vol. 80(3), pp. 381–88.

[62] Hannesen R. 'Improvements of polarimetric radar echo classifications'. *34th Conference on Radar Meteorology*; Williamsburg, VA, 2009.

[63] Schuur T., Ryzhkov A., Heinselman P. *Observations and classification of echoes with the polarimetric WSR-88D radar*. Norman, OK: NOAA National Severe Storms Laboratory Tech Report; 2003.

[64] Rennie S.J., Curtis M., Peter J., Seed A.W., Steinle P.J., Wen G. 'Bayesian echo classification for Australian single-polarization weather radar with application to assimilation of radial velocity observations'. *Journal of Atmospheric and Oceanic Technology*. 2015, vol. 32(7), pp. 1341–55.

[65] Marzano F.S., Scaranari D., Montopoli M., Vulpiani G. 'Supervised classification and estimation of hydrometeors from C-band dual-polarized radars: a Bayesian approach'. *IEEE Trans Geosci Remote Sens*. 2008, vol. 46(1), pp. 85–98.

[66] Yang J., Zhao K., Zhang G., Chen G., Huang H., Chen H. 'A Bayesian hydrometeor classification algorithm for C-band polarimetric radar'. *Remote Sensing*. 1884, vol. 11(16), p. 1884.

[67] Vulpiani G., Montopoli M., Picciotti E., Marzano F.S. 'On the use of a polarimetric X-band weather radar for volcanic ash clouds monitoring'. *35th Conference on Radar Meteorology*; Pittsburgh, PA, 2011.

[68] Lu Y., Kumar J. 'Convolutional neural networks for hydrometeor classification using dual polarization doppler radars'. *IEEE International Conference on Data Mining (ICDM) Workshops*; Beijing, China, 2019.

[69] Liu H., Chandrasekar V. 'Classification of hydrometeors based on polarimetric radar measurements: development of fuzzy logic and neuro-fuzzy systems, and in situ verification'. *Journal of Atmospheric and Oceanic Technology*. 2000, vol. 17(2), pp. 140–64.

[70] NOAA NCEI *NEXRAD data archive, inventory and access* [online]. Available from https://www.ncdc.noaa.gov/nexradinv/

[71] *Google maps* [online]. Available from https://www.google.com/maps/@43.755632,-75.679992,11z

[72] Berenguer M., Sempere-Torres D., Corral C., Sánchez-Diezma R. 'A fuzzy logic technique for identifying nonprecipitating echoes in radar scans'. *Journal of Atmospheric and Oceanic Technology*. 2006, vol. 23(9), pp. 1157–80.

[73] Gourley J.J., Tabary P., Parent du Chatelet J. 'A fuzzy logic algorithm for the separation of precipitating from nonprecipitating echoes using polarimetric radar observations'. *Journal of Atmospheric and Oceanic Technology*. 2007, vol. 24(8), pp. 1439–51.

[74] Beauchamp R.M., Chandrasekar V. 'Dual-polarization radar characteristics of wind turbines with ground clutter and precipitation'. *IEEE Trans Geosci Remote Sens*. 2016, vol. 54(8), pp. 4833–46.

[75] Nai F., Palmer R., Torres S. 'Wind turbine clutter mitigation using range-doppler domain signal processing method'. *27th Conference of Interactive Information Processing Systems (IIPS)*; Seattle, WA, 2011.

[76] Hood K., Torres S., Palmer R. 'Automatic detection of wind turbine clutter for weather radars'. *Journal of Atmospheric and Oceanic Technology*. 2010, vol. 27(11), pp. 1868–80.

[77] Gallardo-Hernando B., Muñoz-Ferreras J.M., Pérez-Martínez F. 'Super-resolution techniques for wind turbine clutter spectrum enhancement in meteorological radars'. *IET Radar, Sonar & Navigation*. 2011, vol. 5(9), p. 924.

[78] Agence Nationale des Fréquences *Perturbations du fonctionnement des radars meteorologiques PAR les eoliennes*. Vol. 1. Rapport CCE5; 2005.

[79] WMO *Guide to instruments and methods of observation volume III – observing systems chapter 7 annex 7.C. WMO guidance statement on weather radar/wind turbine siting [online]*. 2018. Available from https://www.eumetnet.eu/wp-content/uploads/2017/01/OPERA_2010_14_Statement_on_weather_radars_and_wind_turbines.pdf

[80] Isom B.M. 'Characterization and mitigation of wind turbine clutter on the WSR-88D network'. *33rd Conference of Radar Meteorology*; Cairns, AU, 2007.

[81] Nai F., Torres S., Palmer R. 'On the mitigation of wind turbine clutter for weather radars using range-doppler spectral processing'. *IET Radar, Sonar & Navigation*. 2013, vol. 7(2), pp. 178–90. Available from https://onlinelibrary.wiley.com/toc/17518792/7/2

[82] Bachmann S., Al-Rashid Y., Isom B., Palmer R. 'Radar and windfarms – mitigating negative effects through signal processing'. *6th European Conference on Radar in Meteorology and Hydrology (ERAD 2010)*; Sibiu, Romania, 2010.

[83] Gekat F., Teschke B. 'Mitigation of weather radar disturbance caused by wind turbines'. *Conférence sur la cohabitation entre éoliennes et radars*; Berlin, 2015.

Chapter 14

Meteorological polarimetric phased array radar

Igor R. Ivić[1,2], M. David Conway[3], Sebastián M. Torres[1,2], Jeffrey S. Herd[3], Dušan S. Zrnić[2,4,5], and Mark E. Weber[3]

14.1 Motivation for polarimetric phased array radar

Polarimetric meteorological radars have become crucial for the detection and short-term forecast of hazardous weather, as well as quantitative precipitation estimation (QPE). The ability to directly identify hail and gauge its size, detect tornado debris, and anticipate flash floods sets these radars apart from the classical single-polarized ones. Weather agencies in many countries have deployed and/or upgraded their radar systems to dual polarization, including the US National Weather Service (NWS) which completed the upgrade [1] of its Weather Service Radar-1988 Doppler (WSR-88D) network in 2013. Recent experience with the WSR-88D and other operational radars has exceeded expectations, well justifying the upgrade cost, which was about 5% of the initial procurement. The high quality of the quantitative polarimetric measurements has set a standard for any future polarimetric weather radar.

Weather radars typically use narrow "pencil beam" antenna patterns to resolve the three-dimensional structure of storms. They have long (~50 ms) coherent processing intervals (CPI) to mitigate artifacts (e.g., ground clutter, range ambiguous weather) and provide good quality estimates of the meteorological variables. Therefore, the temporal resolution with which weather phenomena are surveilled is low, typically 4–6 minutes for full-volume coverage. As discussed in this monograph, polarimetric phased array radar (PPAR) supports methods to significantly increase the volumetric sampling rate, which provides significant benefits for hazardous weather warnings and QPE. Its agile electronically steered beams can also scan adaptively to increase data quality and/or coverage in volumes of operational interest, improve suppression of ground clutter through adaptive shaping of the

[1]Cooperative Institute for Severe and High-Impact Weather Research and Operations, University of Oklahoma, Norman, OK, USA
[2]NOAA National Severe Storms Laboratory, Norman, OK, USA
[3]Massachusetts Institute of Technology Lincoln Laboratory, Lexington, MA, USA
[4]School of Electrical and Computer Engineering, University of Oklahoma, Norman, OK, USA
[5]School of Meteorology, University of Oklahoma, Norman, OK, USA

lowest elevation beams, and eliminate scan-induced beam broadening. Maintaining the WSR-88D's high-quality polarimetric measurements with PPAR, however, poses major challenges, which are being addressed through the research and risk reduction described herein.

In the remainder of this section, we review the fundamentals of weather radar polarimetry, summarize the expected operational benefits that would be realized with PPAR, and describe existing PPAR demonstration systems.

14.1.1 Polarimetric weather radar

Years of research with single- and dual-polarization weather radars led toward the practical systems in operation today. A substantial literature on the subject is spread among a few specialized journals and summarized in the following books. Doviak and Zrnić [2] have a full chapter devoted to weather radar polarimetry; it highlights the significant and lasting attributes of polarimetry although the text was written in 1993. The first dedicated book is by Bringi and Chandrasekhar [3]; it contains theory and known applications circa the end of the millennium. Zhang [4] discusses scattering, polarimetric signal processing, microphysical applications, and has a section devoted to PPAR. Ryzhkov and Zrnić (2019) concentrate on use of the linear polarization basis with many applications derived from WSR-88D observations.

14.1.1.1 Transmission modes

The polarimetric basis for most polarimetric weather radars is linear horizontal (H) and vertical (V). The horizontally polarized field E_{h} at any observation range is parallel to the ground, while the vertically polarized field tilts away from the true vertical as the elevation of the beam's position increases. The geometric/electric properties of hydrometeors and propagation effects are the main drivers for this choice as the following explains. Gravity shapes and orients the raindrops into oblate spheroids with a vertical symmetry axis. It similarly orients snowflakes. Therefore, the H, V choice accentuates the difference of reflections between the H and the V polarized fields in the rain, snow, and other oriented scatterers like insects, chaff, or smoke particles. Moreover, during propagation through oriented scatterers the H, V polarized waves do not couple. In contrast, circularly polarized waves progressively change polarization as the waves propagate through oriented scatterers, obfuscating meteorological interpretation.

For weather applications, the backscattering matrix $\boldsymbol{S}$ [2–5] concisely describes the polarimetric characteristics that can be sensed by radar. We present an unorthodox definition, which nonetheless is handy for grasping the essence of weather radar polarimetry. The matrix relates the backscattered field $\boldsymbol{E}^b = \begin{bmatrix} E_{\mathrm{h}}^b & E_{\mathrm{v}}^b \end{bmatrix}^T$ with the incident field $\boldsymbol{E}^i = \begin{bmatrix} E_{\mathrm{h}}^i & E_{\mathrm{v}}^i \end{bmatrix}^T$. For a single hydrometeor (scatterer) located at range r_i far from the radar and in free space, the relation is

$$\boldsymbol{E}^b = \frac{\exp(-j\kappa r_i)}{r_i}\boldsymbol{S}\boldsymbol{E}^i = \frac{\exp(-j\kappa r_i)}{r_i}\begin{bmatrix} s_{\mathrm{hh}} & s_{\mathrm{hv}} \\ s_{\mathrm{hv}} & s_{\mathrm{vv}} \end{bmatrix}\begin{bmatrix} E_{\mathrm{h}}^i \\ E_{\mathrm{v}}^i \end{bmatrix}, \tag{14.1}$$

where κ is the wavenumber of free space (i.e., $\kappa = 2\pi/\lambda$). Symbols s_{pq} (p and q are either "h" or "v") are backscattering coefficients, where the first subscript indicates the polarization of the backscattered field and the second of the incident field. The off-diagonal terms in (14.1) are equal because the scatterer is a reciprocal reflector of electromagnetic (EM) waves.

Next, assuming a nonattenuating anisotropic medium, we introduce the propagation matrix $\boldsymbol{T}_{\mathrm{p}}$ ([5], (2.15)):

$$\begin{aligned} \boldsymbol{T}_{\mathrm{p}} &= \begin{bmatrix} \exp\left[-\int_0^{r_1} \kappa_{\mathrm{h}}(r)\,dr\right] & 0 \\ 0 & \exp\left[-\int_0^{r_1} \kappa_{\mathrm{v}}(r)\,dr\right] \end{bmatrix} \\ &= \exp\left[-\int_0^{r_1} \kappa_{\mathrm{v}}(r)\,dr\right] \begin{bmatrix} \exp\left(\Phi_{\mathrm{DP}}(r_i)/2\right) & 0 \\ 0 & 1 \end{bmatrix} \\ &= \exp\left[-\int_0^{r_1} \kappa_{\mathrm{v}}(r)\,dr\right] \boldsymbol{T} \end{aligned} \tag{14.2}$$

where the wavenumbers for the H, V polarizations κ_{h}, κ_{v} depend on the range, and there is no coupling on propagation. The matrix $\boldsymbol{T}$ in (14.2) is the differential propagation matrix often shortened to "propagation matrix" which we use from here on. Furthermore, $\exp\left[-j\int_0^{r_i} \kappa_{\mathrm{v}}(r)\,dr\right]$ will be omitted because it is not known nor a part of computations for any polarimetric variable. The differential phase $\Phi_{DP}(r_i)$ is expressed as follows:

$$\Phi_{\mathrm{DP}}(r_i) = 2\int_0^{r_i} \left[\kappa_{\mathrm{h}}(r_i) - \kappa_{\mathrm{v}}(r_i)\right] dr \tag{14.3}$$

The transmitted field propagates, loses intensity, backscatters off the scatterer, and arrives to the antenna. Voltages in the receiver caused by the backscattered field from a single scatterer are related to the transmitted field via

$$\begin{bmatrix} \delta V_{\mathrm{h}} \\ \delta V_{\mathrm{v}} \end{bmatrix} \propto \begin{bmatrix} \exp\left[-j\Phi_{\mathrm{DP}}(r_i)\right] s_{\mathrm{hh}} & \exp\left[-j\Phi_{\mathrm{DP}}(r_i)/2\right] s_{\mathrm{hv}} \\ \exp\left[-j\Phi_{\mathrm{DP}}(r_i)/2\right] s_{\mathrm{hv}} & s_{\mathrm{vv}} \end{bmatrix} \begin{bmatrix} E_{\mathrm{h}}^t \\ E_{\mathrm{v}}^t \end{bmatrix} \tag{14.4}$$

In presence of many scatterers, the backscattered field is the sum of individual reflections. The largest contribution comes from scatterers in the resolution volume V_6 [2] illustrated in Figure 14.1.

Weather radars generate multiple transmissions (e.g., M) in a given direction that produce multiple returns (e.g., to reduce the errors of estimates) spaced in time by the pulse repetition time (PRT) T_s, where $T_s >> \tau$ (pulse length). The receiver transforms the sum of scatterers' contributions to a voltage pair. Following the mth transmission, the voltage pair is related to the radiation received by the antenna as follows:

$$\begin{bmatrix} V_{\mathrm{h}}(m) \\ V_{\mathrm{v}}(m) \end{bmatrix} \propto \left\{ \begin{bmatrix} S_{\mathrm{hh}}(m) & 0 \\ 0 & S_{\mathrm{vv}}(m) \end{bmatrix} + S_{\mathrm{hv}}(m) \begin{bmatrix} 0 & 1 \\ 1 & 0 \end{bmatrix} \right\} \boldsymbol{E}^t \tag{14.5}$$

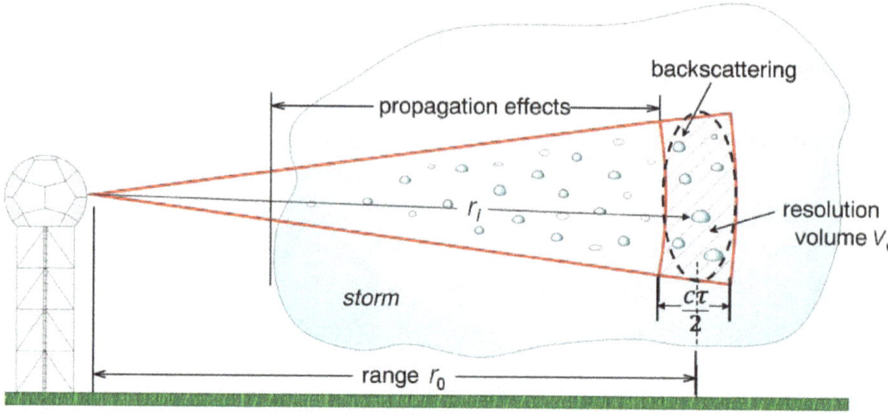

Figure 14.1 Vertical cross section of a beam penetrating a storm. The section at the far end of the cone depicts the vertical cross section of the resolution volume V_6. It is bound by the radiation that is within 6 dB of its peak intensity in azimuth and elevation. In range, V_6 is of length $c\tau/2$ (τ is the transmitted pulse duration) where the range weighting function is maximum in the middle. The elliptical cross section (dashed ellipse) is the cross section modified by the receiver. The range to the center of V_6 is r_0.

where the square matrix multiplying $\boldsymbol{E}^{\,t}$ is decomposed into diagonal $\boldsymbol{S}_{\mathrm{d}} = \mathrm{diag}(\boldsymbol{S})$, and off-diagonal part $\boldsymbol{S}_{\mathrm{off}} = \mathrm{offdiag}(\boldsymbol{S})$ to separate the noncoupled components from the coupled ones. The elements of $\boldsymbol{S}$ are

$$
\begin{aligned}
S_{\mathrm{hh}}(m) &= \exp(-j\Phi_{\mathrm{DP}}) \sum_i s_{\mathrm{hh}}(i) \exp[-j2k_{\mathrm{v}}\Delta r_i(m)] \\
S_{\mathrm{vv}}(m) &= \sum_i s_{\mathrm{vv}}(i) \exp[-j2k_{\mathrm{v}}\Delta r_i(m)] \\
S_{\mathrm{hv}}(m) &= \exp(-j\Phi_{\mathrm{DP}}/2) \sum_i s_{\mathrm{hv}}(i) \exp[-j2k_{\mathrm{v}}\Delta r_i(m)]
\end{aligned}
\tag{14.6}
$$

The Φ_{DP} in (14.6) has the $r_i = r_0$ (r_0 is the range to the middle of V_6, Figure 14.1) so the propagation effects are embedded in terms of $\boldsymbol{S}$. Also, the range $r_{\mathrm{i}}(m)$ depends on m because the ith scatterer moves randomly between successive transmissions. So do other scatterers, and therefore, the $S_{\mathrm{pq}}(m)$ differ from pulse to pulse. The amount of change depends on T_{s} and the relative velocity of scatterers. Consequently, the exponents in the sums cause the S_{pq}s to be random variables with Rayleigh amplitude distribution and uniform phase as well as to have Gaussian distributions of in-phase (I) and quadrature (Q) components [2]. Also, the random motion of scatterers forces the ensemble average of S_{pq} to zero [2], section 4.3). The statistically significant quantities are the second-order moments of the backscattering matrix $\boldsymbol{S}$ terms

$$
\begin{aligned}
&E[|S_{\mathrm{hh}}(m)|^2];\ E[|S_{\mathrm{vv}}(m)|^2];\ E[|S_{\mathrm{hv}}(m)|^2]; \\
&E[S^*_{\mathrm{hh}}(m)\,S_{\mathrm{vv}}(m)];\ E[S^*_{\mathrm{hh}}(m)\,S_{\mathrm{hv}}(m)];\ E[S^*_{\mathrm{hv}}(m)\,S_{\mathrm{vv}}(m)]
\end{aligned}
\tag{14.7}
$$

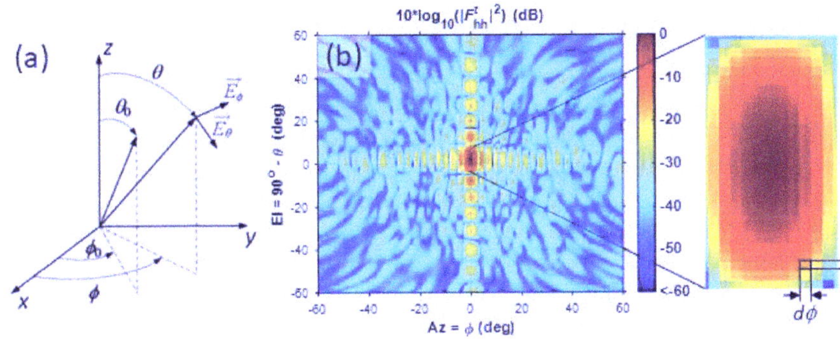

Figure 14.2 *(a) Spherical coordinate system used to plot radiation patterns. It is assumed that the antenna face is aligned with yz-plane. (b) Transmitted antenna pattern of a PPAR system, an example.*

For dual-polarization radars, the antenna patterns are important as they affect the measurements of precipitation. Succinctly, we denote the transmit patterns with F^t_{pq}, where q indicates the intended polarization of the antenna excitation port, and p indicates the polarization of the transmitted field. The patterns are two-dimensional complex functions of θ, ϕ (Figure 14.2a). Unlike parabolic antennas, PPAR antenna patterns are functions of beamsteering directions θ_0, ϕ_0 (Figure 14.2a). Excitation of antenna H port generates a horizontally polarized field (i.e., polarization concomitant with $\vec{E}_\phi$ in Figure 14.2a) described by the pattern denoted as F^t_{hh} (H copolar pattern), but also a vertically polarized field (i.e., polarization concomitant with $\vec{E}_\theta$ in Figure 14.2a) described by F^t_{vh} (H cross-polar pattern). Note that the dependency of antenna patterns on θ, ϕ is implicit (and $\theta = 90° - El$, $\phi = Az$). Likewise, symbols F^t_{vv} and F^t_{hv} denote copolar and cross-polar patterns that describe radiation generated by excitation of the antenna V port. For the purpose of this exposition, all patterns are normalized to the maximum of H copolar pattern at broadside (i.e., $|F^t_{\mathrm{hh}}|$ at $\theta_0 = 90°$, $\phi_0 = 0°$). An example of F^t_{hh} from a PPAR system (described in Reference 6) where the beam is steered electronically at $\theta_0 = 88°$, $\phi_0 = 0°$ is in Figure 14.2b. It demonstrates that the radiated power is concentrated and centered around the beamsteering direction. Furthermore, we show how the space illuminated by the radiation can be sectioned into elemental resolution volumes (V_e) bound by $d\phi$, $d\theta$ in azimuth and elevation; implicit is the extent $c\tau/2$ in range. The scatterers in each V_e are illuminated by approximately constant radiation. Note that V_e is conceptually similar to V_6, and we introduce it to quantify antenna pattern effects on weather radar measurements.

Two polarimetric transmission modes are in use on weather surveillance radars: the simultaneous transmission of H and V polarized fields (SHV) and alternate transmission of these fields (AHV) [5]. The former is the choice of weather services around the world. In both modes, signals are simultaneously received in the H and V channels.

We present the mth transmitted field $\boldsymbol{E}^t = \begin{bmatrix} E_h^t(m) & E_v^t(m) \end{bmatrix}^T$ in a matrix notation as

$$\begin{bmatrix} E_h^t(m) \\ E_v^t(m) \end{bmatrix} = \begin{bmatrix} F_{hh}^t & F_{hv}^t \\ F_{vh}^t & F_{vv}^t \end{bmatrix} \begin{bmatrix} e_h(m) \\ e_v(m) \end{bmatrix} = \boldsymbol{F}^t \boldsymbol{e} \tag{14.8}$$

where F^t is the transmit pattern matrix, and $\boldsymbol{e}$ is the excitation vector that depends on the transmission mode.

The total received voltages are obtained from integrals over the spherical space

$$\begin{aligned} \begin{bmatrix} V_h(m) \\ V_v(m) \end{bmatrix} &= C\int_\Omega \left[\begin{bmatrix} F_{hh}^r & F_{vh}^r \\ F_{hv}^r & F_{vv}^r \end{bmatrix} \left\{ \begin{bmatrix} S_{hh}(m) & 0 \\ 0 & S_{vv}(m) \end{bmatrix} + S_{hv}(m) \begin{bmatrix} 0 & 1 \\ 1 & 0 \end{bmatrix} \right\} \boldsymbol{E}^t \right] d\Omega \\ &= C\int_\Omega \left[\boldsymbol{F}^r \{ \boldsymbol{S}_d + \boldsymbol{S}_{\text{off}} \} \boldsymbol{E}^t \right] d\Omega \end{aligned} \tag{14.9}$$

where the solid angle $d\Omega = \sin\theta d\phi d\theta$, and C is a scalar factor that contains dependence on the range, attenuation, and system parameters. In (14.9), $S_{mn}(m)$ are sums of scatterers contained in V_e (as opposed to sums of scatterers contained in V_6 in (14.6)). The symbols F_{hh}^r and F_{vv}^r are the H and V copolar receive patterns, while F_{vh}^r and F_{hv}^r are the H and V cross-polar receive patterns (where the first letter in the subscript denotes the incident field and the second letter the receiving channel). Note that F_{vh}^r / F_{hv}^r quantify coupling of V/H polarized returns into H/V channel that causes biases in radar polarimetric variable estimates. Later throughout the text, the performance of the PPAR is evaluated on how well it measures the diagonal terms in presence of various couplings.

14.1.1.2 Polarimetric variables

The received voltages are sums of H and V components like in (14.5) and (14.9), and if $S_{hv}(m)$ is significant (14.5) indicates that retrieval of second-order moments would be impossible. Nonetheless, in oriented precipitation (which constitutes most cases) the $S_{hv}(m)$ is very small and routinely neglected. Then, the relation between the voltages and the backscattering terms is expressed with the $\boldsymbol{S}_D$ matrix and it is multiplicative (with no additive terms). The relation (14.9) becomes

$$\begin{bmatrix} V_h(m) \\ V_v(m) \end{bmatrix} = C\int_\Omega \boldsymbol{F}^r \boldsymbol{S}_D \boldsymbol{E}^t d\Omega \tag{14.10}$$

Next, we list the intrinsic polarimetric variables. The reflectivity at horizontal polarization [5]

$$Z_h = \eta \frac{\lambda^4}{\pi^5 |K|^2} = 4E\left[|S_{hh}(m)|^2 \right] \frac{\lambda^4}{\pi^4 |K|^2} \tag{14.11}$$

where η is the cross section per unit volume, and $|K|^2$ quantifies the effect of liquid water's refractive index (at microwave frequencies for water, the value is between 0.91 and 0.93 [2]). The differential reflectivity (in linear units) is

$$Z_{\mathrm{dr}} = \frac{\mathrm{E}\left[|S_{\mathrm{hh}}(m)|^2\right]}{\mathrm{E}\left[|S_{\mathrm{vv}}(m)|^2\right]} \tag{14.12}$$

and the copolar correlation coefficient is

$$\rho_{\mathrm{hv}} e^{j\delta} = \frac{\mathrm{E}\left[S_{\mathrm{hh}}^{*}(m)\, S_{\mathrm{vv}}(m)\right]}{\sqrt{\mathrm{E}\left[|S_{\mathrm{hh}}(m)|^2\right]\mathrm{E}\left[|S_{\mathrm{vv}}(m)|^2\right]}} \tag{14.13}$$

where $\Phi_{DP} = \arg(\rho_{\mathrm{hv}})$ and δ is the differential phase upon scattering (i.e., backscatter differential phase). The δ manifests as a local maximum of Φ_{DP} in range and that is how it can be estimated [5]. It is negligible for Rayleigh scatterers [5]. The differential phase is quantitatively used for estimating precipitation. Attenuation and differential attenuation are additional propagation variables that may affect the measurements. To quantify the effects, relations between attenuation, differential phase, and reflectivity have been successfully used [7]. Finally, the linear depolarization ratio (in linear units) defined as the ratio of cross-polar power to copolar power is

$$L_{\mathrm{dr}} = \frac{\mathrm{E}\left[|S_{\mathrm{vh}}(m)|^2\right]}{\mathrm{E}\left[|S_{\mathrm{hh}}(m)|^2\right]} \tag{14.14}$$

The SHV mode yields the variables (14.11), (14.12), and (14.13) but not the linear depolarization ratio. In the AHV mode, the second-order moments can be retrieved if both the copolar and cross-polar components are processed. We list the attributes of each mode next.

The SHV mode: The standard errors of radar variable estimates are generally smaller [8]. The ρ_{hv} is computed directly. All the operations on the returned signals that are possible (used) in the nonpolarimetric mode can be done with no changes, including filtering ground clutter, phase coding for mitigating ambiguities in velocity and range [9], etc. L_{dr} cannot be computed nor can the other two second-order moments containing the cross-polar components. Compared with the AHV mode, if the same transmitter is operating in the SHV mode, there would be a 3 dB reduction in signal-to-noise ratio (SNR) because the noise powers in the receivers do not change. Combining the autocovariances at lag 1 of the H signal and V signals recovers about 1.5 dB of SNR [10]. Furthermore, coherent addition of the time series data gains 3 dB at the expense of losing the polarimetric information [11]. Another way to increase detectability is by applying a threshold dependent on SNR, autocorrelation, and cross-correlations of H and V returns, which is done on the WSR-88D [12–14]. Also, Keranen and Chandrasekar [15] propose using coherent averages of H and V precipitation signals to enhance detectability.

The AHV mode: The standard errors of polarimetric variable estimates are larger. All the second-order moments can be computed. Special processing must be applied to compute the Doppler spectral moments and the correlation coefficient (including differential phase) because these are coupled. Cross-coupling between the H and V signals by the system as well as via propagation and backscattering is smaller than in the SHV mode. Almost equal biases in the estimates of powers and differential reflectivity in the two modes are produced if the cross-polar isolation in the SHV mode is about twice (in dB) as large as in the AHV mode. For example, the cross-polar isolation of approximately –25 dB in the AHV mode would result in similar biases as a –50 dB cross-polar isolation in the SHV mode. Consequently, the cross-polar isolation requirement in the AHV mode can be relaxed compared with the SHV mode. Nonetheless, the former mode may impede the detectability of useful cross-polar signals along the propagation paths. Furthermore, if the PRTs are chosen to have the same unambiguous range, mitigation of the range and velocity ambiguities is harder because the PRT corresponding to pulses of the same polarization in the AHV mode is twice as large. A succinct summary of these comparisons is in Table 14.1.

14.1.1.3 Applications

Proven benefits of the polarimetric information are as follows [5, 16]: (a) classifying precipitation type (rain, freezing rain, wet snow, dry snow, and hail), (b) gauging hail size, (c) classifying nonmeteorological scatterers (ground clutter, ocean clutter, birds, insects, bats, and chaff), (d) improved quantitative estimation of rain rate and snow rate, (e) detecting debris lofted by tornadoes [17], and (f) detecting icing. Assimilation of polarimetric variables into numerical weather prediction models on scales ranging from storms (tens of kilometers) to mesoscale (hundreds of kilometers) has also been shown to improve model analyses and subsequent forecasts.

Table 14.1 Contrast between the SHV and AHV modes

Feature	SHV	AHV
PRT for pulses of the same polarization	T_s	$2T_s$
Polarimetric information	Partial (no L_{dr} and cross-pol)	Full
The standard error of estimates	Smaller	Larger
Sensitivity (detectability)	1.5 dB loss	
Computation of ρ_{hv} and Φ_{DP}	Direct	More complex estimators
Ground clutter filtering	As on single-pol radar	Somewhat compromised
Mitigation of ambiguities in range and velocity	As on single-pol radar (uniform PRT or staggered)	More complicated and less effective*
Effects of coupling	Two times (in dB) larger	Smaller

*The second-trip returns are orthogonal to the first-trip returns; hence, an automatic reduction of the second trip is achieved.

Apart from the paper by Zrnić *et al.* [18], little information in the literature exists about the relative meteorological (diagnostic) merits of the polarimetric variables. By comparing the classified precipitation field using a full suite of the polarimetric variables with the fields where one fewer variable is used, the authors were able to determine which omitted one did the most damage to the classification outcome. The corresponding rank of the variables and explanation of their potency follows.

Z_h is the most important polarimetric variable. It indicates the presence of scatterers, is used for estimating rain and snow rates, and for identification of hail. In combination with other polarimetric variables, it produces improved rain rate and snow fall estimates and is important for gauging hail size.

After Z_h, Z_{DR} is the most important variable for hydrometeor identification [18]. Z_{DR} contains information about particle shape (a measure of median drop diameter), orientation, and refractive index (density) and is useful for rain/hail/snow discrimination. In combination with Z or K_{DP}, it yields better estimates of rain rate and snow rate. Vertical columns of Z_{DR} indicate updrafts.

The absolute value of ρ_{hv} is the best identifier of nonmeteorological scatterers and is a good indicator of melting precipitation, such as in the melting layer. It detects debris lofted by tornadoes and has the potential for gauging hail size [19].

The backscatter differential phase, δ, depends on the size. It is sensitive to scatterers comparable with or larger than the wavelength; therefore, it is significant in identifying ground clutter, birds, bats, very large insects, hail, and wet snow in the melting layer.

Φ_{DP} is immune to radar miscalibration, to partial beam blockage, and attenuation. It is useful for rainfall estimation via its range derivative. An iterative procedure along range has been proposed to relate Φ_{DP} and Z_h, to specific attenuation and corresponding rain rate, making it among the best polarimetric methods for estimating rain rate [7].

L_{DR} is a good indicator of frozen precipitation, including hail. Its low values also indicate nonmeteorological returns. Nonetheless, the values of $E(|S_{vh}|^2)$ are typically more than 25 dB smaller than the copolar components consequently reducing the useful distance of measurement.

The potency for separating meteorological signals from nonmeteorological is described by Lakshmanan *et al.* [20]. According to these researchers, the variance of Z_{dr}, the vertical reflectivity structure of the storm, and the specific differential phase are the most important ones.

14.1.2 Unique capabilities of PPAR

Research indicates that warning lead times, detection probabilities, and false-alarm rates improve when volumetric storm observations are made approximately once per minute, as opposed to once per 4–6 minutes with current WSR-88D volume coverage patterns (VCP).

Heinselman et al. [21, 22]22, Bowden *et al.* [23], and Wilson *et al.* [24] indicate that rapid update VCPs significantly improve forecasters' cognitive models of severe storm structure and allow them to more quickly identify precursor signatures.

Rapid update VCPs also improve model analysis of storm spin-up [25, 26], which can improve the performance of the warn-on-forecast system [27].

PPAR provides three basic methods for increasing the volumetric scanning rate. First, its beam(s) may be electronically repositioned—essentially instantaneously—allowing for a decrease in average dwell time (e.g., [28], or adaptive reduction in the number of beam positions to be serviced [29]. Second, array geometries with multiple faces (or cylindrical PPAR (CPPAR) sectors) proportionally reduce solid angle coverage for each face with associated scan time reductions. Finally, digital array architectures enable simultaneous active receive beams to be formed from each array face. When coordinated pulse transmission strategies are used (e.g., [30, 31, 32]); scan times can be reduced by a factor equal to the number of receive beams.

PPAR can also improve the accuracy and coverage of "clear-air" wind measurements in the near-storm environment. Weber *et al.* ([33] Appendix A) show that the high average power that can be radiated using modern solid-state power amplifiers and the capability to integrate over long CPIs would enable an operational PPAR to measure winds using Bragg-scattered echoes to an altitude of about 7 km above ground level. Assimilation of these wind observations could further improve numerical model-based severe weather warnings through better analysis of updraft force, size, strength, and the storm track [34].

Finally, PPAR's flexible and adaptive beamsteering and shaping support new methods for reducing data contamination caused by nonstationary ground scatterers such as automobile traffic and wind turbines. Wind turbine clutter will become increasingly problematic as demand for renewable energy drives new deployments. The US Geological Survey estimates that, on average, 3,000 new turbines have been installed each year since 2005.

14.1.3 Existing PPAR demonstrator systems

A number of academic organizations and government research laboratories are investigating PPAR engineering challenges and applications for weather surveillance. The University of Massachusetts Microwave Remote Sensing Laboratory and Raytheon Corporation collaborated to develop a mobile X-band, dual-polarized phased array weather radar. The array is comprised with 2,560 transmit-receive (T/R) elements, operating with relatively low radiated peak power and a long (20%) duty cycle. The radar operates in the AHV mode. Comparisons with precipitation measurements from a colocated reflector antenna weather radar are described by Heberling *et al.* [35], who correct dual-polarization variable estimates using calculations of the ideal radiating patch patterns and vertically pointing measurements in rain. Tanamachi *et al.* [36] present preliminary observations of a nontornadic supercell using this radar.

Morin *et al.* [37] described an X-band PPAR developed at Colorado State University (CSU). The radar scans electronically in one dimension and mechanically in the other. Its antenna can be rotated 90° about its normal so that electronic steering is performed in either the azimuth or elevation principal planes. The authors present measurements of the variation with an electronic steering angle of H and

V amplitude- and differential-phase patterns, using far-field returns from a corner reflector. These measurements are used to correct steering-angle-dependent biases in Z_{DR} and Φ_{DP}. Measurements of dual-polarization variable estimates when the array is electronically steered are compared with near-coincident measurements obtained by mechanical steering the array and with the collocated CSU Chicago-Illinois dual-pol X-band radar. The authors demonstrate that corrections using the far-field measurements improve agreement between electronically scanned Z_{DR} and Φ_{DP} fields and both "truth" fields.

Chinese researchers have developed an X-band [38] and a C-band PPAR [39]. Both utilize a planar array on a rotating platform and operate in the SHV mode. The beams scan electronically in elevation, thus remaining in the principal elevation plane. The beamwidth of the X-band radar is about 2°, and it takes 107 s to complete a volume scan. In 2017, the Guangzhou municipal meteorological department carried out an observation experiment with four X-band dual-polarization phased array radars. The observation domain nearly covered the urban areas of Guangzhou, China. They demonstrate increased temporal resolution of small weather phenomena.

Since the beginning of 2018, an X-band dual-polarization phased array radar has operated in the Tokyo metropolitan area [40, 41]. The radar has a planar array on a rotating platform and uses the SHV mode. The array is one-dimensional and has 12 radiating elements for transmission of several fan beams (four in operational and seven in research mode) to scan elevation in the principal plane; beamwidths are 5° to 27° in elevation and 1.2° in azimuth. At the lowest elevation scan, the radar transmits the 5° beam and completes the highest scan (65° to 90°) with the 25° beam. The receiver has 112 elements (spaced at $\lambda/2$), it records and synthesizes beams 1° wide. The volume scan in typical operation mode takes a minute but in a research mode it can be as short as 10 s. The authors successfully corrected the reflectivity for attenuation by using K_{DP} and have identified K_{DP} columns in the range height indicator (RHI) images from convective cells. For understanding the relations of the column to the updraft and downdraft, they call for three-dimensional analysis of the K_{DP} fields. They compare polarimetric variables computed from the disdrometer (31.2 km away) with radar measured above the disdrometer and note good agreement. However, they report an artifact of high reflectivities near the edges of different fan beams and speculate that "beamforming, pulse compression, and combination of the different fan beams" may be related to the high reflectivity lines.

Zhang *et al.* [42] proposed a CPPAR, where azimuth scanning is accomplished by commutating sectors of the array. Electronic steering in elevation provides volumetric coverage with a beam that is always in the principal elevation plane of the array. This eliminates the need to correct for biases due to strong cross-polar radiation. Li *et al.* [43] provided recent results from the CPPAR demonstrator developed at the University of Oklahoma. The demonstrator is a 2-m diameter, 2-m high cylinder, with 96 columns of radiating elements around its perimeter. A far-field calibration horn installed on a nearby building allowed for a determination of amplitude and phase weights for each column that optimizes the antenna patterns in terms of sidelobe levels and the match between the H and V beams. The authors present measurements of conventional and polarimetric variables, using a mechanically scanned single beam to validate the CPPAR commutating electronic scan concept. Earlier CPPAR studies include Karimkashi and Zhang [44], Lei *et al.* [45], Fulton *et al.* [46], and Saeidi-Manesh and Zhang [47].

The Massachusetts Institute of Technology (MIT) Lincoln Laboratory, National Severe Storms Laboratory, and the University of Oklahoma developed a scaled, S-band PPAR to refine and demonstrate the low-cost panel technology used for the subsequent advanced technology demonstrator (ATD) (see section 14.4). This ten-panel demonstrator (TPD) is comprised with 8×8 T/R-element panels, arranged in a planar configuration with two rows of five such panels. Peak radiated power is 6 W (H and V) per T/R element with a maximum duty cycle of 10%. Effective two-way 3-dB beamwidths (accounting for the amplitude taper used to reduce receive sidelobes) are 2.7° in elevation and 6.7° in azimuth when the array is operated with its rows oriented vertically. Ivić and Byrd [6] presented vertically pointing TPD measurements in stratiform rain, with electronic scanning in the principal elevation plane (i.e., "azimuth" is 0° relative to array broadside) and in planes offset by ±3.5° in "azimuth." Dual-polarization variables differ from the variables obtained by simulating patterns of radiating antenna patches. This is attributed to the significant effects of the active hardware in the antenna back end.

Yeary *et al.* [48] and Fulton *et al.* [49] describe an "all-digital" PPAR prototype being developed by the University of Oklahoma. This PPAR independently controls transmissions and digitizes received signals from each element in the array. As described in section 14.3, an all-digital array enables beams to transmit and receive signals simultaneously from many directions. This capability may be necessary to achieve desired temporal resolution for PPAR. An all-digital PPAR may be continuously calibrated by exploiting mutual coupling between its individual radiating elements. Lujan *et al.* [50] describe mutual-coupling methods and initial analytic and experimental results. All-digital arrays inherently have a high dynamic range since noise from separate elements is uncorrelated, allowing simultaneous observations of low-reflectivity clear-air echoes and high-reflectivity/clutter regions.

14.2 Challenges with PPAR

One of the major challenges for weather surveillance with PPAR is the calibration needed to produce measurements comparable with those of the systems using parabolic-reflector antennas. The first significant issue with PPAR is the scan-dependent measurement bias caused by the H and V copolar antenna patterns, which vary with beamsteering directions (referred to as copolar biases). Second, significant cross-polar antenna patterns on reception induce cross-coupling between returns from the horizontally and vertically oriented fields causing biases of polarimetric variable estimates (referred to as cross-coupling biases). Furthermore, coupling in hardware is likely to exacerbate the cross-coupling effects. These effects on the received voltages are described as

$$\begin{aligned}\begin{bmatrix} V_h(m) \\ V_v(m) \end{bmatrix} &= C\int_\Omega \begin{bmatrix} F^r_{hh} & 0 \\ 0 & F^r_{vv} \end{bmatrix} \mathbf{S}_D \mathbf{E}^t d\Omega + C\int_\Omega \begin{bmatrix} 0 & F^r_{vh} \\ F^r_{hv} & 0 \end{bmatrix} \mathbf{S}_D \mathbf{E}^t d\Omega, \\ &= C\int_\Omega \begin{bmatrix} F^r_{hh} S_{hh}(m) E^t_h(m) \\ F^r_{vv} S_{vv}(m) E^t_v(m) \end{bmatrix} d\Omega + C\int_\Omega \begin{bmatrix} F^r_{vh} S_{vv}(m) E^t_v(m) \\ F^r_{hv} S_{hh}(m) E^t_h(m) \end{bmatrix} d\Omega, \\ &= \begin{bmatrix} V^{co}_h(m) \\ V^{co}_v(m) \end{bmatrix} + \begin{bmatrix} V^{x}_h(m) \\ V^{x}_v(m) \end{bmatrix}\end{aligned} \tag{14.15}$$

where $V^{co}_h(m)/V^{co}_v(m)$ represent voltages caused by H/V polarized returns, while $V^{x}_h(m)/V^{x}_v(m)$ are caused by V/H polarized returns in H/V channel. The latter are the unwanted returns caused by cross-coupling (cross-coupling returns) while the former are the desired returns (copolar returns). Both are affected by the variations of transmit and receive antenna patterns as well as variations of the observed phenomena. The effects on second-order moments are

$$\begin{aligned} V_p(m)\, V^*_q(m+k) = V^{co}_p(m)\, V^{co*}_q(m+k) + V^{co}_p(m)\, V^{x*}_q(m+k) + \\ V^{x}_p(m)\, V^{co*}_q(m+k) + V^{x}_q(m)\, V^{x*}_q(m+k) \end{aligned} \tag{14.16}$$

where all terms, except the first one, are caused by the cross-coupling. The effects on the power estimates in the SHV and AHV modes are in

$$\begin{aligned} \mathbb{E}\left[|V_h(m)|^2\right] &= C^2 \mathbb{E}\left[|S_{hh}(m)|^2\right] \int_\Omega \begin{bmatrix} |F^r_{hh} E^t_h(m)|^2 + \frac{1}{Z_{dr}} |F^r_{vh} R^t_v(m)|^2 + \\ \frac{2|\rho_{hv}|}{\sqrt{Z_{dr}}} \mathrm{Re}\left(\left[F^r_{hh} E^t_h(m)\right]^* F^r_{vh} E^t_v(m) e^{j\Phi_{DP}}\right) \end{bmatrix} d\Omega \\ \mathbb{E}\left[|V_v(m)|^2\right] &= C^2 \mathbb{E}\left[|S_{vv}(m)|^2\right] \int_\Omega \begin{bmatrix} |F^r_{vv} E^t_v(m)|^2 + Z_{dr} |F^r_{hv} R^t_h(m)|^2 + \\ 2\sqrt{Z_{dr}} |\rho_{hv}| \mathrm{Re}\left(\left[F^r_{vv} E^t_v(m)\right]^* F^r_{vh} E^t_h(m) e^{-j\Phi_{DP}}\right) \end{bmatrix} d\Omega \end{aligned} \tag{14.17}$$

where the excitation vector is $\boldsymbol{e} = [A_h(m)\exp[j\alpha_h(m)]\ A_v(m)\exp[j\alpha_v(m)]]^T$ and

$$\begin{aligned} &\text{SHV}: A_h(m) = A_v(m) = 1 \\ &\text{AHV}: A_h(m) = \frac{1+(-1)^m}{2} \text{ and } A_v(m) = \frac{1+(-1)^{m+1}}{2} \end{aligned} \tag{14.18}$$

while $\alpha_h(m)$ and $\alpha_h(m)$ denote the phase shifts imposed on transmission in H and V to implement phase coding techniques [51–53]. The expression demonstrates that the second-order moment estimates are functions of both the antenna patterns and the properties of the observed phenomena.

The conventional estimators of polarimetric variables in the SHV mode are

$$\begin{aligned} &Z_h = 10\log_{10}\left(B\hat{S}_h\right), \hat{Z}_{DR} = 10\log_{10}\left(\frac{\hat{S}_h}{\hat{S}_v}\right) \\ &|\hat{\rho}_{hv}| = \frac{|\hat{R}_{hv}(0)|}{\sqrt{\hat{S}_h \hat{S}_v}}, \hat{\Phi}_{DP} = \arg\left[\hat{R}_{hv}(0)\right] \end{aligned} \tag{14.19}$$

where

$$\hat{S}_{\rm h} = \frac{1}{M}\sum_{m=0}^{M-1}|V_{\rm h}(m)|^2 - N_{\rm h}, \hat{S}_{\rm v} = \frac{1}{M}\sum_{m=0}^{M-1}|V_{\rm v}(m)|^2 - N_{\rm v}$$
$$\hat{R}_{\rm hv}(0) = \frac{1}{M}\sum_{m=0}^{M-1} V_{\rm h}^*(m)V_{\rm v}(m)e^{j[\alpha_{\rm h}(m)-\alpha_{\rm v}(m)]} \tag{14.20}$$

while $N_{\rm h}$ and $N_{\rm v}$ are noise powers in H and V, B is the calibration constant, and M is the number of transmitted pulses in a dwell. In the AHV mode, the estimators are more complex

$$Z_{\rm h} = 10\log_{10}(B\hat{S}_h), \hat{Z}_{\rm DR} = 10\log_{10}\left(\frac{\hat{S}_{\rm h}}{\hat{S}_v}\right),$$
$$|\hat{\rho}_{\rm hv}(0)| = \frac{|\hat{R}_a| + |\hat{R}_b|}{2(\hat{S}_{\rm h}\hat{S}_{\rm v})^{\frac{3}{8}}\,|\hat{R}_{\rm h}(2)\hat{R}_{\rm v}(2)|^{\frac{1}{8}}}, \hat{\Phi}_{\rm DP} = \frac{\arg|\hat{R}_a\hat{R}_b^*|}{2} \tag{14.21}$$

where

$$\hat{S}_{\rm h} = \frac{2}{M}\sum_{m=0}^{M/2-1}|V_{\rm h}(2m)^2|, \hat{S}_v = \frac{2}{M}\sum_{m=0}^{M/2-1}|V_{\rm v}(2m+1)^2|,$$
$$\hat{R}_a = \frac{2}{M}\sum_{m=0}^{M/2-1} V_{\rm h}^*(2m)V_{\rm v}(2m+1), \hat{R}_b = \frac{1}{M/2-1}\sum_{m=0}^{M/2-2} V_h(2m+2)V_{\rm v}^*(2m+1)$$
$$\hat{R}(2) = \frac{2}{M}\sum_{m=0}^{M/2-2} V_{\rm h}^*(2m)V_{\rm h}(2m+2), \hat{R}_{\rm v}(2) = \frac{2}{M}\sum_{m=0}^{M/2-2} V_{\rm v}^*(2m+1)V_{\rm v}(2m+3) \tag{14.22}$$

The measure of the estimates quality is the bias and standard deviation (SD), and herein we focus on the biases induced by the copolar and cross-polar patterns. In this respect, the requirement for the bias limits is ±1 dB for $\hat{Z}_{\rm h}$ and ±0.1 dB for intrinsic (i.e., true) $Z_{\rm DR}$ between 0 and 1 dB (or ±0.2 dB for relaxed requirements) and less than $0.1\times Z_{\rm DR}$ (or $\pm 0.2\times Z_{\rm DR}$ dB for relaxed requirements) for larger $Z_{\rm DR}$ [54]. In the case of $|\hat{\rho}_{\rm hv}|$, the bias of estimates within ±0.01 is deemed sufficient for "*sensing the mixed-phase precipitation and gauging the hail size quantitatively*" [55]. At the same time, National Oceanic and Atmospheric Administration (NOAA)/NWS radar functional requirements (RFR) [56] set the bias limits for $|\rho_{\rm hv}|$ estimates to an even more stringent value of ±0.006. For $\hat{\Phi}_{\rm DP}$, NOAA/NWS RFR bias limit requirements are ±1°. Furthermore, in real systems $\hat{\Phi}_{\rm DP}$ always contains the system differential phase $\Phi_{\rm DPsys}$ $\left(\text{i.e.,}\Phi_{\rm DPsys} \approx arg\left(\left(F_{\rm hh}^{t}F_{\rm hh}^{r}\right)^* F_{\rm vv}^{t}F_{\rm vv}^{r}\right)\right)$, so that,

$$E\left[\hat{\Phi}_{\rm DP}\right] = \Phi_{\rm DPsys} + \Phi_{\rm DP} + \delta \tag{14.23}$$

In operational systems, $\Phi_{\rm DPsys}$ is often measured and subtracted from $\Phi_{\rm DP}$ to extend the interval in which the estimates do not need to be unwrapped.

The copolar biases are summarized as

$$BIAS_{\mathrm{CO}}(\hat{Z}_{\mathrm{h}}) = 20\log_{10}(|F^t_{\mathrm{hh}}F^r_{\mathrm{hh}}|), BIAS_{\mathrm{CO}}(\hat{Z}_{\mathrm{DR}}) = 20\log_{10}\left(\frac{|F^t_{\mathrm{hh}}F^r_{\mathrm{hh}}|}{|F^t_{\mathrm{vv}}F^r_{\mathrm{vv}}|}\right),$$
$$BIAS_{\mathrm{CO}}(\hat{\Phi}_{\mathrm{DP}}) = \arg\left((F^t_{\mathrm{hh}}F^r_{\mathrm{hh}})^* F^t_{\mathrm{vv}}F^r_{\mathrm{vv}}\right) \tag{14.24}$$

Hence, to correct for these (i.e., to calibrate the radar), copolar patterns need to be characterized for beamsteering directions of interest. Note, however, that the copolar pattern characterization only at and around main beam peaks should suffice for correction derivation. In that regard, the following expression describes the limits on two-way copolar power measurement errors ΔP to keep $\hat{Z}_{\mathrm{DR}}$ biases within the desired limits ΔZ_{DR} (e.g., 0.1 or 0.2) as

$$-\Delta Z_{\mathrm{DR}} < 10\log_{10}\left[\left(1+\tfrac{\Delta P}{P_{\mathrm{h}}}\right)/\left(1-\tfrac{\Delta P}{P_{\mathrm{v}}}\right)\right] < \Delta Z_{\mathrm{DR}} \tag{14.25}$$

where P_{h} and P_{v} are true peak powers in H and V. Note that the measurement error is taken to be positive in the numerator and negative in the denominator because the bias in $\hat{Z}_{\mathrm{DR}}$ is the largest in such case. If $P_{\mathrm{h}} = P_{\mathrm{v}} = P$, the expression (14.11) suggests that $10\log_{10}(\Delta P/P)$ must be less than −20/−17 dB if ΔZ_{DR} = 0.1/0.2 dB.

Cross-coupling biases are more difficult to assess due to their dependency on the observed phenomena. To achieve this, we set

$$F^t_{\mathrm{hh}} = F^t_{\mathrm{vv}}, F^r_{\mathrm{hh}} = F^r_{\mathrm{vv}}, F^t_{\mathrm{vh}} = -F^t_{\mathrm{hv}}, F^r_{\mathrm{vh}} = -F^r_{\mathrm{hv}} \tag{14.26}$$

The logic behind this is that the polarimetric patch elements are designed so that the patterns in V can be obtained from H patterns by rotating them by 90° in spherical coordinates. If H pattern is rotated clockwise, relations in (14.12) are valid along principal planes, and their validity diminishes for beamsteering directions away from principal planes. Next, because the cross-coupling biases exhibit a strong dependency on Φ_{DP} but also on $arg\left(F^{t*}_{\mathrm{hh}}F^t_{\mathrm{vh}}\right)$, the two phases are varied from 0° to 360° in 5° steps. At the same time, $20log_{10}\left[|F^t_{\mathrm{vh}}|/|F^t_{\mathrm{hh}}|\right]$ is varied from −54 to −44 dB in the SHV and −26 to −16 dB in the AHV mode. In both modes, Z_{DR} is set to 0 dB and ρ_{hv} to 0.99. Cross-coupling biases are computed using analytical expressions that can be found in [57] for the SHV mode. For the AHV mode, the analytical expressions are derived following the approach in (14.3) but are not given here for brevity.

Details for all variables are given in Table 14.2. The maximum biases in the SHV modes indicate that the powers of cross-polar patterns must be −51/−45 dB below copolar patterns to keep $\hat{Z}_{\mathrm{DR}}$ bias within ±0.1/0.2 dB. In the AHV mode, the cross-polar isolation can be relaxed to −22.5/−19.5 dB.

In general, polarimetric calibration consists of measuring the imperfections (e.g., channel mismatches and leakages) in the radar system and using these to correct (or calibrate) the quantities measured by the radar. Unlike reflector antennas that can be characterized by a single set of antenna radiation patterns and thus require a single set of calibration parameters, PPAR antennas exhibit different copolar and cross-polar patterns for each electronic beamsteering angle and require a different set of calibration parameters for each. Because copolar beam shape mismatches cannot be corrected via polarimetric calibration or through signal processing, copolar

Table 14.2 Estimated biases in the SHV and AHV modes

Estimate	SHV Max. bias at cross-polar isolation −51/−45 dB	AHV Max. bias at cross-polar isolation −22.5/−19.5 dB
$\hat{Z}_h$	−0.05/−0.1 dB @ $arg\left(F'^{*}_{hh}F'_{vh}\right) =$ 0°/180° and Φ_{DP} = 180°/180°	−0.05/−0.07 dB @ $arg\left(F'^{*}_{hh}F'_{vh}\right) =$ 0°/0° and Φ_{DP} = 180°/180°
$\hat{Z}_{DR}$	−0.1/−0.2 dB @ $arg\left(F'^{*}_{hh}F'_{vh}\right) =$ 180°/180° and Φ_{DP} = 180°/180°	−0.1/−0.2 dB @ $arg\left(F'^{*}_{hh}F'_{vh}\right) =$ 45°/45° and Φ_{DP} = 280°/280°
$\left\|\hat{\rho}_{hv}(0)\right\|$	−0.0001/−0.0002 @ $arg\left(F'^{*}_{hh}F'_{vh}\right) =$ 90°/90° and Φ_{DP} = 90°/90°	−0.0025/−0.005 @ $arg\left(F'^{*}_{hh}F'_{vh}\right) =$ 225°/45° and Φ_{DP} = 270°/270°
$\hat{\Phi}_{DP}$	0.65°/1.3° @ $arg\left(F'^{*}_{hh}F'_{vh}\right) =$ 90°/90° and Φ_{DP} = 180°/180°	−0.0025°/−0.005° @ $arg\left(F'^{*}_{hh}F'_{vh}\right) =$ 90°/90° and Φ_{DP} = 180°/180°

beam shape matching must be achieved via precise beamforming. With this assumption, copolar biases can be obtained (and thus corrected) by measuring the transmit and receive copolar beam patterns and cross-coupling biases by measuring both copolar and cross-polar beam patterns. However, if the cross-coupling biases are effectively mitigated using methods like the one discussed in the next section, compensation via polarimetric calibration may not be needed. So, in essence, PPAR calibration requires accurate characterization of copolar and potentially cross-polar beam patterns for each electronic beamsteering direction. This characterization may be obtained using computational EM modeling, measurements in an anechoic chamber, or measurements using external objects such as unmanned aerial systems [58], or a calibration tower [59]. Also, it may require additional infrastructure (e.g., a pedestal and/or a turntable to mechanically reposition the antenna) and measurement systems that introduce minimum distortions (e.g., from multipath or thermal fluctuations). More importantly, to maintain calibration over time, the antenna must be designed so that performance variations occur on time scales that are much longer than the frequency at which beam pattern characterization can be repeated. Although significant progress has been made with the calibration of PPAR (e.g., [49, 59], to date, no polarimetric calibration technique for operational implementation has been demonstrated.

14.2.1 Cross-coupling bias mitigation in the SHV mode

To mitigate the cross-coupling effect in the SHV mode, pulse-to-pulse phase coding in either the horizontal or vertical ports of the transmission elements has been proposed [60, 53, 61, 62]. The basic idea behind this method is to induce the changing phase in the M transmitted pulses, such that,

$$\sum_{m=0}^{M-1} \exp\left[-j2\pi\alpha_c(m)\right] = 0 \tag{14.27}$$

where c is either "h" or "v" (e.g., $\alpha_c(m)$ may be altered between 0° and 180°). The application of the same approach used to assess the cross-coupling biases in the SHV and AHV modes suggests the following for the phase-coded SHV (PCSHV) mode. To keep the cross-coupling $\hat{Z}_{DR}$ biases within ±0.1/±0.2 dB, the power of cross-polar patterns must be more than −22/−19 dB below the copolar pattern powers. Furthermore, to keep the $\hat{\Phi}_{DP}$ cross-coupling biases within ±2.5°, the isolation ought to be at least −20.5 dB. In the case of ρ_{hv}, the isolation of at least −29/−31 dB is needed to keep the $|\hat{\rho}_{hv}(0)|$ bias less than 0.01/0.006. Regarding the latter, one promising method that has the potential to further reduce the biases in ρ_{hv} estimates, applies the filtering in spectral domain for additional reduction of cross-coupling effects [61]. As a result, the required isolation needed to maintain the cross-coupling biases of ρ_{hv} estimates within the desired limits may be relaxed.

The mitigation of cross-coupling effects achieved via pulse-to-pulse phase coding is inversely proportional to the cross-polar pattern levels that increase as beams are steered away from principal planes. Thus, at boresight locations where the cross-coupling mitigation from phase coding is insufficient, one option is to use the correction, which requires the knowledge of both the copolar and cross-polar patterns. A theoretical formulation for such bias correction is given by Zhang *et al.* [63] and Zrnić *et al.* [64] by treating each PPAR radiation element as a pair of crossed Hertzian dipoles or idealized aperture as well as a patch [65]. These works assume that the radar is reciprocal (i.e., transmit and receive patterns are the same), which is usually not the case in phased arrays if active components (e.g., T/R modules) are used. Also, the real patch radiators are likely to behave differently than an ideal Hertzian dipole or idealized aperture/patch. One approach, which is applicable in the SHV mode and makes no assumption on the nature of transmit and receive antenna patterns (but rather requires them to be known) is presented by Fulton and Chappell [66] along with experimental evaluation. This approach requires the manipulation of T/R module amplitude and phase excitations to implement corrections on transmit. However, the limited accuracy of the attenuators and phase shifters in T/R modules may degrade the correction effectiveness. Furthermore, different amplitudes and phases would need to be imposed for each beamsteering angle and would likely complicate the hardware implementation as well as reduce the radar transmit power. Another method that performs corrections during the signal processing stage, and therefore, does not require manipulations in hardware is proposed by [57]. The framework for this approach is based on the analytical derivations, which suggest that the vector of raw (or biased) estimates of powers in H and V as well as cross-correlation is a product of a 4×4 matrix and a vector of unbiased estimates. This implies that the 4×4 matrix can be inverted to produce the correction matrix (CM). This matrix can be multiplied with the vector of biased second-order moment estimates to obtain the corrected estimates. The corrected second-order moments are then used to compute $\hat{Z}_{DR}$, $|\hat{\rho}_{hv}(0)|$, and $\hat{\Phi}_{DP}$. Elements of CM are functions of transmit and receive copolar and cross-polar patterns, so this method requires measurement of both copolar and cross-polar beams (at and around the main beam peaks). Note that this method corrects both copolar and cross-coupling biases at the same time.

In addition to the cross-coupling mitigation approaches that operate strictly in the SHV mode, another technique that requires fully polarimetric measurements (available only in the AHV mode) has been proposed by Galletti *et al.* [67]. This technique requires an extension to the SHV mode that includes transmission in each of the two polarizations (this is referred to as the SHV orthogonal four-pol pulsing scheme in Reference [67]). This enables measurement of second-order moments at lags 1 and 2, which are needed for estimating the copolar correlation coefficient. However, the use of nonzero lags will likely produce correlation coefficient estimates of lower quality for data from scans with long PRTs due to the insufficient correlation in sample time. While promising, this technique has not been tested for application on PPARs.

14.2.2 Cost of PPAR implementation

The implementation of a phased array requires a large number of electronic components, including monolithic microwave integrated circuits (MMICs), multilayer radio frequency (RF) printed circuit boards (PCBs), and RF cables and connectors. These components tend to drive the cost of phased arrays and limit their usage to military applications. The typical cost distribution for military radar is in Figure 14.2, where the phased array represents nearly half of the recurring cost of the system. In the phased array portion of the radar, about half of the cost of the phased array is in the T/R modules. In order to broaden the use of phased arrays, a significant reduction in cost is necessary. Based on the cost distribution (Figure 14.2), the best opportunity for cost reduction is in the T/R modules, the RF boards, and the cabling. Several factors contribute to the high cost of T/R modules. These factors, in order of cost, are the MMICs, housing, ceramic substrates, other electrical/electronic componentry, and assembly, test, and quality control labor. To first order, MMIC costs are driven by total integrated circuits (ICs) area, the choice of IC technology, and market forces. In some cases, where there has been overlap in commercial and military need for MMICs in terms of frequency or function, the defense marketplace has benefited from much lower MMIC costs. Examples of this benefit are predominantly in the commercial wireless bands at L-, S-, and C-bands.

The T/R module housing provides environmental protection and electrical isolation, and the cost is directly proportional to the design sophistication. Traditional module housings are hermetically sealed, gold-plated boxes that incorporate materials with disparate thermal conductivities that must be matched for thermal expansion. Connectors that penetrate the housing walls must also be hermetic. A typical T/R module housing is in Figure 14.3. Most housing designs are tailored to the specific application. Factors that drive housing design include the aperture lattice dimensions, heat dissipation of the active components, hermetic sealing method, connectors, and the method of attaching the module to the next level assembly. The substrates employed in T/R modules are ceramic-based materials with gold or silver trace metals. Common implementations include thin- and thick-film alumina, and low- and high-temperature cofired ceramics. Other electrical components in the module include capacitors, resistors, circulators, power conditioning circuits, and control circuits.

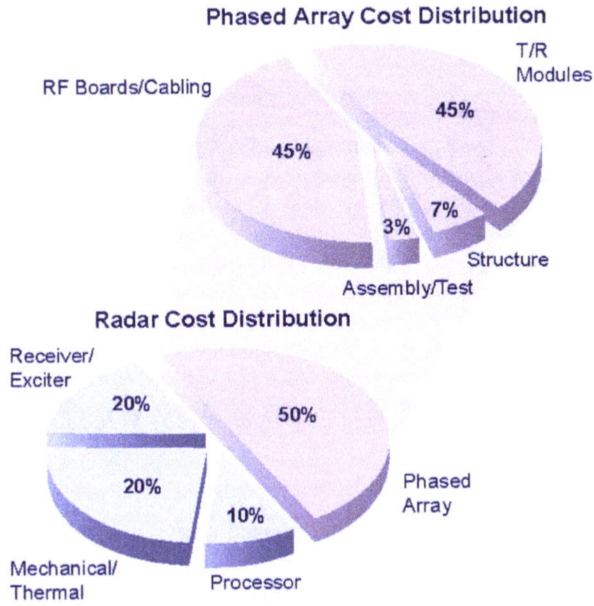

Figure 14.3 *Typical cost distribution for phased array radar [Figure 9[68] © IEEE. Used with permission.]*

The traditional manufacturing technique used to build T/R modules is called ‘‘chip and wire.’’ In chip-and-wire assembly, MMICs and other ICs are handled in bare die form and are done so in a humidity-controlled clean room environment. As opposed to plastic encapsulated MMICs used in commercial manufacture, ICs in die form require additional touch labor before use including on-wafer test, visual inspection, and packaging for shipment. As a result, the cost of an IC in die form can be many times more than its commercially packaged equivalent. Die interconnects are made with gold wire or ribbon, and dies are attached with silver epoxy or gold-tin solder. Resistor and capacitor metallurgy must be compatible with these materials, thus increasing the cost of chip-and-wire assemblies. For example, a capacitor for a chip-and-wire assembly would cost \$0.50–\$1.00, whereas a similar one for commercial (solder-based) manufacture would be \$0.001.

Following the T/R modules, traditional phased array cost is driven by the multi-layer RF boards and cables. This is made clear by considering the current approach to constructing large phased arrays, depicted in Figure 14.4. This architecture is commonly referred to as a slat array, which consists of a series of slats that are oriented perpendicular to the face of the array. The slat approach has the advantage of providing a large surface area for attaching the T/R modules and supporting electronic components. In addition, the high-power amplifier thermal load can be spread out over a large volume – the aperture area times the depth of the slat. A key disadvantage is that it requires a large number of RF boards and cabling to route the RF, DC, and control signals on and off of the slats.

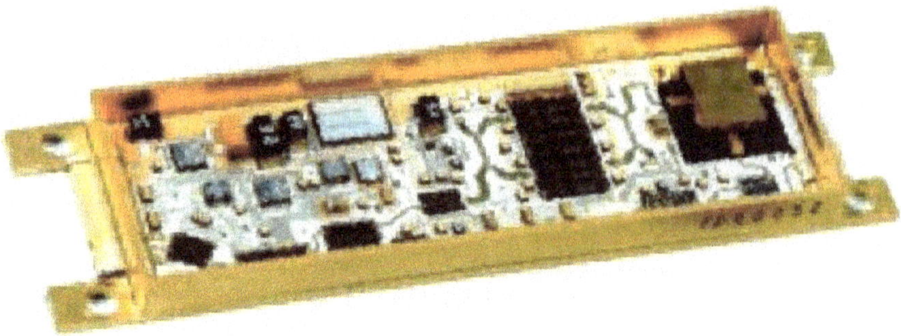

Figure 14.4 Active X-band T/R module [Figure 5[68] © IEEE. Used with permission.]

An alternative approach to building phased arrays is shown in Figure 14.5. In this approach, commonly referred to as a tile architecture, the array is constructed of layers that are oriented parallel to the face of the array. The antenna elements and RF beamformers are integrated into a single-multilayer RF board, with the T/R modules mounted directly onto the back of the board. This approach has the impact of significantly reducing the area of the RF boards and dramatically reduces the number of connectors and cables. For example, assuming a 5-cm element spacing at S-band, a 30-cm deep slat array requires 6 m^2 of board area versus 1 m^2 for a tile array.

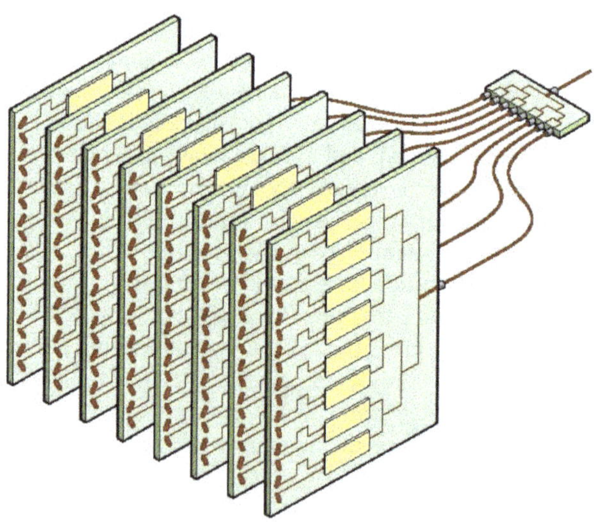

Figure 14.5 Example slat array configuration [Figure 10[68] © IEEE. Used with permission.]

In the low-cost tile approach, the T/R modules are designed and developed to leverage high-volume commercial microwave packaging and manufacturing techniques. Figure 14.6 shows a sample T/R module developed for the multifunction phased array radar (MPAR) program [69, 70] under support from the Federal Aviation Administration and the NOAA. The T/R module MMICs are packaged in low-cost commercial packages, referred to as quad flat no-lead (QFN). The QFN packages are directly soldered onto an inexpensive PCB, which is then soldered directly to the back of the tile. It should be noted that the QFN packages are non-hermetic, and the environmental control for protection of the components must be accomplished at the tile or array level, rather than at the T/R module level. The RF and DC interconnections between the T/R module and the back of the tile consist

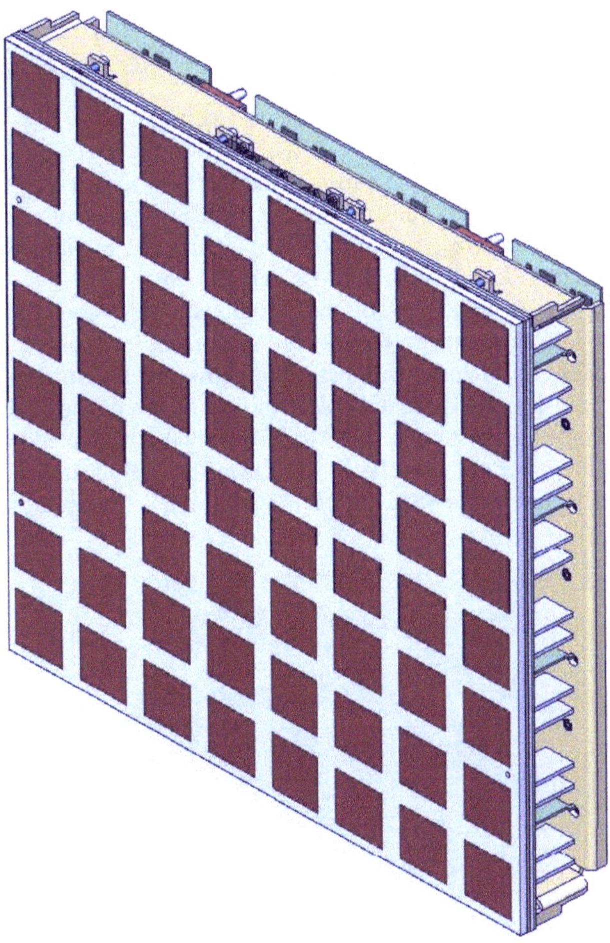

Figure 14.6 Example tile array configuration. [Figure 11[68] © IEEE. Used with permission.]

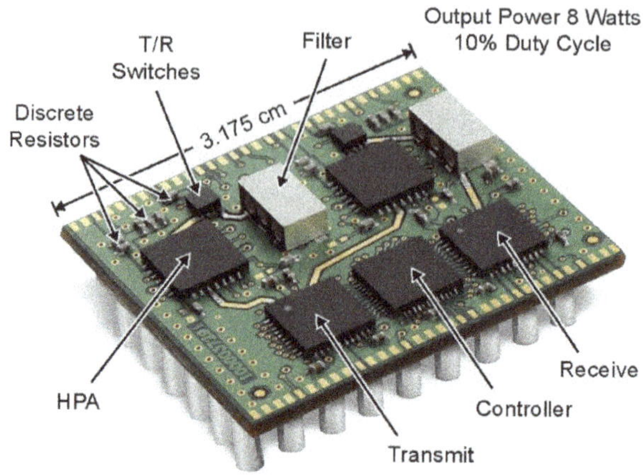

Figure 14.7 Low-cost MPAR T/R module [Figure 12[68] © IEEE. Used with permission.]

of simple metal pads that line the edges of the PCB, as shown in Figure 14.7. The MPAR T/R module also uses transmit-receive switches instead of circulators, which further reduce cost and size. Low-loss T/R switches are implemented using PIN diodes, with insertion loss comparable with a circulator.

Because of the compactness of the design, the RF signal transmission line length in the T/R module substrate is <2 cm. As a result, low-cost, high-loss tangent (>0.01) substrates can be employed instead of high-cost, low-loss tangent (<0.004) materials. The net penalty in added insertion loss of the transmission line is <0.1 dB. In high volume, the cost of the MPAR T/R module is projected to be ~$25, with the bulk of that number going to the MMIC cost. The cost of the substrate, componentry, assembly, and test is <$5. Figure 14.8 shows a fully assembled MPAR tile,

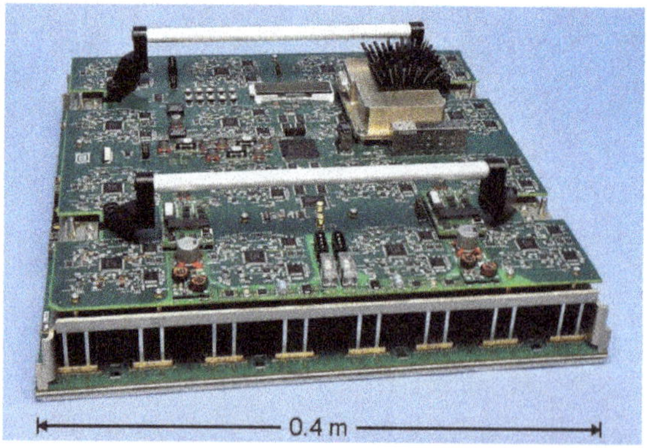

Figure 14.8 Fully assembled MPAR tile. [Figure 13[68] © IEEE. Used with permission.]

which provides a low-cost, scalable building block for large array apertures. As the volume of manufacturing increases, investments in automated assembly and test equipment bring down the cost per unit and improve the yield of the MPAR tiles.

A simple cost model described by Herd and Conway [68] can be used to understand the differences between the slat and tile architectures. A key difference between the slat and tile array architectures is the area of the RF boards. Many of the costs are proportional to the number of RF boards and the total board area. The number of slats per unit area of the array increases directly with frequency because slats are typically spaced at one-half wavelength intervals. As a result, the cost differences between the slat and the tile arrays will be larger at higher frequencies. An example of the cost per unit area for a slat array and a tile array is shown in Figure 14.9. The parameters and assumptions used to calculate these individual costs are given by Herd and Conway [68].

This example assumes an identical RF chipset cost. The higher cost of the slat array is due to the metal T/R module packages, the increased RF board area, and the cooling apparatus (which is proportional to RF board area). Using the cost model by Herd and Conway [68], the relative transmit power per unit area versus cost per unit area for both the slat and tile arrays is plotted for two different transmit power levels. This example shows an improvement ratio of more than 5:1 in cost per unit area for both the low- and high-power cases. For both cases, the dominant cost drivers for the slat array are the RF board area and the T/R module packaging. By minimizing the RF board area, utilizing low-cost T/R packaging, and leveraging high-volume manufacturing and test techniques, the tile array architecture provides a low-cost alternative to slat arrays (Table 14.3).

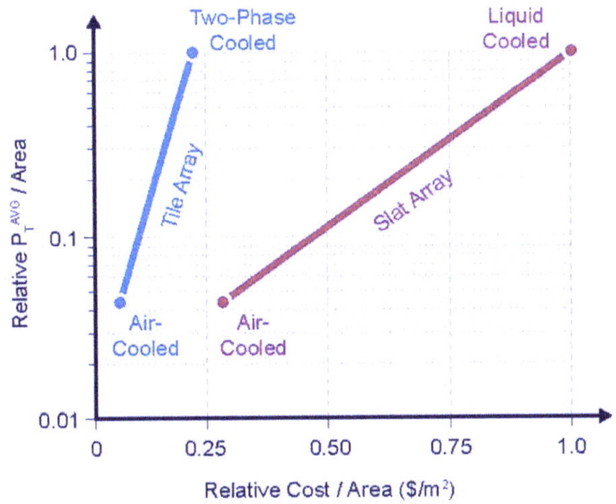

Figure 14.9 Relative transmit power per unit area versus cost per unit area for tile and slat arrays [Figure 15[68] © IEEE. Used with permission.]

Table 14.3 Example cost comparison for slat versus tile arrays

Component	Slat ($/m²)	Tile ($/m²)
T/R chip set	$80,000	$80,000
Die handling	$40,000	N/A
T/R module packages	$400,000	$8,000
RF boards	$150,000	$25,000
Cables and connectors	$10,000	$1,562
Cooling	$180,000	$40,000
Structure	$50,000	$16,250
Assembly and test	$80,000	$12,500
Total	$990,000/m²	$183,312/m²

[Table 1[68] © IEEE. Used with permission]

A critical challenge for tile arrays at high-power densities is the integration of a coldplate into the thin tile structure. In order to maintain low cost at higher power levels, a thin two-dimensional heat exchanger must be incorporated into the PCB stackup using standard high-volume manufacturing techniques. This is a technology that is currently being developed.

14.3 Antennas

14.3.1 Antenna geometry

The geometric configuration of the PPAR has a substantial impact on operational capability, the complexity of the design and manufacturing process, calibration methods, and cost. Here, we summarize three alternatives that have been discussed in the meteorological literature.

A multifaced, planar array is an attractive configuration, in that, it does not require mechanical rotation of the array and facilitates rapid scanning since each array face is responsible for servicing beam positions only within an azimuth interval reduced (relative to 360°) in proportion to the number of faces. For a four-faced array, the time required to replicate current WSR-88D volume-scanning patterns would be reduced to approximately 1 minute. The planar geometry is by far the most common configuration for industry-built military PARs and is compatible with the low-cost tiled architecture discussed in section 14.2.

This geometry, however, requires that beams be electronically steered to angles significantly off broadside to the array. For a four-faced array, steering angles of ±45° in azimuth and 20° or more in elevation will be required, with associated increases in beamwidth and reduction in antenna gain. Given specific requirements for beamwidth and antenna gain, the active aperture area must be more than doubled to replicate broadside performance at these large steering angles.

Calibration and bias compensation for dual-polarization variable estimates are also challenging for this configuration. Required electronic steering angles significantly exceed the envelope within which pulse phase coding is sufficient to mitigate H-V cross-coupling bias in SHV mode [60]. Thus, accurate co- and cross-polar array calibrations will be required for each of the steering angles needed to cover the angular sector assigned to the array faces. For a 1° beamwidth, four-faced array, approximately 1,800 beam positions will need to be calibrated.

The advantages of a cylindrical array geometry have been discussed in section 14.3. Electronic commutation of multiple azimuth sectors and electronic steering in elevation allows CPPAR to provide volumetric coverage with beams that is always in the principal elevation plane of the array. This eliminates the need to correct for polarimetric variable biases due to strong cross-polar radiation and significantly reduces scan-induced beam broadening and antenna gain decrease.

A single-faced array that rotates in azimuth to provide 360° coverage will have significantly lower acquisition and life cycle cost than either of the stationary configurations. As discussed in section 14.2, the area of active aperture is a major component of phased array radar manufacturing cost and will scale with the number of faces. Array power consumption, cooling, routine maintenance, and failed T/R element replacement costs will also increase in proportion to the number of faces. The principal challenge for this configuration is that the multiple, simultaneous transmit/receive beams required to achieve more rapid volumetric scanning must be radiated from the single array face, rather than being distributed across multiple faces (or CPPAR sectors). Accomplishing this requires a highly digital array architecture, variants of which are described in the following subsection.

14.3.2 Array architectures

14.3.2.1 Passive phased arrays

The earliest electronically steered phased arrays utilized a centralized high-power transmit amplifier and receiver as shown in Figures 14.10 and 14.11. Passive arrays are a natural evolution from large parabolic dish antennas, which also use a centralized transmitter source. The passive array architecture requires the use of low-loss analog beamformers and phase shifters, because the RF losses on the output path from the transmit amplifier to free space translate directly into signal loss and, thus, reduced radar sensitivity. High-power centralized transmitters typically utilize klystron vacuum tube amplifiers or traveling wave tube amplifiers, both of which are capable of outputting hundreds of kilowatts of peak power at microwave frequencies. Ferrite phase shifters are commonly used in passive arrays since they are capable of handling very high levels of microwave power with very low insertion loss.

Passive phased array architectures are used in a number of significant US military radar systems, including the S-band Aegis phased array radar developed for the US navy by the Radio Corporation of America (RCA), the C-band patriot radar developed for the US army by Raytheon, and the X-band J-STARS developed for the US air force by Grumman. A number of shortfalls are associated with passive arrays. First, the centralized transmit power amplifier is a single point of failure. Second, the

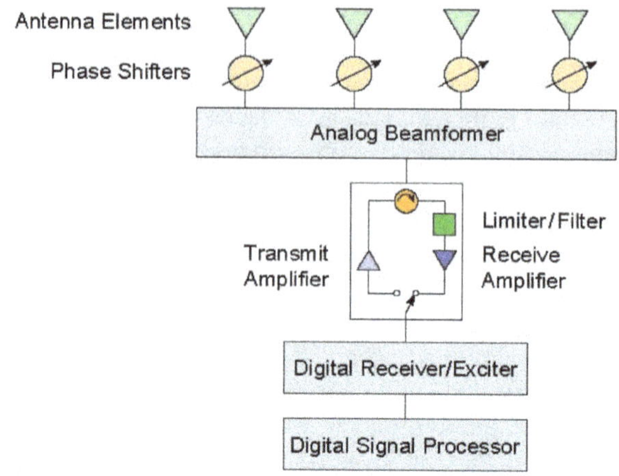

Figure 14.10 *Passive phased array architecture [Figure 2[68] © IEEE. Used with permission.]*

low-loss beamformer networks are large, heavy, and difficult to integrate onto airborne or space-based platforms. Finally, the RF losses incurred in the passive beamformer network and the phase shifters significantly limit radar performance. These factors led to a set of sustained investments by the Defense Advanced Research Projects Agency in the late 1980s and 1990s to develop active phased array technology.

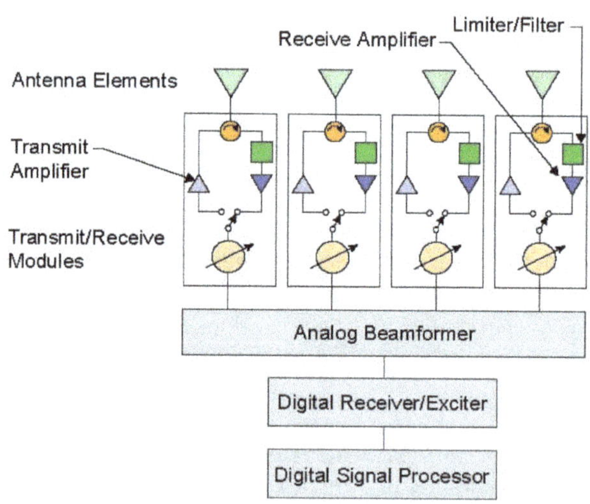

Figure 14.11 *Active phased array architecture [Figure 3[68] © IEEE. Used with permission.]*

14.3.2.2 Active phased arrays

In order to overcome the RF losses imposed by a passive phased array beam-former, the transmit and receive amplifiers can be distributed to each element of the array (Figure 14.11). This approach has the advantage that much lower RF losses incur between the transmit amplifiers and free space. This change reduces the noise figure of the array by a factor of 2 or more, resulting in improved sensitivity and longer operating ranges. A second advantage is that the active phased array architecture "fails" gracefully with the loss of transmit amplifiers at the element level. If the failures are distributed in a random fashion across the array, it is possible to tolerate a loss of 5–10% of the amplifiers, while maintaining acceptable array performance.

An active array T/R module block diagram is shown in Figure 14.12. Low-loss RF switches are incorporated into the module to configure the RF path for either high-power transmit or low-noise receive operation. A circulator, which is a three-port RF device that passes signals in one direction only, is used along with a limiter to protect and isolate the sensitive receive amplifier from the high-power transmit signal. To minimize MMIC area, a common leg topology is often used in T/R module design. The common leg enables the shared use of a phase shifter and gain block for transmit and receive operation. A programmable attenuator is included to provide adjustable amplitude tapers across the array and to compensate for variations in amplitude as a function of phase shifter state and IC manufacture. Active electronically scanned arrays (AESAs) are now used in a number of US military radar systems, including the Theater High Altitude Area Defense ground-based radar, the SPY-3 ship-based radar, the Ground/Air Task Oriented Radar expeditionary radar, military fighter aircrafts (F-15, F-16, F/A-18, F-22, and F-35), and commercial telecommunication system like the Iridium telecommunication satellites [71] and the 5G mobile phone network.

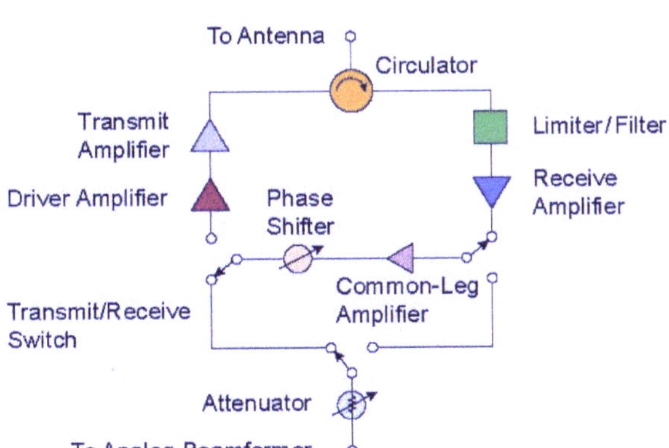

Figure 14.12 Active phased array T/R module block diagram [Figure 4[68] © IEEE. Used with permission.]

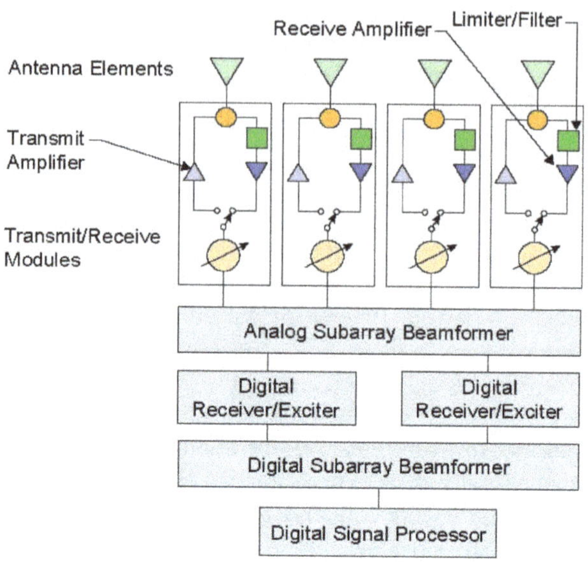

Figure 14.13 Subarray digital phased array architecture [Figure 6[68] © IEEE. Used with permission.]

14.3.2.3 Subarray digital phased arrays

In situations, where it is difficult to achieve radar volume search times with a single beam, a subarray digital beamforming (DBF) architecture has been adopted that provides multiple simultaneous beams over a limited field of view. As shown in Figure 14.13, the subarray digital architecture incorporates a combination of analog and digital beamformers. In this approach, an analog overlapped subarray network produces a flat-topped sector pattern with low sidelobes [72], as shown in Figure 14.14.

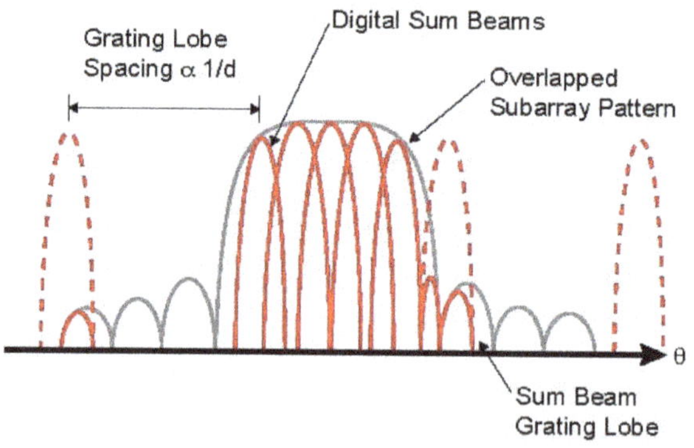

Figure 14.14 Overlapped subarray digital beam patterns (subarray spacing = d) [Figure 7[68] © IEEE. Used with permission.]

The outputs of each analog subarray are downconverted and digitized. Using a digital processor, the subarray outputs are weighted and summed digitally to produce a set of digital beams within the limited angle of the subarray beam pattern. The low sidelobes of the overlapped subarray pattern suppress the unwanted sum beams, referred to as grating lobes, which result from the wide spacing of the digital phase centers. The overlapped digital subarray architecture represents a compromise between the number of simultaneous beams and the number of digital channels required in the system. Depending on the angle extent of the limited scan sector, the analog-overlapped subarray can be complex to build and calibrate [73]. Contiguous randomly sized digital subarrays can also be used to provide digital beam clusters over a limited angle sector [74–76]. All of these digital subarray architectures provide adaptive degrees of freedom that can be used to form pattern nulls and implement sophisticated spatial processing techniques. In cases, where cost is a major consideration and multiple simultaneous beams are required over a limited angle sector, the digital subarray architecture provides an attractive solution [77].

14.3.2.4 Element-level digital phased arrays

As the cost, size, and power consumption of digital receivers/exciters (DREXs) and processors go down, the use of element-level DBF becomes an increasingly realistic approach to implementing large phased arrays. In a DBF array, the received signals are detected and digitized at the element level and then are processed in a special purpose digital computer to form the desired beams. On transmit, the output signals for each element are synthesized digitally and then upconverted to RF. A block diagram of a typical element-level DBF architecture is shown in Figure 14.15.

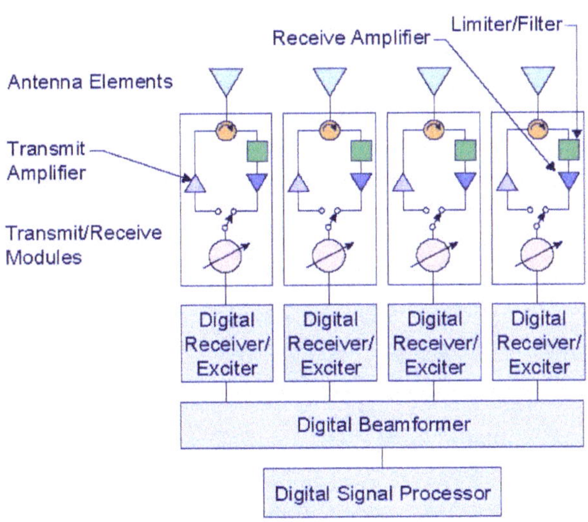

Figure 14.15 Overlapped subarray digital beam patterns (subarray spacing = d) [Figure 8[68] © IEEE. Used with permission.]

Each element of the array has a separate DREX. There are several key features to the element-level DBF approach. First, the total information available at the aperture is preserved, represented by the N individual element signals, in contrast to an analog beamformer, which produces the weighted sum of these signals and reduces the signal dimensionality from N to 1. A second feature is that once the element receive signals have been digitized, they can be manipulated indefinitely because a digital representation of the signal is used rather than the real received signal power. Thus, any number of beams can be formed, and the signal can be subjected to multiple hypothesis testing without degrading signal quality. In DBF, the phase shifting operations of the array are implemented digitally, which allows for arbitrary time delays at each element. This is important for the operation of large arrays over wide bandwidths, where true time delay steering becomes necessary to avoid unwanted beam "squint" effects.

The main operational advantages of DBF are as follows: improved adaptive pattern nulling, multiple simultaneous beams over the full scan volume, antenna self-calibration, ultralow sidelobes, array element pattern correction, and flexible radar power and time management [78, 79, 80]. A key challenge in implementing element-level DBF is the need for an RF transceiver and digitizers at every element of the array. Recent advances in silicon-based RF IC design and fabrication have paved the way toward low-cost element-level digital transceivers with low-power and small form factors [81]. For low-power arrays at millimeter wave frequencies, significant progress has been made in integrating complete phased arrays on RF ICs in SiGe BiCMOS technology [82]. This approach has the potential for extremely low-cost arrays for high-volume commercial applications.

14.3.3 Phased array calibration and thermal management

As noted in section 14.2, maintaining aperture calibration over relevant operational time periods is critical for the overall data quality of a PPAR. Aperture calibration in this case is the element-to-element alignment of transmission gain and phase. The transmission response of the RF hardware that makes up a PPAR element is a function of temperature. General and local drifts in temperature over time will degrade element gain and phase alignment impacting far-field pattern quality and resultant polarimetric variables. Prior to fielding, gain and phase alignment [83] of the aperture take place in a near-field chamber (NFC), an environment with excellent temperature control, no multipath, and near-zero signal interference sources. Aperture alignment is enabled by measuring the transmission response of each element in its reference state to a probe in a common relative position. Differences in phase and amplitude are stored in memory and corrected via phase shifter and variable attenuator setting. Aperture alignment is verified by the subsequent measurement of near-field transformed far-field patterns that samples the operational scan volume. These patterns are evaluated for gain, beam-pointing accuracy, half power beamwidth, cardinal plane sidelobes, and diagonal plane mean square sidelobe level.

If the far-field patterns meet specification and with the PPAR still in the NFC, a second set of measurements is collected and stored by the radar back end. The antenna elements are set to their reference state with alignment corrections applied. Turned on one at a time, the transmission response of the live element is measured via an auxiliary path. This path can take a variety of forms: a wired passive element at the edge of the aperture, an independent feed structure, or a mutual couple to a nearby element. The data collected by this process can be used during fielded operation to monitor (in the background) the alignment of each element. This background process can be used in an adaptive manner to mitigate gain and phase variation as a function of local temperature variation but comes at the cost of time lost for nominal operation. As such, PPAR concepts that minimize local temperature variation would be preferred.

For PPAR operation, liquid cooling is preferred over air. In the latter case, the PPAR must be calibrated in an NFC over the expected environmentally controlled temperature range at the fielded site. This adds time and complexity to the aperture calibration process. Typically, air-cooled systems employ the air under the radome as the heat transfer medium. Due to inefficient mixing, this volume of air can be susceptible to localized hot zones due to solar loading or areas of poor circulation around the PPAR equipment. This in turn leads to a slowly and unevenly varying and imperfectly known thermal environment that drives the overall PPAR pattern performance. In contrast, liquid-cooled systems can be designed to operate at a temperature just above a hot day extreme and be fixed there under all ambient conditions. With proper plumbing design, coolant temperature can be held to within a $<\pm 5^\circ$ range over the entire aperture area. For the balance of this discussion, liquid-cooled PPARs are assumed.

Now turning to PPAR hardware and its susceptibility to temperature variation, we need to separate the hardware suite into domains, the front end and the back end. The front end is made up of T/R modules, analog beamformers, phased array radiators, and their associated infrastructure. The back end comprises the data and RF conversion and local oscillator generation subsystems that collectively make up a DREX. Connecting the front end to the back end is a set of low-loss RF cables that can be several meters in length depending on the PPAR configuration.

To first order, the gain and phase variation of the front-end hardware are driven by the active hardware, i.e., the T/R modules. The passive beamformer and radiators would be fabricated in temperature stable dielectrics, e.g., Tachyon 100G. T/R modules are typically very compact designs populated by SiGe, Si, GaAs, and/ or GaN ICs. Given the nature of the materials and the compactness of the designs, T/R modules exhibit a near constant phase and gain slope versus temperature over (module) population. The temperature the T/R modules "see" is driven to first order by the liquid coolant temperature and the long-term average duty cycle of the waveform employed. For PPARs, this is in the 6–10% range. At these low duties and with current thermal design practices, active control of the liquid coolant temperature, and near-identical phase/gain slope behavior, element-to-element phase and amplitude dispersion will be $<5^\circ$ and <0.3 dB, respectively, over nominal variations in waveform. This level of phase and gain variation would not have a material effect on main beam pointing angle, phase, or amplitude.

Returning to the back-end hardware and its susceptibility to temperature, we need to separate AESA instantiation into two generations. They are delineated by the physical distance between the AESA front end and the back end. In the first generation, roughly 2000 to date, the front end and back end could be separated by several meters of cable and sit in two independent thermal management domains. Considered together, the group delay of the back-end function, i.e., receiver (Rx) channel, and front-end to back-end cable could total many tens of nanoseconds. To first order, the degree of channel-to-channel phase variation is proportional to the nominal group delay of a cable-Rx channel pair. In addition, most modern arrays employ some degree of subarraying, i.e., each cable-Rx channel pair is connected to tens or hundreds of elements. Thus, any phase/gain variation that develops as function of temperature across the Rx channels is correlated over the subarray element count. As a result, the impact on far-field pattern polarimetric metrics is much greater. In AESAs of this generation/configuration, background monitoring of the Rx channel gain and phase and subsequent digital correction is essential.

The second generation of AESAs, roughly 2020 to date, employs RF system-on-a-chip (RFSoC), multichannel (4, 8, 16) devices that integrate fast (>5Gs/s) data converters, analog to digital convertors (ADCs) and digital to analog convertors (DACs), with field programmable gate arrays (FPGAs) and ARM processor circuits. RFSoC technology allows the integration of the back-end function directly onto the AESA [84]. This effectively reduces the group delay from front end to back end to <1 ns. In the work cited, a panel topology is described; the T/R modules and RFSoC share a coldplate, i.e., the thermal management domain is common to the RF hardware elements. The reduced group delay and common thermal management domain effectively limit the magnitude of Rx interchannel gain and phase variation versus temperature. Finally, in PPAR concepts that would employ RFSoCs, subarray element counts drop to <20. The amplified effect of correlated gain and phase errors is reduced.

14.3.4 Rapid scanning for weather observations

In section 14.3, we described how PPAR can increase the frequency of full-volume weather scans. Here we describe specific operational scanning concepts that could realize NWS objectives for approximately once per minute severe weather volume scans. Figure 14.16 summarizes one frequently used VCP for the WSR-88D. This VCP comprises 14 elevation tilts and requires 269 s to complete. Note the lowest three elevation tilts are implemented using a long PRT "surveillance" scan, followed by a short PRT Doppler velocity scan. Thus, a total of 17 physical antenna rotations are required.

We make the conservative assumption that our rapid scan concepts preserve the attributes of the corresponding WSR-88D VCP. Specifically, the number of beam positions serviced and the dwell time, or CPI at each beam position are unchanged. Thus, neither solid angle coverage nor data quality is compromised. Given these assumptions, decreasing the time required to complete the VCP requires that multiple, simultaneously active beams be transmitted and received from the PPAR. The

Elevation Angle (°)	Number of Tilts	CPI Description	Mean Scan Time per Tilt (s)
0.5 – 1.3	3	• Long PRT scan for range-unambiguous surveillance • Short PRT scan for Doppler velocity	41
1.8 – 6.4	6	• Interleaved long PRT surveillance and short PRT Doppler pulses	13
8.0 – 19.5	5	• Short PRT for Doppler and reflectivity	12
Totals	14		269 s

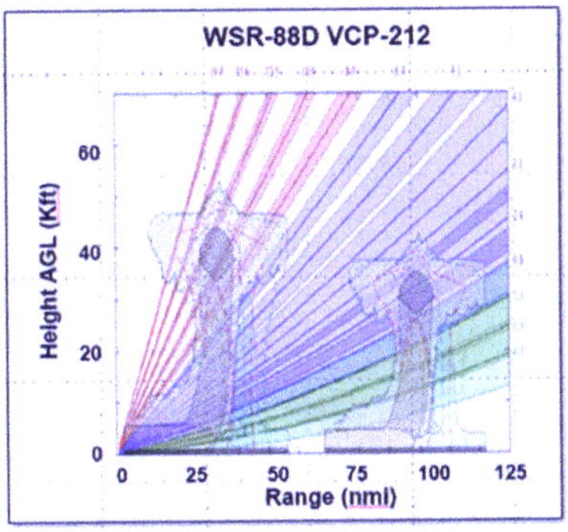

Figure 14.16 WSR-88D volume coverage pattern VCP-212

VCP execution time is reduced proportionally to the number of beams. Here, we assume the use of three or four simultaneous beams, which reduces scan time for VCP-212 to approximately 90 s or 68 s.

For a four-faced planar PPAR realizing the targeted scan time reduction is straightforward. Each array face uses a single, electronically steered beam to surveil all beam positions within its assigned 90° sector. CPI structure at each beam position is identical to that of the WSR-88D so that data quality (clutter suppression, range/Doppler ambiguity mitigation) is not compromised. The beam positions can be surveilled in arbitrary order (e.g., RHIs or 90° sector scans).

Similarly, as its azimuth sectors commutate electronically, CPPAR successively services all elevation beam positions normal to the horizontal tangents to the array at the center of each sector. For VCP-212, each such RHI requires 0.75 s. Thus, after the sectors have commutated through 90° in azimuth, the full volume will have been surveilled in 68 s.

Electronic steering of multiple beams from a single array face can be coordinated with mechanical rotation of the array to achieve rapid scanning. One scanning pattern is a "sector RHI" where limited-in-azimuth sectors (e.g., 30°) are electronically scanned at all elevation tilts as the array rotates slowly in azimuth. The full VCP is completed during one 360° rotation of the antenna. An alternative "multitilt" approach rotates the antenna at rates similar to the WSR-88D (20°–30° s^{-1}), providing full-PPI scans of multiple elevation tilts during a single rotation of the antenna. Both methods are capable of replicating VCP-212 in approximately 90 s, using three simultaneous receive beams, or 68 s using four simultaneous receive beams.

An important consideration is the "cross-talk" between the simultaneously active beams. Each beam will receive echoes from the other transmitted beams, suppressed only by its one-way (receive) antenna pattern at the directions of those beams. If the array's digital architecture supports the use of widely spaced receive beams, this cross-talk can be suppressed to a level close to the array's two-way far antenna sidelobe envelope. As an example, Figure 14.17 shows the two-way response pattern for the center beam of a modeled PPAR, when three simultaneous beams are transmitted/received at broadside and ±30° in azimuth (for details, see

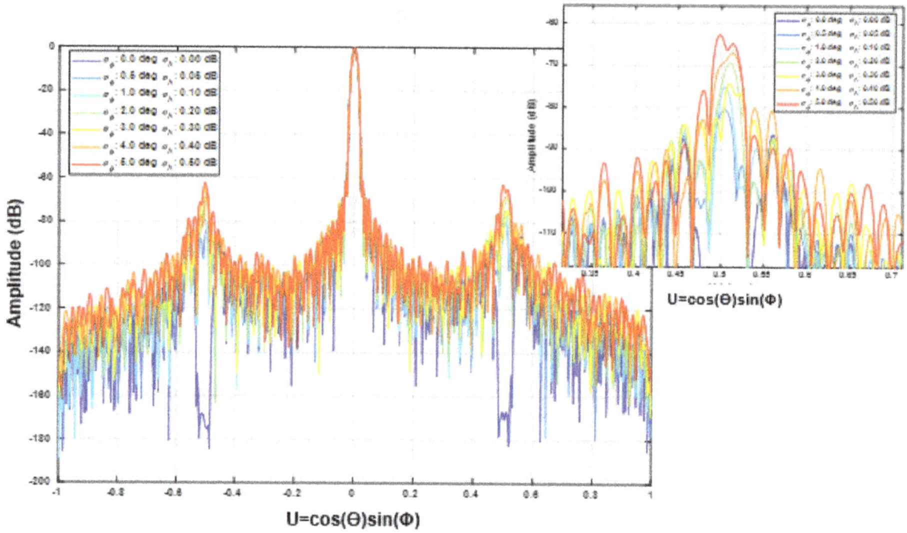

Figure 14.17 Effective two-way antenna response for a modeled PPAR, which transmits three subpulses separated by 30° in azimuth and forms simultaneous receive beams in these directions using its digital degrees of freedom. The different modeled patterns assume the indicated RMS errors in T/R element amplitude and phase calibration across the array.

Reference 85]. Phase-only, deterministic nulling [86] has been used to minimize receive response at the angles of the other transmitted beams. If we assume the array is well calibrated (amplitude/phase RMS errors of 0.5 dB and 5°) and the locations of failed T/R elements are used in computing the beamforming weights, the two-way response in the direction of the other beams can be suppressed more than −60 dB relative to the response in the direction of interest. Additionally, orthogonal phase codes can be applied to the subpulses transmitted in the three beam directions. This “whitens” the interfering returns and facilitates their suppression via Doppler filtering. For expected values of weather echo spectrum width and PPAR unambiguous velocity interval, this filtering would suppress the interference an additional 14 dB below the level shown in Figure 14.17.

14.4 The advanced technology demonstrator

The ATD is a large, S-band, planar, proof-of-concept PPAR at the National Weather Radar Testbed in Norman, Oklahoma. The system serves as a testbed for evaluating the suitability of modern PPAR technology for weather observations [69, 87].

The radar is installed atop a 10-m tall facility and enclosed in a spherical radome (Figure 14.18). The antenna is a 4-m diameter AESA with individual active T/R modules for each of the 4,864 elements. It consists of 76 panels, each with an 8×8

Figure 14.18 The ATD facility and the calibration tower

matrix of dual-polarization radiating elements. The antenna is mounted on an elevation rotator and an azimuth turntable to allow mechanical positioning (Figure 14.19). The elevation rotator provides mechanical positioning of the antenna from −1° to 181°, and the azimuth turntable can rotate the full 360°. These are primarily used to position the antenna prior to scanning using electronic beamsteering, but the radar can also operate while the antenna is moving. Through electronic steering, which is controlled by phase shifters at the element level, the radar beam can be pointed anywhere within a ±52° conical sector with respect to the array normal. Low-level control of the phase shifters and T/R amplifiers on a panel is handled by an FPGA, which receives higher level instructions from the beamsteering generator. On receive, the signals from the 64 elements of each panel are first summed. Each of

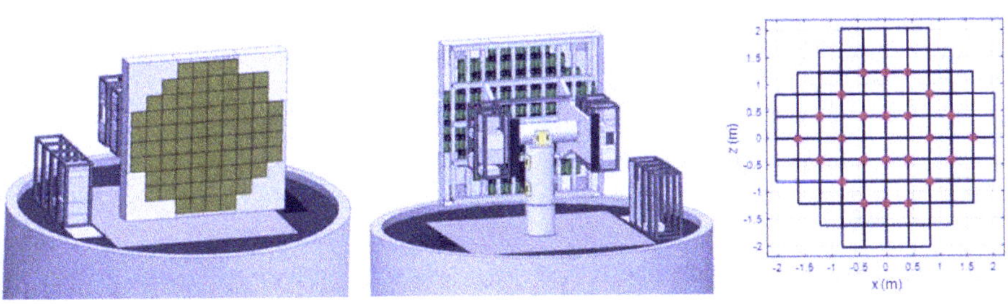

Figure 14.19 (left) Front and (middle) back of the antenna assembly. (right) The antenna layout with subarray centers shown as red dots.

the 76 panels provides four outputs (i.e., two in the H-polarization channel and two in the V-polarization channel) in support of the overlapped subarray architecture [73]. Each subarray consists of eight panels arranged in a 2×4 matrix for a total of 24 subarrays for each polarimetric channel. The configuration chosen for the subarrays supports the digital synthesis of a cluster of three-to-five receive beams. This allows the ATD scan rate to be increased as described in section 14.3. Figure 14.19 shows the antenna layout and subarray centers.

The radar back end is located on the yoke arms of the elevation rotator and comprises all the equipment responsible for waveform generation and downconversion and digitization of the received echoes. The analog receiver handles the analog RF signals from the 48 subarrays and 2 additional sidelobe-blanker channels. These 50 signals are downconverted to an intermediate frequency prior to digitization and processing. The digital beamformer serves as the first level of processing, combining the digitized subarray data, applying DBF, and producing the beam data. To increase the sensitivity, the ATD employs pulse compression and a matched receiver filter that results in range resolution equal to the reciprocal of the transmitted pulse's bandwidth. To mitigate the blind range that results from the transmission of the long, modulated pulse, the ATD transmits a short "fill" pulse immediately after. The two pulses are shifted by ±8 MHz relative to the center operating frequency, and the corresponding returns are separated via filtering in the digital receiver. The application of the pulse compression matched filter results in four streams of in-phase and quadrature signals sampled at 4 MHz that are sent to the signal processor: the H-polarization "long" and "short" pulse data and the V-polarization "long" and "short" pulse data. The application software includes two signal-processing paths – one for aircraft surveillance and one for weather surveillance – and a set of human-machine interfaces for control and monitoring of radar operations.

To support the polarimetric calibration of the ATD, a 45-m tall calibration tower is located approximately 425 m north of the radar. On top of this tower, a radome assembly houses a standard S-band, single-polarization gain horn on a motorized rotating platform that can be remotely controlled to change its orientation. The primary purpose of the calibration tower horn is to transmit and receive test signals to and from the antenna. At the base of the calibration tower, an enclosure containing RF circuitry supports different modes of operation with coherent RF signal paths to and from the radar. It can also transmit an independent, noncoherent, continuous wave signal. Underground single-mode optical fiber transmission lines carry RF and ethernet signals between the ATD and the calibration tower.

The ATD's technical parameters – average radiated power (4.6 kW), beamwidth (1.6°), antenna gain (42 dB), and cross-polarization isolation (−35 to −40 dB) – are sufficient to explore PPAR performance challenges and demonstrate operational benefits. Here, we discuss initial efforts to calibrate the array with sufficient accuracy so as to produce good quality polarimetric variable estimates.

A near-field anechoic chamber at MIT Lincoln Laboratory [83] was used to characterize the array. H and V electric field probe measurements were used to calculate co- and cross-polar transmit and receive patterns at a total of almost 3,000 electronic beamsteering positions. Figure 14.20 shows the ATD antenna in the NFC

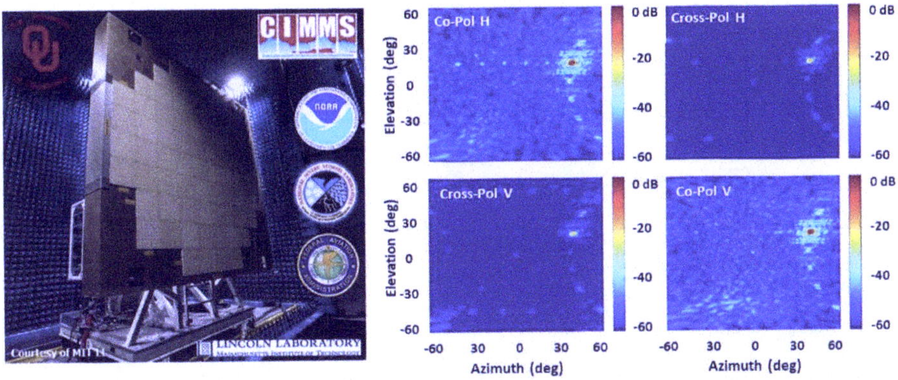

Figure 14.20 *(left) ATD antenna undergoing calibration in an NFC at MIT Lincoln Laboratory. (right) Reconstructed far-field transmit patterns for an electronic beamsteering angle of 35° azimuth and 20° elevation (From Weber* et al. *2021. © American Meteorological Society. Used with permission.)*

and example transmit patterns for an electronic beamsteering angle of 35° azimuth and 20° elevation. Because this is well away from the array normal, cross-polar radiation is significant.

Figure 14.21 shows "beam peak" measurements for each of the steering directions (left and middle panels). These are used to estimate beamsteering copolar biases (relative to 0° azimuth) for Z, Z_{DR}, and Φ_{DP} in the principal horizontal plane (right panels). We smoothed these steering-angle-dependent copolar bias corrections and applied them to ATD data that were collected during a severe weather outbreak in central Oklahoma on 1 May 2019. The ATD polarimetric variable fields and those from the colocated and well-calibrated WSR-88D are compared in Figure 14.22. The ATD estimates prior to correction (top row) exhibit the expected biases in Z_{DR} and Φ_{DP} at the extremes of the ±45° azimuth scan sector. After application of the bias corrections, these are significantly reduced (middle row).

The system Φ_{DP} for both radars is set to about 60° (blue in Figure 14.22). Therefore, 60° is seen where propagation effects are small, such as close to the radar, and may extend in range if precipitation is light. A radial increase (streak) of about 40° over a 6 km distance is seen in the convective cell to the southeast of the radars (white circle). The corresponding Z_{DR} is greater than 4 dB in both the KCRI and ATD data. Some refer to this phenomenon as a "hot spot" and attribute it to the radar signal propagating through heavy rain, possibly containing ice cores ([5], p. 172). Again associated with large *ZDR*, another streak increase occurs at about 190° azimuth and 85 km range. This is also caused by heavy precipitation as the radar signals propagate through a convective cell along this radial. An approximately 100° increase in Φ_{DP} is measured by both radars, and the shapes of the contours are in good agreement; the ATD misses some data at the farthest ranges because it is less sensitive.

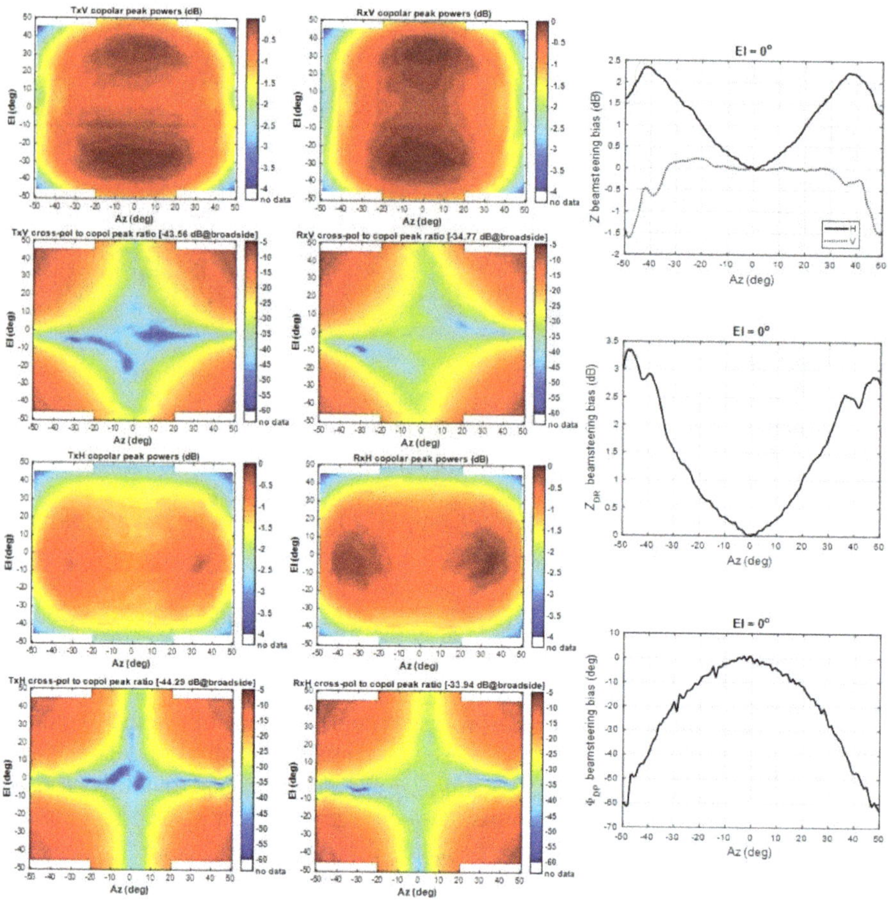

Figure 14.21 (left panels) Transmit and (middle panels) receive beam peak powers of the ATD antenna measured in the NFC. (right panels) Copolar biases as a function of steering angle in the horizontal principal plane. (From Weber et al. *2021. © American Meteorological Society. Used with permission.)*

The ρ_{hv} fields for both ATD and KCRI exhibit a transition to smaller values beginning at a range of approximately 85 km. This may be caused by melting graupel and/or hail below the melting layer in the high-reflectivity convective core. The "pink fringe" on the periphery of the precipitation indicates where ρ_{hv} estimates exceed one and is larger in area in the ATD field than in the KCRI field. The major cause of these positive biases is low SNR (e.g., [88]). This feature is more prominent in the ATD data because the radar has a substantially lower power aperture product than KCRI.

Next, we list the mean differences between the corrected ATD and KCRI polarimetric variable estimates and the SDs of these differences (Table 14.4). We

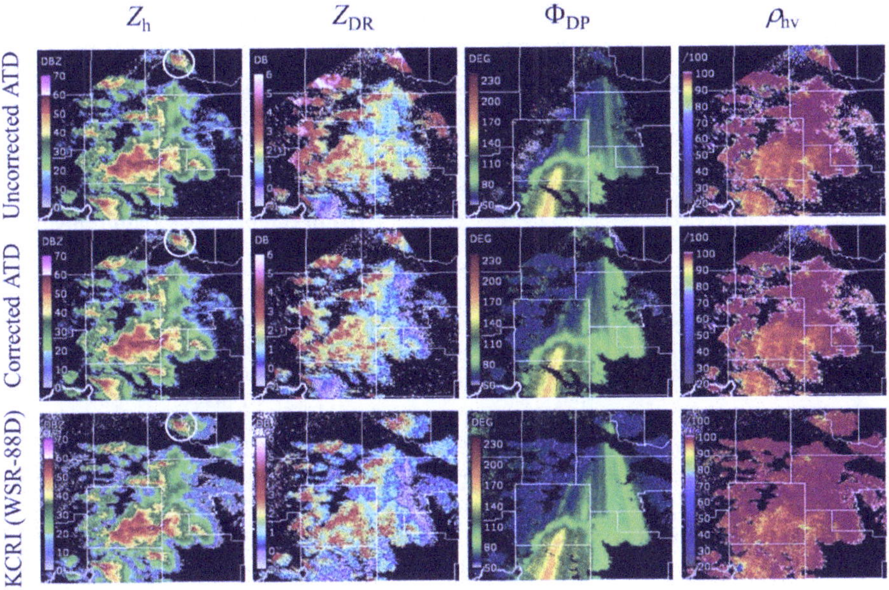

Figure 14.22 ATD reflectivity and polarimetric variable fields on 1 May 2019, (top row) without and (middle row) with bias corrections derived from near-field array characterization. The bottom row is concurrent data from the colocated KCRI radar operated by NOAA's Radar Operations Center. The ATD and KCRI data were collected at the same time (19:57:43 Z) and elevation angle (0.5°). The maximum range in the images is 140 km (From Weber et al. 2021. © American Meteorological Society. Used with permission.)

calculated these by subtracting the KCRI estimates from those of the ATD (Z_{DR} estimates are subtracted in decibel units) and averaging across all-range and azimuth-resolution cells with SNR ≥15 dB, $0.85 \le \rho_{hv} \le 1.00$, and within 140 km of the radar. For reference, we also list the maximum bias and SD requirements from the NWS RFR [56]. Assuming that KCRI is well calibrated, the distributions of the ATD polarimetric variable differences relative to KCRI are consistent with the NWS requirements.

We stress that these observations are very close to the principal horizontal plane of the ATD, where cross-polar radiation is low (see Figure 14.21). Significantly stronger cross-polar radiation at higher elevation angles would likely cause more prominent cross-coupling biases in the estimates of polarimetric variables. Using the ATD radiation patterns and analytical expressions [57], we calculated the cross-coupling bias at the steering angle shown in Figure 14.20. For a true ρ_{hv} of 0.98, the cross-coupling Z_{DR} biases in the SHV mode are predicted to be −0.25 dB and −4.43 dB for $\Phi_{DP} = 0°$ and $\Phi_{DP} = 180°$, respectively. Note that the intrinsic Φ_{DP} strongly affects the cross-coupling biases. These biases can be reduced to 0.057 dB

Table 14.4 Mean differences between ATD and KCRI polarimetric variables and SDs of these differences. NWS RFR maximum bias and SD requirements are subject to the assumptions that true spectrum width ≤2 m/s, SNR ≥20 dB, dwell time = 50 ms, and true ρ_{hv} ≥0.99.

Variable	Mean difference (ATD – KCRI)	SD of differences	NWS RFR bias requirement	RFR SD requirement
Z_{DR}	0.055 dB	0.25 dB	≤0.1 dB for true $\lvert Z_{DR} \rvert \leq 1$ dB ≤10% of true $\lvert Z_{DR} \rvert$ for true $\lvert Z_{DR} \rvert > 1$ dB	≤0.4 dB
Φ_{DP}	0.58°	1.73°	≤1°	≤2.5°
ρ_{hv}	-1.6×10^{-3}	1.2×10^{-2}	$\leq 6 \times 10^{-3}$	$\leq 6.0 \times 10^{-3}$

and −0.026 dB in the SHV mode for the same signal parameters by using the pulse-to-pulse phase coding technique [51, 53]. The Z_{DR} cross-coupling biases in the AHV mode are 0.066 dB and −0.062 dB for the signal parameters above [57]. An important aspect of our future research will be to experimentally evaluate bias corrections in ATD observations well outside the principal plane.

14.5 Outlook

We expect that the use of PPAR for meteorological observations will increase rapidly over the coming decade. In the near term, one of a kind "demonstrators" such as those described in this monograph will be used to drive down the risk for key challenges such as dual-polarization calibration and bias correction, rapid and adaptive scanning methods, and preservation or even improvement of the high data quality provided by today's conventional meteorological radars. Case studies of severe storms and demonstration of benefits for forecasts and warnings will strengthen the "business case" for PPAR. In parallel, critical new technologies will be demonstrated and their manufacturing readiness level (MRL) advanced through interaction with industry. Foundational building blocks of future PPARs will be: (1) highly integrated RFSoC that enables the digital architectures needed for rapid, flexible scanning; (2) efficient high-power amplifiers enabling PPARs that meet or exceed the sensitivity of today's meteorological radars, with modest pulse compression ratios; and (3) commercialization of PPAR building blocks such as the low-cost, tiled panels described herein. This process will leverage rapid progress in microwave systems MRL spurred by the automotive radar and 5G wireless industries.

In the midterm, research and demonstrations using PPAR will proliferate at universities, government research laboratories, and even media outlets, based on the availability of affordable, limited production rate PPAR building blocks or full systems developed by industry. This "community-discovery" phase will be analogous to those that preceded previous incorporation of significant new technologies into operational weather radars, such as Doppler velocity and dual-polarization measurements.

The US NWS is evaluating alternatives, including PPAR, for the eventual replacement of the WSR-88D (NOAA 2020). Its decision will consider both the technical maturity of PPAR and the benefits relative to conventional meteorological radar when integrated with other observing systems and assimilated into human- and model-based warning processes. The large-scale, community-based research described above will be critical in informing this decision.

International partnership will be extremely valuable. Consideration of the distinct meteorological requirements and operational environments of nations other than the United States will enhance the robustness and benefits of PPAR applications. We hope that this monograph will facilitate the international collaborations needed to fully realize PPAR capabilities.

References

[1] Doviak R.J., Bringi V., Ryzhkov A., Zahrai A., Zrnić D. 'Considerations for polarimetric upgrades to operational WSR-88D radars'. *Journal of Atmospheric and Oceanic Technology*. 2000, vol. 17(3), pp. 78–257.

[2] Doviak R., Zrnić D. *Doppler Radar and Weather Observations*. 2nd ed. Mineola, NY: Dover; 2006. p. 562.

[3] Bringi V.N., Chandrasekar V. *Polarimetric Doppler Weather Radar*. Cambridge University Press; 2001. p. 636. Available from https://www.cambridge.org/core/product/identifier/9780511541094/type/book

[4] Zhang G. *Weather radar polarimetry [online]*; 2016. p. 304. Available from https://www.taylorfrancis.com/books/9781439870600

[5] Ryzhkov A.V., Zrnic D.S. *Radar polarimetry for weather observations*. Springer Nature Switzerland AG; 2019. p. 486.

[6] Ivić I.R., Byrd A. 'A first look at the MPAR dual-polarization phased-array-radar mobile demonstrator'. *Proceedings of the 37th Int. Conf. on Radar Meteorology*; 2015.

[7] Ryzhkov A., Diederich M., Zhang P., Simmer C. 'Potential utilization of specific attenuation for rainfall estimation, mitigation of partial beam blockage, and radar networking'. *Journal of Atmospheric and Oceanic Technology*. 2014, vol. 31(3), pp. 599–619.

[8] Melnikov V.M., Zrnić D.S. 'On the alternate transmission mode for polarimetric phased array weather radar'. *Journal of Atmospheric and Oceanic Technology*. 2014, vol. 32(2), pp. 220–33.

[9] Torres S.M., Zrnić D. 'Signal design and processing techniques for WSR-88D ambiguity resolution: evolution of the SZ-2 algorithm'. *NOAA/NSSL Report*. 2006, vol. Part 10, p. 71.

[10] Doviak R.J., Zrnić D.S., Carter J., Ryzhkov A., Torres S., Zahrai A. 'Polarimetric upgrades to improve rainfall measurements'. *Report of the National Severe Storms Laboratory*. 1998, p. 110.

[11] Melnikov V.M., Doviak R.J., Zrnić D.S., Stensrud D.J. 'Mapping Bragg scatter with a polarimetric WSR-88D'. *Journal of Atmospheric and Oceanic Technology*. 2011, vol. 28(10), pp. 1273–85.

[12] Ivić I.R., Zrnić D.S., Yu T.-Y. 'The use of coherency to improve signal detection in dual-polarization weather radars'. *Journal of Atmospheric and Oceanic Technology*. 2009, vol. 26(11), pp. 2474–87.

[13] Ivic I.R., Zrnic D.S., Yu T.-Y. 'Threshold calculation for coherent detection in dual-polarization weather radars'. *IEEE Transactions on Aerospace and Electronic Systems*. 2012, vol. 48(3), pp. 2198–215.

[14] Ivić I.R., Keranen R., Zrnić D.S. 'Assessment of censoring using coherency based detectors on dual-polarized weather radar'. *J. Atmos. Oceanic Technol.* 2014, vol. 31(8), pp. 1694–703.

[15] Keränen R., Chandrasekar V. 'Detection and estimation of radar reflectivity from weak echo of precipitation in dual-polarized weather radars'. *Journal of Atmospheric and Oceanic Technology*. 2014, vol. 31(8), pp. 1677–93.

[16] Park H.S., Ryzhkov A.V., Zrnić D.S., Kim K.-E. 'The hydrometeor classification algorithm for the polarimetric WSR-88D: description and application to an MCS'. *Weather and Forecasting*. 2009, vol. 24(3), pp. 730–48.

[17] Ryzhkov A.V., Schuur T.J., Burgess D.W., Zrnic D.S. 'Polarimetric tornado detection'. *Journal of Applied Meteorology*. 2005, vol. 44(5), pp. 557–70.

[18] Zrnić D.S., Ryzhkov A., Straka J., Liu Y., Vivekanandan J. 'Testing a procedure for automatic classification of hydrometeor types'. *Journal of Atmospheric and Oceanic Technology*. 2001, vol. 18(6), pp. 892–913.

[19] Ortega K.L., Krause J.M., Ryzhkov A.V. 'Polarimetric radar characteristics of melting hail. part III: validation of the algorithm for hail size discrimination'. *Journal of Applied Meteorology and Climatology*. 2016, vol. 55(4), pp. 829–48.

[20] Lakshmanan V., Karstens C., Krause J., Elmore K., Ryzhkov A., Berkseth S. 'Which polarimetric variables are important for weather/no-weather discrimination?'. *Journal of Atmospheric and Oceanic Technology*. 2015, vol. 32(6), pp. 1209–23.

[21] Heinselman P.L., LaDue D.S., Lazrus H. 'Exploring impacts of rapid-scan radar data on NWS warning decisions'. *Weather and Forecasting*. 2012, vol. 27(4), pp. 1031–44.

[22] Heinselman P., LaDue D., Kingfield D.M., Hoffman R. 'Tornado warning decisions using phased-array radar data'. *Weather and Forecasting*. 2015, vol. 30(1), pp. 57–78.

[23] Bowden K.A., Heinselman P.L., Kingfield D.M., Thomas R.P. 'Impacts of phased-array radar data on forecaster performance during severe hail and wind events'. *Weather and Forecasting*. 2015, vol. 30(2), pp. 389–404.

[24] Wilson K.A., Heinselman P.L., Kuster C.M., Kingfield D.M., Kang Z. 'Forecaster performance and workload: does radar update time matter?'. *Weather and Forecasting*. 2017, vol. 32(1), pp. 253–74.

[25] Stratman D.R., Yussouf N., Jung Y., *et al*. 'Optimal temporal frequency of NSSL phased array radar observations for an experimental warn-on-forecast system'. *Weather and Forecasting*. 2020, vol. 35(1), pp. 193–214.

[26] Yussouf N., Stensrud D.J. 'Impact of phased-array radar observations over a short assimilation period: observing system simulation experiments using an ensemble Kalman filter'. *Monthly Weather Review*. 2010, vol. 138(2), pp. 517–38.

[27] Stensrud D.J., Xue M., Wicker L.J, *et al.* 'Convective-scale warn-on-forecast system'. *Bulletin of the American Meteorological Society*. 2009, vol. 90(10), pp. 1487–500.

[28] Yu T.Y., Orescanin M.B., Curtis C.D., Zrnić D.S., Forsyth D.E. 'Beam multiplexing using the phased-array weather radar'. *J. Atmos. Oceanic Technol.* 2007, vol. 24, pp. 616–26.

[29] Torres S.M., Adams R., Curtis C.D, *et al.* 'Adaptive-weather-surveillance and multifunction capabilities of the national weather radar testbed phased array radar'. *Proceedings of the IEEE*. 2016, vol. 104(3), pp. 660–72.

[30] Zrnić D.S., Melnikov V.M., Doviak R.J., Palmer R. 'Scanning strategy for the multifunction phased-array radar to satisfy aviation and meteorological needs'. *IEEE Geosci Remote Sens Lett*. 2015, vol. 12(6), pp. 1204–8.

[31] Weber M.E., Cho J.Y.N., Thomas H.G. 'Command and control for multifunction phased array radar'. *IEEE Trans Geosci Remote Sens*. 2017, vol. 55(10), pp. 5899–912.

[32] Melnikov V.M., Doviak R.J., Zrnic D.S. 'A method to increase the scanning rate of phased-array weather radar'. *IEEE Trans Geosci Remote Sens*. 2015, vol. 53(10), pp. 5634–43.

[33] Weber M., Hondl K., Yussouf N., *et al.* 'Towards the next generation operational meteorological radar'. *Bulletin of the American Meteorological Society*. 2021, vol. 102(7), pp. E1357–83.

[34] Huang Y., Wang X., Kerr C., Mahre A., Yu T.-Y., Bodine D. 'Impact of assimilating future clear-air radial velocity observations from phased-array radar on a supercell thunderstorm forecast: an observing system simulation experiment study'. *Monthly Weather Review*. 2020, vol. 148(9), pp. 3825–45.

[35] Heberling W., Frasier S.J., Wolsieffer C., Adam M. *X-band phase-array weather-radar polarimetry testbed: bias expectation and preliminary results, AMS annual meeting phased array radar symposium* [online]. 2019. Available from https://ams.confex.com/ams/2019Annual/meetingapp.cgi/Paper/350300

[36] Tanamachi R.L., LaFleur A.T., Sharma M., Frasier S.J., Heberling W., Wolsieffer C. 'Observations of severe storms by a low-power, polarimetric, phased array mobile radar'. *AMS Annual Meeting Phased Array Radar Symposium*; Paper 1.3, 2019.

[37] Morin A., George J., Chandrasekar V. 'System description and recent observations from front-X: A dual-polarization X-band phased-array radar'. *AMS Annual Meeting Symposium on Phased Array Radar*; Paper 1.4, 2019.

[38] Wu C., Liu L., Liu X., Li G., Chen C. 'Advances in Chinese dual-polarization and phased-array weather radars: observational analysis of a supercell in Southern China'. *Journal of Atmospheric and Oceanic Technology*. 2018, vol. 9, pp. 1785–806.

[39] Chan D.M. *China gets new 'extreme weather' radar system* [online]. 2019. Available from https://asiatimes.com/2019/09/special-tornado-radar-system-now-monitoring-east-china/
[40] Kobayashi T., Mizutani F., Wada M. 'Development of a multiple-parameter phased-array weather radar'. *AMS Annual Meeting Phased Array Radar Symposium, Paper 744*; 2019.
[41] Ushio T., Kikuchi H., Suezawa T., Onuki A. 'Polarimetric radar observations with multiparameter phased-array radar (MP-PAWR) in the Tokyo area'. *AMS Annual Meeting Phased Array Radar Symposium*; Paper 738, 2019.
[42] Zhang G., Doviak R.J., Zrnić D.S., Palmer R., Lei L., Al-Rashid Y. 'Polarimetric phased-array radar for weather measurement: a planar or cylindrical configuration?'. *Journal of Atmospheric and Oceanic Technology*. 2011, vol. 28(1), pp. 63–73.
[43] Li Z., Zhang G., Golbon-Haghighi M.H., Saeidi-Manesh H., Herndon M., Pan H. 'Initial observations with electronic and mechanical scans using a cylindrical polarimetric phased array radar'. *IEEE Geosci Remote Sens Lett*. 2020, vol. 18(2), pp. 271–75.
[44] Karimkashi S., Zhang G. 'An optimal design of a cylindrical polarimetric phased array radar for weather sensing'. *Radio Science*. 2012, vol. 47(2).
[45] Lei L., Zhang G., Doviak R.J., Karimkashi S. 'Comparison of theoretical biases in estimating polarimetric properties of precipitation with weather radar using parabolic reflector, or planar and cylindrical arrays'. *IEEE Trans Geosci Remote Sens*. 2015, vol. 53(8), pp. 4313–27.
[46] Fulton C., Salazar J.L., Zhang Y., *et al*. 'Cylindrical polarimetric phased array radar: beamforming and calibration for weather applications'. *IEEE Trans Geosci Remote Sens*. 2017, vol. 55(5), pp. 41–2827.
[47] Saeidi-Manesh H., Zhang G. '*CHARACTERIZATION and optimization of cylindrical polarimetric array antenna patterns for multi-mission applications*'. Progress In Electromagnetics Research. 2017, vol. 158, pp. 49–61.
[48] Yeary M., Palmer R., Fulton C., Salazar J., Sigmarsson H. 'Recent advances on an S-band all-digital mobile phased array radar'. *IEEE International Symposium on Phased Array System & Technology (PAST)*. 2019, pp. 1–5. Available from https://ieeexplore.ieee.org/xpl/mostRecentIssue.jsp?punumber=8982099
[49] Fulton C., Palmer R., Yeary M. 'Horus: a testbed for fully digital phased array radars'. *Microwave J*. 2020, vol. 63.
[50] Lujan J., Fulton C., Yeary M. 'A progress report on an iterative algorithm for mutual coupling based calibration for A digital array'. *Proceedings of Radar Sensor Technology XXIV, SPIE Defense, and Commercial*; 2020.
[51] Ivic I.R., Doviak R.J. 'Evaluation of phase coding to mitigate differential reflectivity bias in polarimetric PAR'. *IEEE Trans Geosci Remote Sens*. 2016, vol. 54(1), pp. 431–51.
[52] Sachidananda M., Zrnic D.S. 'Systematic phase codes for resolving range overlaid signals in a doppler weather radar'. *Journal of Atmospheric and Oceanic Technology*. 1999, vol. 16(10), pp. 1351–63.

[53] Zrnić D.S., Doviak R.J., Melnikov V.M., Ivić I.R. 'Signal design to suppress coupling in the polarimetric phased array radar'. *Journal of Atmospheric and Oceanic Technology*. 2014, vol. 31(5), pp. 1063–77.

[54] Zrnić D., Doviak R., Zhang G., Ryzhkov A. 'Bias in differential reflectivity due to cross coupling through the radiation patterns of polarimetric weather radars'. *Journal of Atmospheric and Oceanic Technology*. 2010, vol. 27(10), pp. 1624–37.

[55] Balakrishnan N., Zrnic D.S. 'Use of polarization to characterize precipitation and discriminate large hail'. *Journal of the Atmospheric Sciences*. 1990, vol. 47(13), pp. 1525–40.

[56] *NOAA/national weather service radar functional requirements* [online]. 2015. Available from https://www.roc.noaa.gov/WSR88D/PublicDocs/NOAA_Radar_Functional_Requirements_Final_Sept%202015.pdf

[57] Ivic I.R. 'Options for polarimetrie variable measurements on the MPAR advanced technology demonstrator'. *IEEE Radar Conference (RadarConf18)*; Oklahoma City, OK, 2018.

[58] Umeyama A.Y., Salazar-Cerreno J.L., Fulton C. 'UAV-based antenna measurements for polarimetric weather radars: probe analysis'. *IEEE Access*. 2020, vol. 8, pp. 191862–74.

[59] Ivić I.R., Mendoza R., Schvartzman D., Torres S., Wasielewski D. *Preliminary report on polarimetric calibration for the advanced technology Demonstrator, NOAA/NSSL report* [online]. 2020. Available from https://www.nssl.noaa.gov/publications/par_reports/ Preliminary%20Report%20on%20Polarimetric%20Calibration%20for%20the%20Advanced%20Technology%20Demonstrator.pdf

[60] Ivic I.R. 'Phase code to mitigate the copolar correlation coefficient bias in PPAR weather radar'. *IEEE Trans Geosci Remote Sens*. 2017, vol. 55(4), pp. 2144–66.

[61] Ivić I.R. 'An experimental evaluation of phase coding to mitigate the cross-coupling biases in PPAR'. *38th International Conference on Radar Meteorology, Chicago, IL*; 2017.

[62] Ivic I.R. 'Effects of phase coding on doppler spectra in PPAR weather radar'. *IEEE Trans Geosci Remote Sens*. 2018, vol. 56(4), pp. 2043–65.

[63] GuifuZ.Doviak R.J., Zrnic D.S., Crain J., Staiman D., Al-Rashid Y. 'Phased array radar polarimetry for weather sensing: a theoretical formulation for bias corrections'. *IEEE Trans Geosci Remote Sens*. 2009, vol. 47(11), pp. 3679–89.

[64] Zrnic D.S., Zhang G., Doviak R.J. 'Bias correction and doppler measurement for polarimetric phased-array radar'. *IEEE Trans Geosci Remote Sens*. 2011, vol. 49(2), pp. 843–53.

[65] Lei L., Zhang G., Doviak R.J. 'Bias correction for polarimetric phased-array radar with idealized aperture and patch antenna elements'. *IEEE Trans Geosci Remote Sens*. 2013, vol. 51(1), pp. 473–86.

[66] Fulton C., Chappell W.J. '*IEEE/MTT-S international microwave symposium - MTT 2010*'. Anaheim, CA, USA, 2010.

[67] Galletti M., Zrnic D.S., Gekat F., Goelz P. 'Eigenvalue signal processing for weather radar polarimetry: removing the bias induced by antenna coherent cross-channel coupling'. *IEEE Trans Geosci Remote Sens*. 2014, vol. 52(12), pp. 707–7695.

[68] Herd J.S., Conway M.D. 'The evolution to modern phased array architectures'. Proceedings of the IEEE. 2016, vol. 104(3), pp. 519–29.

[69] Stailey J.E., Hondl K.D. 'Multifunction phased array radar for aircraft and weather surveillance'. *Proceedings of the IEEE*. 2016, vol. 104(3), pp. 649–59.

[70] Weber M.E., Cho J.Y.N., Herd J.S., Flavin J.M., Benner W.E., Torok G.S. 'The next-generation multimission U.S. surveillance radar network'. *Bulletin of the American Meteorological Society*. 2007, vol. 88(11), pp. 1739–52.

[71] Brookner E. 'IEEE International Symposium on Phased ARRAY Systems and Technology (ARRAY 2013)'. 2013. Waltham, MA,

[72] Mailloux R.J. *Phased array antenna handbook*. 2nd ed. Norwood, MA: Artech House; 2005.

[73] Herd J.S., Duffy S.M., Steyskal H. 'Design considerations and results for an overlapped subarray radar antenna'. Proceedings of the IEEE Aerospace Conference, Big Sky, MT, 2005. pp. 1087–92.

[74] Goffer A.P., Kam M., Herczfeld P.R. 'Design of phased arrays in terms of random subarrays'. *IEEE Trans Antennas Propag*. 1994, vol. 42(6), pp. 26–820.

[75] Groger I., Sander W., Wirth W.-D. 'Experimental phased array radar elra with extended flexibility'. *IEEE Aerospace and Electronic Systems Magazine*. 1990, vol. 5(11), pp. 26–30.

[76] Mailloux R.J., Santarelli S.G., Roberts T.M. 'Array aperture design using irregular polyomino subarrays'. *IEEE Int. Symp. Phased Array Syst. Technol.* 2010, pp. 740–44.

[77] Herd J.S., Duffy S.M. 'Overlapped digital subarray architecture for multiple beam phased array radar'. *Proceedings of 5th European Conference Antennas Propagation (EUCAP)*; Rome, Italy, 2011. pp. 3027–30.

[78] Herd J. 'Experimental results from a self-calibrating digital beamforming array'. *IEEE International Symposium on Antennas and Propagation*; Dallas, TX, 1990. pp. 7–384.

[79] Steyskal H. 'Digital beamforming'. *Proc. 18th Eur. Microw. Conf*; 1988. pp. 49–57.

[80] Steyskal H., Herd J.S. 'Mutual coupling compensation in small array antennas'. *IEEE Trans Antennas Propag*. 1990, vol. 38(12), pp. 1971–75.

[81] Chappell W., Fulton C. 'Digital array radar panel development'. *Proceedings of IEEE International Symposium Phased Array System Technology*; Boston, MA, 2013. pp. 50–60.

[82] Kang D.-W., Kim J.-G., Min B.-W., Rebeiz G. 'Single and four element KA-band transmit/receive phased-array silicon RFICS with 5-bit amplitude and phase control, IEEE trans'. *Antennas Propag*. 2009, vol. 57, pp. 3534–44.

[83] Conway M.D., Russel D.D., Morris A., Parry C. 'Multifunction phased array radar advanced technology demonstrator nearfield test results'. Presented at IEEE Radar Conference (Radarconf18); Boston, MA, 2018.

[84] Harger M., Conway M.D., Thomas H, *et al.* 'Fully digital phased array development for next generation weather radar'. *IEEE Radar Conference (RadarConf21)*; Atlanta, GA, 2021.

[85] Schvartzman D., Weber M., Torres S., Thomas H., Zrnić D., Ivić I. 'Scanning concepts and architectures supporting rotating meteorological phased-array radar'. *AMS 37th Conf. Environmental. Information Processing*; Paper 10.8, 2021.

[86] Day D. '*41st Asilomar Conference on Signals, Systems and Computers (ACSSC)*'. 2007. Pacific Grove, CA,

[87] Zrnic D.S., Kimpel J.F., Forsyth D.E, *et al.* 'Agile-beam phased array radar for weather observations'. *Bulletin of the American Meteorological Society*. 2007, vol. 88(11), pp. 1753–66.

[88] Ivić I.R. 'On the use of a radial-based noise power estimation technique to improve estimates of the correlation coefficient on dual-polarization weather radars'. *Journal of Atmospheric and Oceanic Technology*. 2014, vol. 31(9), pp. 1867–80.

Index

www.ingramcontent.com/pod-product-compliance
Lightning Source LLC
Chambersburg PA
CBHW071350220926
51613CB00008B/41
9781839534027